HANDBOOK OF STRATA-BOUND AND STRATIFORM ORE DEPOSITS

Volume 6
Cu, Zn, Pb, AND Ag DEPOSITS

HANDBOOK OF STRATA-BOUND AND STRATIFORM ORE DEPOSITS

Edited by
K.H. WOLF

I
PRINCIPLES AND GENERAL STUDIES

1. Classifications and Historical Studies

2. Geochemical Studies

3. Supergene and Surficial Ore Deposits ; Textures and Fabrics

4. Tectonics and Metamorphism
   Indexes volumes 1—4

II
REGIONAL STUDIES AND SPECIFIC DEPOSITS

5. Regional Studies

6. Cu, Zn, Pb, and Ag Deposits

7. Au, U, Fe, Mn, Hg, Sb, W, and P Deposits
   Indexes volumes 5—7

ELSEVIER SCIENTIFIC PUBLISHING COMPANY
Amsterdam — Oxford — New York   1976

HANDBOOK OF STRATA-BOUND AND STRATIFORM ORE DEPOSITS

II. REGIONAL STUDIES AND SPECIFIC DEPOSITS

Edited by
K.H. WOLF

Volume 6
Cu, Zn, Pb, AND Ag DEPOSITS

ELSEVIER SCIENTIFIC PUBLISHING COMPANY
Amsterdam — Oxford — New York   1976

ELSEVIER SCIENTIFIC PUBLISHING COMPANY
335 Jan van Galenstraat
P.O. Box 211, Amsterdam, The Netherlands

Distributors for the United States and Canada:

ELSEVIER/NORTH-HOLLAND INC.
52, Vanderbilt Avenue
New York, N.Y. 10017

ISBN: 0-444-41406-1

Printed in The Netherlands

LIST OF CONTRIBUTORS TO THIS VOLUME

V.D. FLEISCHER
Mufulira Division, Roan Consolidated Mines Ltd., Zambia

W.G. GARLICK
Magoebaskloof, North Transvaal, Republic of South Africa

R.D. HAGNI
Department of Geology and Geophysics, University of Missouri-Rolla, Rolla, Mo., U.S.A.

R. HALDANE
Chingola Division, Nohanga Consolidated Copper Mines Ltd., Zambia

A.D. HOAGLAND
The New Jersey Zinc Exploration Comp., Eastern Division, Sarasota, Fla., U.S.A.

W. JUNG
V.E.B. Mansfeld Kombinat "Wilhelm Pieck", Eisleben, Germany (G.D.R.)

G. KNITZSCHKE
V.E.B. Mansfeld Kombinat "Wilhelm Pieck", Eisleben, Germany (G.D.R.)

I.B. LAMBERT
Baas-Becking Geobiological Laboratory, Bureau of Mineral Resources, Canberra, A.C.T.,
Australia

N.L. MARKHAM
Department of Mines, Geological Survey of New South Wales, Sydney, N.S.W., Australia

J.C. SAMAMA
Institut National Polytechnique de Nancy, Ecole Nationale Supérieure de Géologie Appli-
quée de Prospection Minière, Nancy, France

D.F. SANGSTER
Geological Survey of Canada, Ottawa, Ont., Canada

E. SCHEIBNER
Department of Mines, Geological Survey of New South Wales, Sydney, N.S.W., Australia

S.D. SCOTT
Geological Survey of Canada, Ottawa, Ont., Canada

G.E. SMITH
The Anaconda Co., Tuscon, Ariz., U.S.A.

M. SOLOMON
Geology Department, The University of Tasmania, Hobart, Tasmania, Australia

F.M. VOKES
Geological Institute, University of Trondhem, Norwegian Technical University, Trondhem, Norway

CONTENTS

Chapter 6. GEOLOGY OF THE ZAMBIAN COPPERBELT
by V.D. Fleischer, W.G. Garlick and R. Haldane

Chapter 7. KUPFERSCHIEFER IN THE GERMAN DEMOCRATIC REPUBLIC (GDR) WITH
SPECIAL REFERENCE TO THE KUPFERSCHIEFER DEPOSIT IN THE
SOUTHEASTERN HARZ FORELAND
by W. Jung and G. Knitzschke

Chapter 8. SABKHA AND TIDAL-FLAT FACIES CONTROL OF STRATIFORM COPPER
DEPOSITS IN NORTH TEXAS
by G.E. Smith

Chapter 9. CARBONATE-HOSTED LEAD—ZINC DEPOSITS
by D.F. Sangster

Chapter 10. TRI-STATE ORE DEPOSITS: THE CHARACTER OF THEIR HOST ROCKS
            AND THEIR GENESIS
by R.D. Hagni

Chapter 11. APPALACHIAN ZINC—LEAD DEPOSITS
by A.D. Hoagland

Chapter 12. THE McARTHUR ZINC—LEAD—SILVER DEPOSIT: FEATURES, METALLO-
           GENESIS AND COMPARISONS WITH SOME OTHER STRATIFORM ORES
by I.B. Lambert

Chapter 1

COMPARATIVE REVIEW OF THE GENESIS OF THE COPPER–LEAD SANDSTONE-TYPE DEPOSITS

J.C. SAMAMA

INTRODUCTION

Among the strata-bound ore deposits, the group of sandstone-type, sometimes called red-bed type, forms a complex case of genetic problems related to the detrital[1] nature of the ore-bearing facies. The group of the so-called red-bed deposits appears to be homogeneous, even if some transitions are frequent, both towards volcano–sedimentary deposits and towards Kupferschiefer[2]. By origin, the term "red beds" is a term from sedimentary geology which is used for continental to subcontinental variegated detrital to ultradetrital series: according to the strict relationship of copper occurrences of Oklahoma with such an environment (Tarr, 1910), the term red beds has been used for the deposits themselves (Rogers, 1916; Finch, 1928). According to the frequent association with other elements (U, V, Ag), the meaning of the term was enlarged and finally it has become more or less synonymous with "sandstone type" because of the preferential localization of ores within the rocks of this group, even if the ore-bearing rock may belong to a different granulometric class (rudites). Of course, placer deposits are not included in this group although implications are evoked by some geologists. Typical economic minerals occurring in placer deposits are chemically stable and mechanically resistant under the physiographic conditions prevailing during weathering and erosion of the parent rocks, as well as during transport and sedimentation of the detrital grains among which the economic ores have become concentrated, chiefly by the effects of gravity. On the contrary, red-bed ore deposits are characterized by the low chemical and/or mechanical resistance of the economic minerals (sulphides, oxides, carbonates) under conditions prevailing during the period of ore concentration; the economic minerals in this case result from processes of deposition and concentration which involve chemical reactions of precipitation.

The genetic problems posed by the deposits of the group are clear in their broad

---

[1] Editorial note: "Detrital" throughout this chapter is used as a synonym for "clastic", i.e. the meaning of terrigenous origin.

[2] Cf. Chapter 7 by Jung and Knitzschke, this volume, for details on the Kupferschiefer ores per se which are the "type" for similar ores elsewhere in the world.

lines but are, according to the various theories presented in the international literature, complicated enough to form the subject of this contribution. Every genetic model has to solve two main aspects: (1) origin of metals (and more precisely, of elements involved by the concentration processes); (2) duration of and process(es) involved in the concentration and (in a few cases) post-concentration events.

As regards the first point, it may be pointed out that — even if this is not so evident from various publications — the same model always cannot be suitable for all the deposits of the group; for instance, it is not because a volcanic origin of copper is obvious for the Cerro Negro deposit (Chile) (Pelissonnier, 1971) that copper also is of a direct volcanic origin in the Fore Urals province (Malyuga et al., 1966). Similarly, for the second point, post-sedimentary copper mineralization may be clear in a detrital environment and a syn-sedimentary origin can be demonstrated in another. Tectonics and metamorphism can be at least as effective as post-concentration processes in one case and may be non-existent in another.

An exhaustive review of the different published conclusions would be of little interest, because every possibility has been evoked from the strictly syn-sedimentary model to the late epigenetic, form the volcanic — hydrothermal origin of metals to that of the pedologic.

In restricting the subject to a comparative review of the genesis of copper and lead sandstone deposits, it appears to be more effective to describe first a simple and complete model of lead–zinc red-bed type (Largentière, France), with some remarks deriving from this model referring to other types, then to consider other models differing from the previous one according to the relative time of deposition, the mechanism of concentration and the origin of the elements (Samama, 1968, 1969; Bernard and Samama, 1970).

DESCRIPTION OF THE Pb–Zn SANDSTONE-TYPE DEPOSIT OF LARGENTIERE (FRANCE)

The Triassic Largentière district is located on the southeastern border of the Hercynian Massif-Central and it appears as the northern part of the "sous-cévenole" province which is characterized by its chemistry (Pb–Zn) and its age (Mesozoic) (Bernard, 1958; Fig. 1). The Largentière deposit and the related occurrences are distributed in a detrital terrigenous complex attributed to the top of Lower Triassic (Buntsandstein) or to the base of the Middle Triassic (Muschelkalk). This Triassic series has been deposited on a stable cratonic area (Hercynian Massif-Central) and it lies unconformably either on Hercynian units or on the Permian basin of Largentière and now forms a faulted monoclinal unit dipping 10–15° southeastwards.

The age of the Pb concentration distributed along four preferential beds among the whole series, has been determined with great accuracy by sedimentological observations mainly based on the relationships of mineralizations with early features such

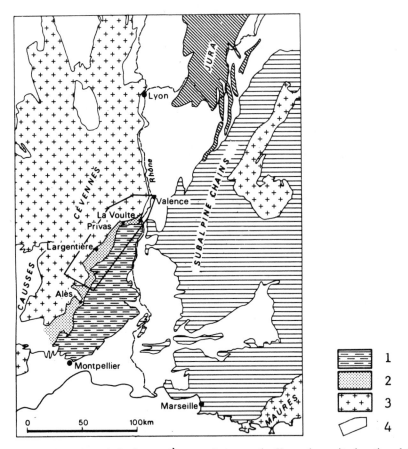

Fig. 1. Situation of the Largentière area between the Hercynian units (west) and the graben of the Rhône valley (east). *1* = Jurassic and Cretaceous units of the "sous-cévenole" border; *2* = Triassic of central and northern Cévennes; *3* = Hercynian basement; *4* = Largentière deposit and its district.

as soft pebbles, synsedimentary and diagenetic faults, slumps and filling of geodes. The result of these observations is that the primary mineralization has taken place after the deposition of the siliceous detrital materials (pebbles, gravels and sand grains) but before the deposition of the next detrital level. More precisely, this primary mineralization originated between the active mechanical sedimentation periods and it appears to be a result from phreatic circulations involving chemical precipitation (internal sediments) (see also the subsection on the outline of the genetic model).

According to this result, the physiographic conditions prevailing during sedimentation are of primary importance: they may be summarized by the consideration of two main units, i.e., the Hercynian basement on the one hand and the pediment area associated with an evaporitic area on the other (Fig. 2).

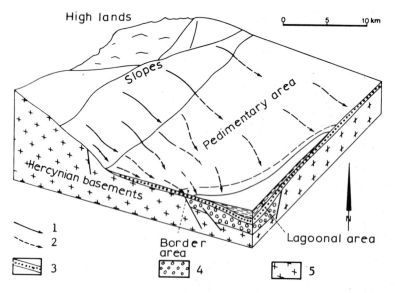

Fig. 2. Sketched block diagram of the distribution of physiographic areas during the Lower Triassic in the Largentière area. *1* = surface water flow; *2* = underground flow; *3* = sedimentary spread;
4 = Permian basin; *5* = Hercynian basement.

The Hercynian basement

The Hercynian basement (granitic and metamorphic rocks) has been deeply eroded
before the Triassic; the pre-Triassic morphology is highly characteristic: it is a nearly
perfect peneplane with a very few irregularities not more than 25 m high. This morphological surface was outlined by slight and constant weathering limited to the destabi-
lization of the most instable mineral species (Na-feldspar, biotite) and the related mechanical disaggregation reaching a few meters, always without any deep rubefaction. The
weathering products are illite, kaolinite, vermiculite and interstratified clays (illite–
montmorillonite). The chemical feature of this weathering phenomenon is the leaching
of sodium and calcium for the major elements and of copper, uranium and partly zinc
for the trace elements.

It appears that the physiographic conditions prevailing between the end of the Permian sedimentation (Saxonian) and before that of the Triassic have induced a drastic separation between copper and uranium on the one hand, completely leached out
from the geochemical landscape, and lead (and partly zinc) slightly enriched in the
residual formation, the reworking of which contributed — on a large scale — to the
Lower Triassic sedimentation.

The geochemical distribution of Pb–Zn–Ba in fresh Hercynian rocks is of great
interest. The lead content is very high (45 ppm) over the whole basement area (2260

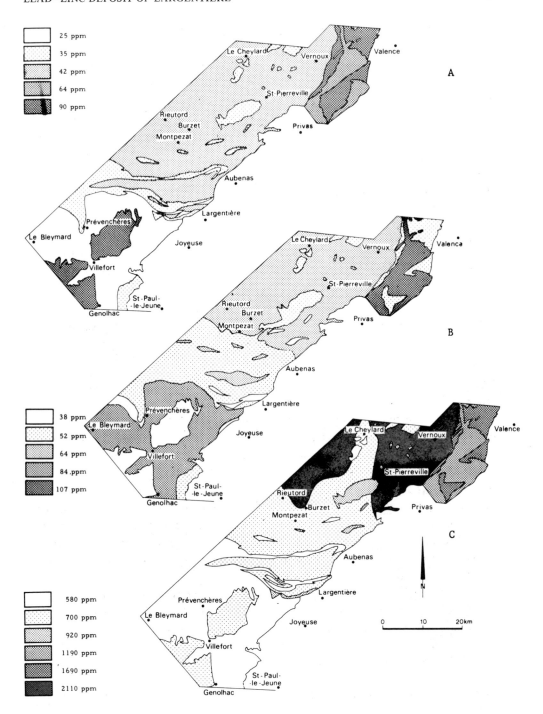

Fig. 3. Geochemical maps of the Hercynian basement. A. Lead. B. Zinc. C. Barium.

km²) and irregular from one lithological unit to another (25–90 ppm) as also is that of zinc (38–107 ppm) or barium (580–2110 ppm) (Fig. 3).

Of course, these irregularities in trace-element distribution have been modified by the pre-Triassic weathering, but we can look for spatial correlations between the Pb-rich areas of the basement and Pb deposits or important occurrences within the Lower Triassic series. From the map (Fig. 4), it appears that all the important occurrences or deposits are less than 20 km from strong geochemical anomalies in the basement. Evidently, this spatial correlation is not so strict, because the true phenomenon of geochemical inheritance from a basement to a basin is modulated by the sedimentary differentiation within the Triassic geochemical landscape.

The pediment area and the evaporitic basin

This second unit, a detrital mantle resulting from the reworking of the pre-Triassic weathering cover is due to the first actions of the Mesozoic distentions between the

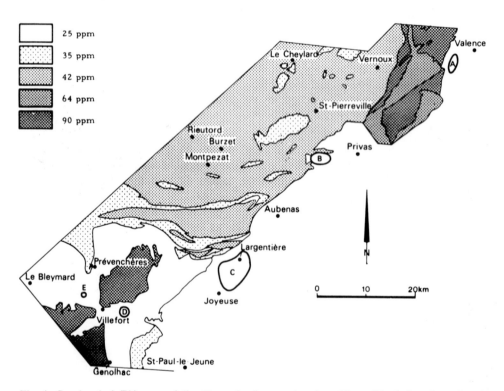

Fig. 4. Geochemical (Pb) map of the Hercynian basement and position of lead deposits and main occurrences within the Lower Triassic. Main Triassic lead occurrences: *A* = Crussol; *B* = Col de l'Escrinet; *C* = Largentière; *D* = Mas de l'Air; *E* = Altier.

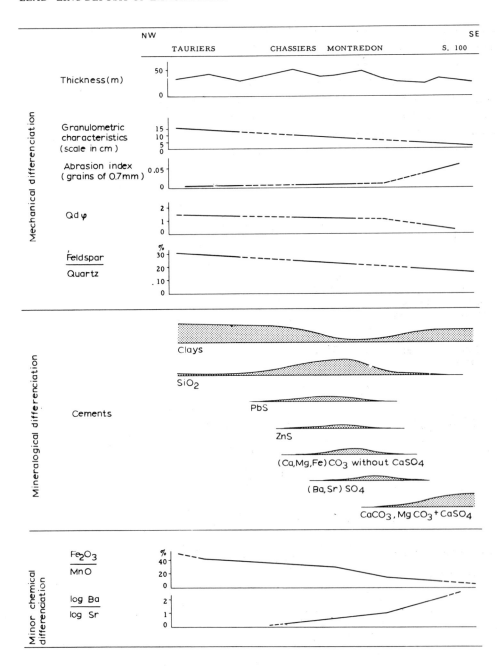

Fig. 5. Table of the main characteristics of the Triassic pediment at Largentière. The total extent from northwest to southeast is 3.5 km.

Hercynian cratonic area of the Massif Central and the mobile zone of the Alps. This detrital mantle forms a thin pediment (30–50 m) formed under a semi-arid climate at the border of an evaporitic basin.

Within this pediment a mechanical and a mineralogical differentiation clearly appear from northwest to southeast, i.e. along the general direction of flow (Fig. 5). Among the different parameters of the chemical differentiation, the occurrence of the sulfide minerals (galena and sphalerite) is constantly in association with silica (quartz overgrowth on detrital grains) between the argillaceous facies, on one hand, and the evap-

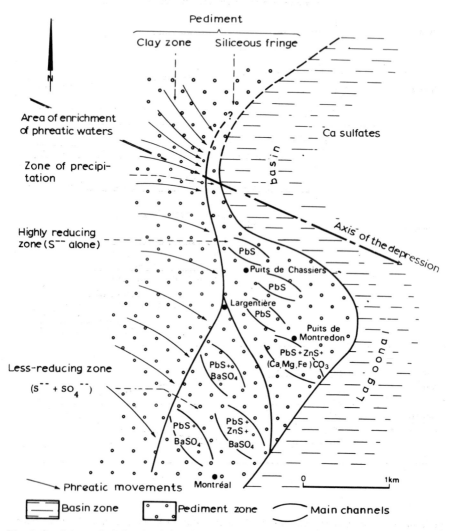

Fig. 6. Semi-theoretical sketch of the distribution of internal chemical sedimentation at Largentière.

oritic facies (dolomite and anhydrite), on the other. This siliceous fringe, located at the border of the evaporitic basin, bears all the lead–zinc occurrences in the Lower Triassic.

A schematic model of the distribution of the main characteristics of the Largentière area is presented in Fig. 6. The zonal distribution shown in Fig. 5, is modified by a flat depression of the basin the axis of which is superposed on that of the local Permian graben. The siliceous fringe itself can be divided into two zones according to their reducing characters and to the mineralized lenses.

Outline of the genetic model

The genetic model proposed for this deposit (Samama, 1969) can be summarized in several steps: (1) geochemical maturation of the basement; (2) progressive formation of the pediment by active mechanical sedimentation; (3) metal enrichment of the waters percolating through the pediment; (4) precipitation of heavy metals at the contact with a geochemical barrier bound to the water interface near the basin (internal chemical sedimentation); (5) diagenetic and epigenetic reorganization.

The geochemical maturation of the basement which is already rich in lead, zinc and barium, induces a chemical separation between elements such as Cu, U and partly Zn leached out from their host minerals by weathering and Pb and Ba which are stable because of their localization within K-feldspar which is stable under the climatic conditions prevailing during this period.

Each stage of the progressive formation of the pediment by active mechanical sedimentation of the weathering mantle is very fast and is the result of a laminar deposition of clastic material; but between each phase, the weathering phenomena continue, being influenced by the circulation of water within the pediment.

Under the new chemical conditions in the pediment, weathering becomes more intense and the residual K-feldspars are destabilized and lead to an enrichment of the percolating waters in silica, lead and barium — almost exactly as established by Helgeson (1966) for deeper subsurface conditions.

The interface between these continental waters and the saline water impregnating the sediments near the limit of the basin acts as a geochemical barrier for silica and heavy metals, because of the differences in chemistry and the reducing properties deriving from the organic matter (Figs. 6, 7 and 8). This geochemical barrier is responsible for the internal chemical sedimentation within the siliceous fringe. Precipitation of silica is probably due to reequilibration of the generally oversaturated water (20–80 ppm SiO_2) in the pediment; in most reducing zone of the fringe, SO_4^{2-} is completely reduced as S^{2-} and SH^- and Pb–Zn are precipitates as sulfides[1]; in the less-re-

---

[1] Due to the different equilibrium and the low content in total sulphur of this water, iron is mainly precipitated with carbonates as ankerite.

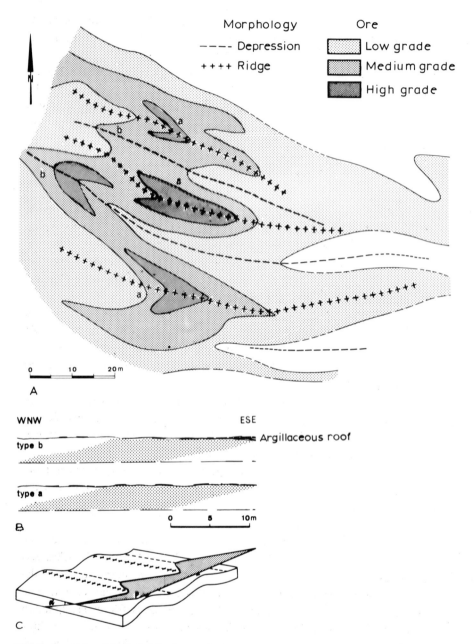

Fig. 7. Example of lead distribution controlled by detailed paleogeography and its interpretation. A. Map of metal accumulation (Largentière – Level 5). Galena forms a croissant-shaped accumulation alternately turned northwest (a) and southeast (b) along the general flow direction (WNW–ESE). B. In both cases of a croissant-shaped body a general vertical distribution can be observed along the flow axis. C. Geometric interpretation of the data. The previous features can be interpreted as the geometrical intersection between an andulated surface and an oblique plane (p) deeping northwestwards (α).

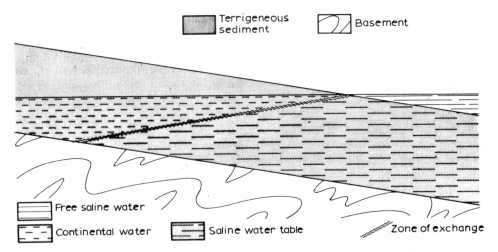

Fig. 8. Theoretical disposition of water tables near the border pediment basin. It has been supposed that equilibrium has been reached, i.e. there is no water flow. This kind of disposition can partly be deduced from the data of Fig. 7.

ducing zone, S^{2-} and SO_4^{2-} are present and Pb–Zn are precipitated as sulfides and Ba as sulfate. In this model, the influence of the S60°E-trending depression of the border of the basin is evident (Fig. 6): in the semi-arid landscape the greater part of the phreatic waters (and of course of the dissolved metals) percolating through the pediment is collected by the lowest part of the landscape and reaches the border of the basin in gulf zones.

The calculation of this hydrogeological model is easy and it proves that with slightly enriched water, under normal conditions, an elementary bed can be mineralized in a few hundred years. Of course, this model requires a great stability of the hydrogeological conditions to induce an intense internal chemical sedimentation: the four mineralized levels of the Largentière deposit would correspond to four main periods of stability during the formation of the pediment.

The diagenetic reorganization of the chemical phase of the sediment is marked by quartz overgrowths affecting detrital grains and the crystallization of sulfides which begin partly during reorganization of silica (syncrystallization of galena and quartz) but it occurs mainly after the quartz with very clear figures of replacement. Later on, under some hundred meters of sedimentary cover and during the Alpine orogeny, the faulting of the whole series induces a local migration of galena in the low-pressure zone, and lower proportions of sphalerite and quartz according to the different molar volumes (galena: 31.5 cm^3/mole; sphalerite: 24 cm^3/mol and silica: 23 cm^3/mole). The recent oxidation by surface waters does not really modify the deposit, except at the level of ancient works attributed to Roman activities.

Relations between Largentière and other sandstone-type deposits

The chemical peculiarities of the Largentière deposit (Pb and Zn without copper, vanadium and uranium) must be looked upon taking a broader view. In this way, and according to the distribution of heavy-metal occurrences and deposits (sandstone type) bound to the Lower Triassic Formation, we can distinguish two geochemical areas: one in which copper and sometimes uranium is present, and another in which copper is unknown except as geochemical anomalies (Fig. 9). Such drastic differences cannot be interpreted as clear differences in the local paleogeographic conditions or in the geochemistry of the basement, but rather as a climatic opposition between an area of pre-Triassic weathering conditions of monosiallitization (*A* on Fig. 9) and another area with conditions of bisiallitization (*B* on Fig. 9). This correlation is of great interest if

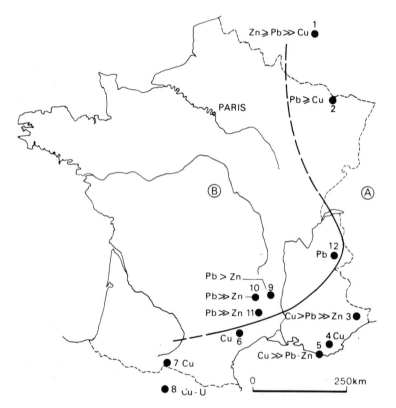

Fig. 9. Map of pre-Triassic weathering facies and of sandstone-type deposits of the Lower Triassic. *A* = area of pre-Triassic monosiallitization process and of sandstone-type deposits with at least small amounts of copper; *B* = area of bibiallitization process and of sandstone-type deposits without any copper. *1* = Maubach; *2* = Saint-Avold; *3* = Le Cerisier; *4* = Le Luc; *5* = Cap Garonne; *6* = Lodève; *7* = Massif de l'Arize; *8* = Plana de Monros; *9* = Largentière; *10* = Mas de l'Air; *11* = Saint-Sébastien; *12* = La Plagne.

we consider the trend of the concentration of these heavy metals according to zonal weathering processes (Bernard and Samama, Chapter 7, Vol. 1): in the first area, pre-Triassic conditions induce copper and partly zinc and lead concentration in the weathered profile, whereas in the second one, lead is more selectively concentrated. *The chemical zonation is influenced by the continental physiographic conditions,* much more so than by the properties of the parent rocks. Based on this remark, a comparison of the environmental characteristics of sandstone-type deposits has been proposed (Bernard and Samama, Vol. 1, table I). Of course, this model must be modified for many deposits according to the influence of volcanic activity as a source for elements and to the relative time of concentration and remobilization.

OTHER TRENDS OF GENETIC MODELS OF SANDSTONE TYPES[1]

In the previous model devoted to a peculiar lead—zinc sandstone-type deposit, the early concentration of heavy metals is well established and the terrigeneous origin of metals highly probable for such a non-volcanic environment. Of course, such a model cannot be generalized and we must review other trends of genetic models established for various deposits.

An analysis of the trends shows major causes of difference, that is to say, the relative time of deposition of the host formation and the first concentration of heavy metal on one hand, the primary source of metals on the other; of course, these two major points are not always independent.

Genetic models and epochs of concentration

Drastic divergences occur between the different models proposed for time relations from the pure synsedimentary ones (synchronism between metal concentration and deposition of the sediment) and late epigenetic models.

Syngenetic and associated models. Many authors have deduced, from various observations, a strict time relationship which is either based on the superposition of particular facies and mineralization or on morphological disposition.

The strict syngenetic model has been evoked by Tixeront (1973) with reference to the Permian formation of Argana (Morocco). According to Tixeront, the copper-uranium occurrences would have been induced by continental metalliferous sources and either a syngenetic concentration in particular sites (controlled by paleomorphology, paleobathymetry) or a general dispersion in a stratigraphic level of the basin.

In the same way, it seems that many sandstone deposits are considered as synchro-

---

[1] Editor's note: See also Chapter 3 by Rackley, Vol. 7, for example.

nous with the host sediment, for instance, the Udokan copper deposit (Grintal, 1968; Bakun et al., 1966; Narkelyun and Yurgeson, 1968), or those of the Dzhezkazgan (Popov, 1953; Shtrem, 1969; Strakhov, 1962; Narkelyun, 1968) or those of the Cretaceous of Morocco (Caia, 1966, 1968, 1969). This model can be summarized by the observation of Shtrem (1969) to explain the conditions of ore concentration: "Only the fore deltaic sediments which ... are framed by sandy sediments of spits and bars are ore bearing; sediments deposited in this facies are a kind of barrier, which in conjunction with the bay-lagoon offshore area, obstructed open flow of riverine arteries to the open sea".

This kind of data leads such authors as Popov (1971) to consider that "the ore matter, regardless of its source, supplied in one form or another into the basin, is an *integral component* of a sediment" or, as Strakhov (1962) who considers numerous deposits as normal components of peculiar lithogenesis and so described with them.

However, the use of the term "syngenetic" is dangerous, even for the primary concentration, because of the sedimentological contradiction between the sedimentation of the detrital phase which involves active depositional currents, and the very calm hydronamic conditions required by the concentration of a very low-grade solution.

For this reason, it seems that the Largentière model is much more suitable and the rough "syngenetic" concept can better be characterized by the "early chemical internal sedimentation" concept.

However, a very clear example such as that of the Holocene deposit at Ray (Arizona) (Phillips et al., 1971) demonstrates that a continental detrital environment can be mineralized nearly syngenetically (less than 7000 years between the deposition of the sediment and the ore concentration) but without any direct relationship between sedimentation and mineralization. This deposit shows that it is evident that other models must be evoked.

Diagenetic models and early fluid migration. According to the common fluid migration related to sediment compaction[1], many authors have built different models based upon this diagenetic process. For instance, such a model has been presented for the White Pine copper deposit (Michigan); for this deposit, White (1971) and White and Wright (1954, 1960) suggested that the copper was carried in connate water squeezed out of the Keweenawan sediments during compaction and sedimentation in the Lake Superior basin. In a similar, but more detailed way, Hamilton (1967) has concluded that the "native copper (of the Upper part of the Harbor conglomerate) is restricted to chloritic sandstone and its presence must be related to conditions that existed in the chloritic layers. The chlorite is diagenetic and formed in a reducing environment. The reducing environment may have been created as water, squeezed out of the overlying Nonesuch Shale during compaction circulated through the sandstone". Because of the obliquity of the limit between pyrite and copper mineralization (chalcocite and native copper), White and Wright

---

[1] Cf. Wolf (1976) for a summary of the influences of compaction on ore genesis.

(1966) concluded on a copper "front" and a replacement of syngenetic/diagenetic pyrite by copper solutions. In 1971, White developed a more sophisticated model involving "lateral migration of fluids through the subjacent Copper Harbor Conglomerate to the site of the deposit and stripping of copper from these solutions where they percolated upward through the overlying Nonesuch Shales". Brown (1971) suggested that the mineralization of the basal Nonesuch "was accomplished by the introduction of copper from the underlying Copper Harbor Conglomerate soon after deposition of the Nonesuch Shale . . . in large part by replacement of original pyrite along an upward advancing mineralization front". For this model, Brown has tested the simple mathematic model of infiltration and diffusion processes.

The great variability of these different models is related to the following three points: (1) origin of the copper; (2) way of migration (and related processes); (3) mechanism of concentration (precipitation).

The two last points are strictly connected, but cannot really be solved without a better chronology of the process, the only way to fix the general conditions of the environment prevailing during concentration.

In these typical models, the fluid migrations are generally related to diagenetic compaction and belong to the sedimentary cycle, but in other models, the fluids are connected with volcanic exhalative activity which is more or less contemporaneous with the sedimentary and diagenetic activity (Morton et al., 1974; Reznikov, 1965).

Nevertheless, this type of genetic model is characterized by early fluid migrations of mineralized waters generated in the ore-bearing series itself. Of course, such processes can act much later and the mineralized waters can originate outside the ore-bearing series.

Epigenetic models. In this group of models, there is no well-defined time relationship between the sedimentary history of the ore-bearing rocks and the metal concentrations except, in particular cases, with some alterations and related selective leaching.

The most typical process is evoked concerning uranium deposits. For instance, for the mineralization of the Shirley Basin area (Wyoming) Harshman (1972) considers that the introduction of the uraniferous solution occurred long after the lithification of the porous ore-bearing levels by the action of precipitation at the contact with an Eh-barrier formed by the reducing water table.

In fact, we must consider the problem of the mobility of the concentrated elements in discussing the epigenetic concentration of a metal. For instance, because uranium is much more mobile than lead in oxydo-reduction processes, it is clear that under similar geological evolutions of lead and uranium sandstone-type districts, uranium concentrations would be remobilized several times (with concentration or dispersion phases) and show clear epigenetic features (Finch, 1967), whereas lead concentrations would show, through its epigenetic aspects, some relics of the previous early phases of mineralization as in the Largentière deposit. In the same way, it is clear that epigenetic reconcentrations can induce high-grade ores (sometimes the only economic occurrences), but this phenomenon must not be confused with an epigenetic concentration of the whole mineralization.

For instance, in the case of copper deposits, epigenetic reconcentrations always occur within or near distension faults, so that evidence of epigenetic controls, such as that presented for instance by Ohle (1968) for White Pine, i.e., control of native copper distribution by post-Nonesuch Shales faulting may be discussed in the light of this point of view.

Nevertheless, and in accordance with this remark, it is evident that epigenetic concentration may act as a metallogenic phenomenon, but chiefly either as a reconcentration process or as a migration process and in this case, the parent concentration generally forms a major metal accumulation.

Chronology of ore concentration. In the previous developments, the primary ore concentration was considered without any consideration of later behaviour. Thus, in the first group of models the epigenetic evolution was neglected, even if epigenetic reconcentration may have been important from an economic point of view. Nearly all deposits attributed to these models have registered such an epigenetic reconcentration, and may be considered as deriving from a sequence of concentration phenomena, that is to say, that a relative chronological classification is far from perfect, except when a process appears as a major one and leads to the characterization of a deposit. In fact, what was developed here is not a classification, but an attempt to characterize the outlines of the genetic models of sandstone-type deposits from a chronological point of view. In this way, models such as those developed by Morton et al. (1974) for copper deposits in the Belt Supergroup (Alberta, Canada) are typically polygenetic with an early diffuse ore concentration, and a later one leading to redistribution and concentration during contact metamorphism and perhaps low-grade regional metamorphism.

Genetic models and the origin of metals

Two aspects of this question must be considered according to the relative time of concentration. If the model belongs to the group involving early phases of the sediment history (syngenetism in its broad meaning), the metal supply may be related either to exhalative activity and to the synchronous volcanic events of the district, or to various geological units of the continental areas. If the model belongs to the group involving later phases of concentration, the source must be looked for either among formations forming the sedimentary pile (or the ore-bearing bed itself) or among particular geologic units within the catchment area.

Origin of metals in "syngenetic" models. From the Largentière model, it appears that the continental catchment area can involve geologic units much older than the sedimentary events and forming an important source of heavy metals when affected by continental weathering. These geologic units could be of igneous, volcanic or sedimentary origin. As well as the Largentière deposit, which has a very high lead anomaly (up to 90 ppm in some Hercynian granites), reference can be made to the Udokan deposit which derives from the copper-rich Archean Aldan Shield (Bakun et al., 1966; Narkelyun and Yurgeson, 1968) or to the Foreland Ural Permian province which is derived from the spilite-keratophyre copper province of the Hercynian Ural region (Malyuga et al., 1966).

The continental origin of metals is not easy to establish, except in a few particular cases such as the Udokan deposit in which the discovery of a very rare mineral (serendibite) has lead to the determination of the source area (Narkelyun and Yurgeson, 1968), but generally the data concerning origin are suggested by considerations as to the metal distributions in the basements and sedimentary basins (Samama, 1969; Narkelyun and Yurgeson, 1968). The metal transportations generally implied for this kind of source are of the order of a few tens of kilometers. In this way, we can determine the criteria for the evaluation of the metal capacity of a basement. Since geochemists (Tauson et al., 1967) consider that for a heavy element, the metal which is concentrated as an ore deposit is estimated at $n \cdot 10^{-4}$ of the metal content of the earth crust, it is evident that a map of the distribution of basement deposits is too inaccurate unless these deposits reflect a large and diffuse geochemical anomaly.

The direct volcanic and exhalative supply of metal in a sedimentary basin is very often evoked by many authors, sometimes with serious arguments and sometimes just because the metal content of basement is considered as being too low. The volcanic exhalative origin is — as for the previous case of continental origin — difficult to prove; to identify volcanic activity during the sedimentary activity is generally possible, but not to decipher whether it is the cause of the metal supply or just a consequence of the instability connected with the subsidence of the basin.

However, the source of sandstone-type deposits occurring in volcano-sedimentary Tertiary basins of the South American Cordilleras is evidently strongly connected with andesitic copper-rich activity (Ruiz et al., 1971; Carter and Nelson Aliste, 1964). Recently, Gossens (1975) has emphasized the "direct" volcanic supply for various deposits such as Kupferschiefer, White Pine, Portage Lake, Corocoro . . . mixing deposits in which the volcanic activity is evident but not always "direct" (continental leaching and weathering of volcanic ashes cannot be considered as a direct supply) and deposits such as Kupferschiefer for which volcanic activity is at least difficult to demonstrate.

In fact, these two types of metal sources are not in opposition: between the early leaching of a volcanic glass deposited on the catchment area of a basin during sedimentation and the deep weathering of volcanic rocks with low-grade mineralization inducing a high metal supply hundreds of millions of years later, the difference is not so drastic and all the intermediaries exist.

Origin of metals in dia- and epigenetic models. In this group of models, the different types of sources have already been mentioned. The most common source to which authors refer is the sedimentary pile and sometimes the ore-bearing bed itself, the metal already being accumulated in well-defined facies or occurring as normal trace content of the sediment. The main difference we can note is the process of metal extraction (sometimes in connection with the time of extraction): diagenetic compaction and expulsion of metals by water or leaching of rocks and minerals (early or late leaching by "crypto-alteration").

In a few cases (see Harshman, 1972 for instance), an external source is required, that is to say that long after the diagenetic period, the leaching of the catchment area and the correlative metal enrichment by infiltration through exposure of rocks of high permeability will induce specially metal-rich aquifers. To find out if the source is really external or not is generally impossible: the way used, for instance, by Harshman is that of the geochemical balance, but to prove that an internal source is inconsistent with geochemical data requires an extensive sampling which is rarely possible. The only clear example is that of the Holocene body at Ray (Phillips et al., 1971), the copper source of which can be connected with the adjacent copper porphyry.

This rough review of the source problem leads to an evident conclusion: the question is one of great difficulty and nearly every kind of model has been proposed except — to the knowledge of the author — that of the marine source.

GENERAL CONCLUSIONS

We will not dwell any longer on the great variety of the genetic models already sketched out in their broad lines. Many of them do exist and the real problem is their metallogenetic effectiveness: among all the processes some can act only under very particular conditions and thus have little importance except for some deposits, others such as the early water deplacement seem to be very positive in the genesis of numerous deposits.

REFERENCES

Bakun, N.N., Volodin, R.N. and Krendelev, F.P., 1966. Genesis of Udokansk cupriferous sandstone deposit (Chitinsk oblast). *Int. Geol. Rev.,* 8 (4): 455–466.

Bernard, A., 1958. Contribution à l'étude de la province métallifère sous-cévenole. *Sci. Terre,* 7 (3–4): 125–403.

Bernard, A. and Samama, J.C., 1970. A process du gîsement de Largentière (Ardèche). Essai méthodologique sur la prospection des "Red-Beds" plombo-zoncifères. *Sci. Terre,* 15 (3): 209–264.

Brown, A.C., 1971. Zoning in the white Pine Copper deposit, Ontonagon country, Michigan. *Econ. Geol.,* 66: 543–576.

Caia, J., 1966–1968. Découverte d'une minéralisation plombifère stratiforme dans le Crétacé inférieur de la bordure septentrionale du Haut-Atlas. *C.R. Acad. Sci., Sér. D,* 267: 282.

Caia, J., 1968. Minéralisations plombo–cupro–zincifères dans les formations griso-conglomérati-

ques continentales du début de l'Infracénomanien en bordure septentrionale du Haut-Atlas oriental (Maroc). *C.R. Acad. Sci. Sér. D*, 267: 283–286.

Caia, J., 1969. Les minéralisations plombo–cupro–zincifères stratiformes de la région des plis mar-
ginaux du Haut-Atlas oriental: un exemple de relations entre des minéralisations et une sédi-
mentation détritique continentale. *Notes Serv. Géol., Maroc*, 29 (213): 107–120.

Carter, W.D. and Nelson Aliste, T., 1964. Paleo-channels at the Guayacan copper mine, Cabildo
district, Aconcagua Province, Chile. *Econ. Geol.*, 59: 1283–1292.

Chernyshev, N.J., 1967. Peculiarities of copper occurrences in the Upper Permian deposits of the
Permian region of the Urals. *Lithol. Miner. Resour., (USSR)*, 1967 (4): 517–521.

Finch, J.W., 1928. Sedimentary metalliferous deposits of the Red Beds. *Trans. Am. Inst. Min. Metall.
Pet. Eng.*, 76: 378–392.

Finch, W.I., 1967. Geology of epigenetic uranium deposits in sandstone in the United States. *U.S.
Geol. Surv., Prof. Pap.*, 538: 121 pp.

Gossens, P.J., 1975. L'apport métallifère "direct" du volcanisme continental. *Miner. Deposita*, 10:
43–45.

Grintal, E.F., 1968. Some distribution patterns of mineralization in the Udokan deposit applicable
to the sedimentary hydrothermal interpretation. *Lithol. Miner. Res.*, 1968 (3): 280–287.

Grip, E., 1960. The lead deposits of the eastern border of the Caledonides in Sweden. *Int. Geol.
Congr., 21st., Copenh., Rep. Sess., Norden*, 16: 149–159.

Grip, E., 1967. On the genesis of the lead area of the eastern border of the Caledonides in Scandinavia. *Econ. Geol., Monogr.*, 3: 208–218.

Hamilton, S.K., 1967. Copper mineralization in the upper part of the Copper Harbor Conglomerate
at White Pine, Michigan. *Econ. Geol.*, 62:; 885–904.

Harshman, E.N., 1972. Geology and uranium deposits, Shirley Basin area, Wyoming. *U.S. Geol. Surv.,
Prof. Pap.*, 745: 82 pp.

Helgeson, H.C., 1966. Silicate metamorphism in sediments and the genesis of hydrothermal ore
solutions. *Econ. Geol., Monogr.*, 3: 333–341.

Malyuga, V.I., Proskuryakov, M.I. and Sokolova, T.N., 1966. Distribution of exogenic copper concentrations in the Ural region. *Lithol. Miner. Resour., (USSR)*, 1966 (6): 765–774.

Morton, R.D., Goble, R.J. and Fritz, P., 1974. The mineralogy, sulfur isotope composition and origin
of some copper deposits in the Belt Supergroup, Southwest Alberta, Canada. *Miner. Deposita*,
(9): 223–242.

Narkelyun, L.F. and Yurgenson, G.A., 1968. Copper sources in the formation of deposits of the
cupriferous sandstone type. *Lithol. Miner. Resour. (USSR)*, 1968 (6): 739–747.

Ohle, E.L., 1968. Copper mineralization in the upper part of the Copper Harbor Conglomerate at
White Pine, Michigan. Discussion. *Econ. Geol.*, 63: 190–195.

Pelissonnier, H., 1971. Le gîsement de cuivre stratiforme de Cerro Negro (Aconcagua, Chili). *Bull.
Bur. Rech. Géol. Minières (Fr.), Sect. 2*, 6: 43–50.

Phillips, C.H., Cornwall, H.R. and Meyer, R., 1971. A Holocene ore body of copper oxides and carbonates at Ray, Arizona. *Econ. Geol.*, 66: 491–498.

Popov, V.M., 1953. The intraformational conglomerates of Dzhezkazgan and the nature of their
mineralization. *Dokl. Akad. Nauk S.S.S.R., N. S.*, 1953 (4).

Popov, V.M., 1970. Origin of stratiform deposits and ways of solving this question. *Lithol. Miner.
Resour. (USSR)*, 1970 (2): 198–211.

Popov, V.M., 1971. On the anisotropy of ore-bearing series in stratiform deposits. In: Amstutz and
Bernard (Editors), *Ores in Sediments*. Springer, Berlin, 1973, pp. 221–225.

Raby, J.A., 1833. Sur le gisement des divers minerais du cuivre de Sain-Bel et de Chessy (Rhône).
Ann. Mines, 4: 393–408.

Reznikov, I.P., 1965. The problem of the origin of Udokan deposit. *Lithol. Polezn. Iskop*, 1965 (2).

Rogers, A.F., 1916. Origin of copper ores of the "Red-Beds" type. *Econ. Geol.*, 11: 366.

Ruiz, C., Aguilar, A., Egert, E., Espinosa, W., Peebles, F., Quezada, R. and Serrano, M., 1971. Stratabound copper sulphide deposits of Chile. *Proc. IMA–IAGOD Meet., 1970, Min. ,Geol. Soc.,
Jap., Spec. Issue*, 3: 252–260.

Samama, J.C., 1968. Contrôle et modèle génétique des minéralisations en galène de type "Red-Beds" – Gisement de Largentière, Ardèche, France. *Miner. Deposita,* 3 (3): 261–271.

Samama, J.C., 1969. *Contribution à l' Etude des Gisements de Type Red-Beds. Etude et Interprétation de la Géochimie et de la Métallogénie du Plomb en Milieu continental. Cas du Trias ardéchois et du Gisement de Largentière.* Thesis, Nancy, 450 pp.

Shtrem, E.A., 1969. Mineral zoning in seams of the Dzhezkazgan deposit. *Lithol. Miner. Resour. (USSR),* 1969 (3): 316–325.

Strakhov, N.M., 1962. *Principles of Lithogenesis.* Trad. Cons. Bur., New York, 1967–1970, 3 vols: 245; 609 and 577 pp.

Tarr, W.A., 1910. Copper in the "Red-Beds" of Oklahoma. *Econ. Geol.,* 5: 221–226.

Tauson, L.V., Kozlov, V.D. and Kuzmin, M.I., 1967–1970. The potential ore bearing of intrusives and the sources of the ore matter. In: Z. Pouba and M. Stemprok (Editors), *Problems of Hydrothermal Ore Deposition.* Schweizerbart, Stuttgart, pp. 25–27.

Tixeront, M., 1973. Lithostratigraphie et minéralisations cuprifères et uranifères stratiformes syngénétiques et familières des formations détritiques permo-triasiques du couloir d'Argana, Haut-Atlas occidental (Maroc). *Notes Serv. Géol., Maroc,* 33 (249): 147–177.

Vasil'eva, E.G., 1973. Simulation of depositional processes of uranium, selenium and molybdenum during the interaction between metal-bearing oxygenated waters and a counterflow of gazeous reducing agents. *Lithol. Miner. Res. (USSR),* 1973 (6): 703–713.

White, W.S., 1971. A paleohydrologic model for mineralization of the White Pine copper deposit, Northern Michigan. *Econ. Geol.,* 66 (1): 1–13.

White, W.S. and Wright, J.C., 1954. The White Pine copper deposit, Ontonagon Country, Michigan. *Econ. Geol.,* 49: 675–716.

White, W.S. and Wright, J.C., 1960. Lithofacies of the Copper Harbor conglomerate, Northern Michigan. *U.S., Geol. Surv., Prof. Pap.,* 400 B: 5–8.

White, W.S. and Wright, J.C., 1966. Sulfide mineral zoning in the basal Nonesuch Shale, Northern Michigan. *Econ. Geol.,* 61: 1171–1190.

Wolf, K.H., 1976. Ore genesis influenced by compaction. In: G.V. Chilingarian and K.H. Wolf (Editors), *Compaction of Coarse-Grained Sediments,* II. Elsevier, Amsterdam, in press.

Zhurbitskii, B.I. and Marichev, K.I., 1968. Certain geological features of cupriferous sandstones of Dzhezkazgan and questions of copper ore prospecting. *Lithol. Miner. Res. (USSR),* 1968 (3): 259–268.

Chapter 2

"VOLCANIC" MASSIVE SULPHIDE DEPOSITS AND THEIR HOST ROCKS — A REVIEW AND AN EXPLANATION

M. SOLOMON

INTRODUCTION

"Volcanic" massive sulphide deposits[1] are composed almost entirely of iron sulphides and varying amounts and proportions of copper, lead, and zinc sulphides. They are generally stratiform within volcanic rocks, are commonly layered and zoned, and in many cases overlie low-grade mineralization. They are essentially the same as the "volcanogenic" massive sulphides of Sangster (1972), Hutchinson (1973) and Sangster and Scott (Chapter 5 in this volume), but the term "volcanic" is used in this paper simply to indicate that they occur in volcanic rocks and to avoid the genetic implications of "volcanogenic".

Difficulties always arise in defining this group (e.g., Gilmour, 1971; Chapter 4, Vol. 1; Sangster, 1972) because of the number of deposits that depart only marginally, but possibly significantly, from the definition. For example, the small but important group of Pb–Zn-rich deposits including Sullivan, Mt. Isa, MacArthur River, and Faro is commonly included in discussions of "volcanogenic massive sulphides" (e.g., Hutchinson, 1973), but the deposits occur in sedimentary successions with a relatively minor proportion of volcanic rocks. Their tendency to be laminated, their narrow age range (Early Proterozoic), relatively large size, S-isotope distributions (cf. Chapters in Vol. 2), and tectonic setting are other factors which set these deposits apart, even though their genesis may well be fundamentally similar to that of the "volcanic" types.

With increasing magnetite content, volcanic massive sulphide ores grade to massive oxide ores dominated by magnetite and/or hematite, e.g., Savage River, Tasmania (Coleman, 1975), Fosdalen and Tverrfjellet, Norway (Waltham, 1968), and Kiruna, Sweden (see Parak, 1975). A small marginal group that should also be included in this discussion is that of the stratiform sulphur, and sulphur-iron sulphide, deposits such as those in Pleistocene andesites and dacites of northeastern Honshu (Mukaiyama, 1970) and northern Taiwan (Tan, 1959).

---

[1] Numerous chapters discuss volcanogenic deposits, e.g., those by Sangster and Scott, Lambert, Gilmour, Vokes, Ruitenberg.

Some polymetallic massive sulphide deposits occur in wholly sedimentary successions and are thereby excluded from this discussion but a few of these (e.g., the Cobar district, Kappelle, 1970) are otherwise identical in composition (including S-isotope distributions) and any genetic model should be flexible enough to account for such deposits. Others have S-isotope trends indicating biogenic influences (e.g., Rammelsberg; and Anger et al., 1966) and closer genetic associations with the Sullivan-Mt. Isa Group.

Another complication in identifying and grouping deposits is that they may contain massive sulphide orebodies, but derive most of their production from disseminated, low-grade mineralization (e.g., Mt. Lyell, Tasmania; Reid, 1975).

The nature of the host rocks to massive sulphide deposits, and the probability of there being associations of certain ore-types and host-rock types, are clearly important points for investigation by theoretical and applied geologists (cf. Chapter 4 by Gilmour, Vol. 1). It will be shown in this chapter that clear-cut associations are difficult to detect on present information but that this result is compatible with recent theories on the genesis of massive sulphide deposits.

VOLCANIC HOST ROCKS

Felsic vs. mafic

In a preliminary study comparing host-rock and ore-type, Anderson (1969) was unable to find any well-defined relationship between the metal contents of massive sulphides and the proportions of felsic and mafic volcanic rocks (and sediments). The writer has undertaken a similar study of 50 fairly well documented deposits (Fig. 1), initially subdividing the rocks 100 m or so stratigraphically below the orebodies into rhyolite, dacite, andesite, basalt and sediment. Difficulties in the definition and recognition of dacite and andesite finally reduced the subdivision to felsic (rhyolite and dacite), mafic (basalt and andesite), and sediment, or combinations of these. The deposits range from Archaean to Tertiary in age, and may be divided into the following types on the basis of their overall Cu/Pb/Zn ratios (wt %): (a) Zn–Pb–Cu (generally Zn > Pb > Cu), (b) Zn–Cu (generally Zn ⩾ Cu), and (c) Cu (cf. also Chapter 12 by Lambert).

The mineralogy of the three groups is as follows:

(a) Pyrite–sphalerite–galena–chalcopyrite (in order of abundance) and minor tetrahedrite, plus or minus pyrrhotite, bornite, etc., with both gold and silver recovered (e.g., Rosebery, Kosaka, Shakanai, Buchans).

(b) Pyrite–pyrrhotite–sphalerite–chalcopyrite, plus or minus arsenopyrite, magnetite, etc. (see Sangster, 1972, p. 36), with both gold and silver recovered (Timmins, Noranda district, West Shasta district).

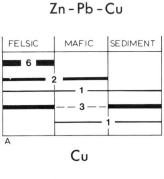

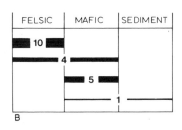

Fig. 1. Diagram to illustrate the nature of the footwall rocks to massive sulphides, from a study of the following 50 deposits:

America: Mammoth, Iron Mountain, Bully Hill, and Keystone-Union (California); Iron King and United Verde (Arizona).

Australia: Mt. Lyell, Rosebery, Captains Flat and Mt. Morgan.

Canada: Horne, Quemont, Vauze, Lake Dufault, Flin Flon, Kidd Creek, Mattagami Lake, Orchan, Hidden Creek, Weedon, Anaconda Caribou, Brunswick Mining and Smelting No. 6 and No. 12, Buchans, Stirling and Betts Cove.

Cyprus: Skouriotissa and Mavrovouni.

Eire: Avoca.

Fiji: Undu (Vanua Levu).

Japan: Four Kuroko deposits of the Hokuroko district; Besshi, Sazare and Ikadatsu of Shikoku; Yanahara of the Chugoku district.

Norway: Skorovass, Lokken and Sulitjelma.

Philippines: Barlo.

Russia: Blyavinskia, Urals.

Spain: Rio Tinto, San Miguel, San Paulo, and Tharsis.

Sweden: Boliden.

Turkey: Ergani and Lahanos.

(For further explanation, see text.)

(c) Pyrite—chalcopyrite, plus or minus pyrrhotite, sphalerite, marcasite, etc., with gold recovered (Cyprus mines, Rio Tinto, Tharsis). These deposits are relatively low in silver.

Only a few deposits fail to fit into these broad groups, e.g., Bawdwin in Burma with Pb > Zn and Cu (Sommerlatte, 1958; H.W. Walther, personal communication, 1973), and some pyrite deposits in volcanic rocks with little or no zinc, lead, or copper. Difficulties occur with mines that contain more than one type of ore and it is emphasised that the subdivision refers only to the total-metal ratio. The method of subdivision is different from that used by Hutchinson (1973) which includes the nature of the host rocks in the definition of each group, e.g., his pyrite—chalcopyrite types include only those associated with ophiolites. As the purpose of this chapter is to investigate ore—host-rock relationships, such a classification is unsuitable here.

The principal conclusions from the first stage of the study are as follows (Fig. 1):
50% overlie only felsic volcanic rocks (8% include sedimentary rocks);
30% overlie only mafic volcanic rocks (12% include sedimentary rocks);
20% overlie mixed volcanic rocks (4% include sedimentary rocks);
25% overlie volcanic rocks with more than about 5% of sediments (mainly shales).

A wider view of about 300 deposits, based on more limited data, revealed similar findings. Another variation of the first study, in this case to include rocks 1—2 km below the ore, was also hampered by poor data, but indicated that more than half of the deposits overlie mixed volcanic terrains with or without sediments (notable exceptions being the Cyprus deposits, which overlie entirely mafic rocks).

The importance of rhyolites as host rocks (particularly fragmental types) has been noted by many geologists (e.g., Sangster, 1972), and appears to be maintained even in terrains dominated by andesites and basalts. For example, the Abitibi volcanic belt in the Canadian Archaean consists of 44% basalt and 26% andesite (Descarreaux, 1973) but almost all the orebodies overlie rhyolitic rocks. In the Iberian pyritic belt, rhyolite is the dominant footwall rock to the major deposits, although the terrain contains basalts, andesites and sediments (Schermerhorn, 1970), and a similar pattern is found in the Palaeozoic volcanic belts of the Tasman Orogenic Zone of eastern Australia.

No account in these studies has been taken of the possibility of the deposits under review being allochthonous as a result of processes such as gravity sliding (Schermerhorn, 1970; Jenks, 1971).

The second stage of the 50-sample study involved comparisons of host-rock type and ore type, and some of the conclusions are as follows (Fig. 1):

(a) 70% of the Zn—Pb—Cu types overlie only felsic rocks (half including some sedimentary rocks) and 92% have abundant rhyolites in the footwall;

(b) 50% of the Zn—Cu types overlie only felsic rocks, and 30% only mafic rocks;

(c) 35% of the Cu types overlie only felsic rocks (a few with sedimentary rocks) and 47% overlie only mafic rocks (half including some sedimentary rocks); and

(d) there are no Zn—Pb—Cu ores overlying wholly basaltic terrains.

The most interesting results involve the distribution of lead. This is almost invariably associated with felsic rocks and never with ophiolitic terrains, which contain only Cu or Zn—Cu types of ore. The distributions are complicated by the low levels of lead (<1%) in Archaean deposits which are almost entirely of the Zn—Cu type (Sangster, 1972; Hutchinson, 1973). As these are numerous and mainly overlie rhyolites, the Zn—Cu ores as a whole emerge as a rhyolite-oriented group despite the fact that Phanerozoic Zn—Cu ores occur mostly with intermediate and mafic rocks. It is not surprising to find that Cu and Zn—Cu deposits overlying ophiolites have significantly lower silver contents compared to those in other terrains, and this provides a rough basis for subdividing each of these groups.

A crude estimate of the proportions of each class of ore indicates that the Zn—Cu total is about equal to Zn—Pb—Cu plus Cu. This result is again influenced by the abundance of Zn—Cu ores in the Archaean, almost to the exclusion of the other two groups, even though large areas are dominantly basaltic and might therefore be expected to carry some Cu deposits.

Magma-type and ore occurrence

Most of the volcanic rocks associated with massive sulphide deposits appear to belong to the calc-alkaline, basalt—andesite—rhyolite suite of Andean and island arcs, but there are important occurrences in tholeiitic basalts of oceanic origin, and possibly some in low-K tholeiites of island arcs. Few, if any, are associated with alkaline volcanic rocks in island arcs or rifting zones, or with "continental" volcanic rocks. In many areas, there are insufficient chemical data to classify the rocks, making this part of the review particularly difficult.

Calc-alkaline association. Most of the deposits in calc-alkaline rocks are Zn—Pb—Cu or Zn—Cu types, though pyrite—chalcopyrite types are known. If, as suspected, the Iberian pyritites are associated with calc-alkaline rocks, then the Cu types are fairly common.

Terrains of clearly established calc-alkaline type containing massive sulphide deposits occur around the Pacific margin (Fiji, Japan, North American Cordillera) and in Canada, eastern Australia and the Urals. One of the areas worked out in some detail is the Abitibi region in Canada, where the deposits lie exclusively in calc-alkaline (mainly rhyolitic) rocks (Descarreaux, 1973). In a pattern resembling Pacific island arcs, the rhyolite rocks occupy only 15% of the succession and occur with basalts and andesites of calc-alkaline and tholeiitic affinities (Baragar, 1968; Descarreaux, 1973). The localization of ores mainly in felsic rocks within mafic-dominated terrains is the trend for most of the Canadian Archaean, but does not apply generally in the Phanerozoic. For example, in the basalt- and andesite-dominated terrains of the island arcs from New Guinea to Fiji, massive sulphide deposits occur in dacites and rhyolites (Undu, Vanua Levu; Rickard, 1970; Colley and Rice, 1975) and in andesites (Wainivesi, Viti Levu; Rodda, 1963) while in the Solomon Islands pyrite—chalcopyrite mineralization is known in basalts of ocean-floor or low-K tholeiite type (Taylor, 1974).

Similarly in New South Wales, Australia, Ordovician volcanic successions that appear to be dominated by mafic rocks (e.g., the Cargo and Walli Andesites, Packham, 1969) carry small massive sulphide deposits (Stanton, 1955), and younger Silurian volcanic rocks in the same region (which generally contain a much higher proportion of rhyolites) are hosts to the Captains Flat, Woodlawn, Mineral Hill and many minor deposits. A comparable situation exists in the East Shasta and West Shasta districts in California.

In New South Wales the mineralization appears to be much more intense in the rhyolite-rich terrain than the mafic terrain, and the deposits in the island arc chain from New Guinea to Fiji have to date proved uneconomic. This raises the possibility that the intensity of mineralization is related to magma-type. Jakes and White (1972) indicated there were major chemical differences between the rhyolite-rich, Andean calc-alkaline rocks and those from island arcs, and on a selected range of samples, Solomon and Griffiths (1974) found statistically significant differences in some trace element contents between the types. In fact, there may well be a range in composition from the rocks of primitive arcs to those formed in mature arcs developed over substantial thicknesses of continental crust, as indicated by Miyashiro (1974). Existing information on the intensity of mineralization seems to be inadequate to pursue comparative studies and there may be many complications. For example, the lack of mineralization in mafic terrains could reflect unsuitable physical conditions for solution flow or sulphide preservation rather than different magmatic processes, and in the case of the Pacific Islands may reflect the relative lack of mineral exploration and development.

Other associations. Another volcanic rock type that appears to be rarely mineralized is the island-arc tholeiite, though Horikoshi (1972) has concluded from chemical data that metabasalts of this type are associated with the Besshi pyrite—chalcopyrite orebodies in Shikoku, Japan (see also Mitchell and Garson, 1973). Horikoshi (1975) also noticed that some Kuroko deposits and all the sulphur—pyrite deposits of northern Honshu lie within the tholeiitic magma zone at the arc "front", and the time of initiation of the tholeiite (about 13 m.y.) is very close to the time of the Kuroko mineralization. Though all the Kuroko and sulphur—pyrite orebodies are associated with calc-alkaline volcanic rocks and many Kuroko deposits lie in the high-Al basalt zone, it may be that the beginning of tholeiite volcanism triggered or accompanied the ore-forming processes.

The Cyprus pyrite—chalcopyrite and pyrite—chalcopyrite—sphalerite deposits are often quoted as examples of ophiolite-associated deposits. However, there is considerable confusion as to the nature of the pillow lavas which overlie the sheeted dyke complex and gabbro (Fig. 2). The footwall rocks to the deposits are either the basal group lavas (Searle, 1972) or the lower pillow lavas (Hutchinson and Searle, 1971; Adamides, 1975) and Pearce (1975) showed that these rocks have island-arc and ocean-floor tholeiite characteristics (though he found it difficult to isolate the chemical effects of sea-water—basalt reactions). He suggested that the post-ore, upper pillow lavas have affinities with low-K tholeiites. Though detailed studies of similar deposits (e.g., Betts Cove; Upadhyay and Strong, 1973) may well reveal similar complications, the association of pyrite—chal-

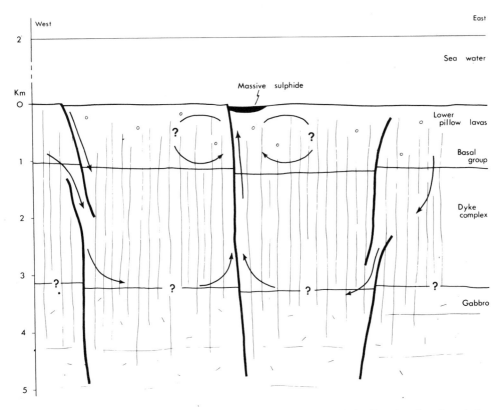

Fig. 2. Hypothetical east—west cross-section of the northern part of the Troodos Complex at the time of formation of the massive sulphide deposits, showing possible ore solution flow-paths (after Vine and Moores, 1972). The dykes are oriented north—west (Adamides, 1975) and the faults trend north.

copyrite deposits and pillow lava—dolerite—gabbro—ultramafic complexes seems well established.

No mention has been made so far of the association of massive sulphides with spilites and keratophyres (Amstutz, 1968). Studies of altered basalts appear to have led to a generally accepted view that spilites and keratophyres represent a facies of altered volcanic rocks of various types (Cann, 1969; Vallance, 1969) and this view is adopted here. In most cases, alteration has not totally obscured the nature of the original rock, and major and minor element analyses may be used to classify magma-types. In western Tasmania, for example, Solomon (see Solomon and Griffiths, 1974) and White (1975) have been able to show that a suite of keratophyres and spilites (the Mt. Read Volcanics) containing the orebodies at Mt. Lyell and Rosebery (Fig. 3) has calc-alkaline affinities, and that adjacent basalts have compositions close to those of low-K tholeiites.

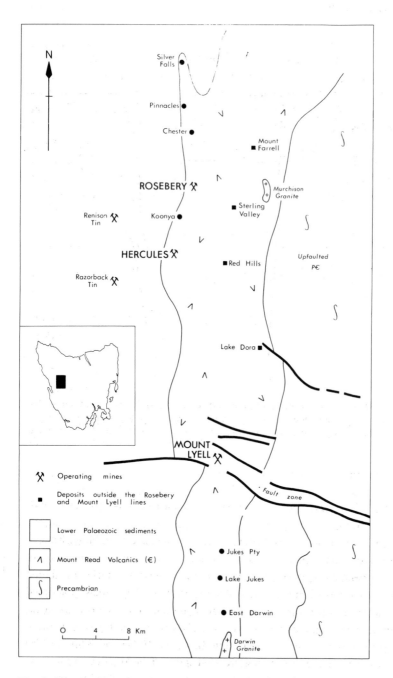

Fig. 3. Western Tasmania showing the distribution of Cambrian calc-alkaline volcanic rocks and the Rosebery and Mt. Lyell "lines" of ore deposits.

Local host-rock—ore relationships

Massive sulphide deposits commonly occur in groups and in any one area they tend to occur at one, or more than one, specific horizon within the succession (Stanton, 1972, p. 508). This horizon may represent a change in composition of the volcanic rocks, e.g., from felsic to mafic (Horne and Vauze mines, Noranda), or a change from volcanism to sedimentation (Rosebery, Tasmania), or simply a pause in volcanism without any distinctive features. Spence and De Rosen-Spence (1975) have outlined five successive zones of rhyolite and andesite in the Noranda area, and though sub-economic deposits occur in the fourth and fifth zones, over 90% of the metal production was derived from deposits within the third zone. Though the nature of the footwall and hanging wall rocks is variable, all the major deposits formed during a short time interval that in some places coincided with the intrusion of small rhyolite domes. The third-zone rhyolites do not appear to have anomalous chemical compositions.

In the Buchans area of Newfoundland, Thirlow et al. (1975) have recognized four cycles of calc-alkaline volcanism, each cycle beginning with andesite and passing through to dacite and felsic breccia. Only the first two of these are mineralized, there being two massive Zn—Pb—Cu orebodies near the top of each cycle, but the rocks of these cycles appear to be similar in primary composition to those of the remaining cycles. High barium and base-metal contents near the ore bodies may well be developed during mineralization (cf. Thirlow et al., 1975).

In the Iberian pyrite belt, ore occurs on at least two stratigraphic horizons within the Lower Carboniferous in association with altered rhyolitic and dacitic rocks of various types and also with shales (Schermerhorn, 1970; Bernard and Soler, 1974). Despite the possible complexities resulting from gravity sliding of the pyrite to exotic environments it seems highly likely that ore deposition took place more than once and on more than one rock type.

The stratigraphic position of the Cyprus deposits is uncertain; if they overlie the basal-group basalts there is little or no difference between the composition of pre- and post-ore rocks (the case of Searle, 1972), but if they overlie the lower pillow lavas (as in Fig. 2) there is a pronounced change (Pearce, 1975) that may be related to the intrusion of relatively siliceous dykes (Govett and Pantazis, 1971). Upadhyay and Strong (1973) have suggested that the ophiolitic pyrite—chalcopyrite deposits in Newfoundland (and also the Oman Mountains) overlie the sheeted dyke complex.

In the Mt. Lyell district of western Tasmania, mineralized calc-alkaline volcanic rocks are overlain by unmineralized rocks and though it is possible to identify the two groups in the field by the proportions of sedimentary and acid and basic volcanic rocks, the whole succession belongs to one petro-chemical suite (White, 1975). Similar conclusions have been drawn for the succession around the Rosebery Mine (W.B. Anderson, personal communication, 1973).

In the Hokuroko district of northtern Honshu, all the major Kuroko deposits occur in the Upper Nishikurosawa formation of Middle Miocene age, though a few minor deposits lie at higher stratigraphic levels (Sato et al., 1974). The deposits commonly overlie fragmental rocks around the margins of rhyolitic domes, and Horikoshi (1969) suggested the ores formed in the waning stages of a volcanic cycle. Tatsumi and Clark (1972) suggested the plugs were the most fractionated and latest phase of such a cycle and attempted to establish this by comparing the differentiation index (normative Q + Or + Ab) of dome rocks and surrounding volcanic rocks. The main study, involving 19 samples from the Kosaka mine, failed to yield significant differences (cf. Tatsumi and Clark, 1972). A subsidiary study of rocks from the Tsuchikata mine showed a very large difference, but only five samples were involved in this comparison (2 dome rocks and 3 others), making the test inconclusive. The Kuroko orebodies of the Aizu district of Honshu occur at three horizons in Miocene volcanics and appear to be associated with rhyolitic flows as well as domes (Hayakawa et al., 1974).

A close association between coarse-grained fragmental rocks and ore occurrence appears to be fairly common though by no means universal. Sangster (1972) and Sangster and Scott (Chapter 5, this volume) have noted that massive sulphides in the Canadian Archaean are commonly associated with coarse-grained pyroclastic rocks, and commonly the most coarsely grained in any one area (usually greater than 64 mm diameter and locally termed "mill rock"). This is also true for the Kuroko ores of northern Honshu, where many of the orebodies overlie the explosive products of the adjacent rhyolite domes. In western Tasmania, the Rosebery Zn—Pb—Cu orebody occurs in tuffs and shales(?) overlying a rhyolitic pyroclastic rock (Brathwaite, 1974), and the largest orebody at Mt. Lyell, the Prince Lyell, occurs mainly in highly altered fragmental rhyolite. At the Lake Jukes deposit, 10 km south of Mt. Lyell (Fig. 3), bornite veins occur in a rhyolite breccia containing blocks over 50 cm in diameter. However, there are many deposits without coarse-grained volcanic rocks and it appears that though explosive volcanism is commonly a related process it is not necessary pre-requisite to mineralization.

RELATED PLUTONIC ROCKS

In an important paper that began a revival of syngenetic theories for massive sulphide deposits, Stanton (1955) noted that a number of small Upper Ordovician and Lower Devonian (Packham, 1969) deposits in the Bathurst area of New South Wales occurred within or close to certain sedimentary beds, and did not appear to be related to the local granitic plutons of probable Devonian age that cropped out within 2—3 km of the deposits. Similar patterns may be seen in other Phanerozoic orogenic belts containing zones of calc-alkaline volcanic rocks intruded by syn- and/or post-volcanic granitic plutons, e.g., Newfoundland and New Brunswick, southern Spain, Californian Cordillera, etc. In the Tasman Orogenic Zone in west Tasmania the Mt. Read Volcanics were intruded by several sills and plugs of granitic composition (Fig. 3) and these appear to have been intruded to shallow depths during the growth of the volcanic pile; the Darwin Granite at least could well be synchronous with massive sulphide mineralization (Williams et al., 1975; White, 1975) but none of them occur close to, or have any systematic relationship to, the ore deposits. In contrast, there are cases in other areas of more or less synchronous mineralization and intrusion where the intrusives have metamorphosed the ores. For example, at Mt. Morgan the pyrite—chalcopyrite orebody lies within Middle (?) Devonian rhyolites and has been metamorphosed by an Upper Devonian granitic pluton (Lawrence, 1972). In the Weedon Mine, in Quebec, a pre-Middle Ordovician Cu—Zn deposit has been metamorphosed and disrupted by an albite granite plug of pre-Normanskill (Middle Ordovician) age (McAllister and Lamarche, 1972). In eastern Viti Levu, Fiji, the Wainivesi deposits appear to be massive sulphides lying in Miocene andesitic tuffs and sediments that have been intruded by several small tonalite—diorite intrusions of similar age (Hirst, 1965).

Relationships of this nature raise queries concerning the time relationships between porphyry copper and massive sulphide mineralization. In the two cases known to the writer, where the two occur in the same region, the massive sulphide deposits are older. Firstly, in Central Bulgaria, east of Sofia, a Cretaceous volcanic sequence containing several pyrite—chalcopyrite deposits (e.g., Radka in Bogdanov and Bogdanova, 1974) has been deformed and intruded by Laramide plutons carrying porphyry—copper mineralization, e.g., Medet (Angelkov, 1974). Secondly, in Fiji the Wainivesi deposits occur in Miocene andesites while 50 km to the southwest, in the Namosi area, similar volcanic rocks have been intruded by a suite of granitic plutons bearing porphyry—copper mineralization of probable Pliocene age (Geological Map of Fiji). There is thus some support for the time separation proposed by Hutchinson and Hodder (1972).

TECTONICS

Previous workers have pointed out that massive sulphide deposits occur in Andean volcanic belts, island arcs, island arc—trench zones, and ocean basins (probably formed at spreading centres), e.g., Sillitoe, 1972; Mitchell and Garson, 1972, 1973; and others). They also appear to have formed in shales of the continental rise (Cobar, Meggen), and the magnetite—pyrite ores at Savage River probably developed near a hinge line at a shelf-rise junction of Late Proterozoic age (Coleman, 1975). Thus massive sulphide deposits may be found throughout the classical orogenic belt and also in upthrust masses of oceanic material. They are not found in the stable regions of the continental plates though it is possible that deposits like MacArthur River and Mt. Isa have formed during tensional faulting (rifting?) within large intracontinental basins, with or without significant vulcanism (see also Sawkins, 1974).

A feature of many massive sulphide deposits is their obvious spatial relationship to faults. Sangster (1972) drew attention to some Archaean examples in relatively undeformed rocks and postulated pre-mineralization movements. Similarly for the Cyprus ores, Kortan (1970) noted their association with N–S fracture zones and suggested that the basins of ore deposition were developed by N–S, pre-mineralization faulting. At Mt. Lyell, the ore deposits are concentrated at the intersections of prominent N–S and E–W fractures, both on regional and local scales. Though the fractures are clearly Devonian in age, movements along some fractures in the Late Cambrian–Early Ordovician are indicated by local unconformities and variations in thickness of Ordovician sediments (Solomon, 1969). Mineralized rocks are overlain with local unconformity by volcanics dated as late Middle/early Upper Cambrian (Jago et al., 1972), and it seems likely that fault movement occurred about the same time as mineralization.

In the Hokuroko district of northern Honshu, many of the Kuroko orebodies and vein-type deposits lie on N–S to NNE fractures (Fujii, 1974; Sato et al., 1974). In this district it is possible to account for the major structures (NNE and N–S faults, NNW folds and faults, NE and ENE fractures) in terms of a NNE dextral shear system. This interpretation predicts strong WNW structures and in the Aizu district further south, such fractures are prominent and appear to exert some control over the distribution of ore deposits (Hayakawa et al., 1974).

Tensional fracturing parallel to the regional structural grain is a feature of Andean and island arc terrains (e.g., the Taupo volcanic rift in New Zealand) and mid-ocean ridges, and the fracture systems must significantly increase the effective permeabilities of the volcanic successions.

INTERIM SUMMARY OF FIELD AND LABORATORY DATA

This review has highlighted a number of features of massive sulphide deposits, e.g.:

(a) They occur most commonly in calc-alkaline volcanic rocks, and particularly, but not exclusively, in rhyolitic fragmental types.

(b) They also occur in oceanic and low-K tholeiites, and in addition are found in sedimentary successions containing little or no evidence of volcanic activity.

(c) There are no exclusive relationships between ore-type (defined by metal ratio) and host-rock type, except that lead-rich deposits most commonly overlie rhyolites and never overlie wholly basaltic, or ophiolitic, terrains.

(d) The orebodies in any one field tend to lie on one (in some areas more than one) stratigraphic horizon. This commonly has no unusual features apart from the mineralization and associated exhalites, and the nature of the rocks above and below may vary within one orefield.

(e) The volcanic rocks intimately associated with the ores generally have no distinctive primary features compared to those in the rest of the succession.

(f) The deposits commonly occur in terrains intruded by acidic plutons, but appear to have no systematic relationships to the plutonic rocks. These may have intruded close to or at some distance from the deposits and be more or less the same age, or later.

(g) The limited evidence available indicates that deposits in ophiolites occur at or near the base of the pillow lava sequence and overlie dolerite dykes and gabbroic intrusives.

(h) They commonly lie on major faults or within fault zones.

Before attempting to explain these features, it is necessary to summarize some salient characteristics of massive sulphide deposits that have not been reviewed here, and the list is continued as follows:

(i) The lack of detritus and massive nature of the orebodies, and the lack of major changes in patterns of vulcanicity and/or sedimentation before and after mineralization, all point to rapid deposition of the sulphide masses. This contrasts with the field evidence for the Sullivan-Mt. Isa group of deposits.

(j) Many deposits have pipe-like stockwork zones beneath the massive ore that extend downwards for several hundred metres and have diameters less than that of the orebodies (e.g., Sangster, 1972). They are marked by low-grade mineralization and hydrothermal alteration. Other deposits overlie much wider zones of alteration that extend beyond the ore limits (e.g., Mt. Lyell, and the Hokuroko area; Iijima, 1974).

(k) Massive sulphide deposits seldom occur singly and have characteristic grouping patterns. For example, there are twelve major orebodies in a 50 km zone along the northern flank of the Troodos complex (Fig. 4) and their spacings and the approximate ton-

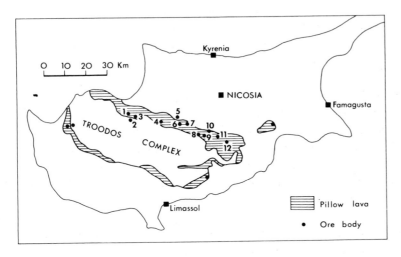

Fig. 4. Map of Cyprus showing the massive sulphide deposits. Those referred to in Fig. 5 are numbered as follows (after Bear, 1963; Kortan, 1970; Constantinou and Govett, 1972): *1* = Mavrovouni; *2* = Apliki; *3* = Skouriotissa; *4* = Memi; *5* = Kokkinoyia; *6* = Kokkinopezoula; *7* = Agrokopia; *8* = Kampia; *9* = Kapedhes; *10* = Kambia; *11* = Mathiati North; *12* = Sha.

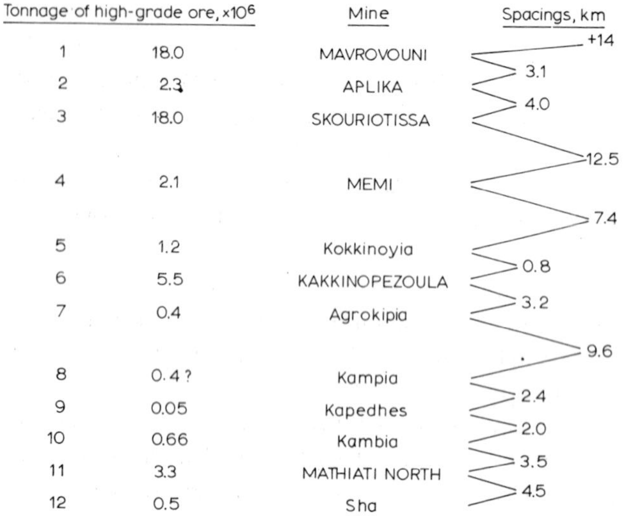

Tonnage of high-grade ore, ×10⁶		Mine	Spacings, km
1	18.0	MAVROVOUNI	+14
2	2.3	APLIKA	3.1
3	18.0	SKOURIOTISSA	4.0
			12.5
4	2.1	MEMI	
			7.4
5	1.2	Kokkinoyia	
6	5.5	KAKKINOPEZOULA	0.8
7	0.4	Agrokipia	3.2
			9.6
8	0.4 ?	Kampia	
9	0.05	Kapedhes	2.4
10	0.66	Kambia	2.0
11	3.3	MATHIATI NORTH	3.5
12	0.5	Sha	4.5

Fig. 5. Spacings and approximate tonnages of the massive sulphides on the northern flank of the Troodos Massif (from Bear, 1963; Kortan, 1970; data from the Cyprus Mines Corporation, 1974; and the geological Map of Cyprus; see Fig. 4). The "plus" sign indicates the exposed extent of the ore zone that has been proposed without success.

nages (production and reserves) of high-grade ore are given in Fig. 5. The orebodies clearly form groups, with intra-group spacings mainly between 2.0 and 4.5 km (average 3.2, plus one at 0.8) and spacings between groups of 7.4 and 12.5 km. Each group contains at least one substantial (>1 m tons) deposit. No deposits (other than minor showings; Kortan, 1970) have been found around the Mavrovouni–Apliki–Skouriotissa group for at least 14 km to the west and 12.5 km to the east, despite intensive exploration.

A similar pattern to that on Cyprus is seen in the Iberian pyritic belt, where about 60 deposits occur in a zone of volcanic rocks some 220 km long (Bernard and Soler, 1974, fig. 1). The three major deposits (Tharsis, Zarsa and Rio Tinto), each containing at least $100 \cdot 10^6$ tons of pyrite, are spaced 25 km apart while the smaller deposits occur at much smaller intervals. This mineral province has more massive sulphide deposits than any other and also contains the largest in the world, viz, the Rio Tinto pyritite-stockwork ores, with about $525 \cdot 10^6$ tons of pyrite (D. Williams, personal communication, 1974).

In the Hokuroko district of Japan, the massive and major vein deposits tend to form groups with internal spacings of 1–2 km while the groups themselves (e.g., Hanaoka–Matsumine–Shakanai) have considerably greater, though rather variable, spacings ranging from about 8 km (e.g., Furotobe–Kosaka) to 16.8 km (e.g., Hanaoka–Furotobe) (Sato et al., 1974, figs. 1, 3). If the spacings in this area are measured along NNE lines

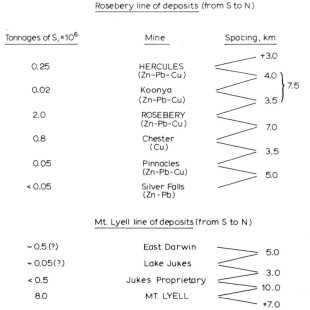

Fig. 6. Spacings and approximate tonnages of the deposits in the Rosebery and Mt. Lyell "lines" (see Fig. 3).

(see previous discussion) they become much more uniform around 14–15 km, e.g., Hanaoka–Yunosawa; Furotobe–Fukazawa; and Kosaka–Ginzan.

In western Tasmania, the Rosebery and Mt. Lyell deposits lie on separate N–S zones containing several smaller deposits (Figs. 3, 6) and again there is a relationship between size and spacing for the larger deposits.

This brief review of the distribution pattern indicates a tendency for massive sulphide deposits to form clusters, and that large deposits and whole groups tend to be spaced roughly according to their size.

PROBLEMS OF GENESIS

Theories as to the origin of massive sulphides fall fairly clearly into two groups: (a) solutions derived largely from magmas (e.g., Tatsumi and Watanabe, 1971) and (b) solutions of mainly surface origin leaching ore components from the local rocks (e.g., Corliss, 1971). The magmatic theories have difficulty in explaining the grouping and spacing of the deposits, the stratigraphic control, the wide range of tectonic settings and associated igneous rocks, and in particular the lack of systematic spatial relationships with plutons. In some cases, there is a size problem because the spatially related intrusives could not act as major sources of the ore solutions. For example, the Kuroko ores commonly occur around the perimeters of rhyolitic plugs (Horikoshi, 1969) which have helped to shape the topography of the ore basins and also act as conduits for the ore solutions. A plug about 600 m in diameter, extending vertically for say, 8 km, and containing 2% of water in solution could yield about 10^{14} g of water. If the solubility of Fe is 20 ppm (Sato, 1972) then the solution could yield $2 \cdot 10^9$ g Fe. However, Kuroko deposits usually contain 10^{11} to 10^{12} g Fe (Lambert and Sato, 1974), leaving a large shortfall.

It would appear that the most satisfactory answer to this problem is provided by convective circulation of surface water, as proposed by Ellis (1967), Corliss (1971), Henley (1973), Spooner and Fyfe (1973) and others. As massive sulphide deposits appear to be formed mainly in marine environments (Anderson, 1969), sea water should be the principal surface contribution.

The probable importance of sea water in some ore solutions is indicated by $\delta^{18}O$ and δD data from fluid inclusions in the Matsumine and Shakanai deposits (Ohmoto and Rye, 1974; see also Chapter 4 by Roedder, Vol. 2) and from hydrous minerals in alteration zones of the Cyprus deposits (Heaton and Shepherd, 1974). The role of sea water in the alteration of the Cyprus basalts had already been indicated by $\delta^{18}O$ studies (Spooner et al., 1974).

Before considering the evidence further, it is proposed to examine the convection or geothermal model in greater detail.

THE GEOTHERMAL MODEL

Conditions for convection

Considerable success has been achieved in the theoretical modelling of the Wairakei system in which the water is driven upwards towards the surface by buoyancy effects (Wooding, 1957; Donaldson, 1962; Elder, 1965). Lapwood (1948) was able to show that in a fluid-saturated, horizontal, permeable slab under the influence of a vertical temperature gradient, convection is only possible if the Rayleigh number of the system exceeds a certain value. This dimensionless number, R, is given by

$$\frac{k \, \alpha \, g \, H \, \Delta T}{\kappa_m \nu}$$

where k = permeability; α = coefficient of cubical expansion; g = gravitational constant; H = thickness of the medium; ΔT = temperature difference between top and bottom of the permeable slab; κ_m = thermal diffusivity of the water-saturated medium; ν = kinematic viscosity.

For a medium of infinite extent bounded above and below by impermeable layers, Lapwood (1948) showed that for convection to take place, R must be greater than $4\pi^2$, and this has been confirmed by experiments (see Elder, 1965). For a situation where the upper surface is bounded by free-standing fluid, which is presumably somewhat closer to the massive sulphide situation, the critical Rayleigh number is 27.1 (Lapwood, 1948). Using this model, it can be shown that for resonable values of permeability, etc., convection seems likely in a wide range of geological situations. For example, drilling at the Mavrovouni and Skouriotissa mines on Cyprus has shown that hydrothermal alteration extends downwards for at least 700 m, and that mineralization occurs near the base of the pillow lavas, so the solutions probably rose through a vertical distance approaching 1 km. $\delta^{18}O$ date indicate that the rocks for at least 2 km depth have been involved in reactions involving high water/rock ratios (Spooner et al., 1974) and in this review H is taken as 2 km. The coefficient of cubical expansion is taken at a pressure of 400 bars, assuming a depth of water of 2 km. Though the temperatures are unknown, indications from fluid inclusions in the Kuroko deposits (e.g., Tokunaga and Honna, 1974; Chapter 4 by Roedder, Vol. 2) indicate a temperature range for the solutions near the base of the ore bodies of 200–250°C, and modelling of the Wairakei deposit suggests the temperature at the bottom of the convection cell could be 400°C (to give $\Delta T = 200$). The other variables are estimated in c.g.s. units for pure water as follows: $\alpha \approx 1.6 \cdot 10^{-2}$ at 400°C/ 400 bars (Burnham et al., 1969); $\Delta T = 200$; $H = 2 \cdot 10^5$; $\kappa_m = \kappa/\rho c \approx 6 \cdot 10^{-3}$, estimated from $\kappa = (1-e)K_s + eK_f$ (Elder, 1965), where e = porosity, taken as 0.005 (see below), K_s = thermal conductivity of solid (basalt), taken as $4 \cdot 10^{-3}$ (Kappelmeyer and Heanel, 1974). K_f = thermal conductivity of fluid (sea water) taken as $4 \cdot 10^{-4}$ (Toulmin and Clark, 1967), ρ = density of system (~2.8), and c = heat capacity of system (~0.25); $\nu = 3 \cdot 10^{-3}$ (Toulmin and Clark, 1967).

The permeability is difficult to estimate and there is likely to be a wide variation between the dyke complex and the pillow lavas. A conservative estimate would allow 1 crack/mm (10^3 cracks/m^3) and each crack 5 μm wide, to yield a permeability of about 10^{-2} darcys (1 darcy $\approx 10^{-8}$ cm^2) and a porosity of 0.005 (Henley, 1973). Values averaging 10^{-1} darcys were obtained from near the surface at Wairakei (Elder, 1965, 1966). Using the figures given above yields a Rayleigh number of $3.5 \cdot 10^3$, well above the lower stability limit of convection.

At high Rayleigh numbers, the rising fluid below an upper impermeable boundary tends to mushroom out just below that boundary (Donaldson, 1962) and this approximates the flow pattern at Wairakei (Fig. 7a), where the ground–air interface acts as a fairly impermeable medium. In the case with free-standing fluid above the surface of the permeable medium, the rate of discharge and recharge and the ability of the overlying fluid to participate in the convection are all critical factors in determining the flow pattern, and particularly the size or even absence of secondary circulation patterns near the discharge area. Elder (1965) has simulated the ideal situation in a heated Hele-Shaw cell with hot water withdrawn at the top of the rising column, and cold water

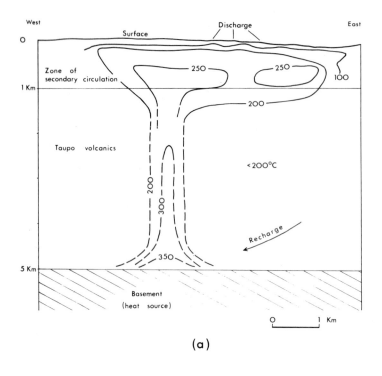

(a)

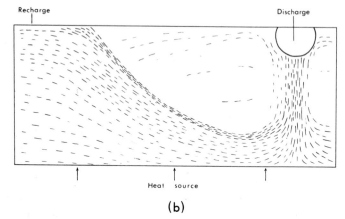

(b)

Fig. 7. a. Possible temperature distribution in the Wairakei convection cell (after Elder, 1965). b. Flow pattern as illustrated by aluminium particles in a Hele-Shaw cell with point discharge, drawn from a photograph in Elder (1965, fig. 19).

recharged at the sides of the cell (Fig. 7b). This probably represents the simplest geological situation with a well-defined vertical feeder pipe below a massive orebody, and attention is drawn to the sharp upper boundary of the cool, recharging water as it trav-

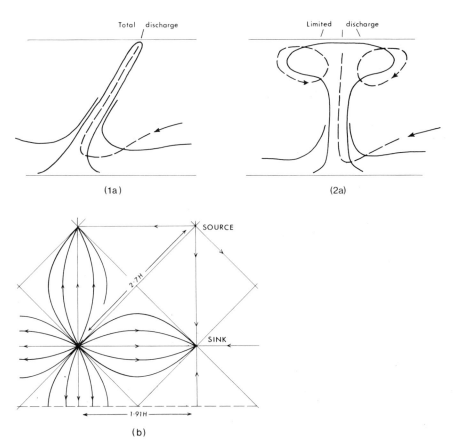

Total   discharge

Limited   discharge

(1a)

(2a)

SOURCE

2·7H

SINK

1·91H

(b)

Fig. 8. a. Possible flow patterns for (1) discharge = recharge, high flow rate and fault control (2) limited discharge, no fault control. Solid lines are isotherms. b. Theoretical spacings of sink and source in a saturated porous slab with impermeable material below, and water above (after Lapwood, 1948).

els downwards towards the hot column. An application to a geological situation involving a dipping fracture is shown in Fig. 8a.

Wooding (1963) has shown that with high flow rates the rising column behaves as a jet, having a very small diameter compared to the size of the cell. Applying the Wairakei flow rate (W) of 10^{-4} cm/sec to the Cyprus model indicates a jet diameter (X) of about 80 m, using the expression $X = (4 H \kappa_m/W)^{1/2}$ given by Elder (1965). The limited information available from below the Skouriotissa orebody indicates an alteration pipe narrowing upward from 175 m to 100 m (Kortan, 1970), and pipes of similar size appear to have existed below several Canadian deposits (Sangster, 1972).

If, as shown in Fig. 8a, secondary circulation is set up near the surface (due to low pipe velocity or slow discharge) the ore fluids may be discharged at several points at the surface in a pattern approaching that at Wairakei, and this could lead to formation

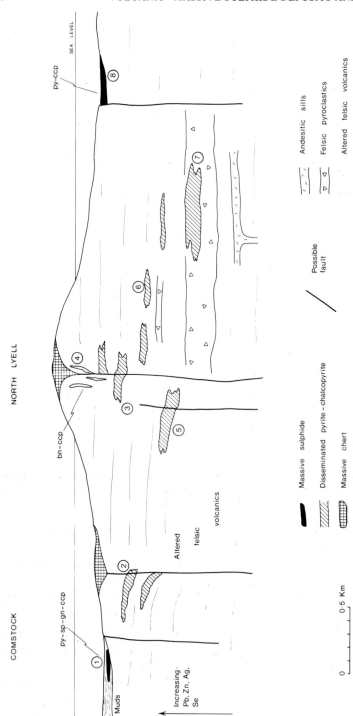

Fig. 9. Speculative north–south cross-section of the Mt. Lyell field at the time of mineralization, allowing for 25% shortening. Numbers refer to the following orebodies: *1* = Tasman and Crown Lyell; *2* = Comstock; *3* = Crown Lyell; *4* = North Lyell; *5* = Cape Horn; *6* = Royal Tharsis; *7* = Prince Lyell; *8* = Blow (or Mt. Lyell).

of several small, closely-spaced, deposits. Another corollary of this situation is that the alteration halo will be much wider than both the main jet and the diameter of the individual orebodies. In parts of the Hokuroko district of northern Honshu, there are several groups of deposits, the largest being that of Hanaoka–Matsumine–Shakanai–Matsuki, with over fourteen individual orebodies in an area of about 12 km² (Lambert and Sato, 1974; Sato et al., 1974, fig. 1). Judging from the cross-sections of Iijima (1974), sericite–chlorite alteration extends even beyond this area, indicating that a single mushroom-like flow pattern could be involved.

The several types of orebody at Mt. Lyell have been severely folded and faulted (Reid, 1975), but it is possible to make a fairly satisfactory reconstruction of the geology at the time of mineralization (Middle–Upper Cambrian). The surface area that originally contained the orebodies was at least 9 km² (allowing 25% N–S shortening during cleavage folding in the Devonian). As shown in the N–S cross-section of Fig. 9, there were two massive sulphide deposits, one of pyrite–chalcopyrite type and one of Zn–Pb–Cu type, and others may have been removed by Late Cambrian or Devonian erosion. There were also bornite–chalcopyrite orebodies formed very close to the Cambrian surface and related to extensive silica and barite deposition. The bulk of the known mineralization is in the form of disseminated, low grade, pyrite–chalcopyrite–sericite–quartz–chlorite lenses that probably formed by subsurface precipitation in, and replacement of, felsic fragmental volcanic rocks. The E–W faults indicated in Fig. 9 are demonstrably Devonian in age, but the possibility of Cambrian movements has already been indicated and it is suggested that these fractures may have modified local flow patterns. The estimated 1–1.5 km depth to the mineralization (and alteration) indicates that the size of the upper part of the flow system could have been similar to, or slightly larger than, the Wairakei system, and this would require a convection cell occupying at least 100 km² and extending in depth for 4–5 km (Elder, 1965).

In some geological situations, much of the solution flow could be in fault channels, resulting in high flow rates, less contact with surrounding rock, and considerable modifications to the size and shape of the cell. However, buoyancy effects due to heating will still be the main driving force.

Size of the convection cell

Lapwood (1948) calculated that in an infinite medium with impermeable upper and lower layers the spacing of rising streams of fluid would be from $2(2H)^{1/2}$ to $(4H)^{1/2}$, and in the case of the medium covered with free-standing fluid the spacing would be $2.7H$ to $3.8H$ (Fig. 8b). This gives an overall range of diameter/H ratios of 1.3 to 3.8 for cells with depths ranging from 1 to 10 km.

For Wairakei, Elder (1966) has calculated an area for the convection cell of 40 km², though his model (op. cit., fig. 19) indicates an area of 100 km² and a depth of 5 km (giving a diameter/H of 2.0). The geothermal districts of the Taupo zone are about 10 km apart, according to Henley (1973), which makes the larger cell area more probable. The 5 km depth is to some extent taken from the estimated thickness of the Taupo volcanics in this area.

If the smaller spacings on Cyprus (Figs. 4, 5) reflect the widths of the convection cells, then depths of, say, 1.0—1.5 km are indicated. This immediately raises problems of metal supply for the large deposits (e.g., Mavrovouni contained about $9 \cdot 10^{11}$ g Cu) and to supply this by leaching copper from the basalts and dolerite of the convection cell requires more copper than the rocks ever contained. A more likely approach is to regard the Apliki—Skouriotissa—Mavrovouni complex as a single complex system with three discharge points, a diameter of about 10 km and a depth of about 3.5 km. Possible flow patterns, modified by N—S faults, are indicated in Fig. 2.

Similar problems arise if the metals of the individual Kuroko deposits are derived by extraction in cells with similar diameters to their local spacings. For example, in the Hanaoka district the main orebodies could be regarded as spaced at 1 km intervals (Takahashi and Suga, 1974; Ohtagaki et al., 1974). A typical orebody contains $2 \cdot 10^{11}$ g Zn (Lambert and Sato, 1974), so a cell 1 km wide and 0.5 km deep would have to yield more than five times its total content of zinc to produce the orebody. It seems much more likely that the groups of deposits (such as Hanaoka—Shakanai and Kosaka) are derived from much larger, mushroom-type cells with multiple discharge points, as indicated by the pattern of subsurface alteration. The spacings between the major groups indicate cell diameters of 10 km or so and hence depths of 4—5 km, thus requiring extensive circulation of ore fluids in the Palaeozoic basement (cf. Ohmoto and Rye, 1974).

Life of the convection cell

Data from Wairakei allow order-of-magnitude calculations concerning the time of ore deposition for comparison with the geological evidence which strongly favours rapid formation of the ore.

The Mavrovouni and Skouriotissa orebodies, both massive and stockwork, contained about $1.4 \cdot 10^{12}$ g of copper and about $32 \cdot 10^{12}$ g of pyrite in massive and disseminated form (data from Cyprus Mines Corporation, 1974). If the solubility of copper in the ore solution is taken as 1 ppm (from the models of Kuroko ore solutions developed by Sato, 1972), $1.4 \cdot 10^{18}$ g of solution are required to form the orebody. The solubilities of iron and sulphur required to yield the same answer are about 15 ppm, similar to those used by Sato (1972). Though the salinity of the sea-water-derived ore solutions circulating in the Cyprus basalt and dolerite may well have been no higher than that of sea water (~0.5 m NaCl), it is unlikely on present information that these solubilities are out by much more than one order of magnitude. The rate of discharge at Wairakei is estimated to be 600 kg/sec (Elder, 1966), so that the time required for ore deposition could be approximately 70,000 years, which seems a geologically reasonable figure. For the largest pyrite—chalcopyrite orebody in the world, the Rio Tinto deposit, the time would be about $1.25 \cdot 10^6$ years at the Wairakei rate of discharge, though multiple outlets could well reduce this figure.

To satisfy the suggestion of Stanton (1974) that massive sulphide deposits form in weeks or months requires a flow rate of about 10^8 kg/sec for Mavrovouni, which seems unreasonable.

For the Mavrovouni–Skouriotissa cell to pass 10^{18} g of fluid indicates an overall water content of about 3 g/cm^3 (or at NTP for pure water, an overall water/rock ratio by volume of 3 : 1). If the porosity is 0.005, as used previously, then the pore fluid must be replaced 600 times. In investigating extraction of metals, isotope exchange mechanisms, etc., it may be significant that the experiment of Fig. 7b shows the sub-surface flow to be concentrated in the lower $\frac{1}{3}$ or so of the cell. Thus the actual water/rock ratio in this part of the cell could be considerably greater than 3 : 1. These conclusions are compatible with minimum estimates based on $\delta^{18}O$ data for the pillow lavas and feeder pipes on Cyprus by Spooner et al. (1974) and Heaton and Shepherd (1974), and also on those based on $^{87}Sr/^{86}Sr$ ratios (Spooner, 1974). The study of these systems is complicated by the extensive post-ore (and probably pre-ore) hydrothermal alteration.

Heat supply[1]

Crude estimates of the heat budget can be made for the Mavrovouni–Skouriotissa system, based on previous calculations. To raise $1.4 \cdot 10^8$ g of fluid from 20°C to say, 400°C at the base of the cell, requires about $3.8 \cdot 10^{20}$ cal., assuming no recirculation of warm fluid (i.e., a single-pass system). Over 70,000 years, this would require a heat discharge of about $2.5 \cdot 10^8$ cal./sec, similar to that at Wairakei, and for the whole cell this represents a heat flow at the base of about 250 H.F.U. This is well above any heat flow rates found on mid-ocean ridges (the Cyprus analogues), but these results fail to account for heat transfer by circulating sea water (see references in Anderson and Halunen, 1974). For the Cyprus situation, it is clear that there is no difficulty in supplying heat to the convection cells whether they be 2 km or 4 km deep. The pillow lavas are underlain by sheeted dykes (with only thin screens), and mafic and ultramafic intrusions (Fig. 2). If a slab 3 km thick and 100 km^2 in area is cooled from 1000°C to 400°C by convecting fluid at the upper surface, the heat removed would be about $1.6 \cdot 10^{20}$ cal. (using $c = 0.3$ cal./g and $\rho = 3.0$). Additional sources of heat are the latent heat of crystallization (providing another $0.6 \cdot 10^{20}$ cal.), and the latent heat of alteration in the overlying rock which at 40 cal./g (see Ellis, 1967) could provide another $0.2 \cdot 10^{20}$ cal. In addition, heat is stored in rocks within the cell and is also provided by upward transfer by conduction from magma below the cooling slab. Cells for Mavrovouni and Skouriotissa based on their 3–4 km spacings would transfer only a tenth of

---

[1] See also "Note added in proof", p. 49.

the heat estimated above, confirming that the larger cell size is more reasonable.

For non-ophiolite situations, the heat-supply situation is less favourable and substantial plutons must exist to provide sufficient energy, as postulated by Elder (1966, fig. 2) for the Wairakei system. A 100 km^2 cell in a volcanic arc might require an intrusive complex occupying some 25% of its base, extending downward for 10 km or so and maintaining the heat flow by repeated intrusions of magma during the life of the cell.

These requirements seem quite reasonable for orogenic belts, judging by the nature of the Tasman Orogenic Zone of eastern Australia, in which the area of Palaeozoic outcrop occupied by granitic plutons is about 35%. Some of these plutons are much younger than the massive sulphide deposits, but a gross compatability with the requirements of the model is indicated. In addition, this particular orogenic belt is not deeply dissected, so that as erosion proceeds the area of outcrop occupied by plutons may well increase. It is noted that Taylor (1974) encountered heat energy problems for the porphyry–copper hydrothermal systems and had to conclude that the intrusive complexes enlarged in area with increasing depth. The plutons of eastern Australia are not confined to the volcanic belts and thus provide a suitable source of heat for the massive sulphide deposits that are found in sedimentary rocks in central and western New South Wales (e.g., Cobar).

Metal and sulphur supply

It has already been noted that unreasonable extractions are required for cell sizes based on local spacings on Cyprus. However, using the cell size indicated by the intergroup spacing for the Mavrovouni–Skouriotissa complex (10 km diameter and 3.5 km deep) requires an overall extraction of 1.5 ppm of copper. Of only a third of the cell is participating significantly in the reactions (see Fig. 7b), then 4.5 ppm is required. Similar or lower extraction rates apply to zinc, which was present in some orebodies (e.g., Agrokipia, Mavrovouni; Bear, 1963). Such levels of extraction are compatible with the availability of metals in the mafic rocks (Table I) and also with experiments on water (and sea water)–rock interactions which indicate that copper and zinc readily transfer to the aqueous phase (Ellis, 1968; Bischoff and Dickson, 1974). The low extractions required are also compatible with the results of an extensive geochemical survey of the pillow lavas and dykes by Govett and Pantazis (1971). Most of the samples contained "normal" copper abundances (see comparative oceanic basalt in Table I), but minor depletion (10 ppm?) near some of the open-pit mines; no evidence of severe depletion was found.

Similar low extraction rates are required for deposits in other terrains. For example, the Rosebery deposit probably contained $2 \cdot 10^{12}$ g Zn, $1 \cdot 10^{12}$ g Pb and $5 \cdot 10^{10}$ g Cu and is underlain by 3 to 4 km of Andean-type calc-alkaline volcanic rocks dominated by rhyolites. If the convection cell had an area of 49 km^2 (see Figs. 3, 6) and a depth of 4 km, the average yield from a third of the volume would have to be 12 ppm Zn, 6 ppm Pb and 0.3 ppm Cu.

TABLE I

Average Pb, Zn and Cu contents of volcanic rocks, shales, and greywackes (ppm)

		Cu	Pb	Zn	References
Cenozoic island arc volcanics	basalt	80	3.5	–	Solomon and Griffiths (1974)
	andesite	52	5	–	
	rhyolite	20	3	–	
Cenozoic Andean volcanics	basalt	47	3.5	74	Solomon and Griffiths (1974)
	andesite	30	12	59	
	rhyolite	60	35	48	
Mesozoic ocean floor basalts (Cyprus)	means of groups, 2000 analyses	64–82	–	51–74	Govett and Pantazis (1971)
Cenozoic oceanic tholeiite	range of 12 analyses	58–72	3–<2	–	Thompson et al. (1972)
Archaean volcanics	mafic (andesite and basalt), 254 analyses	92	7.0 (4.7)*	101	Baragar and Goodwin (1969)
	felsic (dacite and rhyolite), 60 analyses)	44	7.6 (6.9)*	72	
	calc-alkaline basalt	116	–	99	Descarreaux (1973)
	calc-alkaline andesite	96	–	85	
	calc-alkaline rhyolite	~43	–	~55	
Shales	average	45	20	95	Wedepohl (1968)
Greywackes	average	45	–	60	Wedepohl (1968)

* Result after excluding Yellowknife Belt.

The whole field at Mt. Lyell is estimated to have contained $3 \cdot 10^{12}$ g Cu and the volcanic geology is similar to that at Rosebery except for a greater proportion of andesitic and basaltic rocks. Using a cell width of 10 km (Fig. 6) and a depth of 4 km, an extraction of 9 ppm Cu is required for a third of the cell volume. These figures are compatible with abundances typical of Andean volcanic terrains (Table I).

The cycle of sulphur from cold sea water to hot ore solution is complex, probably involving initially loss to, and then gain from, the cell rocks. For Cyprus, the early, relatively low-temperature stages of basalt–sea-water interaction probably involved precipitation of calcium minerals, mainly anhydrite, and Bischoff and Dickson (1974) recorded only 60 ppm of sulphate in sea water reacted with basalt at 200°C and 500 bars for 6 months. At considerable depths (say > 2 km) and nearer the hot zone (temperatures 200–300°C) reaction between iron silicates and solution probably resulted in precipitation of pyrite (Spooner and Fyfe, 1973), leaving a few ppm of reduced sulphur in solution. With rising temperature (300–400°C) around the base of the buoyancy column, continuing reaction probably dissolved primary rock-sulphide to increase the reduced-sulphur content of the solution. In felsic rocks the relatively high content of barium could be important in precipitating sulphate from sea water in barite. Even in a 1 M NaCl solution, the solubility of barite at 300°C is only about 100 ppm (Uchameyshvili et al., 1966). In western Tasmania, barite veins are found in volcanic rocks away from the ore fields and one example near Mt. Lyell indicated S- and O-isotope compositions similar to that of Cambrian sea-water sulphate (Solomon et al., 1969).

In Archaean situations, the content of oxidised sulphur species in the sea water is presumably very low, so that the proportion of sulphur supplied by dissolution of rock-sulphide could be significantly higher than for younger systems.

DISCUSSION

General

This brief continuation of studies commenced by Ellis (1967), Henley (1973), Spooner and Fyfe (1973) and others, indicates that at the reconnaissance level the geothermal model is a viable theory for the genesis of massive sulphide deposits. Convective circulation of water (probably sea water) in columns of volcanic or sedimentary rocks is feasible in the presence of a heat source, and the provision of sufficient metals and sulphur in reasonable times presents no problems at this stage, even ignoring contributions from magmatic sources.

The metal ratios of the ore solutions are probably determined by the bulk composition of the cell-rocks, particularly those at the base of the cell near the hot zone. Hence, there are only a few special cases where the solutions will not contain substantial lead, zinc, and copper. The main exceptions will be those involving lead-poor rocks, e.g., Archaean volcanic rocks, ophiolites and volcanic rocks typical of immature island arcs (Table I). The lack of lead and zinc and the presence of iron and copper minerals in some deposits probably reflects the influence of variables such as temperature, oxygen fugacity and pH in surface or near-surface situations where the differing solubilities of lead and zinc compared to copper and iron are important.

Other features predictable by the geothermal model are the lack of relationships between footwall rocks and ore types, the wide variety of magma-types and tectonic situations, the spatial distribution patterns, and the common association with major faults. However, possible weaknesses of the model are its inability to explain completely the tendency of the deposits to occur at specific horizons, and for rhyolitic rocks to be common in the footwall. The commencement and continuation of convection requires a heat supply at a suitable depth and lack of physical disturbance such as might be caused by vigorous eruption or repeated shallow intrusions. Faulting and the presence of a thick succession of permeable volcanic rocks probably assist in producing favourable conditions.

The similar stratigraphic position of massive sulphides in several mineralized ophiolites (Upadhyay and Strong, 1973) indicates a common process in ophiolite development and this may be the emplacement of a substantial, shallow heat source (gabbroic sills?) that causes uplift and faulting.

In orogenic belts presumably the later stages of tectonic cycles provide optimum conditions for ore formation (a thick volcanic pile, rifting, and rising granitic plutons) and this may explain the abundance of felsic rocks in the footwall. The plutons responsible for heating the circulating ore solutions are likely to be frozen at some depth below the ore horizon, making it difficult to trace any relationship between magmatic activity and mineralization. However, this will be more obvious if later magma rises through the frozen plutons to reach, or pass, the level of the ore deposits. Variations in the amount of plutonism, depth of crystallization, and timing with respect to ore formation, account for the lack of general systematic relationships between granitic plutons and massive sulphide deposits.

Sulphur and lead isotopes

The variation in isotopic composition of the sulphur in massive sulphide deposits is too complex to discuss here (see Chapter 9 by Sangster, this volume), but it is apparent from the discussion on sulphur supply that the initial $\delta^{34}S$ values of reduced sulphur within the buoyancy-driven jet will tend to vary between the composition of the major sources of sulphur, i.e., coeval sea water (providing an upper limit) and cell-rock sulphide (providing a lower limit). In at least the Cyprus case, the latter is probably close to zero (Kanehira et al., 1973) and it is most likely not far from zero for most volcanic rocks. Almost all "volcanic" massive sulphide deposits have $\delta^{34}S$ values between these limits (Sangster, 1968 and Chapter 9, this volume). The variation in $\delta^{34}S$ values, if any, within and near the orebody is probably a function of oxygen fugacity, Ph and temperature (e.g., Ohmoto and Rye, 1974).

The isotopic composition of ore-lead will be controlled entirely by the composition of the cell-rocks, as sea water will make a negligible contribution (cf. Kajiwara, 1973). Thus the ore-lead should represent a weighted average of the isotopic compositions of lead in the lead-bearing cell-rocks. The lead in the Kuroko ore deposits has a well-homogenized isotopic composition similar to that in Japanese Cenozoic calc-alkaline and tholeiitic volcanic rocks, and also Mesozoic granitic rocks (Sato and Sasaki, 1973; Sato et al., 1973). These results are as anticipated, but do not represent confirmation of the

model because no data are available for the remaining basement rocks of the Hokuroko region which are also likely to have been involved (see preceding discussion).

Evolutionary trends

Hutchinson (1973) noted that there are very few massive sulphide deposits in the
Middle and Late Proterozoic and Solomon (1974) pointed out that there are relatively
few deposits of Permo—Triassic age compared to the rest of the Phanerozoic. If the
Sullivan-Mt. Isa Group is removed from consideration then the whole Proterozoic is
poorly endowed with massive sulphides. Both gaps probably reflect the apparent paucity of orogenic, calc-alkaline volcanism and associated plutonism at these times.

Hutchinson (1973) also suggested an evolutionary trend in Phanerozoic orogenic
belts of Zn—Cu, then Zn—Pb—Cu, then Cu. Part of this trend is probably a function of
the composition of the volcanic rocks. The early volcanic rocks in volcanic belts are
commonly largely basalt and andesite of both calc-alkaline and low-K tholeiite types.
These rocks, and the underlying oceanic crust, are likely to be low in lead and are thus
associated with Cu or Zn—Cu orebodies. For instance, as Hutchinson (1973) noted,
the West Shasta deposits are low in the Devonian volcanic sequence and are Zn—Cu
types. They overlie the Copley Greenstone, reputed to be over 1 km thick, and to consist of basic and intermediate volcanics (Kinkel et al., 1956). On the other hand, the
Triassic deposits of East Shasta, which contain 1—2% Pb, occur in felsic volcanic and
sedimentary rocks overlying several km of mixed volcanic and sedimentary rocks (Albers and Robertson, 1961).

In eastern Australia, the Upper Ordovician deposits near Bathurst are of Zn—Cu
type, but the Lower Devonian deposits contain lead (Stanton, 1955). A similar explanation is likely here though the geology is less well known.

Hutchinson (1973) also suggested that Cu-type ophiolite-associated orebodies formed
late in the orogenic cycle, although he found variable time relationships in his examples. Solomon and Griffiths (1974) have noted that ophiolitic complexes in eastern
Australia may be tectonically re-intruded and thus be older than their host rocks, and
the relative ages of ophiolite complexes, particularly the low-viscosity, serpentinite-
rich examples, may thus be difficult to determine. Presumably in any one tectonic
cycle, massive sulphides in ophiolites could form earlier than, or at much the same
time as, massive sulphide deposits in associated Andean or island-arc volcanic rocks.

The geothermal model, and the low lead contents in Archaean rocks, account for
the change in massive sulphide mineralization from Zn—Cu in the Archaean to Zn—Pb—
Cu in the Late Proterozoic and Phanerozoic (Hutchinson, 1973). However, there is no
obvious explanation for the dearth of Cu types in the older rocks and their relative
abundance in the Phanerozoic.

SUMMARY AND CONCLUSIONS

This review of the host rocks to "volcanic"massive sulphides has found lack of clear-cut relationships between ore occurrence and volcanic rock type, and between the composition of the ore and that of the footwall rocks (apart from lack of lead ore in ophiolites and Archaean rocks). Though the majority of massive sulphides occur in felsic calc-alkaline rocks, they are also associated with mafic types and also with low-K tholeiites, oceanic basalts and in some cases, sedimentary rocks with minor volcanic component. In nearly all cases, there are plutonic rocks in the coeval rock column, whether they be grabbros of the ocean floor or granitic plutons of orogenic belts, but the spatial relationships between these rocks and the deposits is extremely variable except for those deposits in ophiolites. Single deposits are unusual and they commonly form groups or clusters that may be aligned along local structures. The spacing between deposits in any one group is commonly fairly regular and the spacing between groups and between major deposits appears to be related to the size of the deposits.

The only genetic model compatible with all the available information appears to be one involving convective circulation of surface water (mainly sea water) as suggested by Ellis (1967), Corliss (1971) and later workers. This geothermal model does not preclude, but does not rely on, a contribution from magmatic solutions. The order-of magnitude calculations by Henley (1973) and Spooner and Fyfe (1973) have been extended and so far appear to be compatible with the field data with respect to the times required for ore deposition, the availability of heat sources, and the distribution patterns of the deposits. The clusters of deposits that are seen in the Hokuroko and other orefields may have been formed above a wide area of secondary circulation, such as is developed above the rising jet at Wairakei.

If the conclusions concerning the sizes and forms of the convection cells are correct, they should allow more sophisticated predictions as to the locations of new deposits and a better understanding of alteration patterns.

NOTE ADDED IN PROOF

Studies of the cooling rates of plutons (see for example, Kappelmeyer and Haenel, 1974, p. 72) indicate that transfer of heat from the pluton to the convection cell solely by conduction would be too slow to maintain the convection. For example, the slab suggested as a heat source for Mavrovouni would take about 200,000 years to cool from 1000°C to 400°C solely by conduction. If the estimated life of the cell is even approximately correct it would be necessary for the fluids of the convection cell to maintain intimate contact with the heat source.

ACKNOWLEDGEMENTS

Many geologists have contributed to this paper by discussion but special thanks are due to M.J. Russell (University of Strathclyde), F.J. Sawkins (University of Minnesota) and R. Varne (University of Tasmania), and to many graduate students at the University of Tasmania, particularly J. Foden, G.R. Green, A.G. Stevens and J. Walshe. I am grateful for the hospitality of the Department of Geology, Imperial College of Science and Technology, London, during preparation of this chapter.

REFERENCES

Adamides, N.G., 1975. Geological history of the Limni concession, Cyprus, in the light of the plate tectonics hypotheses. *Trans. Inst. Min. Met.,* 84: B17–B23.
Albers, J.P. and Robertson, J.F., 1961. Geology and ore deposits of East Shasta copper–zinc district, Shasta County, California. *U.S. Geol. Surv. Prof. Pap.,* 338.
Amstutz, G.C., 1968. Les laves spilitiques et leurs gîtes mineraux. *Geol. Rundsch.,* 57: 936–954.
Anderson, C.A., 1969. Massive sulphide deposits and volcanism. *Econ. Geol.,* 64: 129–146.
Anderson, R.N. and Halunen, A.J., 1974. Implications of heat flow for metallogenesis in the Bauer Deep. *Nature,* 251: 473–475.
Angelkov, K., 1974. The Medet molybdenum–copper deposit. In: P. Dragov and B. Kolkovski (Editors), *Twelve Ore Deposits in Bulgaria. Field Excursion Guideb., IAGOD Meet., 4th, 1974, Sofia,* pp. 102–113.
Anger, G., Nielsen, H., Puchelt, H. and Ricke, W., 1966. Sulfur isotopes in the Rammelsberg ore deposit (Germany). *Econ. Geol.,* 61: 511–536.
Baragar, W.R.A., 1968. Major-element geochemistry of the Noranda volcanic belt, Quebec–Ontario. *Can. J. Earth Sci.,* 5: 773–790.
Baragar, W.R.A. and Goodwin, A.M., 1969. Andesites and Archean volcanism of the Canadian Shield. In: A.R. McBirney (Editor), *Proceedings of the Andesite Conference. Oreg., Dep. Geol. Miner. Ind., Bull.,* 65: 121–142.
Bear, L.M., 1963. The mineral resources and mining industry of Cyprus. *Bull. Geol. Surv. Cyprus,* 1.
Bernard, A.J. and Soler, E., 1974. Aperçu sur la province pyriteuse Sud-Ibérique. In: *Gisements Stratiformes et Provinces Cuprifères. Cent. Soc. Geol. Belg.,* pp. 287–315.
Bischoff, J.L. and Dickson, F.W., 1974. Seawater–basalt interaction at 200°C and 500 bars: implcations for origin of sea-floor heavy-metal deposits and regulation of seawater chemistry. *Earth Planet. Sci. Let.,* 25: 385–397.
Bogdanov, B. and Bogdanova, R., 1974. The Radka copper-pyrite deposit. In: P. Dragov and B. Kolkovski (Editors), *Twelve Ore Deposits of Bulgaria. Field Excursion Guideb., IAGOD Meet., 4th, 1974, Sofia,* pp. 114–133.
Brathwaite, R.L., 1974. The geology and origin of the Rosebery ore deposit, Tasmania. *Econ. Geol.,* 69: 1086–1101.
Burnham, C.W., Holloway, J.R. and Davis, N.F., 1969. Thermodynamic properties of water to 1,000°C and 10,000 bars. *Geol. Soc. Am. Spec. Pap.,* 132.
Cann, J.R., 1969. Spilites from the Carlsberg Ridge, Indian Ocean. *J. Petrol.,* 10: 1–19.
Coleman, R.J., 1975. Savage River magnetite deposits. In: C.L. Knight (Editor), *Economic Geology of Australia and Papua–New Guinea.* Australasian Institute of Mining and Metallurgy, Melbourne.
Colley, H. and Rice, C.M., 1975. A kuroko-type ore deposit in Fiji. *Econ. Geol.,* in press.
Constantinou, G. and Govett, G.J.S., 1972. Genesis of sulphide deposits, ochre and umber of Cyprus. *Trans. Inst. Min. Met.,* 81: B34–46.

Corliss, J.B., 1971. The origin of metal-bearing submarine hydrothermal solutions. *J. Geophys. Res.*, 76: 8128–8138.

Descarreaux, J., 1973. A petrochemical study of the Abitibi volcanic belt and its bearing on the occurrence of massive sulphide ore. *Can. Inst. Min. Met., Bull.*, 66: 61–69.

Donaldson, I.G., 1962. Temperature gradients in the upper layers of the earth's crust due to convective water flows. *J. Geophys. Res.*, 67: 3449–3459.

Elder, J.W., 1965. Physical processes in geothermal areas. *Am. Geophys. Union, Monogr. Ser.*, 8: 211–239.

Elder, J.W., 1966. Heat and mass transfer in the earth: hydrothermal systems. *N.Z. Dep. Sci. Ind. Res., Bull.*, 169.

Ellis, A.J., 1967. The chemistry of some explored geothermal systems. In: H.L. Barnes (Editor), *The Geochemistry of Hydrothermal Ore Deposits*. Holt, Rinehart and Winston, New York, N.Y., pp. 465–514.

Ellis, A.J., 1968. Natural hydrothermal systems and experimental hot-water–rock interactions: reactions with NaCl and trace metal extraction. *Geochim. Cosmochim. Acta*, 32: 1356–1363.

Fujii, K., 1974. Tectonics of the Green Tuff region, northern Honshu, Japan. In: S. Ishihara, K. Kanehira, A. Sasaki, T. Sato and Y. Shimazaki (Editors), *Geology of the Kuroko Deposits. Soc. Min Geol. Jap., Spec. Issue*, 6: 251–260.

Gilmour, P., 1971. Strata-bound massive pyritic sulfide deposits – a review. *Econ. Geol.*, 66: 1239–1249.

Govett, G.J.S. and Pantazis, Th.M., 1971. Distribution of Cu, Zn, Ni and Co in the Troodos pillow lava series, Cyprus. *Trans. Inst. Min. Met.*, 80: B27–46.

Hayakawa, N., Shimada, I., Shibata, T. and Suzuki, S., 1974. Geology of the Aizu metalliferous district, north-east Japan. In: S. Ishihara, K. Kanehira, A. Sasaki, T. Sato and Y. Shimazaki (Editors), *Geology of the Kuroko Deposits. Soc. Min. Geol. Jap., Spec. Issue*, 6: 19–28.

Heaton, H.E. and Sheppard, S.M.F., 1974. Hydrogen and oxygen isotope evidence for the origins of fluids during the evolution of the Troodos igneous complex (Cyprus). *Meet. Geochem. Group, Mineral. Soc., Lond., November 1974*. (Abstr.).

Henley, R.W., 1973. Some fluid dynamics and ore genesis. *Trans. Inst. Min. Met.*, 82: B1–8.

Hirst, J.A., 1965. Geology of east and north-east Viti Levu. *Geol. Surv. Fiji, Bull.*, 12.

Horikoshi, E., 1969. Volcanic activity related to the formation of the Kuroko-type deposits in the Kosaka district, Japan. *Miner. Deposita*, 4: 321–345.

Horikoshi, E., 1972. Orogenic belts and plate tectonics of the Japanese Islands. *Kagaku*, 42: 665–673 (in Japanese).

Horikoshi, E., 1975. Development of late Cenozoic petrogenic provinces and metallogeny in northeast Japan. In: D.F. Strong (Editor), *Metallogeny and Plate Tectonics. Nato Symp.*, in press.

Hutchinson, R.W., 1973. Volcanogenic sulfide deposits and their metallogenic significance. *Econ. Geol.*, 68: 1223–1246.

Hutchinson, R.W. and Hodder, R.W., 1972. Possible tectonic and metallogenic relationships between porphyry copper and massive sulphide deposits. *Can. Inst. Min. Met., Trans.*, 75: 16–22.

Hutchinson, R.W. and Searle, D.L., 1971. Stratabound pyrite deposits in Cyprus and relations to other sulphide ores. In: Y. Takéuchi (Editor), *Papers and Proceedings of the IMA–IAGOD Meetings, 1970. Soc. Min. Geol. Jap., Spec. Issue*, 3: 198–205.

Iijima, A., 1974. Clay and zeolitic alteration zones surrounding Kuroko deposits in the Hokuroko district, northern Akita, as submarine hydrothermal–diagenetic alteration products. In: S. Ishihara, K, Kanehira, A. Sasaki, T. Sato and Y. Shimazaki (Editors), *Geology of the Kuroko Deposits. Soc. Min. Geol. Jap., Spec. Issue*, 6: 267–289.

Jago, J.B., Reid, K.O., Quilty, P.G., Green, G.R. and Daily, B, 1972. Fossiliferous Cambrian from within the Mt. Read Volcanics, Mt. Lyell mine area, Tasmania. *J. Geol. Soc. Aust.*, 19: 379–382.

Jakes, P. and White, A.J.R., 1972. Major and trace element abundances in volcanic rocks of orogenic areas. *Geol. Soc. Am. Bull.*, 83: 29–40.

Jenks, W.F., 1971. Tectonic transport of massive sulfide deposits in submarine volcanic and sedimentary host rocks. *Econ. Geol.*, 66: 1215–1224.

Kajiwara, Y., 1973. Chemical composition of ore-forming solution responsible for the Kuroko type mineralization in Japan. *Geochem. J.*, 6: 141–149.

Kanehira, K., Yui, S., Sakai, H. and Sasaki, A., 1973. Sulphide globules and sulphur isotope ratios in the abyssal tholeiite from the Mid-Atlantic Ridge near 30°N latitude. *Geochem. J.*, 7: 89–96.

Kappelle, K., 1970. Geology of the C.S.A. mine, Cobar, N.S.W. *Proc. Australas. Inst. Min. Metall.*, 233: 79–94.

Kappelmeyer, O. and Haenel, R., 1974. Geothermics with special reference to application. *Geoexplor. Monogr., Ser. 1*, no. 4.

Kinkel Jr., A.R., Hall, W.E. and Albers, J.P., 1956. Geology and base-metal deposits of the West Shasta copper–zinc district, Shasta County, California. *U.S. Geol. Surv. Prof. Pap.*, 285.

Kortan, O., 1970. *Zur Bildung der Schwefelkies–Kupferkies-Vorkommen Cyperns unter besonderer Berücksichtigung der Lagerstätte Skouriotissa.* Unpub. Doctorate thesis, Technischen Universität Clausthal-Zellerfeld.

Lambert, I.B. and Sato, T., 1974. The Kuroko and associated ore deposits of Japan: a review of their features and metallogenesis. *Econ. Geol.*, 69: 1215–1236.

Lapwood, E.R., 1948. Convection of a fluid in a porous medium. *Proc. Cambridge Philos. Soc.*, 44: 508–521.

Lawrence, L.J., 1972. The thermal metamorphism of a pyritic sulfide ore. *Econ. Geol.*, 67: 487–496.

McAllister, A.L. and Lamarche, R.Y., 1972. Mineral deposits of southern Quebec and New Brunswick. *Int. Geol. Congr., 24th, Guideb., Field Excursion*, A58.

Mitchell, A.H.G. and Garson, M.S., 1972. Relationship of porphyry copper and circum-Pacific tin deposits to palaeo-Benioff zones. *Trans. Inst. Min. Met.*, 81: B10–25.

Mitchell, A.H.G. and Garson, M.S., 1973. Relationships of porphyry copper and circum-Pacific tin deposits to palaeo-Benioff zones. Discussion. *Trans. Inst. Min. Met.*, 82: B43–44.

Miyashiro, A., 1974. Volcanic rock series in island arcs and active continental margins. *Am. J. Sci.*, 274: 321–355.

Mukaiyama, H., 1970. Volcanic sulphur deposits in Japan. In: T. Tatsumi (Editor), *Volcanism and Ore Genesis*. University of Tokyo Press, Tokyo, pp. 285–294.

Ohmoto, H. and Rye, R.O., 1974. Hydrogen and oxygen isotopic compositions of fluid inclusions in the Kuroko deposits, Japan. *Econ. Geol.*, 69: 947–953.

Ohtagaki, T., Tsukada, Y., Hirayama, H., Fujioka, H. and Miyoshi, T., 1974. Geology of the Shakanai mine, Akita Prefecture. In: S. Ishihara, K. Kanehira, A. Sasaki, T. Sato, Y. Shimazaki (Editors), *Geology of the Kuroko Deposits. Soc. Min. Geol. Japan, Spec. Issue*, 6: 131–139.

Packham, G.H., 1969. Ordovician System. In: T.G. Vallance (Editor), *The Geology of New South Wales. J. Geol. Soc. Aust.*, 16: 76–103.

Parak, T., 1975. The origin of the Kiruna iron ores. *Sver. Geol. Unders., Ser. C*, no. 1.

Pearce, J.A., 1975. Basalt geochemistry used to investigate past tectonic environments on Cyprus. *Tectonophysics*, 25: 41–67.

Reid, K.O., 1975. Geology of the Mt. Lyell copper ore deposits, Tasmania. In: C.L. Knight (Editor), *Geology of Australia and Papua–New Guinea*. Australasian Institute of Mining and Metallurgy, Melbourne.

Rickard, M.J., 1970. The geology of north-eastern Vanua Levu. *Bull. Geol. Surv. Fiji*, 14.

Rodda, P., 1963. Wainivesi zinc property, Tailevu. Diamond drilling, 1961–2. *Geol. Surv. Fiji, Econ. Invest.*, 2.

Sangster, D., 1968. Relative sulphur isotope abundances of ancient seas and stratabound sulphide deposits. *Proc. Geol. Assoc. Can.*, 19: 79–91.

Sangster, D., 1972. Precambrian volcanogenic massive sulphide deposits in Canada: a review. *Geol. Surv. Can., Pap.*, 72.

Sato, K. and Sasaki, A., 1973. Lead isotopes of the black ore ("Kuroko") deposits from Japan. *Econ. Geol.*, 68: 547–552.

Sato, K., Slawson, S.F. and Kanasewich, E.R., 1973. Additional isotopic measurements on Japanese ore leads. *Geochem. J.*, 7: 115–122.

Sato, T., 1972. Model for ore-forming solutions and ore-forming environments: Kuroko vs. veins in Miocene "Green Tuff" region of Japan. *Bull. Geol. Surv. Jap.*, 23: 457–466.

Sato, T., Tanimura, S. and Ohtagaki, T., 1974. Geology and ore deposits of the Hokuroko district, Akita Prefecture. In: S. Ishihara, K. Kanehira, A. Sasaki, T. Sato and Y. Shimazaki (Editors), *Geology of the Kuroko Deposits. Soc. Min. Geol. Jap., Spec. Issue*, 6: 11–18.

Sawkins, F.J., 1974. Massive sulphide deposits in relation to geotectonics G.A.C./M.A.C. Meet., 1974, St. Johns, Newfoundland, Abstr. Vol., p. 81 (Abstr.).

Schermerhorn, L.J.G., 1970. The deposition of volcanics and pyritite in the Iberian pyrite belt. *Miner. Deposita*, 5: 273–279.

Searle, D.L., 1972. Mode of occurrence of the cupriferous pyrite deposits of Cyprus. *Trans. Inst. Min. Met.*, 81: B189–197.

Sillitoe, R.H., 1972. Formation of certain massive sulphide deposits at sites of sea-floor spreading. *Trans. Inst. Min. Met.*, 81: B141–148.

Solomon, M., 1969. The copper–clay deposits at Mount Lyell, Tasmania. *Proc. Australas. Inst. Min. Metall.*, 230: 39–47.

Solomon, M., 1974. Massive sulphides and plate tectonics. *Nature*, 249: 821–822.

Solomon, M. and Griffiths, J.R., 1974. Aspects of the early history of the southern Tasman Orogenic Zone. In: A.K. Denmead, G.W. Tweedale and A.F. Wilson (Editors), *The Tasman Geosyncline – A Symposium*. Geol. Soc. Aust., Queensland Div., Brisbane, pp. 19–46.

Solomon, M., Rafter, T.A. and Jensen, M.L., 1969. Isotope studies on the Rosebery, Mount Farrell and Mount Lyell ores, Tasmania. *Miner. Deposita*, 4: 172–199.

Solomon, M., Groves, D.I. and Klominsky, J., 1972. Metallogenic provinces and districts in the Tasman Orogenic Zone of eastern Australia. *Proc. Australas. Inst. Min. Metall.*, 242: 9–24.

Sommerlatte, H., 1958. Die Blei–Zink-Erzlagerstätte von Bawdwin in Nord-Burma. *Z. Dtsch. Geol. Ges.*, 110/111: 491–504.

Spence, C.D. and De Rosen-Spence, A.F., 1975. The place of sulfide mineralization in the volcanic sequence at Noranda, Quebec. *Econ. Geol.*, 70: 90–101.

Spooner, E.T.C., 1974. Water/rock ratios during hydrothermal metamorphism of the ophiolitic rocks of East Liguria (Italy) and Troodos (Cyprus). *Meet. Geochem. Group, Mineral. Soc. Lond., November, 1974*.

Spooner, E.T.C. and Fyfe, W.S., 1973. Sub-sea-floor metamorphism, heat and mass transfer. *Contrib. Mineral. Petrol.*, 42: 287–304.

Spooner, E.T.C., Beckinsale, R.D., Fyfe, W.S. and Smewing, J.D., 1974. ^{18}O enriched ophiolitic metabasic rocks from E. Liguria (Italy), Pindos (Greece) and Troodos (Cyprus). *Contrib. Mineral. Petrol.*, 47: 41–62.

Stanton, R.L., 1955. Lower Palaeozoic mineralisation near Bathurst, N.S.W. *Econ. Geol.*, 50: 681–714.

Stanton, R.L., 1972. *Ore Petrology*. McGraw-Hill, New York, N.Y., 713 pp.

Stanton, R.L., 1974. The development of ideas on the evolution of mineralization in the Tasman Geosyncline. In: A.K. Denmead, G.W. Tweedale and A.F. Wilson (Editors), *The Tasman Geosyncline – A Symposium*. Geol. Soc. Australia, Brisbane, pp. 185–219.

Takahashi, T. and Suga, K., 1974. Geology and ore deposits of the Hanaoka Kuroko belt, Akita Prefecture. In: S. Ishihara, K. Kanehira, A. Sasaki, T. Sato and Y. Shimazaki (Editors), *Geology of the Kuroko Deposits. Soc. Min. Geol. Jap., Spec. Issue*, 6: 101–113.

Tan, L.P., 1959. The sulfur-melnikovite deposits of the Szehuangtzeping area, Taipeihsien, Taiwan. *Proc. Geol. Soc. China*, 2: 123–145.

Tatsumi, T. and Clark, L.A., 1972. Chemical composition of acid volcanic rocks genetically related to formation of the Kuroko deposits. *J. Geol. Soc. Jap.*, 78: 191–201.

Tatsumi, T. and Watanabe, T., 1971. Geological environment of formation of the Kuroko-type deposits. In: Y. Takéuchi (Editor) *Papers and Proceedings of the IMA – IAGOD Meetings, 1970. Soc. Min, Geol. Jap., Spec. Issue*, 3: 216–220.

Taylor, G.R., 1974. Volcanogenic mineralization in the islands of the Florida Group, B.S.I.P. *Trans. Inst. Min. Met.*, 83: B120–130.

Taylor Jr., H.P., 1974. The application of oxygen and hydrogen isotope studies to problems of hydrothermal alteration and ore deposition. *Econ. Geol.*, 69: 843–883.

Thirlow, J.G., Swanson, E.A. and Strong, D.F., 1975. Geology and lithogeochemistry of the Buchans polymetallic sulfide deposits, Newfoundland. *Econ. Geol.*, 70: 130–144.

Thompson, G., Shido, F. and Miyashiro, A., 1972. Trace element distribution in fractionated oceanic basalts. *Chem. Geol.*, 9: 89–97.

Tokunaga, M. and Honma, H., 1974. Fluid inclusions in minerals from some Kuroko deposits. In: S. Ishihara, K. Kanehira, A. Sasaki, T. Sato and Y. Shimazaki (Editors), *Geology of the Kuroko Deposits. Soc. Min. Geol. Jap., Spec. Issue*, 6: 385–388.

Toulmin, P., III, and Clark, Jr., S.P., 1967. Thermal aspects of ore formation. In: H.L. Barnes (Editor), *Geochemistry of Hydrothermal Ore Deposits*. Holt, Rinehart and Winston, New York, N.Y., pp. 437–464.

Uchameyshvili, N.Ye., Malinin, D. and Klutarov, N.I., 1966. Solubility of barite in concentrated chloride solutions of some metals at elevated temperatures in relation to problems of the genesis of barite deposits. *Geochem. Int.*, 3: 951–963.

Upadhyay, H.D. and Strong, D.F., 1973. Geological setting of the Betts Cove copper deposits, Newfoundland: an example of ophiolite sulfide mineralization. *Econ. Geol.*, 68: 161–167.

Vallance, T.G., 1969. Spilites again: some consequences of the degradation of basalts. *Proc. Linn. Soc. N.S.W.*, 94: 8–51.

Vine, F.J. and Moores, E.M., 1972. A model for the gross structure, petrology and magnetic properties of oceanic crust. *Geol. Soc. Am., Mem.*, 132: 195–205.

Waltham, A.C., 1968. Classification and genesis of some massive sulphide deposits in Norway. *Trans. Inst. Min. Met.*, 77: B153–161.

Wedepohl, K.H., 1968. Chemical fractionation in the sedimentary environment. In: L.H. Ahrens (Editor), *Origin and Distribution of the Elements*. Pergamon, Oxford, pp. 999–1016.

White, N.C., 1975. *Cambrian Volcanism and Mineralization in South-west Tasmania*. Unpub. Ph.D. Thesis, University of Tasmania.

Williams, E., Solomon, M. and Green, G.R., 1975. The geological setting of metalliferous ore deposits in Tasmania. In: C.L. Knight (Editor), *Economic Geology of Australia and Papua–New Guinea*. Australasian Institute of Mining Metallurgy, Melbourne.

Wooding, R.A., 1957. Steady state free thermal convection of liquid in a saturated permeable medium. *J. Fluid Mech.*, 2: 273–285.

Wooding, R.A., 1963. Convection in a saturated porous medium at large Rayleigh number or Peclet number. *J. Fluid Mech.*, 15: 527–544.

Chapter 3

TECTONIC SETTING OF SOME STRATA-BOUND MASSIVE SULPHIDE DEPOSITS IN NEW SOUTH WALES, AUSTRALIA[1]

E. SCHEIBNER and N.L. MARKHAM

INTRODUCTION

The tectono-genetic approach to the study of metallogeny is becoming very important in the rational search for mineral deposits. In the context of plate-tectonics theory it is generally accepted that cupriferous massive sulphide deposits of the Cyprus type are associated with ophiolites and were generated in connection with the formation of oceanic crust (cf. Sillitoe 1972a, and others). The tectonic setting of Kuroko-type massive sulphide deposits is generally assumed to be sites of orogenic volcanism in connection with subduction zones (cf. Sawkins, 1972; Lambert, 1973; and others).

The distribution of these two main types of strata-bound massive sulphide deposits, as well as other types of deposits and their tectonic settings in New South Wales, have been discussed in a recently published Centenary Volume of the New South Wales Department of Mines entitled *The Mineral Deposits of New South Wales* (Markham and Basden, 1974). The present requested paper has a review character based on the above publication.

TECTONIC SETTING OF CYPRUS-TYPE AND KUROKO-TYPE STRATA-BOUND MASSIVE SULPHIDE DEPOSITS

The recent application of plate tectonics to metallogeny (Sawkins and Peterson, 1969; Guild, 1971, 1972, 1973; Hutchinson, 1971; Hodder and Hollister, 1972; Hutchinson and Hodder, 1972; Sawkins, 1972; Sillitoe, 1972a,b,c, 1973; Wright and McCurry, 1973; Tarling, 1973; and others) has been very fruitful. It is believed that natural tectonic units (stratotectonic) distinguished on the basis of plate tectonics (Scheibner, 1972a, 1974a,b) are also, to a certain extent, natural metallogenic units. Therefore, the application of plate tectonics to metallogeny should eventually lead to an understanding of the genetic aspects of metallogenic (really genetic) provinces. It has been suggested that practical (from the point of view of prognosis) basic genetic entities or *basic metallogenic units* can be defined by genetic type(s) or groups of metallogenic deposits (cf. Scheibner, 1974b).

---

[1] Permission to publish granted by the Under Secretary, New South Wales Department of Mines and Energy.

A basic metallogenic unit has been defined as an entity of mineral deposits comprising part of a metallogenic stage, and originating in a specific, well-defined area of a metallogenic province (Scheibner, 1974b).

Oceanic metallogenic unit

One of the most typical basic metallogenic units of the pre-cratonic or orogenic metallogenic province is the oceanic metallogenic unit. An oceanic metallogenic unit represents an entity of mineral deposits formed at centres of lithospheric growth by sea-floor spreading. According to plate-tectonics theory, at the margins of diverging major plates oceanic lithosphere forms by axial sea-floor spreading at true mid-oceanic ridges. At the margins of diverging small plates, oceanic lithosphere of small ocean basins and marginal seas forms by axial or non-axial (diffused) sea-floor spreading. Generally, there is no substantial difference in material composition of the sea floor at different centres of lithospheric growth, but differences in structure have been observed, i.e., fast spreading mid-oceanic ridges are smooth. However, recently Nisbett and Pearce (1973) have pointed out that the TiO_2 content of abyssal tholeiitic basalts appears to be proportional to the rate of sea-floor spreading. Slow spreading oceanic crustal basalts contain less than 1% TiO_2. It might be coincidental, but the abyssal tholeiites of Cyprus have 1% of TiO_2, and similarly the well-documented Palaeozoic examples discussed below in the Girilambone Beds have a lower than 1% content of TiO_2. It is possible that formation of Cyprus-type massive sulphide deposits is associated only with slow sea-floor spreading, while fast spreading, with higher content of TiO_2 in abyssal tholeiites, is characterised by lesser metallic concentrations.

During sea-floor spreading (Hess, 1955; Kay et al., 1970), oceanic crustal layers 1 to 3 and the depleted upper mantle originate, and within them characteristic mineral deposits.

The expression *ophiolite* was coined by alpine geologists for a distinctive assemblage of mafic to ultramafic rocks occurring in orogenic belts, and this term should be used in this sense according to the recommendations of the Ophiolite Conference (Coleman, 1973). Ophiolites occurring in present continental areas represent slices or whole slabs of oceanic lithosphere tectonically emplaced during orogenies (plate convergence). Oceanic lithosphere is incorporated as slices in trench complexes or flysch wedges (slope and basin in Scheibner, 1972b, 1974a), or it is emplaced by obduction (Coleman, 1971), i.e., thrusting. Normally, oceanic lithosphere sinks and is subducted, or in marginal seas it is partially melted and fused if a proper subduction zone does not develop. Oceanic crust in these cases probably contributes substantially to the formation of lower continental crust. Special circumstances must exist for oceanic lithosphere to crop out in presently continental orogenic areas.

The oceanic metallogenic unit is characteristic of several basic tectonic units: oceanic basin; small ocean basin, including marginal seas; slope and basin of converging oceanic and continental or two oceanic plates, i.e., trench complex or flysch wedge; convergence zone of two plates where obduction or upthrusting of oceanic lithosphere occurs.

The oceanic metallogenic unit can be subdivided into: oceanic crustal, and oceanic (upper) mantle metallogenic subunits.

An *oceanic crustal metallogenic subunit* comprises the entity of mineral deposits formed in an oceanic crustal and supracrustal environment in oceans and small ocean basins (including marginal seas).

The pyrite—copper—zinc strata-bound deposits or massive sulphides of the Cyprus type associated with tholeiitic basalts (part of the ophiolite complex: Gass, 1968; and others) are the most characteristic and common genetic type of deposits of this metallogenic subunit. These deposits have been described in a voluminous literature, cf. review by Sillitoe (1972a).

Massive sulphide deposits of the Cyprus type were episodically generated during the formation of layers 2 and 1 of the oceanic crust, and also during later hydrothermal activity, and are distributed as irregular elongate, lenticular bodies in pillow lavas. Massive sulphides at the base of sediments forming part of oceanic layer 1 presumably formed as metal-rich brines and muds, filling depressions on the sea floor. Mineralising fluids were supplied through subjacent faults and fractures which now possess epigenetic mineralisation (Sillitoe, 1972a).

Schneiderhöhn (1955) recognised three types of submarine exhalative deposits connected with "initial magmatism" (i.e., sea-floor spreading): oxidic siliceous red iron ores of the Lahn-Dill type, sulphidic iron—copper—zinc ores, and exhalative sedimentary ores. The Lahn-Dill ores are connected with mid-Devonian ophiolites exposed today in the form of spilite, keratophyre, and diabase, i.e., they are strongly altered and metamorphosed. Examples of exhalative sedimentary manganese ores are occurrences in Graubünden in Switzerland in the Jurassic radiolarian cherts closely associated with the classical ophiolites of Steinmann (1905). Further examples occur in the Woolomin Beds in the New England Fold Belt of New South Wales. An example of the sulphidic iron—copper—zinc deposits would be the Cyprus occurrences mentioned above, and similar ones occurring elsewhere, again associated with ophiolite.

The oceanic crustal metallogenic subunit also contains accumulations of manganese nodules, which form in connection with specific micro-organisms capable of manganese oxidation. Enrichment in metallic cations occurs in such manganese nodules.

Evaporites, formed especially during the initial stages of ocean-basin development, could also be characteristic of this metallogenic subunit.

It is necessary to stress that massive sulphide concentrations of the oceanic crustal metallogenic subunit differ from the massive pyrite—zinc—lead—copper sulphides of volcanic rifts discussed below. The main difference lies in their igneous rock association and in the proportion of metallic elements.

Volcanic-rift metallogenic unit

A volcanic rift (volcano-tectonic depression of Van Bemmelen, 1931; Karig, 1970) is a basic tectonic unit characterised by the orogenic volcanism, transitional regressive tectonic realm (extended continental-type crust) and upwelled asthenosphere, and forms in a setting of lithospheric tension in orogenic areas (Scheibner, 1974a). It represents a combination of two basic tectonic units: an inter-arc basin, and a volcanic arc or arch. When extension of continental-type crust occurs, sudden upwelling of the asthenosphere causes high heat-flow, to which the thermal magmatic diapirs rising from the associated Benioff zone contribute substantially. In this setting sudden partial melting of continental-type crust takes place, and is expressed in the rapid eruption of large volumes of ignimbrites. Ignimbrites prevail over other calc-alkaline volcanics in volcanic rifts. Most volcanics are submarine, as is evidenced by intercalated marine sediments. A certain sedimentary starvation, i.e., low rate of deposition of clastics, is a local requirement for development of mineral deposits in volcanic rifts.

From evidence available from the Palaeozoic complexes in eastern Australia, it is suggested that Kuroko-type strata-bound sulphide deposits are associated with volcanic rifts, and this conclusion is probably valid also for the deposits in Japan, and elsewhere. It is quite clear that Kuroko-type strata-bound sulphide deposits are not associated with all types of orogenic volcanic chains (Scheibner, 1974b).

A volcanic-rift metallogenic unit can be defined as an entity of mineral deposits formed in a volcanic rift (basic tectonic unit). The most characteristic genetic type of deposits here are strata-bound massive pyrite–zinc–lead–copper sulphides of the Kuroko type (Matsukuma and Horikoshi, 1970; Lambert, 1973). Normal faults bounding volcanic rifts penetrate the whole thinned lithosphere, as indicated by basic eruptives. These faults are important for the circulation of descending meteoric and mainly sea water and ascending thermal waters charged with metals in the form of complex soluble compounds. The metals appear to be derived from the upwelled mantle, from magmatic calc-alkaline diapirs rising from the Benioff zone, and by leaching out of metallic elements from the volcanic pile. Probably the partial melting of oceanic lithosphere, including subducted sediments having a relatively high content of metals, is the most important source of metals (Sawkins, 1972; Wright and McCurry, 1973); however, the upwelled asthenosphere could contribute also (copper). For effective formation of Kuroko-type strata-bound sulphide deposits a marine environment appears to be essential (cf. Lambert, 1973).

In classical Kuroko-type deposits, underneath the massive black "Kuroko" ore, which consists essentially of pyrite–zinc–lead sulphides, there is yellow "Oko" ore, rich in chalcopyrite (Matsukuma and Horikoshi, 1970; Lambert, 1973). Similar metal zoning occurs in sulphide veinlets probably representing "plumbing" systems for the strata-bound sulphides. In some strata-bound massive sulphide deposits outside Japan, such zoning is present (Hutchinson and Hodder, 1972), but in many Palaeozoic (including eastern Australian) and Precambrian deposits the yellow ore is absent; however, all other properties are identical. The absence of "yellow" ores is not fully understood. If the copper has been derived from upwelled asthenosphere and is connected with the formation of tholeiitic oceanic crust, then the presence or absence of yellow ore could indicate the amount of rifting in the volcanic rift. Where rifting was intensive and led to the formation of some oceanic-type crust, even if in a very restricted amount, then yellow ore is present. This hypothesis could be proved or disproved by the study of basic rocks which are present in some volcanic rifts.

During orogenic deformation, volcanic rifts become part of synclinorial zones and blocks in fold or orogenic belts. As such they are less exposed to erosion than volcanic arches which become parts of anticlinorial zones and blocks (cf. Scheibner, 1974b). This is probably the reason why, since Archean time, volcanic-rift metallogenic units have been preserved and not eroded, as opposed to porphyry copper deposits characteristic of volcanic arches which are prone to erosion, and no complicated explanation is required to explain their occurrence since the early stages of development of the earth's crust.

This is very important in exploration activity, for synclinorial zones should contain stratiform massive sulphides and anticlinorial zones should contain porphyry copper and subvolcanic deposits.

DEVELOPMENT OF SOME TECTONIC UNITS IN WHICH STRATA-BOUND MASSIVE SULPHIDE DEPOSITS OCCUR IN THE PALAEOZOIC SEQUENCES OF NEW SOUTH WALES

The best examples of Cyprus-type strata-bound massive sulphide deposits in New South Wales occur in trench complexes of various ages: in the Girilambone Beds and in the Woolomin Beds. The Girilambone Beds represent probably a remnant of a Cambrian trench and the Woolomin Beds form part of a Middle to Late Palaeozoic trench complex (Scheibner, 1974a).

Kanmantoo pre-cratonic metallogenic province (Fig. 1)

The character of Early Cambrian rock complexes in eastern Australia indicates that decoupling of the Palaeo-Pacific oceanic lithosphere from the continental lithosphere occurred, or a subduction zone developed within the oceanic plate, and a converging plate margin developed. A typical pacific-type marginal mobile zone developed rapidly, characterised by marginal seas (the Kanmantoo Through in South Australia and New South Wales, the Bancannia Trough in New South Wales, the Dundas Trough in Tasmania, and a similar feature in Queensland), volcanic chains (the Mount Wright Volcanic Arc in New South Wales and the Mount Read Volcanic Arch in Tasmania; the Argentine Metamorphics in Queensland, as mentioned by Heidecker (1972), probably represent a volcanic arc), microcontinents (the Wonominta Block in New South Wales, the Tyenna Block in Tasmania, and Proterozoic elements of the Ravenswood-Lolworth Block in Queensland) and a trench complex or flysch wedge (the Girilambone Beds in New South Wales and metamorphics in the Anakie High in Queensland).

From the Girilambone oceanic metallogenic unit the *Girilambone oceanic crustal sub-*

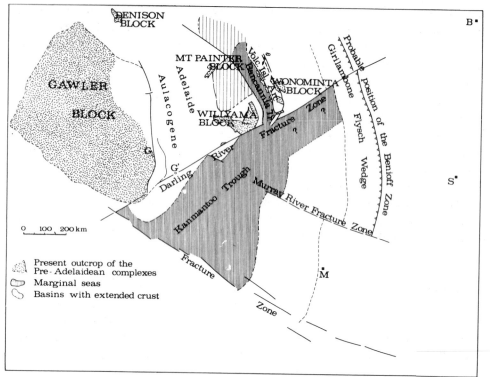

Fig. 1. Palinspastic reconstruction for the area of southeastern Australia for Early Cambrian time. (Reprinted from Scheibner, 1974c, in: A.K. Denmead, G.W. Tweedale and A.F. Wilson–Editors: *The Tasman Geosyncline – A Symposium.* Geol. Soc. Austr. Qld. Div., Brisbane, pp. 65–92.)

unit is best developed. In several locations in the Girilambone Beds, which represent a trench complex or flysch wedge and tectonically are classified as the Girilambone Slope and Basin (Scheibner, 1974a), strata-bound pyrite–copper–zinc sulphide deposits occur. The best known are the deposits around Tottenham (Suppel, 1971, 1974) and Girilambone (Smith, 1974). Sulphides occur in close association with basic volcanics (oceanic layer 2), especially at Tottenham where their abyssal tholeiitic character has been established. The sulphide deposits occur in rocks which represent oceanic layers 1 and 2 and form regional anticlinal structures. These deposits formed in the Palaeo-Pacific Basin and were incorporated into a trench complex probably during the Early Palaeozoic, perhaps latest Precambrian. The Girilambone Beds do not contain recognisable fossils, but they occur unconformably under a mid-Ordovician sequence.

Based on the presence of typical quartz-magnetite rocks, interbedded basics and ultra-basics, and style of deformation, the Jindalee Beds east of Cootamundra are also considered to be equivalent to the Girilambone Beds (Basden, 1974) and to belong to the Girilambone oceanic metallogenic unit.

After orogenic deformation the Girilambone Beds were involved in a mid-Palaeozoic orogenic episode, they formed an internal massif in the Lachlan mobile zone, and at present they form an anticlinorial zone in the Lachlan Fold Belt.

Lachlan pre-cratonic metallogenic province (Fig. 2)

This metallogenic province represents the entity of deposits which formed in the Lachlan mobile zone during its orogenic development. The tectonic history of the Lachlan mobile zone has been described recently from a plate-tectonics point of view (Scheibner, 1974a,b).

After the Benambran-Quidongan tectonic stage, which includes the Ordovician to Early Silurian development, the Bowning tectonic stage started in the late Early to early Late Silurian and continued till the end of the Silurian. During this tectonic stage a new episode of tensional stresses was expressed in the formation of new tensional features, marginal seas and volcanic rifts. Intensive volcanic arch and rift volcanism and orogenic granite plutonism commenced in the Lachlan marginal mobile zone. Tensional stresses were caused perhaps by formation of a secondary Benioff zone in the Cowra Trough (a marginal sea), and by formation of a new primary Benioff zone, or intensive retrograde movement of it. Recently, Uyeda and Miyashiro (1974) have suggested that subduction of a mid-oceanic ridge and of the hot parts of an oceanic plate can cause ubiquitous tension and a wide zone of orogenic igneous activity in the leading plate. Perhaps this is applicable to certain episodes of the Palaeozoic development of eastern Australia; however, it is difficult to accept that oceanic plates could be so strong that they could be subducted at very low angles for distances up to three thousand kilometres (cf. Uyeda and Miyashiro, 1974).

On the west of the Lachlan mobile zone, in the present Cobar area, rifting occurred during the Bowning tectonic stage, and probably at this time the Cobar Trough was formed. It was a volcanic rift extending into a narrow marginal sea. Volcanic-rift volcanism (Mount Hope Volcanics) occurred in an area to the south of the Lachlan River fracture zone (Fig. 2), but not north of it. In the Cobar synclinorial zone (which developed from the Cobar Trough) no volcanics of Silurian age are known north of the Lachlan River fracture zone as yet, despite the fact that the same strata-bound sulphide deposits occur in both segments of the synclinorial zone. Volcanic rifting also occurred in the present Mineral Hill synclinorial zone during the Bowning tectonic stage. The entity of mineral deposits characterised by strata-bound sulphide deposits which developed in this area form the Cobar volcanic-rift metallogenic unit. The large size of copper deposits probably indicates intensive rifting leading to oceanic crustal development.

The strong extension widened the Late Ordovician initial Cowra Trough into the relatively wide marginal sea (Cowra Trough), the character of which is evidenced by oceanic lithosphere forming the Coolac serpentinite belt. A secondary Benioff zone dipping to the west probably developed in the Cowra Trough, and it is suggested that all the

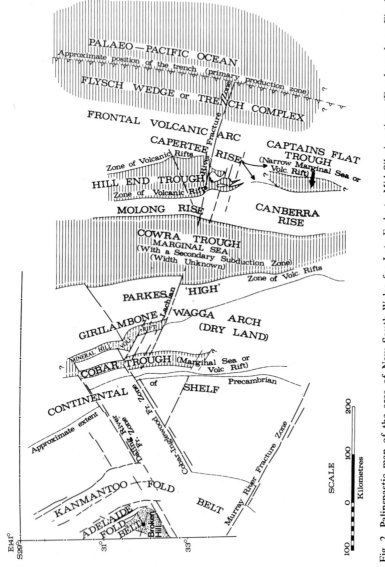

Fig. 2. Palinspastic map of the area of New South Wales for Late Early to Mid Silurian time. (For legend, see Fig. 3.)

Silurian to Early Devonian volcanic-arch volcanics and associated orogenic granites in the area west of the Cowra Trough were produced in connection with the mentioned subduction zone (Scheibner, 1974a).

The subduction in the Cowra Trough apparently terminated in Early Devonian time, because the upthrust subduction zone (Coolac serpentinite belt) is intruded by the Bogong Granite (Ashley et al., 1971). It is interesting to note that massive post-kinematic orogenic granites were emplaced for a further 10 m.y. or even longer, during the Tabberabberan tectonic stage.

The Bowning orogenic granite plutonism and the Bowning volcanic-arch volcanism (calc-alkaline orogenic volcanism) were closely associated. Many of these orogenic granites still have volcanic ejecta associated with them. Mineral deposits which formed in this volcanic arch setting are today mostly eroded because of the continuous rise of the anticlinorial zone where they formed. However, the entity of mineral deposits which survived erosion forms the Bowning volcanic-arch metallogenic unit.

The Cowra Trough, from a metallogenic point of view, can be subdivided into the Cowra oceanic-crustal metallogenic subunit, which includes a few mineral deposits associated with the oceanic crust (Ashley, 1974a), and the Cowra oceanic-mantle metallogenic subunits, with deposits in the Coolac serpentinite belt (Ashley, 1974b). The Tumut Pond serpentinite belt probably represents upper mantle material of the Cowra Trough emplaced along a major fault, but no deposits are known to be associated with this belt.

Around the Cowra Trough, like around other marginal seas, localised volcanic rifts developed as a result of tensional stresses. The entity of mineral deposits formed in connection with these volcanic rifts can be grouped into the Cowra volcanic-rift metallogenic unit. The recently described Basin Creek copper deposit near Tumut (Nethery and Ramsden, 1973) is an example, and generally this province is a good exploration target.

After Early Silurian Campbells Group time, tensional stresses caused formation of the Hill End Trough. Its en-echelon equivalent to the south was the Captains Flat Trough. It has been suggested that the Hill End Trough, probably a marginal sea (Packham, 1973), originated by a new splitting or rifting of the Ordovician Molong Volcanic Rise (Scheibner, 1972b, 1974a). The eastern separated segment built by the Sofala and Rockley Volcanics and underlying Ordovician rocks of the Monaro trench complex all contributed to the formation of the Capertee Rise. At present, oceanic rocks have not yet been described from the Hill End Trough. Basic sills occurring in the Chesleigh Formation in the Hill End synclinorial zone (Dickson, 1962) have not been studied in detail and their geochemical character is unknown. The large thickness of sediments, rapid subsidence, high-temperature/low-pressure metamorphism, style of deformation, and presence of gold deposits, the gold of which was probably extracted from the underlying oceanic crust, favour the marginal-sea interpretation of the Hill End Trough.

Assuming continental separation and rifting along the Lachlan River and Pyramul Creek fracture zones, a very impressive fit of the pre-Hill End Trough complexes can be achieved (Scheibner, 1974c; Scheibner and Stevens, 1974). Opening of the segment of the

Hill End Trough north of the Lachlan River fracture zone took place around a pole of rotation positioned on the Darling River fracture zone. This means that the Hill End Trough was restricted to the area south of the Darling River fracture zone. The southern segment of the Hill End Trough opened around a pole of rotation positioned on the projection of the Murray River fracture zone in the present south coast area of New South Wales or northeastern Victoria, or even further to the south. The opening of the southern segment was limited and this has important metallogenic implications because moderately rifted areas have the character of volcanic rifts, and strata-bound sulphide deposits are associated with them.

The entity of deposits associated with the volcanic rifts which formed around the Hill End Trough, and especially in its southern segment, form the Hill End volcanic-rift metallogenic unit.

The Captains Flat Trough was probably a narrow marginal sea, which initially had a volcanic-rift character, or a volcanic rift in which intensive rifting took place. In connection with this rifting, strata-bound sulphide deposits of the Kuroko type formed (Felton et al., 1974; cf. Lambert, 1973). They constitute an entity of mineral deposits named the Captains Flat volcanic-rift metallogenic unit. Into this province it is possible to group localised volcanic rifts situated in the Canberra-Yass Rise area (Gilligan, 1975).

During the Late Silurian a new inter-arc basin can be identified east of the Capertee Rise, the Murruin Basin. Unfortunately, most of the rocks of this tectonic unit are concealed under the Sydney Basin, but an oceanic metallogenic unit is envisaged in the area concerned.

Based on the character of sediments and clastics occurring in the Tamworth synclinorial zone, it has been suggested that a frontal volcanic arc existed perhaps in Late Silurian times but certainly in the Devonian east of the Murruin Basin. This volcanic arc is concealed at present, but the hypothetical deposits formed in this setting constitute an unnamed volcanic-arc metallogenic unit.

The area east of the frontal volcanic arc has been described as the Tamworth Frontal Arc Area (Scheibner, 1972b, 1974a), in which relatively unstable continental-shelf to continental-slope conditions prevailed. A relatively narrow trough can be distinguished within the above tectonic unit, the Tamworth Trough (Fig. 3), characterised by spilite volcanism (Vallance in Packham, 1969). This volcanism indicates tensional stresses and possibly restricted sea-floor spreading. Some oceanic metallogeny could have been associated, but it is not known as yet. Further to the east was the Woolomin slope and basin, a flysch wedge or trench complex which probably accumulated in connection with a primary Benioff zone. Oceanic deposits characterised this tectonic unit and they form an entity of mineral deposits, the Woolomin oceanic metallogenic unit. This province can be subdivided into the Woolomin oceanic crustal metallogenic subunit which is represented by deposits associated with oceanic layer 2 occurring sporadically in the basal sections of the oceanic Woolomin Beds and in the obducted oceanic lithosphere along the Peel Thrust. As yet no deposits have been described from oceanic layer-3 rocks which occur in

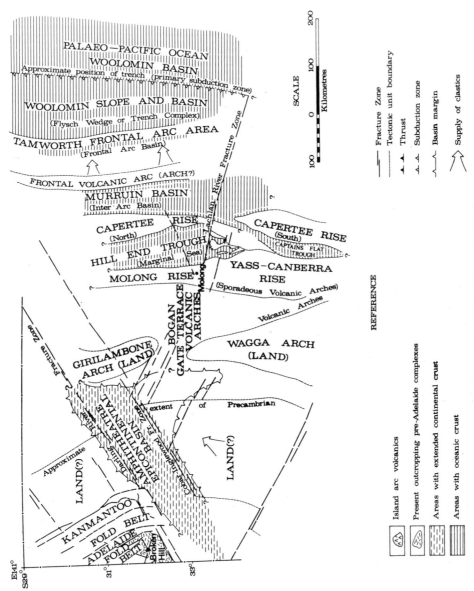

Fig. 3. Palinspastic map of the area of New South Wales for Early Devonian time.

the obducted oceanic lithosphere (Scheibner and Glen, 1972). The Woolomin oceanic mantle metallogenic subunit is represented by mineral deposits associated with obducted oceanic lithosphere along the Peel Thurst, including those deposits at Port Macquarie.

During the late Late Silurian and Early Devonian the Australian plate rotated oceanward. The oceanic crust of the Cowra Trough was probably completely subducted, while the oceanic crust of the narrow Cobar Trough was deformed and eventually partly fused to give rise to orogenic plutonism. Also the axis of sedimentation moved into the Amphitheatre Basin. In the terminal episodes of orogeny, upthrust of oceanic lithosphere during collision of small plates resulted in the development of obduction zones such as the Coolac serpentinite belt and the Tumut Pond serpentinite belt (Scheibner, 1972a, 1974a). Shelf sedimentation reflecting the progress of cratonization spread over the areas of deformed troughs north of the Murray fracture zone, but not south of it, because in the Melbourne Trough orogenic sedimentation (flysch) continued. Similarly, in the Hill End Trough and Murruin Basin orogenic sedimentation continued. In the Tamworth Frontal Arc Area further large quantities of sediments and volcanic debris accumulated. The Woolomin flysch wedge (trench complex) continued to accrete.

The orogenic development of the Lachlan mobile zone continued during the Tabberabberan metallogenic stage. In Middle Devonian times, during the strong Tabberabberan orogeny, pre-cratonic development was terminated and a transitional tectonic province (Lambian transitional province) developed. In the Middle Carboniferous the Kanimblan orogeny terminated the sedimentation and the Lachlan Fold Belt was formed (cf. Scheibner, 1972a, 1974a,b).

The eastern part of the former Lachlan mobile zone was later involved in a remobilisation and became part of the Late Palaeozoic New England mobile zone. During the Hunter-Bowen orogeny the New England Fold Belt developed from this mobile zone.

It is not intended to discuss the tectonic development of the Tasman mobile zone further, because this has been done recently in several publications. The above brief review serves to illustrate how the more important strata-bound massive sulphide deposits of the region can be related to the tectonic development of the Tasman mobile zone.

DESCRIPTION OF SOME EXAMPLES OF STRATA-BOUND SULPHIDE DEPOSITS IN NEW SOUTH WALES

Cyprus-type deposits

Stratiform massive sulphide deposits of Cyprus type are known from both the New England and Lachlan Fold Belts. Their characteristic features are outlined in the following summary description taken largely from "The Mineral Deposits of New South Wales" Centenary Volume (Markham and Basden, 1974).

Within the New England Fold Belt a long narrow arcuate belt of sediments and

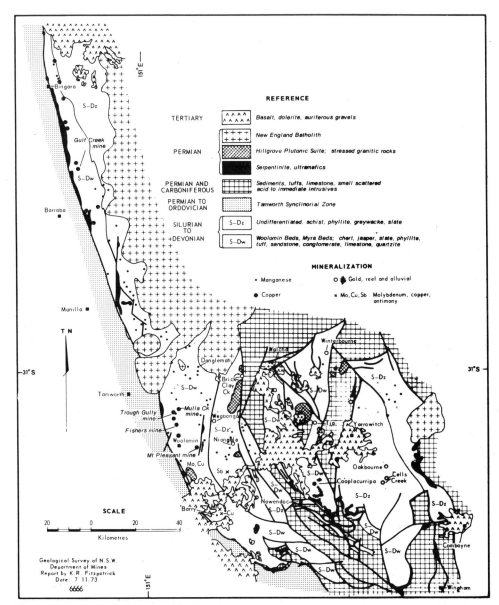

Fig. 4. Mineralisation within the Woolomin Beds and associated sediments of western and southern New England. (Geology modified after Pagson, 1972. Reprinted from Fitzpatrick, 1974, in: Markham and Basden, 1974. Courtesy of Geol. Survey of New South Wales.)

volcanics (designated the Woolomin Beds) is well developed on the western and south-western margin (Fig. 4). The Woolomin Beds, together with the related Myra Beds, comprise a sequence of argillite, siltstone, and greywacke with variably developed jasper, chert, and altered basic lavas. These rocks range in age from Silurian to Devonian but

could be as old as Ordovician. According to Fitzpatrick (1974) jasper, chert, and tholeiitic basaltic lavas show a close spatial association and are especially well developed along
the western margin of the belt near their contact with the Peel Thrust and Great serpentinite belt. Mineralisation that occurs within the Woolomin Beds comprises both stratiform massive sulphide deposits and siliceous manganese ores. The latter may be regarded
as banded manganese formations and are clearly chemically precipitated sediments. Lusk
(1964) and Fitzpatrick (1974) have described the main features of the massive sulphide
mineralisation. Some twenty-one occurrences of this type are known within the Woolomin Beds over a strike length of about 230 km (Fig. 4). Individual sulphide bodies are
lenticular in form, conformable to cleavage and probably also to layering of the host
rocks, and show an intimate spatial relationship to jasper and altered basalt. Most of the
known deposits are quite small, the maximum recorded production, from the Gulf mine,
being some 25,000 tons of ore.

Mineralogically, the ores are fine-grained and pyritic with evidence of primary mineralogical banding. Chalcopyrite may be present in amounts ranging from 5 to 10% and
sphalerite is the next most abundant sulphide. The elements titanium, manganese, lead,
and cobalt are present in decreasing order of abundance. The ores show clear evidence of
post-depositional deformation and recrystallisation. On the basis of observed features,
Fitzpatrick (1974, p. 344) concluded that the above deposits within the Woolomin Beds
"originated as chemical precipitates on a deep (?) ocean floor at a time when little or no
clastic material was being supplied". The metals are believed to be related to fumarolic
activity connected with "outpourings of submarine tholeiitic basalt lava" which supplied
not only the copper, iron, silver, zinc, and other metals but were also responsible for the
chemically precipitated siliceous iron and manganese formations.

Within the Lachlan Fold Belt a similar type of mineralisation is known from the
Tumut (Ashley, 1974a) and also from the Tottenham-Girilambone region (Suppel, 1974).
According to Ashley (1974a), the Tumut occurrences are developed within metabasaltic
rocks of the Honeysuckle Beds of possible Early Silurian age. These, together with the
neighbouring Coolac serpentinite, from a typical ophiolite assemblage, the basalts being
of abyssal tholeiitic composition. The Tumut gold mine and Snowball copper mine are
two specific examples of stratiform, cupriferous pyrite mineralisation quoted by Ashley.
Neither deposit is well exposed and hence precise details of the mineralisation are not
known, but as far as can be ascertained the lodes are conformable to layering.

The sulphide bodies, of tabular to lenticular shape, comprise both massive and disseminated mineralisation, the dominant mineral being pyrite with subordinate chalcopyrite,
sphalerite, and other sulphides. The sulphide ores are locally banded, pyrite-rich bands
alternating with silicate-rich bands containing such minerals as albite, epidote, actinolite,
and chlorite. These latter minerals are an expression both of hydrothermal alteration of
the basalts and their subsequent deformation and regional metamorphism. Ashley
(1974a) has proposed a volcanic exhalative origin for the Tumut mineralisation, the
source of the metals being submarine, tholeiitic volcanism.

The Tottenham and Girilambone deposits are localized within the Girilambone Beds, a largely undifferentiated sequence of metasediments and volcanics that crop out over a wide area of central New South Wales. The Girilambone Beds make up the northern part of the Girilambone-Wagga anticlinorial zone and according to Suppel (1974) their presently known extent is encompassed within a belt some 360 km long and locally up to 120 km wide. The Girilambone Beds comprise a highly deformed sequence of siliceous sediments and associated basic volcanic rocks now represented by phyllite, quartz-mica schist, quartzite, metagreywacke, metabasalt, and banded magnetite quartzite. The sequence, which is largely undifferentiated on account of limited outcrop and complex structure, has been intruded by older mafic rocks represented by serpentinite, peridotite, or dunite as well as intrusive complexes of basic to intermediate composition. Stratiform cupriferous pyrite mineralisation is developed at a number of localities within this broad belt, the more important and best studied being at Tottenham and in the Girilambone-Hermidale district. At Tottenham, Suppel (1974) has recorded the presence of tabular massive sulphide layers from 0.9–1.5 m thick occurring in a sequence of quartz-mica schists, "basic" schists (altered basalt lavas), and banded magnetite quartzite. The sulphide bodies appear to be conformable to layering and show a close spatial relationship to the magnetite quartzites. In addition, most occur stratigraphically above a prominent band of "basic" schist (Fig. 5). Suppel (1974) recorded the presence of at least three separate lode horizons. The sulphide layers comprise essentially massive and sometimes disseminated sulphides that are rich in pyrite with subordinate chalcopyrite and lesser sphalerite and other sulphide minerals. The ores are granular, fine-grained, and exhibit typical metamorphic recrystallisation and deformation textures. Transgressive vein-type mineralisation is also represented in the Tottenham area but is thought to be the result of remobilisation of material from the layered ores.

In the Girilambone-Hermidale region to the north of Tottenham, similar mineralisation has been described by Smith (in Suppel, 1974). Stratiform cupriferous mineralisation is developed within fine-grained schistose and quartzitic sediments closely associated with or immediately overlying metavolcanics or their pyroclastic derivatives. The mineralisation occurs in both massive and disseminated form. Locally, the ore could best be described as laminated with alternating layers of sulphide-rich material and sulphide-free host rock. Pyrite is again the most abundant primary sulphide with minor chalcopyrite, sphalerite, and traces of pyrrhotite. At the Girilambone mine itself, the best mineralisation is developed within a pink quartzite layer which appears to be a chemically precipitated chert.

In both the Tottenham and Girilambone areas the mineralisation is related to rocks that form part of a typical ophiolite assemblage. The origin of such mineralisation appears to be the result of volcanic exhalative processes related to submarine basaltic volcanism. Suppel (1974) has suggested that the local occurrence of such volcanic rocks with their associated sediments and sulphide mineralisation as at Tottenham and Girilambone-Hermidale, may be due to their exposure in eroded anticlinal structures.

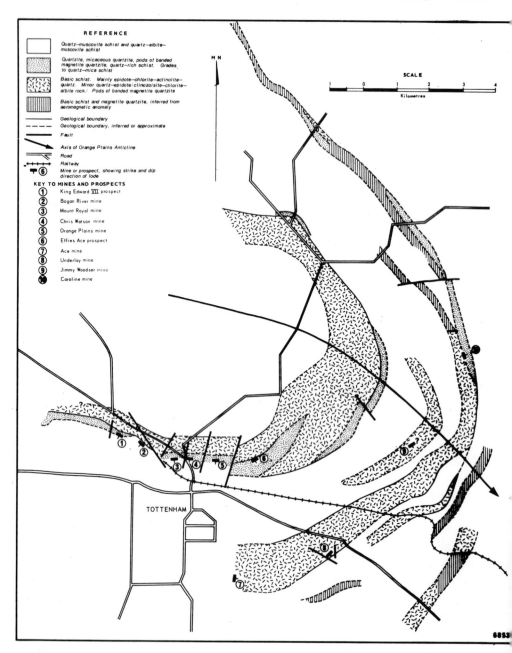

Fig. 5. Geological plan of the Tottenham area showing the location of copper deposits. (Reprinted from Suppel, 1974, in: Markham and Basden, 1974. Courtesy of the Geol. Survey of New South Wales.)

With the exception of the Girilambone deposit itself, which contains significant reserves of copper mineralisation, all of the known examples of Cyprus-type massive sulphide deposits in New South Wales are comparatively small in size and of limited economic importance. However, the areas in which they occur are clearly of considerable exploration interest and it is possible that further exploration may disclose significantly larger tonnages.

Kuroko-type deposits

Stratiform massive sulphide deposits of Kuroko type differ from those described above in that they are associated with volcanic rocks of acidic rather than basic composition. Moreover, the ores themselves show differing metal ratios, zinc being characteristically more abundant than both lead and copper, as opposed to $Cu > Zn > Pb$ abundances in the Cyprus-type mineralisation.

Kuroko-type mineralisation is especially well developed in acid volcanic rocks of the Lachlan Fold Belt, particularly those of Middle–Late Silurian age. Most of the known deposits occur within comparatively narrow, elongate fault-bounded blocks which represent ancient volcanic rift zones. Within these rift zones acid volcanism commenced at an early stage in their evolution, resulting in the accumulation of thick volcanic piles both under submarine and subaerial conditions. As noted in an earlier section of this paper, volcanic rifts became incorporated into synclinorial zones during subsequent deformation. Hence the volcanic rocks and their enclosed mineral deposits are comparatively well preserved.

With the eastern part of the Lachlan Fold Belt, deposits of this type occur in the Hill End, Captains Flat-Goulburn, and Cowra-Yass synclinorial zones (Stevens, 1974; Gilligan, 1974a,b; Felton et al., 1974). Most are localised along the margins of these zones.

Although each deposit tends to show some individual features, most are remarkably similar in character and the following characteristics apply to the majority of known deposits of this type in the Lachlan Fold Belt.

(1) Stratiform massive sulphide deposits of Kuroko type are localised within belts of acid volcanic rocks of generally Middle–Late Silurian age. They occur along the margins of volcanic rifts in close proximity to a faulted boundary with the neighbouring structural unit.

(2) Volcanic rocks associated with the mineralisation are dominantly rhyolitic-dacitic in composition and comprise lavas, ignimbrites, tuffs and agglomerate. These are interbedded with sediments including shale, greywacke, and limestone. The well-stratified nature of the volcanic rocks, the abundance of pyroclastics, and the presence of interlayered sediments is indicative of explosive volcanism under essentially submarine conditions.

(3) The volcanic host rocks are strongly cleaved and have suffered low-grade, regional metamorphism. They are composed of quartz, feldspar, and white mica, with lesser

chlorite and carbonate minerals. Individual rock units can be delineated on the basis of abundance and grain size of quartz and feldspar "phenocrysts" relative to the schistose groundmass. Rapid facies changes characterise the volcanic sequences.

(4) The orebodies themselves are lenticular in form and are conformable with layering of the host volcanics and sediments. Orebodies tend to be localised in a fine-grained shale (?tuffaceous) unit in close proximity to an underlying coarse-grained pyroclastic rock. The host shale is often highly siliceous and may be described as chert.

(5) Individual deposits comprise one or more lenses of massive sulphide, the size and tonnage of these lenses varying greatly from one deposit to another. Mineralisation may be present intermittently over a strike length of up to 1200 m with individual lenses from 1–20 m in thickness and up to 200 m in length. Measured tonnages range from a few tens of thousands of tonnes up to 10 million tonnes in one deposit.

(6) The orebodies are composed of fine-grained massive sulphides which commonly display a well-developed mineralogical banding. The banding is a primary depositional feature and is expressed by differences in grain size and abundance of the constituent sulphide and gangue minerals. Sulphide minerals, in order of abundance, comprise pyrite, sphalerite, galena, and chalcopyrite with lesser tetrahedrite and arsenopyrite. Silver and gold are generally present in amounts of the order of 60 g/t and 2 g/t, respectively.

(7) The orebodies show mineralogical zoning. Copper, as the mineral chalcopyrite, is enriched towards the base of the lens, zinc and lead towards the top. Where present, barite is associated particularly with lead at the top of the ore sequence, but barite is present in only a few deposits. (A number of deposits are also known in which stratiform barite occurs without the presence of associated massive sulphides.) Overall metal abundances are in the ratio $Fe > Zn > Pb > Cu$.

(8) The ores show clear evidence of post-depositional deformation and recrystallisation.

(9) Most stratiform massive sulphide deposits of this type within the Lachlan Fold Belt do not show an underlying zone of "yellow" ore. As such, they differ from the classical Kuroko-type deposit of Japan. However, as noted by Gilligan (1974b), strata-bound deposits rich in copper and exhibiting disseminated to stockwork vein mineralisation are known from within the Cowra-Yass synclinorial zone.

Economically, the two most important Kuroko-type deposits from within the Lachlan Fold Belt in New South Wales are those of Captains Flat and Woodlawn. The Lake George mine at Captains Flat has yielded in excess of 4.5 million tonnes of ore while the newly discovered deposit of Woodlawn near Tarago contains reserves presently estimated at some 10 million tonnes. Brief descriptions of these two deposits are given below, this information being summarised from Gilligan (1974a).

At Captains Flat the host Silurian Kohinoor Volcanics comprise some 800 m of strongly foliated volcanic rocks of mostly rhyolitic to dacitic composition but with some andesitic tuffs represented (Fig. 6). Crystal tuffs are the most common rock type. These contain typically phenocrysts of quartz and feldspar (both plagioclase and alkali feldspar)

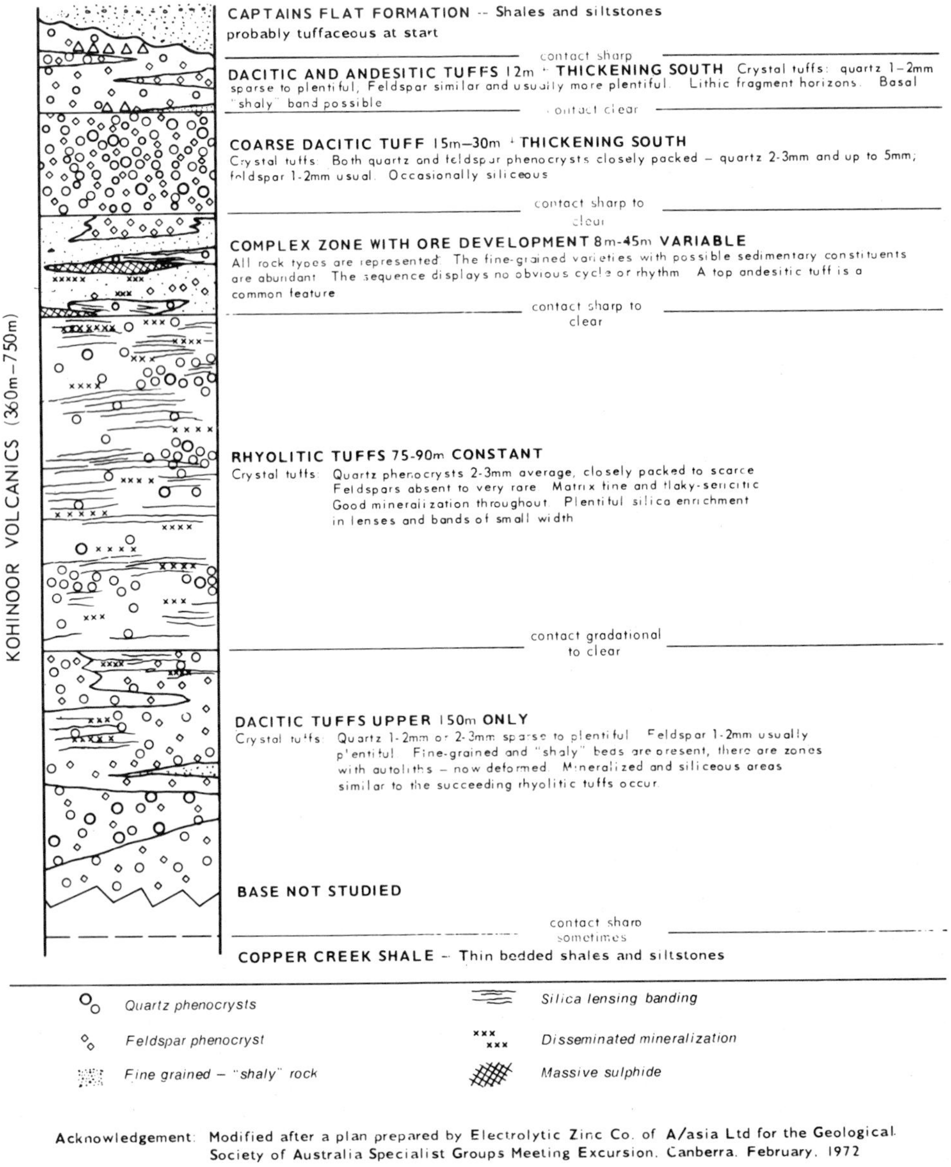

CAPTAINS FLAT FORMATION -- Shales and siltstones
probably tuffaceous at start
_____ contact sharp _____
DACITIC AND ANDESITIC TUFFS 12m + THICKENING SOUTH Crystal tuffs: quartz 1-2mm
sparse to plentiful. Feldspar similar and usually more plentiful. Lithic fragment horizons. Basal
"shaly" band possible
_____ contact clear _____
COARSE DACITIC TUFF 15m—30m + THICKENING SOUTH
Crystal tuffs. Both quartz and feldspar phenocrysts closely packed — quartz 2-3mm and up to 5mm;
feldspar 1-2mm usual. Occasionally siliceous
_____ contact sharp to _____
clear
COMPLEX ZONE WITH ORE DEVELOPMENT 8m-45m VARIABLE
All rock types are represented. The fine-grained varieties with possible sedimentary constituents
are abundant. The sequence displays no obvious cycle or rhythm. A top andesitic tuff is a
common feature
_____ contact sharp to _____
clear

RHYOLITIC TUFFS 75-90m CONSTANT
Crystal tuffs. Quartz phenocrysts 2-3mm average, closely packed to scarce
Feldspars absent to very rare. Matrix fine and flaky-sericitic
Good mineralization throughout. Plentiful silica enrichment
in lenses and bands of small width

_____ contact gradational _____
to clear

DACITIC TUFFS UPPER 150m ONLY
Crystal tuffs. Quartz 1-2mm or 2-3mm sparse to plentiful. Feldspar 1-2mm usually
plentiful. Fine-grained and "shaly" beds are present, there are zones
with autoliths — now deformed. Mineralized and siliceous areas
similar to the succeeding rhyolitic tuffs occur.

BASE NOT STUDIED
_____ contact sharp _____
sometimes
COPPER CREEK SHALE — Thin bedded shales and siltstones

KOHINOOR VOLCANICS (360m—750m)

Quartz phenocrysts   Silica lensing banding

Feldspar phenocryst   Disseminated mineralization

Fine grained — "shaly" rock Massive sulphide

Acknowledgement: Modified after a plan prepared by Electrolytic Zinc Co. of A/asia Ltd for the Geological
Society of Australia Specialist Groups Meeting Excursion. Canberra. February. 1972

Fig. 6. Detailed stratigraphic succession the the Laek George mine, Captains Flat. (Reprinted from
Gilligan 1974a, in: Markham and Basden, 1974. Courtesy of Geol. Survey of New South Wales.)

set in a sheared matrix of quartz, feldspar, sericite, and chlorite. The phenocrysts are of
variable grain size. Other rock types represented include agglomerate, rhyolitic lavas, and
fine-grained tuffaceous sediments, the latter being the host rock for the Lake George
mine orebody.

The mineralisation occurs in the form of a number of discrete lenses of massive sulphide which extend over a strike length of 1300 m and have been traced down dip for about 1000 m. These lenses are localised within the Elliot's and Keatings shale lenses which occur stratigraphically above a rhyolitic crystal tuff and are, in turn, overlain by a dacitic crystal tuff. The orebodies comprise massive banded sulphides containing pyrite, sphalerite, galena, chalcopyrite, arsenopyrite, tennantite, and minor gold. Some indication of grade and metal ratios is afforded by the average grade of ore mined over the period 1937–1962, viz.,

10% Zn,   6% Pb,   0.67% Cu,   54 g/t Ag,   1.6 g/t Au

The following pattern of mineralogical zoning across the orebody has been described by Davis (in Gilligan, 1974a, p. 298).

"(1) A broad zone of disseminated pyritic mineralisation 75–90 m thick with minor chalcopyrite towards the top of the zone. This zone lies stratigraphically below the main sulphide ore body.

(2) Massive and semi-massive pyrite with occasional chalcopyrite and rare sphalerite and galena occurring in narrow lenses and bands. This zone is about 15 m thick and merges with zone (1) above.

(3) Above zone (2) the bulk of the massive sulphides is represented by a banded pyrite-sphalerite zone which can be up to several metres thick.

(4) At the stratigraphic top of the ore body the sulphide ore is rich in sphalerite and galena.

The broad zone of disseminated pyrite described above is also characterised by intense silicification of the volcanic host ..."

"Down dip form the massive sulphides the ore horizon is represented by chert and dolomite."

The newly discovered deposit at Woodlawn is similar to Captains Flat in many respects but does show important differences. At Woodlawn the massive sulphide orebody is some 200 m long and extends down dip for about 330 m. The main ore lens or series of lenses is composed of banded, fine-grained sulphide ore containing pyrite, sphalerite, galena, chalcopyrite, and tetrahedrite. In the stratigraphic footwall of the deposit, lenses of chlorite schist ore, containing up to 50% of pyrite, chalcopyrite, and sphalerite, immediately underly the massive sulphide body and are also conformable. The chloritic schist itself is an intensely sheared rock composed of both chlorite and talc with subordinate quartz and is characterised chemically by a high magnesium content. Petrographic work suggests that it represents an original magnesium-rich sediment and is not the product of chloritic alteration of an original acid volcanic rock.

A full suite of acid lavas, crystal tuffs, agglomerates, and related pyroclastic rocks is also present at Woodlawn but the orebody itself is situated in a dominantly fine-grained tuffaceous sequence marginal to the above rocks and does not immediately overly them, as at Captains Flat and other deposits of this type.

As a result of detailed studies of massive sulphide deposits of the Kuroko type in Japan, eastern Australia and in many parts of the world there is now general agreement that such deposits are volcanic exhalative in nature and owe their origin to fumarolic activity connected with explosive, submarine acid volcanism. The evidence from the New South Wales deposits suggests that the sulphide ores were chemically and rhythmically precipitated on the sea bottom in periods of relative quiescence such that only a minimum quantity of clastic material was being supplied. The observation that most massive sulphide deposits are located within fine-grained tuffaceous shale horizons immediately overlying a coarse-grained crystal tuff suggests that this period of quiescence followed a period of explosive volcanism and corresponded to a time when fumarolic activity was at a maximum. The fumaroles rising to the surface would also be responsible for some disseminated and stockwork type mineralisation in the underlying pyroclastic rocks.

CONCLUSIONS

The above brief paper illustrates the tectono-genetic approach to the study of metallogeny being applied in New South Wales by the State Geological Survey. Only two tectonic settings were discussed, but a more complete analysis has recently been made (Markham and Basden, 1974), and detailed systematic study is under way. The tectono-genetic approach, combined with the study of structural control of the distribution of mineral deposits, enables prognosis of mineral deposits in areas where as yet no deposits are known. This has a special bearing on the exploration for concealed deposits and in areas relatively unexplored, and also calls for reassessment of well-known areas.

REFERENCES

Ashley, P.M., 1974a. Strata-bound pyritic sulphide occurrences in an ophiolite assemblage near Tumut, New South Wales. *J. Geol. Soc. Austr.,* 21: 53–62.
Ashley, P.M., 1974b. Southern serpentinite belts. In: N.L. Markham and H. Basden (Editors), *The Mineral Deposits of New South Wales.* Geol. Surv. N.S.W., Sydney, pp. 184–194.
Ashley, P.M., Chenhall, B.E., Cremer, P.L. and Irving, A.J., 1971. The geology of the Coolac Serpentinite and adjacent rocks east of Tumut, New South Wales. *J. Proc. R. Soc. N.S.W.,* 104: 11–29.
Basden, H., 1974. Preliminary report on the geology of the Cootamundra 1 : 100,000 sheet. *Q. Notes Geol. Surv. N.S.W.,* 15: 7–18.
Coleman, R.G., 1971. Plate-tectonic emplacement of upper mantle peridotites along continental edges. *J. Geophys. Res.,* 76: 1212–1222.
Coleman, R.G., 1973. Ophiolite Conference combines field trips, seminars. *Geologist (Newslett. Geol. Soc. Am.),* 8: 1, 4.
Dickson, T.W., 1962. *The Geology of Parishes of Guroba and Merinda.* Thesis Univ. Sydney, Sydney (unpublished).
Felton, E.A., Gilligan, L.B., Matson, C.R. and Stevens, B.P.J., 1974. Base-metal mineralization associated with Silurian acid volcanics in the eastern part of the Lachlan Fold Belt. *Rec. Geol. Surv. N.S.W.,* 16 (in press).

Fitzpatrick, K.R., 1974. Woolomin Beds and associated sediments. In: N.L. Markham and H. Basden (Editors), *The Mineral Deposits of New South Wales*. Geol. Surv. N.S.W., Sydney, pp. 339–349.

Gass, I.G., 1968. Is the Troodos Massif of Cyprus a fragment of Mesozoic ocean floor? *Nature,* 220: 39–42.

Gilligan, L.B., 1974a. Captains Flat-Goulburn synclinorial zone. In: N.L. Markham and H. Basden (Editors), *The Mineral Deposits of New South Wales*. Geol. Surv. N.S.W., Sydney, pp. 294–306.

Gilligan, L.B., 1974b. Cowra-Yass synclinorial zone. In: N.L. Markham and H. Basden (Editors), *The Mineral Deposits of New South Wales*. Geol. Surv. N.S.W., Sydney, pp. 217–230.

Gilligan, L.B., 1975. A metallogenic study of the Canberra 1 : 250,000 Sheet. Geol. Surv. N.S.W., (in preparation).

Guild, P.W., 1971. Metallogeny: a key to exploration. *Min. Eng.,* Jan. 1971: 69–72.

Guild, P.W., 1972. Metallogeny and the new global tectonics. *Rep. Int. Geol. Congr., 24th, Montreal,* 4: 17–24.

Guild, P.W., 1973. Massive sulfide deposits as indicators of former plate boundaries. (Open-File Rep. U.S. Geol. Surv., unpublished.)

Heidecker, E., 1972. Evolution of the Ravenswood-Lolworth Block: influence upon Devonian tectonism in northeastern Queensland. *Abstr. Joint Specialist Groups Meet., Geol. Soc. Austr., Canberra,* pp. F7–F11.

Hess, H.H., 1955. Serpentinites, orogeny and epirogeny. *Geol. Soc. Am., Spec. Pap.,* 62: 391–408.

Hodder, R.W. and Hollister, V.F., 1972. Structural features of porphyry copper deposits and the tectonic evolution of continents. *Can. Min. Metall. (C.I.M.) Bull.,* Febr. 1972: 41–45.

Hutchinson, R.W., 1971. Volcanogenic sulphide deposits and their metallogenic significance. (Abstr.) *Min. Eng.,* 23: 71.

Hutchinson, R.W. and Hodder, R.W., 1972. Possible tectonic and metallogenic relationships between porphyry copper and massive sulphide deposits. *Can. Min. Metall. (C.I.M.) Bull.,* Febr. 1972: 34–40.

Karig, D.E., 1970. Ridges and basins of the Tonga-Kermadoc island arc system. *J. Geophys. Res.,* 75: 239–254.

Kay, M., Hubbard, N.J. and Gast, P.W., 1970. Chemical characteristics and origin of oceanic-ridge volcanic rocks. *J. Geophys. Res.,* 65: 1585–1613.

Lambert, I.B., 1973. The features and genesis of the Kuroko-type Cu-Zn-Pb-Ag-Au deposits of Japan, with comments on some other ore deposits of volcano-sedimentary sequences. *CSIRO Div. Miner., Invest. Rep. Canberra,* 98: 53.

Lusk, J., 1964. *Copper Ores and Their Distribution in Western New England*. Thesis, Univ. New England, Armidale (unpublished).

Markham, N.L. and Basden, H., (Editors), 1974. *The Mineral Deposits of New South Wales*. Centenary Volume. Geol. Surv. N.S.W., Sydney, 682 pp.

Matsukuma, T. and Horikoshi, E., 1970. Kuroko deposits in Japan, a review. In: T. Watanabe (Editor), *Volcanism and Ore Genesis*. Univ. Tokyo Press, Tokyo, pp. 153–179.

Nethery, J.E. and Ramsden, A.R., 1973. Geological setting of the Basin Creek copper mineralization. (Paper presented to Geol. Soc. Austr., Sydney, 13 Sept., 1973.)

Nisbett, E. and Pearce, J.A., 1973. TiO$_2$ a possible guide to past oceanic spreading rates. *Nature,* 246: 468–470.

Packham, G.M. (Editor), 1969. *The Geology of New South Wales–J. Geol. Soc. Austr.,* 16: 654 pp.

Packham, G.H., 1973. A speculative Phanerozoic history of the southwest Pacific, In: P.J. Coleman (Editor), *The Western Pacific Island Arcs, Marginal Seas, Geochemistry*. Univ. West. Austr. Press, Perth, pp. 369–388.

Sawkins, F.J., 1972. Sulfide ore deposits in relation to plate tectonics. *J. Geol.,* 80: 377–397.

Sawkins, F.J. and Petersen, U., 1969. A tectonic-genetic classification of sulphide ore deposits. (Abstr.) *Geol. Soc. Am. Ann. Meet., Atlantic City,* pp. 197–198.

Scheibner, E., 1972a. Tectonic concepts and tectonic mapping. *Rec. Geol. Surv. N.S.W.,* 14: 37–83.

Scheibner, E., 1972b. A model of the Palaeozoic tectonic history of N.S.W. *Abstr. Joint Specialists Groups Meet., Canberra,* pp. F12–F16.

Scheibner, E., 1974a. A plate-tectonic model of the Palaeozoic tectonic history of New South Wales. *J. Geol. Soc. Austr.,* 20: 405–426.

Scheibner, E., 1974b. An outline of the tectonic development of New South Wales with special reference to mineralization. In: N.L. Markham and H. Basden (Editors), *The Mineral Deposits of New South Wales.* Geol. Surv. N.S.W., Sydney, pp. 1–39.

Scheibner, E., 1974c. Fossil fracture zones (transform faults), segmentation, and correlation problems in the Tasman Fold Belt System. In: A.K. Denmead, G.W. Tweedale and A.F. Wilson (Editors), *The Tasman Geosyncline – a Symposium.* Geol. Soc. Austr., Qld. Div., Brisbane, pp. 65–92.

Scheibner, E. and Glen, R.A., 1972. The Peel Thrust and its tectonic history. *Q. Notes Geol. Surv. N.S.W.,* 8: 2–14.

Scheibner, E. and Stevens, B.P.J., 1974. The Lachlan River lineament and its relationship to metallic deposits. *Q. Notes Geol. Surv. N.S.W.,* 14: 8–18.

Schneiderhöhn, H., 1955. *Erzlagerstätten* (Kurzvorlesungen zur Einführung und zur Wiederholung). VEB Gustav Fischer Verlag, Jena, 3rd ed., 375 pp.

Sillitoe, R.H., 1972a. Formation of certain massive sulphide deposits at sites of sea-floor spreading. *Trans. Inst. Min. Metall.,* 81: B141–B148.

Sillitoe, R.H. 1972b. A plate tectonic model for the origin of porphyry copper deposits. *Econ. Geol.,* 67: 184–197.

Sillitoe, R.H., 1972c. Relation of metal provinces in western America to subduction of oceanic lithosphere. *Geol. Soc. Am. Bull.,* 83: 813–817.

Sillitoe, R.H., 1973. Tops and bottoms of porphyry copper deposits. *Econ. Geol.,* 68: 799–815.

Smith, E.A., 1974. The geology and genesis of the Girilambone cupriferous pyrite deposits. *Abstr. Joint Specialists Groups Meet., Geol. Soc. Austr., Brisbane,* pp. 13–15.

Steinmann, G., 1905. Geologische Beobachtungen in den Alpen, 2. Die Schardt'sche Überfaltungstheorie und die geologische Bedeutung der Tiefseeabsätze und der ophiolitischen Massengesteine. *Ber. Natl. Ges. Freiburg,* 26: 44–65.

Stevens, B.P.J., 1974. Hill End synclinorial zone. In: N.L. Markham and H. Basden (Editors), *The Mineral Deposits of New South Wales.* Geol. Surv. N.S.W., Sydney, pp. 278–293.

Suppel, D.W., 1971. Preliminary report on copper lodes in the Tottenham and Albert areas, central western New South Wales. *Q. Notes Geol. Surv. N.S.W.,* 4: 11–18.

Suppel, D.W., 1974. Girilambone anticlinorial zone. In: N.L. Markham and H. Basden (Editors), *The Mineral Deposits of New South Wales.* Geol. Surv. N.S.W., Sydney, pp. 119–131.

Tarling, D.H., 1973. Metallic ore deposits and continental drift. *Nature,* 243: 193–196.

Uyeda, S. and Miyashiro, A., 1974. Plate tectonics and the Japanese islands: a synthesis. *Geol. Soc. Am. Bull.,* 85: 1159–1170.

Van Bemmelen, R.W., 1931. Is de Oeloebeloe een vulkaan? *Mijningenieur,* 3: 30–32.

Wright, J.B. and McCurry, P., 1973. Magmas, mineralisation and sea-floor spreading. *Geol. Rundsch.,* 62: 116–125.

Chapter 4

CALEDONIAN MASSIVE SULPHIDE DEPOSITS IN SCANDINAVIA: A COMPARATIVE REVIEW

F.M. VOKES

INTRODUCTION

This contribution to the present volume will attempt to give a general review of an important group of sulphide-ore deposits occurring within the metavolcanic—metasedimentary sequence of the Caledonian (Early Palaeozoic) fold mountain belt of Scandinavia (Fig. 1). Attention will be directed mostly towards the ore occurrences on the Norwegian side of the international boundary, due to the greater abundance of deposits here, the more abundant literature concerning them, and the writer's exclusive engagement in the Norwegian sector.

Within the Late Precambrian and Early Palaeozoic geosynclinal sequences of the Scandinavian Caledonides, two main groups of generally strata-bound sulphide mineralization may be distinguished: (1) deposits of mainly lead sulphide in arenites and other clastic rocks of Eocambrian and Lower Cambrian age along the Caledonian front zone (the so-called "Laisvall Horizon"); and (2) deposits of generally massive, bi- or polymetallic pyritic and pyrrhotitic deposits in the Upper Cambrian—Ordovician and possibly Silurian metavolcanic and metasedimentary sequences.

The present contribution considers only the deposits of the second group, i.e. the massive sulphide ores. It will deal with the general aspects of these deposits, their economic importance, mineralogy, geochemistry, problems of genesis, their metamorphism and deformation, lithostratigraphical associations, and possible relation to plate tectonics.

The Scandinavian, especially Norwegian, massive sulphide ores have long been recognized as "classic" examples of their type and have long figured in the literature concerning it. However, modern descriptions of individual deposits and mining-districts are generally few and far between, especially in the world literature, though the situation is changing for the better at the present time.

The deposits belong undoubtedly to a world-wide class of ores which has received increasing attention from both "academic" and "practical" geologists over the last decade or so, as an increasing number of examples has been discovered or recognized in widely separated parts of the world. On the academic side, the origin of these massive sulphide deposits, has been perhaps one of the most widely and hotly argued subjects in the whole

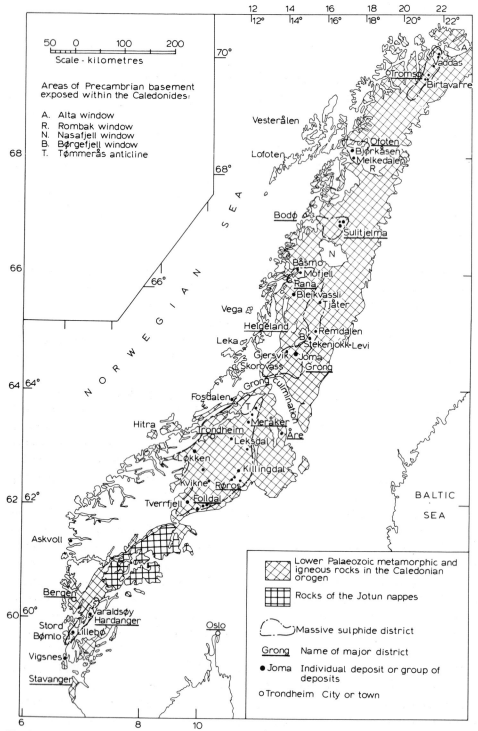

Fig. 1. Map of part of the Scandinavian peninsula showing the distribution of massive sulphide deposits within the Lower Palaeozoic Caledonides. Simplified from the tectonic and metallogenetic maps of Europe. Original scale 1 : 2,500,000.

of ore-geology within the last ten to fifteen years, and it is only relatively recently that a "break-through" has taken place in this field which gives promise of settling the controversy once and for all. However, as parts of the following account will try to show, the problem of the genesis and subsequent history of the Scandinavian massive sulphide ores is by no means resolved and many geologists prefer to keep what they refer to as an open mind on the subject.

Massive sulphide ores are generally characteristic of the geosynclinal belts of the earth's crust and have been found on, and described from, practically all the continents, from areas ranging in age from Archaean to Tertiary.

Whatever their exact origin may ultimately prove to be, they seem undoubtedly closely related spatially to the effusive and pyroclastic rocks of the so called "initial volcanism" of the geosynclinal stage of mobile-belt development. In the last decade or two, there has grown up a world-wide opinion among ore-petrologists that the relationship between the massive sulphide deposits and the geosynclinal volcanism is not only spatial, but temporal, that is to say, genetical. The ore-forming "fluids", using this term in its widest sense, may have originated within the magma, which itself gave rise to the volcanic rocks and to related, penecontemporaneous, intrusive rocks. On the other hand, the volcanism may have had a role as a source of heat which initiated circulation of connate and meteoric water, leading to leaching, transport and redeposition of the ore-components.

Because of this apparent genetical relationship to the geosynclinal volcanism, the terms "synvolcanic" or "volcanogene" have been used in describing these deposits. This is undoubtedly a more satisfactory term than the various alternatives which have been applied from time to time to denote this ore type; for example, "exhalative–sedimentary" or "syngenetic" on the one hand, and "hydrothermal–epigenetic" or "replacive" on the other. (See also Anderson and Nash, 1972, p. 862, and Sangster, 1972, p. 1.)

The now numerous, well-described examples of massive sulphide bodies of the type here under consideration from areas with little or no subsequent deformation and metamorphism, such as parts of the Canadian Shield (Roscoe, 1965 and Sangster, 1972)[1], Cyprus (Hutchinson and Searly, 1971; Constantinou and Govett, 1972 and Searle, 1972) and Japan (Matsukuma and Horikoshi, 1970 and Horikoshi, 1972), seem to point towards a common model, or a very limited number of models, which show that a fully developed ore body of the volcanogene type contains elements which have involved formation by replacement and open-space filling (with accompanying "classical" wall-rock alteration) in close relationship to other elements where deposition has taken place on top of consolidated or nearly consolidated rocks (often pillow lavas) and most probably in varying depths of seawater. Fig. 2 shows a general model of a volcanogene massive sulphide ore body, incorporating features from various published models (see above).[1]

There is being gathered a considerable weight of evidence to show that also features

---

[1] Editor's note: See Chapter 5 by Sangster and Scott, this volume, for a summary on ores of the Canadian Precambrian Shield and genetic models.

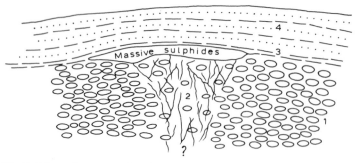

Fig. 2. Schematic cross-section through an idealized volcanogene sulphide deposit: *1* = older lava units, often pillowed, *2* = stringer zone (feeder channel or pipe); quartz-sulphide veins and disseminations, wall-rock alteration (silicification, argillization, chloritization, etc.), *3* = bedded chert unit, often ferruginous and/or manganiferous (iron formation), *4* = younger volcanic, volcanic—sedimentary or sedimentary unit.

arising from post-depositional, pre-consolidational, processes, such as sedimentary reworking, turbidity currents and slumping, are present in many of the deposits (Schermerhorn, 1971; Horikoshi, 1972 and Searle, 1972). Whether all or only some of the above features are present in any deposit, or group of deposits, depends on many variables which will not be discussed here, but the general model seems to be well-substantiated from places as far apart in space and time as the Canadian Shield (Archaean), Cyprus (Mesozoic) and Japan (Tertiary).

Scandinavian "volcanogene" deposits

If one considers the Scandinavian deposits of this type, it can be said that, although their general spatial relationship to the early geosynclinal volcanism of the Caledonian orogen is now well-established, there is available on the whole very little of the detailed evidence which would allow them to be fitted into a model of the type just mentioned. The main bulk of the Norwegian Palaeozoic massive sulphide mineralization seems related to the Støren group of Early Ordovician age and its supposed equivalents, though Wolff (1967) has pointed out that in the eastern Trøndelag (Meråker) region (Fig. 1), sulphide deposits of this type are also related to an upper greenstone horizon of middle Upper Ordovician age. Recent mapping in the southcentral Trondheim region also indicates that minor volcanic episodes, both younger and older than the Støren event, have associated sulphide deposits (e.g., Nilsen and Mukherjee, 1972). In Sweden, the important deposit at Stekenjokk in Västerbotten (Fig. 1), lies quartz-keratophyres of the Lasterfjäll group, the age of which is given as Silurian (Zachrisson, 1969 and 1971).

The ores are generally of the strata-bound type (and sometimes they can be shown to be stratiform and wholly conformable with their country rocks) though in the majority of cases the ore-body morphology has been determined by later deformational events, so that original relations are difficult to decipher. The majority of the ore bodies have been described, variously, as plates, lenses, pencils, rulers, etc., and most are elongated to

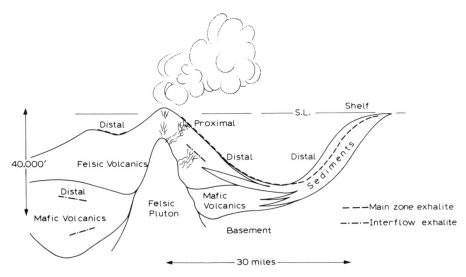

Fig. 3. Idealized section through a volcano–sedimentary complex to illustrate the various sites of exhalative deposition. (After Ridler, 1973.)

varying degrees along axes which may or may not coincide with a fold-axial, or other linear, direction in the country rocks (see Th. Vogt, 1944, 1952).

The sulphide bodies would, therefore, appear to represent the massive sulphide, supralaval elements of the general models mentioned above. Evidence for both feeder channels and reworked or resedimented elements is less obvious, though it is not altogether lacking in the deposits in the lower metamorphic areas, e.g., Løkken and Skorovass (Fig. 1).

The apparent lack of, say, feeder-channel elements in the majority of Scandinavian massive sulphide deposits may be relatable to one or more of the following reasons:

(1) original deposition at a point away from the feeder channel (cf. the "distal exhalities" of Ridler, 1971 and 1973), Fig. 3; (2) slumping and turbidity-current transport of the precipitated, but unconsolidated sulphides down the palaeoslope away from the feeder channels (Schermerhorn, 1971 and Horikoshi, 1972); (3) tectonic transport of the consolidated sulphides away from their footwall zone during tectonic movements (Jenks, 1971); and (4) folding and other deformational movements producing a new spatial relationship between massive ore and feeder zone (cf. Sangster, 1972, p. 24).

One or more of these mechanisms has probably been operative in the case of the Scandinavian Caledonian sulphide ores, making the original depositional morphological picture difficult to interpret. In particular the writer believes that the effects of the polyphase metamorphism and deformation during the Caledonian orogeny cannot be stressed too strongly when trying to reconstruct the original nature of the deposits. This subject is treated in rather more detail in a subsequent section, but at this point the writer would like to state his considered view that the ore-forming event, as indicated from detailed studies of the metamorphic and deformational effects seen in the ores, occurred at an early, preorogenic, geosynclinal stage, consistent with the synvolcanic model for ores of the same general type to which reference has already been made.

ECONOMIC IMPORTANCE

The Caledonian massive sulphide deposits have played an important economic role, in Norwegian metal-mining especially, over many generations. Mines working this type of ore were among some of the earliest to be brought into production in Scandinavia. In the beginning, the value of the ores lay in their copper content and some of Scandinavia's oldest copper-producing mines worked bodies of massive sulphide ore (Kvikne, 1631; Løkken, 1634, Røros, 1644; Folldal, 1748 and Åreskutan, 1750's). The ores were normally roasted and smelted locally to produce blister copper and nearly all their other potentially economic components were lost. In the middle of the 19th century, the sulphur contents of the ores began to be of importance for the production of sulphuric acid. This heralded a long period of intense exploitation of the massive sulphide ores, many of which had proved to be too low in copper to smelt directly for this metal. By the end of the century, the Norwegian pyrite production had passed 100,000 metric tons (tonnes) per year and it continued to rise rapidly, with some setbacks due to world economic conditions, until, immediately preceding World War II, it reached over 1,000,000 tonnes per year.

Following the general adoption of differential-flotation methods for treating such ores, concentrates of copper (chalcopyrite), zinc (sphalerite) and sulphur (pyrite) became the normal products from most of the Caledonian massive sulphide ores, with lead (galena) concentrates from one or two special types. Some especially fine-grained types, notably the Løkken and Skorovass ores, were either shipped directly as lump material or, for a time, treated pyrometallurgically (Orkla Process, see Aanerud, 1954). At the time of writing, these two mines are in the process of adapting to differential-flotation processes to produce copper and zinc concentrates. These steps have been more or less forced on the companies due to the near impossibility of selling lump pyritic ores. In fact, the world sulphur situation has in recent years become such that, with one or two notable exceptions, most of the Norwegian mines have given up producing pyrite concentrates as these are unsaleable.

By world standards, the Norwegian output of the products of the massive sulphide ores is not at all significant. The Norwegian production data for 1969, the last year for which detailed figures are available, are represented in Table I.

The domestic importance of the Caledonian sulphide-ore production can be illustrated by the data for 1970. In that year, the total primary mineral production in the country had a first-hand value of 892 million Norwegian Kroner (N.Kr.), representing about 2% of the Norwegian Gross National Product. Of the former figure the metal mines were responsible for 500 million Norwegian Kroner, or 56%. Within this group the sulphide-ore products amounted to 180 million Norwegian Kroner, or 36%.

TABLE I

Norwegian production figures in relation to world-production figures of massive sulphide ore products
(1969)

Product	Norwegian production (metric tons)	Relation to world production (%)
S (in pyrite concentrate)	352,000	3.4
Cu (in concentrate)	21,000	0.3
Zn (in concentrate)	11,000	0.2
Pb (in concentrate)	3,750	0.1

At the present time, there are nine companies working Caledonian massive sulphide
ores in Norway. Table II gives the production figures for 1972 for these companies. At
present there is no production from massive sulphide ores in the Swedish Caledonides,
but the large deposit at Stekenjokk (see below) is being made ready for production in the
near future. Prospecting for ores of the massive-sulphide type is being actively pursued
nowadays by private companies and state institutions in both countries, and an increase
in production seems assured within the next few years.

TABLE II

Norwegian massive sulphide ore production 1972

Company (mine)	Ore broken (tonne)	Lump ore exported (tonne)	Concentrates produced			
			S (tonne)	Cu (tonne)	Zn (tonne)	Pb (tonne)
A/S Folldal Verk, Tverrfjell	600,000	–	232,260	22,980	9,735	–
Orkdla A/B, Løkken	440,000	284,100	–	–	–	–
A/S Sulitjelma Konnerverk	435,600	–	88,200	24,490	2,107	–
Elhem-Spigerverket A/S, Skorovass	201,800	165,190	–	–	–	–
A/S Bleikvassli Gruber	127,745	–	17,940	–	9,630	5,470
A/S Grubeselskapet Nord-Norge, Mofjell	103,000	–	5,222	1,000	3,640	725
A/S Grong Gruber, Joma	75,850*	–	–	3,250	2,942	–
A/S Killingdal Gruveselskap	41,490	–	–	2,967	2,942	–
A/S Røros Kobberverk	?	–	–	3,868	–	–

* Planned yearly production for 1973: 300,000 tonnes of ore.

MINERALOGY

The main sulphide minerals comprising the Caledonian ores are relatively few in number and of simple composition. They are reflected directly in the chemistry of the ores and in the nature of the concentrates produced from them (Table III).

Mineralogical modal analyses of the ores are not too numerous, but Table IV gives some figures which are available.

TABLE III

Main sulphide minerals of Caledonian ores

Concen-trate	Essential component	Chemical composition
S	pyrite	FeS_2 ($\pm$pyrrhotite, $Fe_{1-x}S$)
Cu	chalcopyrite	$CuFeS_2$
Zn	sphalerite	$(Zn,Fe)S$
Pb	galena	PbS

TABLE IV

Mineralogical modal analyses of some Caledonian sulphide ores

Deposit	Mineralogical analysis (wt. %)					Reference
	cp[*]	sl[*]	gn[*]	py[*]	po[*]	
Skorovass	6.4	2.8	–	67.0	–	T. Malvik, unpublished results
Joma (Grong)	9.6	1.8	–	?	–	T. Malvik, unpublished results
Sulitjelma	6.4	0.7	–	22.9	present	T. Malvik, unpublished results
Tverrfjellet (Folldal)	4.7	2.2	–	50.1	?	T. Malvik, unpublished results
Nordre Gjetryggen, (Folldal)	5.5	7.5	trace	55.6	5.3	Page, 1963, average figures
Lergruvbakken, Røros	2.1	14.4	trace	3.7	9.9	T. Malvik, unpublished results
Mofjell	1.8	5.5	2.5	12.0	–	T. Malvik, unpublished results
Bleikvassli pyritic ore	0.8	17.4	6.1	55.6	5.4	Vokes, 1963, average figures
Bleikvassli pyrrhotitic ore	1.8	19.3	6.5	2.5	49.0	Vokes, 1963, average figures
Killingdal, main ore body	5.5	9.9	0.46	81.15	trace	Rui, 1973a, average figures
Killingdal, north ore body	3.0	17.6	–	67.2	–	Rui, 1973a, average figures
Vaddas	5.1	0.1	–	27.2	23.2	I. Lindahl, unpubl. thesis, Norges Tekniske Høgskole

[*] cp = chalcopyrite, sl = sphalerite, gn = galena, py = pyrite, po = pyrrhotite.

Major minerals

Iron sulphides. Pyrite and pyrrhotite are always the most abundant sulphides present in the typical, massive Caledonian ores, though their relative proportions vary greatly from deposit to deposit, and often between different parts of the same deposit. This aspect of the ores has been discussed by, e.g., Vokes (1962), who distinguished two main types of ore, the pyritic and the pyrrhotitic, based on the predominance of, respectively, pyrite or pyrrhotite as the iron sulphide. The two types of ore can occur separately as independent ore bodies or as distinctive parts of the same ore body. There are examples of pyritic deposits with no, or practically no, pyrrhotite and where the sulphur content approaches 50%, as, for example, in the Løkken and Skorovass ores. Other deposits which must be classified, mineralogically, as belonging to the same type, show sulphur contents of only 20–30%.

At the other mineralogical extreme, there occur bodies of quite massive sulphides where pyrite is totally lacking, its place having been taken by pyrrhotite. Examples of such ores include the Birtavarre ores, northern Norway (Vokes, 1957) and several of the ore bodies in the Røros mining-district, e.g., the Olav and Mugg deposits.

Intermediate between these two extremes are mixed ore bodies consisting partly of pyritic and partly of pyrrhotitic ore. In such cases, pyrrhotite is often present in notable amounts in the matrix to the pyrite grains in the pyritic portions. An example of such a mixed ore body is the Bleikvassli Mine in Nordland. Vokes (1963) has presented modal analyses for the two types of ore at Bleikvassli. His average figures are reproduced in Table IV. They show, among other things that the pyrrhotitic type of ore has over twice the amount of chalcopyrite than the pyritic type. This is a general characteristic of the sulphide ores of the Scandinavian Caledonian belt and will be commented on further in the following.

Geological evidence suggests that, in at least some of the composite bodies, the pyrrhotite-rich type of ore has been emplaced in its present position at a later stage than the pyritic ore. For example, at the Jakobsbakken Mine in the Sulitjelma district, chalco-pyrite–pyrrhotite mineralization could be seen clearly veining a massive, granular pyritic ore with less chalcopyrite and variable quantities of sphalerite. At the Joma deposit in the Grong area, rich chalcopyrite–pyrrhotite ore cements a breccia of the banded, fine-grained pyritic ore that constitutes the bulk of the deposit.

However, although the age relations between the two ore types seem clear in these deposits, there are other examples where the evidence is not so unequivocal, e.g. Bleik-vassli. The origin of the pyrrhotite-rich type of ore is also far from being settled. In most cases where it appears to be epigenetic with respect to the pyritic ore, an origin by metamorphic mobilization of pre-existing ore, possibly from the same body, seems plausi-ble. In other instances, the pyrrhotite-rich ore type may be the result of metamorphism of pre-existing pyrite ore, without any noticeable mobilization having taken place. In this respect, it is relevant to note that most of the ores occurring in the low-metamorphic areas are solely of the pyritic type and that in general, the relative content of pyrrhotite in the ores increases with increasing metamorphic grade in their country rocks. The already-mentioned relative abundance of chalcopyrite in the pyrrhotitic ore bodies is not at all easy to explain.

One possibility, following from the above suggested explanation of the type as a whole, is that copper was extracted from silicate and other minerals in the vicinity of the ore by a process of sulphurization during the metamorphic breakdown of pyrite to pyrrhotite.

Chalcopyrite. This is undoubtedly the economically most important base-metal sulphide in the majority of the Caledonian massive sulphide ores. It occurs in practically all the normal types of deposits in varying proportions. Table IV shows that in the "normal", lead-free, ore types, chalcopyrite predominates over sphalerite as the main base-metal sulphide present. However, in ores where galena is abundant, the relative positions of chalcopyrite and sphalerite are reversed and chalcopyrite is relegated to being the least abundant of the three base-metal sulphides. As already mentioned, chalcopyrite is usually more abundant in the pyrrhotite-rich ores than in the pyritic ores (see the Bleikvassli data).

Chalcopyrite is a typical mineral of the matrix or groundmass between the pyrite

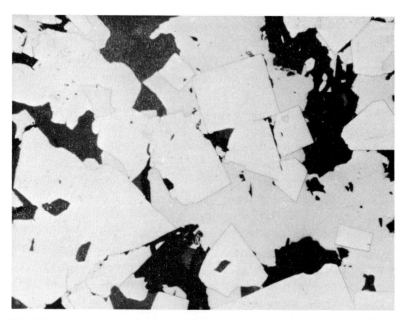

Fig. 4. Photomicrograph of pyritic ore, Skorovass Mine. Euhedral to subhedral pyrite grains (white) in matrix of anhedral chalcopyrite (off-white) sphalerite (grey) and gangue minerals (black). Polished section. Oil immersion (×200).

Fig. 5. Photomicrograph of pyrrhotitic ore, Moskogaissa, Birtavarre. Directed mosaic intergrowth of chalcopyrite (lighter) and pyrrhotite (darker); gangue and pits (black). Polished section, reflected light. Oil immersion, nicols partly + (X200).

granules of the pyritic ores, occurring in a polygonal intergrowth together with the other sulphides and the gangue quartz, etc. The matrix sulphides as a whole exhibit the typical triple junction of recrystallized or annealed ores (e.g., Stanton, 1964. Fig. 4, this chapter). In the pyrrhotitic-ore types, chalcopyrite occurs as often coarse-grained, irregular, allotriomorphic intergrowths with the other sulphides, mainly pyrrhotite and to a lesser extent sphalerite. The textures of these pyrrhotite-rich ores are typically "kneaded" or "durchbewegt", less often gneissose (Fig. 5).

Sphalerite. As shown by the few figures in Table IV, this is generally the second most abundant base-metal sulphide of the Caledonian massive sulphide ores, but becomes important in certain Zn–Pb types. Sphalerite follows chalcopyrite closely in its textural features. It is normally a mixed Zn–Fe sulphide with very varying iron content from deposit to deposit, and often within the same deposit.

Vokes (1962) has published figures for the Fe-contents of sphalerites separated from selected Norwegian Palaeozoic massive sulphide ores (Table V).

More recently, Ljøkjell (1972) has carried out a large number of electron-microprobe analyses of sphalerite in zinc concentrates from Norwegian Caledonian sulphide ores (Table VI).

In the past, attempts have been made to interpret the Fe-content of the (ZnFe)S

TABLE V

Fe-contents of sphalerites from selected Norwegian Palaeozoic massive sulphide ores (Vokes, 1962)

Ore deposit	Fe in (Zn, Fe)S (%)
Skorovass	1.50
Løkken	3.50
Mofjell	4.76
Mofjell	5.97
Bleikvassli	7.50
Killingdal	7.27
Nordre Gjetryggen, Folldal	7.65
Jakobsbakken, Sulitjelma	8.07

mix-crystals, following Kullerud (1953), in terms of the temperature of deposition or of the metamorphic recrystallization of the ores (Kullerud et al., 1955; Vokes, 1962 and Page, 1963). Recent work on the Zn—Fe—S system has shown the doubtfulness of such interpretations (Barton and Toulmin, 1966; Boorman, 1967; Boorman et al., 1971 and Scott and Barnes, 1971). The figures presented in Table VI are not easily interpreted in terms of the metamorphic grade of the rocks surrounding the various deposits. The

TABLE VI

Electron microprobe analyses of sphalerite from Norwegian Palaeozoic sulphide ore (flotation concentrates)

Deposit	Contents (wt. %)[*]			
	Zn	Fe	Cd	Mn
Skorovass	64.9	1.41	0.12	0.01
Vigsnes, Karmøy	64.3	3.29	0.13	0.03
Løkken	62.5	3.65	0.18	0.01
Mofjell	61.4	4.76	0.22	0.08
Tverrfjell	58.4	6.87	0.19	0.01
Bleikvassli	58.4	7.00	0.11	0.16
Sulitjelma	58.5	7.05	0.19	0.05
Lergruvbakken, Røros	58.7	7.10	0.11	0.24
Killingdal	58.2	7.13	0.14	0.12
Joma, Grong	58.2	7.23	0.30	0.01
Gamle Storwartz, Røros	58.7	7.48	0.10	0.06
Rieppe, Vaddas	58.5	7.82	0.14	0.06

[*] Average figures from Ljøkjell, 1972.

TABLE VII

Ore deposits, metamorphic grade of rocks and iron-sulphide contents of the associated sphalerite

Deposit	Metamorphic grade	Iron sulphide
Skorovass	upper greenschist	pyrite
Vigsnes	upper greenschist	pyrite (? pyrrhotite)
Løkken	greenschist	pyrite
Mofjell	amphibolite	pyrite
Tverrfjell	amphibolite	pyrite; pyrrhotite
Bleikvassli	amphibolite	pyrite; pyrrhotite
Sulitjelma	amphibolite	pyrite; pyrrhotite
Lergruvbakken	biotite grade	pyrrhotite; pyrite
Killingdal	upper greenschist	pyrite
Joma	greenschist	pyrite (pyrrhotite)
Storwartz	amphibolite	pyrrhotite; pyrite
Rieppe	amphibolite	pyrrhotite

presence or absence of pyrrhotite in the ores does, however, seem to show a relationship to the Fe-contents of the associated sphalerite, as is indicated by Table VII. An apparent exception to this is the Killingdal ore (Rui, 1973a), which does not contain pyrrhotite, but where the sphalerite is comparatively Fe-rich.

Pyrrhotite is itself, of course, mainly a product of the increasing metamorphism of the ores, so that the two factors, grade of metamorphism and presence of pyrrhotite, are really interdependent. It would seem that, coinciding with the appearance of pyrrhotite (at about the beginning of amphibolite-facies conditions?), the Fe-content of the sphalerites increases rapidly to the order of 7–8 wt.%. The Mofjell ore, which does not show pyrrhotite, even though it is in amphibolite-grade rocks (itself a puzzling feature) contains sphalerite with an intermediate Fe-content (4–5%).

Galena. This is normally absent, or present only in trace amounts, in many of the Scandinavian ores of this type. In a number of others, galena is present only in small amounts (tenths to a few percent) so that it cannot be regarded as an economically important mineral. However, in a few ore bodies, mainly restricted to the central Nordland district of the Caledonides in Norway and in adjacent areas in Sweden, galena is present in amounts such as to make it an economically important mineral (see Table IV). In Norway, two mines in the Rana area of Nordland recover lead concentrates from ores of this type (Bleikvassli and Mofjell).

Magnetite. This is the only significant oxide ore mineral that can be present in the polymetallic massive sulphide ores. In the majority of the ores it is absent or only present in minor amounts. However, certain ores, such as parts of the important new deposit at

Tverrfjell, Dovre, and lesser deposits in the southern Trondheim region, show considerable magnetite contents, though up to now the mineral has not been economically recovered from Norwegian massive sulphide ores. On the other hand there are, especially in the Fosen area, northwest of Trondheimsfjord, economically important magnetite-iron ores with up to 3—4% of pyrite and minor though recoverable, quantities of chalcopyrite. Such types seem to represent transitions between dominantly oxidic facies iron deposits and dominantly sulphide facies iron deposits (H. Carstens, 1955).

Minor minerals

In addition to the above major minerals, the Palaeozoic massive sulphide ores contain a variety of minor sulphides and sulphosalts which do not figure in the economic products, some of which, however, are beneficial to the value of the various concentrates or potential concentrates. These latter include native gold and various silver minerals, such as dyscrasite, ruby-silver minerals, silver-bearing fahlore and, extremely rarely, minerals of the Ag—Cu—S system.

Silver minerals. On the whole, the silver contents of the Caledonian massive sulphide ores are quite low, but they show an increase in those ores where lead enters as an economically important mineral. Table VIII shows that the Pb-bearing ores can contain up to twice as much silver as those which have only trace amounts of, or no, lead. However, it is

TABLE VIII

Noble-metal assays for a number of Caledonian massive sulphide ores

Mine	Cu (%)	Zn (%)	Pb (%)	Ag (ppm)	Au (ppm)	S (%)
Lillefjell, Meråker	5.89	4.76	tr	15	0.5	26.8
Løkken	2.3	1.8	0.02	16	0.2	42.0
Ankarvatnet[**]	0.5	5.5	0.4	17	0.2	
Tverrfjell	0.81	1.20	0.06	18	0.1	n.d.
Levimalmen[*]	1.16	1.55	<0.1	20	<0.1	16.1
Mofjellet	0.47	5.00	1.12	20	n.d.	n.d.
Bleikvassli	0.24	5.12	2.75	30	n.d.	22.77
N. Gjetryggen, Folldal	1.18	3.35	0.37	31	0.2	n.d.
Killingdal 1954	1.50	7.40	0.61	39	0.1	43.4
Rikarbäcken[**]	0.8	4.3	1.1	43	0.2	35
V. Storbäcksdalen[**]	1.2	6.3	2.4	49	0.5	16
Tjåter[**]	1.0	4.8	1.9	49	0.5	13
Stekenjokk[*]	1.46	3.03	0.3	53	0.25	20.1

[*] After Zachrisson, 1971.
[**] After Grip and Frietsch, 1973.

difficult to show any significant correlation between the Pb- and Ag-contents of these ores on the basis of present data and more research is required on this aspect of their geochemistry. In most cases the exact mineralogical natures of the "Ag-bearers" in these ores have not been determined. A certain amount is certainly present in the lattice of the PbS, though this cannot always account for the total quantities reported.

Silver contents of allegedly pure galena concentrates from Caledonian polymetallic massive sulphide ores in Norway are usually of the order of some few hundred parts per million (ppm or g/tonne) to a maximum of perhaps 1,000 g/tonne.

What proportion of these contents can be ascribed to silver in true solid solution in the galena and what proportion to extremely finely divided silver minerals has very seldom been investigated. However, it would seem from the reported results of laboratory attempts to determine the miscibility of Ag in PbS that lattice silver can only account for a small proportion of the reported contents of the galena concentrates. Boyle (1968, pp. 37–40) cites figures which show that the amount of Ag_2S soluble in galena varies from 0.03 mole% at 300°C to 0.52 mole% at 600°C.

Vokes (1963) has determined the presence of dyscrasite and a ruby-silver mineral in small quantities in the Bleikvassli ore, though others may be present. Dyscrasite has also been reported from the Mofjell deposit by Saager (1966, p. 54) and from some of the Røros ores by Jøsang (1964, pp. 187 and 193), while Lindahl (1968, p. 90) has tentatively identified this mineral in specimens from low-grade mineralization of the same general type on Kongsfjell, east of the Bleikvassli Mine.

Hessite is reported from Mofjell (Saager, 1966, p. 52) and from Kongsfjell (Lindahl 1968, pp. 85–86), while Saager (1966, p. 42) reports the presence of an Ag—Te sulphosalt with Ag : Te : S approximately 6 : 2 : 1 from Mofjell. Jøsang (1964, p. 196) has described an Ag—Te mineral, which he identifies tentatively as naumannite from the Røros ores.

Pyrargyrite has also been reported from the small prospect at Finnsaeter near Mofjell by Saager (1966, p. 43).

Minerals of the "fahlore group" (tennantite—tetrahedrite) are well-known "Ag-bearers" in many types of deposits and their presence has been recorded from several massive sulphide ores in the Scandinavian Caledonides. Little has been reported regarding the Ag-contents of these fahlore minerals.

Vogt (1894a, p. 43) reported fahlore carrying 0.5% Ag and occurring together with "Ag-rich galena" (?) from the Fløttum deposit, south of Trondheim. Ramdohr (1938) in his description of the antimony-rich mineral association at Jakobsbakken, Sulitjelma, reported the presence, from ore-microscopical observations, of an Ag-rich tetrahedrite. Fahlore from the Bleikvassli Mine (Vokes, 1963, pp. 62–63) has been analyzed chemically (Analyst: B. Bruun, Oslo) and shown to be a tennantite with 1.3% Ag. However, electron microprobe analyses of tennantite from Bleikvassli by C. Wheatley (personal communication, 1970) failed to detect any Ag-content (detection limit 0.5% Ag). The same worker reported Ag at or just below this detection limit in tennantite from the Skorovass deposit

in the Grong area. Thus, the role played by fahlore minerals as Ag-bearers in the Caledonian sulphide ores is not at all clear.

Recently Bergstøl and Vokes (1974) have discovered stromeyerite and mckinstryite in
the outcrop exposures of a small pyrite–sphalerite–chalcopyrite–(galena) deposit as the
Godejord's prospect in the southern part of the Grong distrct. The Cu–Ag sulphides in
this deposit appear to be replacing pre-existing sulphides in the ore, and may well be
formed by supergene cementation.

Undesirable minerals. Other minor minerals present may be regarded as undesirable in
many of the commercial concentrates, if not downright damaging to their values. Minerals, such as arsenopyrite (deleterious in S-concentrates) and the Sn–Cu sulphide stannite
(deleterious in zinc concentrates), may be included in this group.

These minerals appear to be more abundant and in greater variety in Pb-bearing ores
than in Pb-free ones, though insufficient detailed studies of the minor minerals of the
polymetallic ores have been carried out as yet. Of those which have been undertaken may
be mentioned those of Vokes (1963) on the Bleikvassli ore, Saager (1966) on the Mofjell
and adjacent ores, Jøsang (1964) on samples from the Røros ores and Juve (1975) on the
Stekenjokk ore.

SIZE AND GRADE OF ORE BODIES

With one or two notable exceptions, the ore bodies are not, by world standards, large
ones. The largest single ore body to be discovered so far is probably of the order of 20
million tonnes, but the majority are much smaller and many are of the order of 2 million
tonnes or less. Table IX lists some selected ore bodies, with tonnage and grade, without
attempting to be statistically representative.

Elemental composition

As will be apparent from the foregoing, and especially from Table IX, the main
base-metal elements of the Scandinavian Palaeozoic massive sulphide ores are copper and
zinc, with lead a poor third. The first five examples in Table IX are of bimetallic Cu–Zn
ores, where the Cu/Zn ratio is greater than unity. As has already been discussed, such ores
often show pyrrhotite as the main iron sulphide in preference to pyrite, though Løkken
and Båsmo are two notable exceptions which are dominantly pyritic in type.

Lead is absent, or present only as a trace component, in these ores. The other deposits
listed in Table IX are also, with five exceptions, lead-free, or practically so. However, in
these deposits the Cu/Zn ratios (wt.%) are less than unity, varying from 0.83 to 0.09,
with an average value of 0.39.

Five deposits in Table IX may be classified as polymetallic Zn–Cu–(Pb) or

TABLE IX

Tonnage and grades of some selected Scandinavian Palaeozoic massive sulphide deposits

Deposit*	Tonnage (10⁶ tonnes)	Grade (wt.%)				Source
		Cu	Zn	Pb	S	
Vaddas, Troms	0.5	1.60	0.01	–	25.0	I. Lindahl, unpublished thesis, Norges Tekniske Høgskole
Gjersvik, Grong	1.7	1.35	0.30	–	28.0	H. Bjørlykke, unpublished report, 1960
Olavs, Røros	2.5	1.80	1.44	tr	10.53	108 monthly mill head samples
Løkken	20.0	2.00	1.80	tr	42.00	"Typical ore". Company pamphlet
Båsmo, Rana	2.0	0.42	0.40	tr	49.21	3 bulk ore shipments
Tverrfjell, (Folldal)	10.0	1.00	1.20	tr	36.00	Company literature
Joma, Grong	17.0	1.27	1.70	–	35.00	Company reserves figures
Levi, Västerbotten	4.6	1.16	1.55	<0.1	16.1	Zachrisson, 1971
Bjørkåsen, Ofoten	4.7	0.46	0.69	–	27.0	130 monthly mill head samples
Kongens, Røros	2.5?	2.74	4.17	–	43.82	Foslie, 1925
Skorovass, Grong	7.3	0.73	1.65	–	35.00	300 drill core samples
Rødkleiv, Vigsnes	2.5	0.97	2.85	–	34.4	7 yearly production averages
N. Gjetryggen, Folldal	1.5	1.25	5.25	tr	35.64	Page, 1963 (200 mine samples)
Killingdal, (North)		1.03	10.61	–	42.14	Rui, 1973
Lergruvbakken, Røros	0.9	0.7	7.8	–	?	Company reserves figures
Stekenjokk, Västerbotten	15.1	1.46	3.03	0.30	20.1	Zachrisson, 1971
Killingdal, (Main)	2+	1.90	5.89	0.40	48.15	Rui, 1973a
Mofjell, Rana	3	0.36	4.66	0.98	9.67	67 drill core samples
Bleikvassli, Rana	3.5	0.42	6.81	3.67	25.0	Vokes, 1963
Lillebø, Stord	2.5	0.027	0.030	0.018	38.8	4 yearly bulk shipments

* The deposits, producers or past or prospective producers, are listed generally in order of decreasing Cu/Zn weight ratios and of increasing Pb/(Cu + Zn) weight ratios.

Zn–Pb–(Cu) types, with Zn the most abundant of the three base metals. In the case of the Killingdal and Stekenjokk deposits, the Pb-content is of little or no economic value. However, Bleikvassli and Mofjell are, as we have already seen, lead producers, in fact the only deposits of this class in Scandinavia which do yield lead concentrates at present. The polymetallic Zn–Pb–Cu ores are concentrated mainly in a rather restricted subdistrict of the central Caledonides in the Rana–Helgeland district of Norway and adjacent parts of Sweden (Tjåter, Joefjället, Rikarbäcken and others, see Grip and Frietsch, 1973). Otherwise, Pb-bearing deposits, some of potential economic importance, occur singly, at several places along the length of the Caledonides.

One deposit listed in Table IX warrants special mention, that is the so-called "Leksdals-type" or "vasskis" deposit (see p. 109) at Lillebø on the island of Stord in southwest Norway (see Fig. 1). The analysis in Table IX shows that it is, relatively speaking,

lacking in base metals. The ore was worked extensively for its pyrite (sulphur) content for many years, the only example of its type in Scandinavia to be so worked.

As can be seen from the analysis data for sulphur in Table IX, the "massiveness" (expressed as total sulphides) in the Caledonian ores varies considerably. Other factors being equal, the pyrrhotitic ores are, of course, lcwer in sulphur than the pyritic types, but within the latter class the sulphur content appears to bear no relation to any of the base-metal ratios.

The relative proportions of the three main base metals have formed the basis of various attempts to classify the ores into compositional types using triangular plots of the type developed originally by Stanton (1958 and 1962) and others. Vokes (1963), Saager (1967) and Waltham (1968) have dealt with this problem, based on studies in different parts of the Norwegian Caledonides and the last two have proposed classifications based on geochemistry and mineralogy. These attempts have on the whole met with little acceptance, mainly due to the often great variations in base-metal contents in closely related groups of deposits and, often, within one and the same deposit.

We have already noted that within the Caledonides, pyritic ores of the same general geological class are dominantly zinc—copper or less often copper—zinc types in which lead is absent or present in only minor to trace amounts. Most of the ores plot along the Cu—Zn join of the Cu—Zn—Pb compositional diagram. The deposits with notable Pb-contents, such as Bleikvassli and Mofjell and others mentioned above, of course, plot substantially away from the Cu—Zn join (Fig. 6).

It is also worth mentioning in this context that iron sulphide-free, Pb—Zn ores, having a negligible Cu-content, also occur in the Caledonides. They are mostly small and of no economic significance. The best-known examples are to be found in the Ofoten basin in Nordland and Troms (Juve, 1967).

Geological differences between the Pb-rich and Pb-poor pyritic ores are not readily apparent. There is some evidence that there may be a correlation between ore chemistry and the lithology of the country rocks in which the ores occur. However, many of the Pb-rich ores are situated in areas of high-metamorphic grade where it is very difficult to determine exactly the original character of the enclosing rocks. All that can be said, perhaps, is that there is usually an absence or paucity of amphibolites, which often are the metamorphosed representatives of basic volcanics, in the vicinity of the Pb-bearing ores. The lithology comprises mainly schists and other metamorphic rocks, some of which could have been acid volcanics originally. On the other hand, there seem to be large proportions of metasedimentary rocks characterizing the successions enclosing may of the Zn—Pb—Cu ore types, so that we are certainly dealing with mixed volcanic—sedimentaty environments, perhaps even with purely sedimentary ones in places.

Similar relations to hold for the Pb-free types of pyritic ores in that those deposits enclosed in massive greenstone lavas (pillowed and non-pillowed) tend to show a higher Cu/Zn ratio than those lying within a more mixed volcanic—sedimentary sequence. These differences are well shown in the Trondheim region of the central Caledonides,

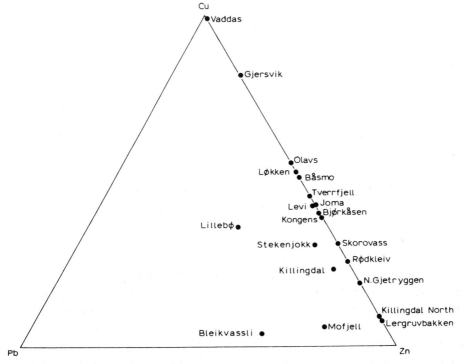

Fig. 6. Triangular plot (wt.%) of the relative proportions of Pb, Cu and Zn in the Scandinavian massive sulphide deposits listed in Table IX.

where deposits such as Løkken, occurring in the Lower Ordovician Støren greenstone lavas in the west of the region, are of the Cu–Zn type. Along the eastern volcanic belt (Fundsjø group, supposedly time-equivalent with the Støren) the rocks are more often mixed volcanic–sedimentary greenstones and here the ore types are definitely Zn-rich in comparison to those in the western belt. Pb may even appear in low amounts in this eastern zone.

It has been noted above that those sulphide deposits, or parts of deposits, where pyrrhotite is the predominant or sole iron sulphide, show a marked increase in Cu-contents, both absolutely and in relation to the other base metals, as compared with those where pyrite is the main or sole, iron sulphide. In areas where the deposits are Pb-poor or Pb-free, the pyrrhotite-ore types tend to be significantly enriched in copper, such that the average Cu/Zn ratio becomes substantially greater than unity. In extreme examples, zinc may be reduced to almost trace concentrations (e.g., Vaddas, Table IV). In the Pb-bearing ores, the pyrrhotitic types also show a significant increase in Cu-contents, as Table X shows for the Bleikvassli ore.

However, the pyrrhotitic-ore type at Bleikvassli is still a decidedly Zn–Pb-rich type in keeping with the general geochemistry of the ore as a whole. There is no question of the

TABLE X

Cu-, Zn- and Pb-contents for the Bleikvassli ore (after Vokes, 1963, p. 120)

Sample	Cu (%)	Zn (%)	Pb (%)
Mean of 90 underground samples	0.50	8.06	3.81
Mean of 12 pyrrhotitic samples	1.03	6.98	4.11
Mean of 78 pyritic samples	0.42	8.22	3.76

Cu-enriched pyrrhotitic portions being completely without zinc and lead as, for example, is the case in the Black Star, Mt. Isa, ore bodies investigated by Stanton (1962).

Comparisons with other ores

It seems relevant to compare the Scandinavian Caledonian ores geochemically with other undoubtedly (or at least widely accepted as such) volcanogene ores. In this respect, the Japanese Neogene Kuroko-type ores come to mind immediately. Matsukuma and Horikoshi (1970) characterize a fully developed Kuroko ore body by, e.g., the ore types, in stratigraphic succession, listed in Table XI. From this table it appears that the Keiko zone apparently represents the feeder-channel zone, already mentioned in the discussion of models of volcanogene ore bodies. The reasons for the absence of feeder-channel zones in the majority of Scandinavian massive sulphide ores have already been outlined on pp. 82–84.

The Oko zone has a composition which in fact corresponds well with that of some of the Scandinavian Palaeozoic pyritic ores of the Cu—Zn type. Specimens of Oko ore

TABLE XI

A fully developed Kuroko ore body (after Matsukuma and Horikoshi, 1970)

Stratigraphic succession	Composition
Top	
Barite ore zone	usually monomineralic barite ore
Kuroko zone	barite-Zn-Pb-Cu-Ag-bearing polymetallic sulphide ore
Oko zone	cupriferous pyritic ore
Bottom	
Keiko zone	copper-bearing siliceous ore, some disseminated, others stockwork type

collected by the present writer in 1970 from the Shakanai mine near Odate, Akita Prefecture, are, for example, macroscopically identical with ore from the relatively little metamorphosed Løkken deposit. At higher metamorphic grades the comparison is less easy. The Kuroko zones of the Japanese ores seem to find their geochemical and mineralogical equivalents in the Pb-bearing ores already mentioned, e.g., Bleikvassli, Mofjell and others of lesser economic importance. In this case, comparison is made less easy due to the often highly recrystallized nature of the Scandinavian Pb-bearing ores, especially in northern Norway. Barite is present up to a few % in the Mofjell ore, but at Bleikvassli this mineral appears to be lacking. There appear to be no Scandinavian equivalents of the pure barite zones of the Japanese deposits.

The evidence so far is thus that some of the ore types represented as zones in the Japanese Kuroko ore bodies occur as individual ore bodies in the Scandinavian Caledonides. There appears to be no clear-cut example of a composite "Kuroko-type" ore body in Scandinavia, though again the subsequent deformation and recrystallization undergone by the Caledonian ores makes it difficult, if not impossible, to reconstruct the original zoning, if any. Isolated examples can, however, be cited where it appears that minor, incomplete examples of Kuroko-zone development may be present at the extreme upper (?) portions of Caledonian massive sulphide ores, where the ore type is predominantly "Oko".

In parts of the Skorovass ore body, the extreme hanging wall contact of the so-called ore "lenses" consisted of a zone or layer, of the order of thickness of 5—10 cm, within which sphalerite and tetrahedrite were concentrated, in places making up almost 100% of the zone.

The portions consisting almost wholly of tetrahedrite were along the actual ore contact. This example is hard to interpret as it is by no means certain that the hanging wall in question is the original hanging wall of the ore body; it may well be structural, since it seems that the lens-like internal structure of the Skorovass ore may well be the result of isoclinal folding along flat-lying or gently dipping axial planes (A. Reinsbakken, personal communication, 1973).

Other non-metamorphic, volcanogene ores of interest in the present context are those occurring in the Mesozoic Troodos volcanics of Cyprus. These ores, with one or two exceptions, are Cu-bearing pyritic ores, resembling texturally and mineralogically the "Oko" type ore of the Japanese Kuroko deposits. Only in a few small ore bodies do there appear appreciable quantities of a second base metal, namely zinc. Galena is for all practical purposes lacking in the Cyprus ores. In this respect they show marked similarities with Scandinavian Cu-rich ores such as Løkken.

The Scandinavian Caledonian ores thus may be said to include deposits which are geochemically and mineralogically similar to both the Japanese Kuroko and the Cyprus types. However, the majority of the Scandinavian ores seem to be of an intermediate Zn—Cu type which is, seemingly, not represented in either of the above mentioned examples. It may be that these geochemical and mineralogical differences reflect the

TABLE XII

Summary of chemistry and lithological associations of Japanese Kuroko, Cyprus and Norwegian Caledonian massive sulphide ores

Ore deposits	Age	Chemistry	Lithology
Kuroko ores, Japan	Tertiary	Zn−Pb−Cu−Ag−Ba	felsic volcanics (lavas, pyroclastics; dacites−rhyolites); siliceous sediments
Troodos massive sulphides, Cyprus	Mesozoic	Cu(±Zn)	pillowed basaltic lavas, ochres, umbers
Caledonian massive sulphides, Norway	Palaeozoic	Cu−Zn	basic lavas (often pillowed), greenstones, greenschists, cherts, iron formation
		Zn−Cu(±Pb)	greenstones, greenschists, metasedimentary schists, quartzites, quartz−keratophyres
		Zn−Pb−(Cu)	metasedimentary (?metavolcanic) schists, gneisses; minor amphibolites

varying geological environments in which the ores were deposited in the three areas. Table XII presents a summary of the chemical and lithological features of the ores just discussed.

SURVEY OF HISTORY OF GENETICAL CONCEPTS

A summary of the history of ideas concerning genesis of the Scandinavian Palaeozoic massive sulphide deposits is of relevance in the present context. It reveals a cycle of changing views which seems to have had a rotation period of something over 100 years, so that many of the present ideas concerning this problem are very similar to those advanced in the very earliest accounts (Fig. 7 and Tables XIII and XIV). What were probably the first geological observations concerning these ores, appear to have been made by Vargas Bedemar (1819) who described the various ore bodies of the Røros mining-area in south-central Norway as being disrupted parts of one and the same *layer* or *bed* of sulphides. However, no view was expressed concerning the origin of this layer or bed.

The first definite genetical view regarding the Norwegian sulphide ores of this type seems to have been that of Strøm (1825) who expressed the opinion that the ore bodies, as well as their enclosing wall rocks, were of a sedimentary origin. The majority of Norwegian ore geologists and mining engineers of the period appear to have been in agreement with Strøm's views.

The syngenetic view of genesis was also strongly championed in the second half of the century by Helland (1873) in an important work which dealt with all the then known

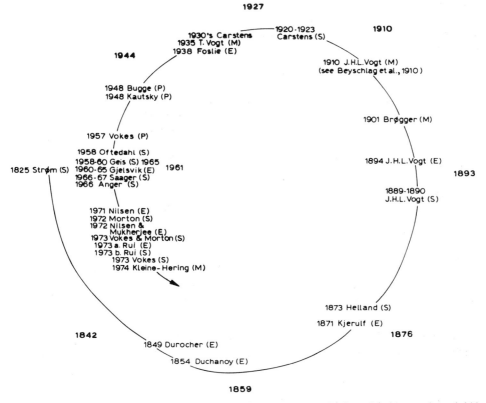

Fig. 7. To illustrate the cycle of thought regarding the genesis of the Scandinavian massive sulphide deposits. S = syngenetic hypothesis; E = epigenetic; M = sulphide melt and P = palingenic origin.

deposits between Hardanger in the southwest of Norway and the northern part of the centrally situated Trondheim district. Helland observed that the sulphide ores occurred very often as massive layers intercalated in the surrounding metavolcanic and metasedimentary schists. There could nearly always be observed a strict parallelism between the ore walls and the layering or schistosity in the country rocks. Helland concluded that the sulphides represented chemically precipitated sediments and he assumed that the heavy metal sulphides were supplied by volcanic exhalations on the floor of the Palaeozoic geosynclinal sea.

This syngenetic–exhalative view of sulphide genesis – almost identical to more recently held views in certain quarters – was initially supported by the "Grand Old Man" of Norwegian ore geology, J.H.L. Vogt, in publications in 1889 and 1890, though in works published in 1894, Vogt (1894a and b) rejected the sedimentary origin of the ores in favour of an epigenetic one.

Earlier, other workers, such as Durocher (1849) and Duchanoy (1854) had expressed the view that the sulphide ores were related genetically to late-orogenic eruptive (intru-

TABLE XIII

Summary of the literature concerning the modes of origin of the Scandinavian massive sulphide deposits

Author	Year	Theory of genesis					subsequent metamorphism (ores recrystallized, deformed, partly mobilized)
		syngenetic	epigenetic				
		sedimentary (volcanic–exhalative)	hydrothermal (general and or metasomatic)	injected sulphide melt	metamorphic–palingenic		
Strøm	1825	considered ores and wall rocks sedimentary					
Durocher	1849		ores related to late-orogenic intrusives				
Duchanoy	1854		ores related to late-orogenic intrusives				
Kjerulf	1871 (1876)		ores related mainly to gabbros, to a lesser extent to granites				
Helland	1873	chemically precipitated heavy-metal sulphides supplied by sea-floor volcanism					
J. H. L. Vogt	1889, 1890	advocated volcanic–exhalative processes					
J. H. L. Vogt	1894a,b		rejected syngenesis in favour of epigenesis				
Brøgger	1901			result of study of Sulitjelma ores			

J. H. L. Vogt	1910			view expressed in text book of Beyschlag, Krusch and Vogt	
Carstens	1920, 1923	considered applicable to deposits of the "vasskis" type			
Foslie	1925		most ores originate from gabbroic differentiates; some from granites		
Berg	1928				described pyrite porphyroblasts at Sulitjelma
Ramdohr	1928				Sulitjelma ores subjected to katazonal metamorphism
Carstens	1931				Porsanger Pre-C ores considered recrystallized "vasskis"
Kurek	1931	"vasskis" type deposits on Stord			
Carstens	1932, 1936		hydrothermal–metasomatic origin related to gabbroic and granitic magma		
T. Vogt	1935			theory of injected pyrite deposits	
Foslie	1938, 1939		advocated views identical to those of Carsten's		considered ores and wall rocks had been regionally metamorphosed
Carstens	1941, 1944		hydrothermal–epigenetic synorogenic ores		regionally metamorphosed by late-orogenic phase

TABLE XIII (continued)

Author	Year	Theory of genesis				
		syngenetic	epigenetic			subsequent metamorphism (ores recrystallized, deformed, partly mobilized)
		sedimentary (volcanic–exhalative)	hydrothermal (general and or metasomatic)	injected sulphide melt	metamorphic–palingenic	
Kautsky	1948				volcanic pre-concentration; mise en place due to palingenic processes	
Bugge	1948 (1954)				S derived from sulphidic sediments by palingenesis; heavy metals from magmatic emanations	
Kautsky	1952				Emphasized contributions from graphitic sediments	
Krause	1956					discussed metamorphism of Sulitjelma ores
Vokes	1957				palingenic mobilization of components into breccia zones at Birtavarre; Mg–Fe metasomatism	
Oftedahl	1958	reintroduced general theory of volcanic–				

		genesis	
Gjelsvik	1960	considered most likely origin of Skorovass ore	discussed certain metamorphic effects on sulphide minerals
Vokes	1962		pyritic-type ores completely recrystallized during metamorphism
Vokes	1963		study of metamorphic sulphide fabrics at Bleikvassli
Page	1963		study of metamorphic fabrics at Folldal
Geis	1965	Vigsnes ore as volcanic–exhalative deposit	
Anger	1966	comparison of German and Norwegian volcanogene ores	
Saager	1966, 1967	ores of Mofjell area considered syngenetic	discussion of metamorphic effects at Mofjell
Anger and Eisbein	1968	Joma as volcanic–exhalative ore	
Gjelsvik	1968	origin of Skorovass ore; emphasized wall-rock alteration	
Waltham	1968	deposition in marine environment; close association with volcanics	pyrite ores regionally metamorphosed
Vokes	1968	ores regarded as volcanogene	reviewed effects of regional metamorphism
Nilsen	1971	Rødhammer late orogenic; Mg–Fe metasomatism emphasized	
Zachrisson	1971		considered relation of Stekenjokk ore to regional fold-phase; ore folded

TABLE XIII (continued)

Author	Year	Theory of genesis				
		syngenetic	epigenetic		metamorphic—palingenic	subsequent metamorphism (ores recrystallized, deformed, partly mobilized)
		sedimentary (volcanic—exhalative)	hydrothermal (general and or metasomatic)	injected sulphide melt		
Morton	1972	refuted Nilsen's views on Rødhammer; general volcanogene view				ores considered regionally metamorphosed
Nilsen and Mukherjee	1972		Kvikne deposits considered epigenetic			
Vokes and Morton	1973	discussion of origin of the Kvikne deposits				Kvikne and other ores metamorphosed volcanogene
Rui	1973a				palingenic origin for Killingdal ores	ore folded and recrystallized in orogeny
Rui	1973b	Røstvangen ore of submarine volcanic origin				ore subsequently metamorphosed and deformed
Vokes	1973					discussed origin of "ball-ores" by deformation of pre-existing deposit
Kleine-Hering	1973			Moskodal, Troms ores due to injection of sulphide melt		ore subsequently deformed

TABLE XIV

Chronological review of the main publications on the genesis of Scandinavian massive sulphide deposits

Year	Primary origin				Ores later metamorphosed and deformed
	syngenetic	epigenetic			
	sedimentary (volcanic–exhalative)	hydrothermal (open-space filling and or replacement)	injected sulphide melt	metamorphic–palingenic	
1830	Strøm (1825)				
1840					
		Durocher (1849)			
1850					
		Duchanoy (1854)			
1860					
1870					
		Kjerulf (1870's)			
	Helland (1873)				
1880					
	J. H. L. Vogt (1889, 1890)				
1890					
		J. H. L. Vogt (1894a,b)			
1900					
			Brøgger (1901)		
1910			J. H. L. Vogt (see Beyschlag et al., 1910)		
1920	Carstens (1920, 1923)				
		Foslie (1925)			
					Berg (1928) Ramdohr (1928)
1930		Carstens (1930's)			
	Kurek (1931)				Carstens (1931)[**]
			T. Vogt (1935)		
1940		{ Foslie (1938, 1939)			Foslie (1938)[*]
					Carstens (1941, 1944)[*]
			Kautsky (1948, 1952) Bugge (1948, 1954)		

TABLE XIV *(continued)*

Year	Primary origin				Ores later metamor-phosed and deformed
	syngenetic	epigenetic			
	sedimentary (volcanic–exhalative)	hydrothermal (open-space filling and or replacement)	injected sulphide melt	metamor-phic-pa-lingenic	
1950					Krause (1956)
				Vokes (1957)	
	Oftedahl (1958) Geis (1958, 1960, 1965)				
1960		Gjelsvik (1960)			Gjelsvik (1960)[*] Vokes (1962)[***] Vokes (1963)[***] Page (1963)[***]
	Saager (1966, 1967) Anger (1966) Anger and Eisbein (1968)	Gjelsvik (1968)			Walthan (1968)[**] Vokes (1968)[**]
1970		Nilsen (1971) Nilsen and Mukherjee (1972)			Zachrisson (1971)[***] Sen and Mukherjee (1972)
	Morton (1972)				
	Vokes and Morton (1973) Rui (1973b) Vokes (1973)			Rui (1973a)	Wilson (1973) Rui (1973a)[*] Rui (1973b)[**] Vokes (1973)[**]
		Kleine-Hering (1973)			

[*] Primary origin considered epigenetic; [**] primary origin considered syngenetic; [***] question of primary origin left open.

sive) rocks. These views were supported and elaborated on by Kjerulf (1871 and 1876), who opined that the majority of the deposits were genetically connected to gabbroic intrusions, while some examples seemed more to be connected to granites. Kjerulf may thus be said to have been one of the earliest exponents of an epigenetic–hydrothermal view of the origin of the Caledonian massive sulphide deposits in Scandinavia.

As already mentioned, this view received the powerful support of J.H.L. Vogt in 1894 and subsequently little was heard of the syngenetic–sedimentary view of origin until its revival in the 1950's (see below).

The eruptive view of origin was again supported and modified by Brøgger (1901) in a publication dealing with the ore body of the Charlotta mine at Sulitjelma in northern Norway. Brøgger regarded the rounded megascopic pyrite crystals to be found in certain parts of the deposit, enveloped in a chalcopyrite groundmass, as being due to resorption in a sulphide melt which had been injected along the schistosity of the country rocks. Vogt also came to accept the idea of an injected sulphide melt of magmatic origin and in the text book of Beyschlag et al., (1910) not only the Scandinavian, but also the Spanish and other similar deposits are referred to this class.

Th. Vogt, the son of J.H.L. Vogt, carried the idea of a sulphide melt further and as late as 1935 published a paper concerning the "injected pyrite deposits". However, as early as the 1920's, other workers were beginning to advance different ideas concerning sulphide genesis in the Norwegian Caledonides. C.W. Carstens (1920 and 1923) studied a minor class of sulphide deposits which occur as relatively thin, extensive sheets or layers, often alternating with layers of magnetite—quartz rock and chlorite metasediments, the whole resembling closely a type of iron formation showing both sulphidic and oxidic facies. This type of deposit, called "Leksdals-type" after the type locality, or "vasskis" (Weisskies) after the rather pale appearance of the almost solely iron sulphide, was considered by C.W. Carstens as of undoubted sedimentary origin, being formed by biochemical precipitation of iron supplied to the seawater by volcanic exhalations.

This view of the Leksdals-type deposits, generally of little economic significance, was also supported by Kurek (1931) in a study of the largest deposit of this class, situated on the island of Stord in southwest Norway. The views of C.W. Carstens on the "vasskis" have received general approval from Scandinavian ore geologists over the years and have never been seriously disputed.

The vast majority of the Norwegian massive sulphide deposits, however, comprising those of the greatest economic value, due to their contents of base-metal sulphides, have continued to be a subject of scientific controversy up to the present day. Following the general acceptance of the "injected sulphide melt" theory of Brøgger and J.H.L. Vogt, Foslie (1925) referred the vast majority of the massive sulphide deposits to "gabbroidal magmatic differention products". He did, however, distinguish a minor group of ores, apparently spatially related to granitic rocks, which he considered as being of hydromagmatic or hydrothermal, epigenetic character.

The most eloquent proponent of the epigenetic—hydrothermal replacement origin of the Caledonian sulphide ores was undoubtedly C.W. Carstens, who in a series of publications (1932, 1937 and 1944) put forward his evidence for the view that the ores were formed by "hydrothermal—metasomatic processes" due to solutions originating either in a granitic or a gabbroid magma. C.W. Carstens pointed to certain relict textures in the ores (for instance, alleged replaced pillows from the original pillow lavas) as well as a form of "wall-rock alteration", in support of his views. C.W. Carstens' views received the support of Foslie (1938 and 1939), in the years immediately preceding World War II, and it may be said that the hydrothermal-replacement theory, with variations, was accepted in Scandinavia with no dissent until at least the end of the 1950's.

One variant of the epigenetic–hydrothermal view of deposition was first put forward in the immediate post-war years by Bugge (1948 and 1954). Bugge considered that the base-metal bearing massive sulphide deposits had crystallized from solutions generated by metamorphic palingenic processes acting on sedimentary sulphides of the Leksdals-type. The base-metal contents of the ores were, in Bugge's view, added to the products of palingenesis by solutions "ascending from magmatic sources".

At about the same period Swedish geologists were entertaining similar views on the origin of the massive Caledonian sulphide deposits. Foremost among these was Kautsky (1948), who in a discussion of a lecture delivered by Magnusson (1948), put forward the following conclusions regarding the ores in Västerbotten and Norrbotten counties. The *mineralization* is for the greater part referable to the intrusions of greenstones and to the basic volcanicity occurring in the geosynclinal belt. The *concentration* of the sulphides to form the present day ore-deposits occurred later, in connection with metamorphic palingenic processes. Thus, the formation of sulphides was considered by Kautsky to be a function of both the presence of the geosynclinal intrusives and extrusives and of the operation of the palingenic processes. By means of these processes, the sulphides originating from the basic rocks were concentrated in certain tectonic zones in their near vicinity, in a zone parallel to the eastern border of the "zone of plastic folding".

In a publication in 1952 Kautsky refers briefly to palingenic processes in connection with the formation of the deposits of the Sulitjelma mining-district. In this work he seems to assign a more structural role to the greenstones and suggests that the mineralization could have, in part, originated from the ubiquitous black (carbonaceous) schists of the area. The present writer (Vokes, 1957) also adhered to a late-orogenic, palingenic theory of genesis to explain the small but Cu-rich ore bodies of the Birtavarre mining-district in northern Norway, while as late as 1973, Rui (1973a) invoked the process to explain the formation of the Killingdal ore bodies in the northern part of the Røros mining-district.

About the middle of the 1950's, however, the cycle of genetical thought appears to have turned a full circle (Fig. 7) and advocates of genetic theories resembling those of Helland and other earlier workers appeared. In 1955, H. Carstens, the son of C.W. Carstens, presented an important paper – unfortunately, perhaps, in Norwegian – in which he proposed that the magnetite–pyrite–chalcopyrite ores of the Fosen peninsula to the northwest of Trondheim, are best explained by an original volcanic-exhalative, syngenetic deposition, followed by recrystallization and deformation in the subsequent Caledonian orogeny. H. Carstens also showed the close genetic relationship between these sulphide-bearing magnetite ores and the sedimentary pyrite ores of the Leksdals-type which, as already mentioned, often show interlayers of magnetite–quartz composition. The two types, the Fosen iron ores and the Leksdal pyrite deposits, were considered to represent, respectively, an oxidic and a sulphidic facies of iron deposition under differing sedimentary environments in the same geosyncline.

H. Carstens implicitly excluded the massive base-metal bearing sulphide deposits of the central Caledonian geosynclinal zone from his genetical model. Three years later, however, Oftedahl (1958) in a paper which aroused interest far beyond the countries of Scandinavia, proposed a volcanic exhalative–sedimentary mode of formation for all the massive sulphide ores of the Caledonian belt. As Oftedahl himself made clear (1961) the theory was not an original one. Apart from the early Norwegian workers already mentioned, many German geologists had long regarded exhalative–sedimentary syngenesis as the true origin of ores of the same type ("Kieslagerstätten") in central Europe. (A German geologist – Geis, 1958 and 1960 – working in Norway, advocated an exhalative–sedimentary hypothesis for the Caledonian sulphides at about the same time as Oftedahl.) It was not without considerable opposition that Oftedahl reintroduced the syngenetic theory of massive sulphide genesis, but eventually his paper gave rise to a valuable rethinking of ideas, not only within Scandinavia, but also among English-language geologists in many parts of the world. Also other workers, such as for instance Stanton (1959, 1960), were almost simultaneously arriving at similar conclusions regarding massive sulphides in Australia, Canada and elsewhere.

The world-wide appreciation of the role played by volcanic and sedimentary processes in the formation of massive sulphide ores of the type here under consideration may be said to date from this period of the late 1950's. The implications of the syngenetic, exhalative–sedimentary mode of formation are still being worked out (see also Chapter 5, by Sangster and Scott, this volume), but the rethinking involved has already stimulated considerable advances in the geological knowledge concerning this important, and controversial type of sulphide deposit.

That the controversy over the origin of Caledonian massive sulphide deposits in Scandinavia was by no means settled, can be seen from the variety of views expressed in publications appearing since Oftedahl's 1958 paper. Advocates of practically all the genetical processes recorded in the above summary have appeared in the last one and a half decades. The late-postorogenic, epigenetic–hydrothermal view has been advocated by Gjelsvik (1960 and 1968) for the Skorovass deposit in the Grong area; by Nilsen (1971) for the Rødhammer deposit in the central Trondheim district; and by Nilsen and Mukherjee (1972) for the Kvikne deposits in the same general area. As already mentioned, Rui (1973a) has maintained the paligenic–epigenetic view with regard to the Killingdal deposits, a view which has been criticized by Vokes (1974).

The volcanic–exhalative sedimentary view of deposition has been preferred in papers by Geis (1965) for the Vigsnes deposits in the extreme southwestern end of the Caledonian metallogenic belt; by Saager (1966 and 1967) for the Mofjell and similar deposits in the Mo i Rana area of central Nordland; and by Anger (1966) and Anger and Eisbein (1968). Morton (1972) and Vokes and Morton (1973) have questioned the views of Nilsen (1971) and Nilsen and Mukherjee (1972) on the origin of the Rødhammer and Kvikne deposits. Rui (1973b) has also recently come out in favour of a submarine volcanic origin (followed by metamorphic deformation and recrystallization) for the Røstvangen deposit in the Kvikne mining district (see Fig. 1). The "injected sulphide melt" hypotheses has been recently advocated for the pyrrhotitic ores of the Moskodal area, Troms, by Kleine–Hering (1973).

The present writer, as the result of studies of the metamorphic effects produced in the ores during the Caledonian orogeny, has come to regard these ores in general as being pre-orogenic in origin and most probably related to the Early Palaeozoic volcanic activity of the geosynclinal stage of the development of the Caledonides. (See also Vokes and Gale, 1975.)

METAMORPHISM OF THE ORES

Whatever view one may hold regarding the primary origin of strata-bound massive sulphide ores of the type represented by the Scandinavian Caledonian deposits, it is clear that their environment of deposition is very typically "geosynclinal" and that they occur interlayered, and very often conformable, with the metavolcanic and mixed metavolcanic—metasedimentary rocks deposited during the pre-orogenic stage. It is becoming increasingly recognized through mineralogical, textural and structural studies, that such ores have been, as a rule, affected, to greater or lesser degree, by the regional dynamothermal metamorphism of the orogenic stage of mobile-belt development. It now seems clear that these "sulphidites" have been affected by this metamorphism to the same extent as have the other types of rocks in the enclosing or surrounding geosynclinal pile.

The metamorphic effects on the sulphidites generally parallel those that can be observed in the enclosing rocks; in addition there may be found unusual features which result from the sulphidites' special physical and chemical properties. Many of these deposits have been subjected to orogenic deformation and recrystallization to such an extent that their original natures have been more or less obscured. Indeed, one may say that much of the controversy which has gone on over the primary origin of the Scandinavian Caledonian sulphide ores has been the result of insufficient recognition in the past of the part played by the subsequent regional dynamothermal metamorphism. The increased understanding attained in recent years of the features which metamorphism can produce (see Vol. 4, Chapter 5, by Mookherjee) in ores of all-types has greatly contributed to a truer understanding of their original mode of formation. Many apparently epigenetic or magmatic features in ores can be produced by metamorphic recrystallization, deformation and mobilization and these features may completely dominate the primary, possibly non-magmatic, features of the deposits.

The Scandinavian Caledonides are especially well-suited for the study of the effects of metamorphism on massive sulphide deposits. The ore bodies occur along practically the whole length (some 1,500 km) of the orogenic belt, while the metamorphic grade varies considerably from district to district, from lower greenschist to almandine—amphibolite facies. It is thus possible to study deposits of apparently the same origin, chemistry, mineralogy and original form in areas showing very different metamorphic conditions. By this means it has been possible to build up a fairly complete picture of the effects of increasing regional metamorphism on the deposits.

While attention will be here concentrated on the Scandinavian Palaeozoic sulphide ores, it must be mentioned that similar considerations apply to many other regionally metamorphosed areas in the world (see Vol. 6, Chapter 5, by Sangster and Scott). The reader is referred to a series of more general treatments of the problem, such as Ramdohr (1953, 1960), Domarev (1956), Schneiderhöhn (1962, pp. 286–330), McDonald (1967), Vokes (1969), Shadlun (1970) and others for more general discussions, as well as Vol. 4, Chapter 5 by Mookherjee and, Chapter 6, by Both and Rutland.

Although the metamorphic aspects of strata-bound ores of the type represented by the Caledonian deposits have perhaps received special attention only in recent years, the general concept is by no means new, having been accepted by geologists in many parts of the world for several decades now. Meanwhile it seems that such ideas have undergone changes in awareness from time to time. After an initial period of lively investigation and description earlier this century, the question of the metamorphism of massive sulphide ores tended to be disregarded to a considerable degree in many places and only in the last two decades has a greatly revived interest in the subject occurred.

The deposits of the Scandinavian Caledonides have been recognized as having been affected by metamorphism since the 1920's. Berg (1928) as a result of his studies of the large, rounded pyrite crystals in parts of the Sulitjelma ore bodies, expressed the opinion that these were porphyroblasts developed during the Caledonian metamorphism. The Sulitjelma ores were also discussed by Ramdohr (1928) in a study of the mineralogical and textural changes brought about in ores of the *Kieslagerstätten*-type during regional metamorphism. Ramdohr considered these ores to have been subjected to fairly high-grade (Katazonal) metamorphism, as compared, for example, with the Rammelsberg ores of Germany. Norwegian geologists took up the problem of sulphide ore metamorphism in the 1930's. The earliest published reference to this, however, appears to be a paper dealing, not with the Caledonian ores, but with the Precambrian deposits of the Porsanger area of Finnmark in northern Norway. C.W. Carstens (1931) regarded the Cu-free strata-bound pyritic deposits of this area as the metamorphic products of a Precambrian equivalent of the Palaeozoic Leksdals-type of deposit.

The first printed reference to the metamorphism of the Palaeozoic sulphide deposits themselves is the report of a discussion at a meeting of the Norwegian Geological Society in 1937, when Foslie (1938) forwarded the view that the sulphides, in common with their enclosing rocks, had been subjected to regional metamorphism. Foslie pointed to the fact that those deposits which occur in rocks showing low-grade metamorphism (greenschist facies) are as a rule characterized by a very fine-grained texture and very intimate intergrowth of their mineral constituents. In regions of higher metamorphic grade, where the rocks have often been metamorphosed under conditions of the amphibolite or higher facies, the massive sulphides everywhere show a coarse grain size.

The theme was later taken up and enlarged upon by C.W. Carstens in papers published in 1941 and 1944. Among other things, C.W. Carstens dealt with the rounded pyrite cubes of Sulitjelma, as well as similar ones from the Foldal deposits of central Norway. Carstens interpreted these crystals as being "typically developed porphyroblasts with inclusions of several different groundmass minerals". In particular, the Folldal crystals appeared to have been rotated, since the inclusions showed a circular or bow-shaped arrangement, in the same manner as do inclusions in rotated garnets. Following C.W. Carstens' work, the metamorphic nature of the Norwegian Caledonian sulphide ores has been more or less tacitly accepted by Norwegian geologists and, for example, incorporated into standard lecture courses in the Universities.

Of the newer publications referring to the effects of metamorphism on the Norwegian Palaeozoic sulphide deposits may be mentioned that of Krause (1956) on the deposits of the Sulitjelma district, and Gjelsvik (1960) who discussed in a general way the effects of the Caledonian metamorphism on the Skorovass ore in the Grong area. Gjelsvik's conclusion was that certain replacement phenomena between the euhedral—subhedral pyrite grains and the base-metal sulphides in the matrix could be ascribed to this metamorphism.

Vokes (1962) in a general review of the sulphide parageneses of the Caledonian ores, refocussed attention on the probability that the pyrite-rich ore types had completely recrystallized during the regional metamorphism. In a detailed study of the Bleikvassli ore body, the same writer (Vokes, 1963) described and discussed the various fabrics resulting from the metamorphism. The question of the timing of the ore-forming event at Bleik-vassli in relation to the orogenic events in the area formed the subject of a later, shorter publication (Vokes, 1966). Page (1963), in an account of the Nordre Gjetryggen deposit at Folldal, concluded that whatever the origin of the ore, its structural and textural features were the results of the process of regional metamorphism and that the various sulphide geothermometers gave temperatures of recrystallization consistent with the highest possible temperatures of metamorphism. The non-pyritic, stratabound zinc—lead deposits of the Håfjell syncline in the Ofoten area have been shown by Juve (1967) to have been deformed and partly remobilized during the Caledonian orogeny.

Waltham (1968) described certain metamorphic features of the ores of the Folldal area, including apparent redistribution of metal values within the bodies due to folding. A general review of the metamorphic effects of the Caledonian orogeny on the Norwegian Palaeozoic stratabound sulphides as a whole, has been given by Vokes (1968). It was here concluded that the effects in the ore bodies appear to correspond closely to those in their enclosing rocks and that systematic changes can be traced from areas of one regional metamorphic facies to another. Three main effects were recognized. *Recrystallization* of the ores during progressive regional metamorphism resulted in changes mainly in the fabrics, but also in the mineralogy, of the ores. *Deformational* effects present varied from zero, through a brittle cataclasis of the sulphide mass at low metamorphic grades, to a thorough plastic deformation ("Durchbewegung") of the whole mass at higher meta-morphic grades. *Mobilization* at high metamorphic grades has produced irregular, pegma-tite-like bodies of vein quartz and ore minerals, either within the massive ores or in their immediate country rocks. A summary of the effects of metamorphism is given in Table XV.

Some years ago Th. Vogt (1944, 1952) pointed to the parallelism between the longitu-dinal axes of the Caledonian massive sulphide bodies and certain linear structures in the

TABLE XV

Summary of effects of metamorphism on massive sulphide deposits, as exemplified by the Norwegian Caledonian deposits

Changes in fabric

Due to recrystallization
1. General increase in grain size with metamorphic grade.
2. Changes in mineral morphology.
    Growth of minerals as porphyroblasts. Idioblastesis.
    Sulphides : pyrite, arsenopyrite.
    Oxide   : magnetite.
    Silicates   : garnet, staurolite, amphiboles.
    Carbonate : dolomite.
    Development of polygonal mosaic in matrix minerals.
    Triple junction point texture.
    Sulphides : pyrrhotite, sphalerite, chalcopyrite, galena.
    Others   : quartz, calcite.

Due to deformation
1. Brittle deformation (cataclasis) of ore prior to or subsequent to recrystallization.
    Prior: Brecciation and mylonitization of fine-grained sulphide mass.
    Subsequent: fracturing and brecciation of brittle porphyroblasts, infilling with ductile matrix sulphides, oriented deformational textures may be preserved in ductile matrix minerals, stress twinning or cataclasis of softer sulphides, e.g., "steel galena".
2. Plastic (ductile) deformation, usually accompanying or succeeding recrystallization, at elevated temperatures.
    Folding and later disruption of banded textures, formation of detached cores of silicate layers, through-going kneading and rotation of sulphide mass (Durchbewegung).

Changes in structure of ore bodies

Folding, thickening of sulphides in fold-hinges, disruption, boudinaging, "stretching".
Resulting in elongated bodies, often with marked axial elongations, which may coincide with a regional fold- or lineation direction. Shapes described as plates, lenses, pencils, cigars, rulers.

Changes in mineralogy

Increase in pyrrhotite content due to progressive breakdown of pyrite at higher metamorphic grades. Fe content of sphalerite increases on appearance of pyrrhotite.
Sulphide–silicate (–oxide,–carbonate) reactions.
Sulphurization of Fe-bearing minerals.

Mobilization

Selective mobilization of ore-forming components by creep, fluid-phase transport, or as melts.
Limited mobilization distances, either within parent body or in restricted halo surrounding it (up to some tens of metres).
General order of increasing mobility apparent:
Most mobile : galena, sulphosalts, chalcopyrite.
Mobile   : pyrrhotite, sphalerite.
Least mobile : pyrite, magnetite.

surrounding metamorphic rocks. This feature has been later commented on by, i.a., Vokes (1957), Page (1963) and Rui (1973a, b). Increasing attention is now being paid to the detailed structural geology of the sulphide bodies and their surrounding rocks. Results so far published show that in many cases the sulphides have taken part in the earliest fold phase decipherable in the area. Zachrisson (1971), as the result of a study of the structural setting of the Stekenjokk ore bodies in the central Swedish Caledonides concluded that the sulphides are to be dated before what he terms the F_1 folding in the area. However, since there is the possibility of an important earlier deformation phase having occurred, Zachrisson was unable to do more than to point to two possibilities: (1) that the Stekenjokk ore is epigenetic and related to the early, pre-F_1, fold phase; and (2) that the ore is syngenetic.

Sen and Mukherjee (1972) in a study of the structural evolution and metamorphism of the Bleikvassli ore body have reached similar conclusions regarding the relation between the ore layers and the two folds phases recognized in the area. Comparison of the petrofabrics of the ore lenses with biotite cleavage poles suggests, according to Sen and Mukherjee (1972) that the sulphides are premetamorphic in nature.

Wilson (1973) has studied the structural setting of the Sulitjelma ore bodies in northern Norway and concluded that they show evidence of being involved in "most if not all of the structural phases in this part of the Caledonides", although he was not able to show that the ores are involved in the early folds seen in the surrounding schists. Rui (1973a), on the other hand, was able to state that the Killingdal ore bodies in the Trondheim district "have been subjected to isoclinal folding which in style is consistent with early F_1 folds", and that "sulfide ores formed prior to the folding may thus have been largely remoulded during periods of intense folding and regional metamorphism". In spite of this, Rui prefers an origin for the Killingdal ores based on the generation of solutions within the eugeosynclinal, supracrustal rocks "during a late stage of regional Caledonian metamorphism and simultaneous strong deformation by folding and thrusting" (Rui, 1973a, p. 882).

There is no doubt that there is a considerable need for more detailed modern structural studies of the Caledonian ores of Scandinavia to fix more exactly the time relations of the sulphides to the deformational and metamorphic events. There seems to be little doubt among the majority of geologists who have studied the ores in recent years that they *have* been involved in the Caledonian metamorphism. Until recently agreement has been lacking as to how early the sulphides were present in the metamorphic-rock pile and, as a consequence of this, on their primary or initial mode of genesis.

LITHOSTRATIGRAPHICAL RELATIONS OF THE ORES

The Lower Palaeozoic sequence

The Lower Palaeozoic volcanic rocks, with which the massive sulphide deposits are intimately associated, occur extensively over a length of over 1,200 km of the Caledonian orogen in Scandinavia (Fig. 1). The width and thickness of the volcanic rocks as now exposed are very variable, due to tectonic deformation and to later erosion. In the northern part of the Caledonides, these volcanics are largely represented by relatively thin, discontinuous bodies of amphibolite. Primary volcanic textures and structures are very seldom discernible, though pillow structures are reported from the greenstone volcanics of the Sulitjelma area (E. Horikoshi, personal communication, 1973) and the Vaddas district (I. Lindahl, unpublished thesis).

The volcanic rocks attain their maximum development in the central and southern Caledonides where the geosynclinal pile is exposed over a width of about 150 km at right angles to the axis of the orogen. Considerable repetition due to folding has taken place, but it has been estimated that the volcanics attained a maximum thickness of 2—3 km in this region (Oftedahl, 1968; Furnes, 1972; Furnes and Skjerlie, 1972).

In the Trondheim region, where detailed stratigraphic and structural studies are being carried out, the submarine Cambro-Ordovician basic lavas and pyroclastics of the Støren group (and supposed local equivalents) overlie the Gula schists group, which is a thick sequence of clastic sediments with volcanic horizons in the upper part. Overlying the Støren group is a thick pile of middle to upper Ordovician flysch-type lithologies with thin horizons of acidic volcanics and minor intermediate intrusive and extrusive volcanic rocks. (Hølonda porphyry.) These are in turn overlain by lower Silurian sediments which are the youngest exposed rocks in the geosynclinal pile.

A climactic Caledonian deformation and metamorphism took place in middle Silurian times (Roberts, 1967 and Wilson et al., 1973).

More detailed descriptions of the geosynclinal sequences can be found in the publications of Roberts (1967), Wolff (1967), Skjerlie (1969), Roberts et al. (1970) and Rui (1972). Table XVI presents a general stratigraphical column which has up to now been generally accepted as being valid throughout central and southern Norway.

Recently Gale and Roberts (1974) have reinterpreted the tectonics and stratigraphy within the Trondheim region in terms of a plate tectonic model for the central Scandinavian Caledonides. Their scheme of relationships is reproduced as Table XVII.

Nature of the volcanic activity

Investigations of Lower Palaeozoic rocks in southern Norway at present in progress (G.H. Gale, personal communication, 1973) have shown that the Cambro-Ordovician volcanics of the Løkken, Støren and Stavfjord (Askvoll) areas are predominantly basic in character and were extruded into a submarine environment, as evidenced by the presence of abundant pillow lavas and hyaloclastites. In the Grong area the same volcanics appear to have been emplaced in a mixed submarine—subaerial environment, and consist of basic pillowed and non-pillowed lavas, with minor andesites, silicic lavas and pyroclastics. Basic and acidic plutonic intrusives are especially abundant in the Grong area.

TABLE XVI

Trondheim region: stratigraphy and lithology*

Litho-/chrono-stratigraphic unit	West (Løkken–Stjørdal–Ekne–Snåsa)	East (Meråker–Tydal–Røros)
?Horg Group (west)/ Slågån Group (east) (Silurian)	Sandstone and shale (possibly Upper Ordovician, see below) basal conglomerate	Phyllites, sandstone (Llandoverian)
Upper Hovin Group (Middle–Upper Ordovician)	metagraywackes, polygenous conglomerates, volcaniclastic sandstones, silicic tuffs basal conglomerate	metagraywackes, phyllites, conglomerates calc-schists in the Røros area
Lower Hovin Group (Lower–Middle Ordovician)	rhyolite tuffs pelites, sandstones, limestones, green schists, locally porphyritic andesites basal conglomerate	greenstone/greenschist locally at top phyllite, calc-sandstone, polygenous conglomerates, local limestones basal conglomerates
Støren Group (Tremadocian)	greenstones, greenschists, quartz-keratophyres, tuffs, ?hyaloclastites, some sediments banded quartzites locally at base	mostly greenstones and amphibolites (with quartz-keratophyres) basal quartzite-conglomerate, schists
Gula Group (Cambrian)	medium-grade schists, calc-schists, phyllites, amphibolite bands	various med.-high grade schists and gneisses, locally phyllites, some quartzite, amphibolite
Eocambrian	quartzite, arkose, semi-pelites metadolerite sheets	quartzite, arkose, schists some amphibolites
Precambrian	various gneisses, schists, leptites metadolerite sheets locally	granite, dolerite, porphyry

* Compiled by D. Roberts, NGU, Trondheim, from unpublished work (in west), plus data from publications by: Carstens (1960), Peacey (1963, 1964), Wolff (1967), Siedlecka (1967), Chaloupsky (1970), Roberts, Springer and Wolff (1970), Roberts (1967), Vogt (1945) and Rui (1972). According to Chaloupsky (1970) the Horg Group is probably Upper Ordovician, not Silurian as suggested by Vogt (1945).

The Bømlo–Stord area in southwest Norway shows dominantly basaltic to andesitic lavas and pyroclastics with minor basaltic pillow lavas, rhyolitic lavas and ignimbrites. The predominantly subaerial volcanics in this area have also been intruded by basic and acidic plutonic masses.

Data on the geochemistry of the basic volcanics (G.H. Gale, unpublished results) indicate that those of the Grong area are low in potassium ($<0.3\% K_2O$) and that some of the basalts are chemically similar to low-potassium tholeiites (cf. Jakeš and White, 1972

TABLE XVII

Stratigraphical relationships in the Trondheim region interpreted in terms of a plate-tectonic model by Gale and Roberts, 1974

Chronostratigraphic unit	Eugeosynclinal		Miogeosynclinal	
	western areas	eastern areas		
Llandoverian	Horg Group	Slågån Group		
Upper Ordovician	U. Hovin Group	Kjølhaug Group		
Lower–Middle Ordov.	L. Hovin Group	Sulåmo Group		
Tremadocian	Støren Group	Fundsjø Group	Gula Group	Trondheim Supergroup
Cambrian				
Eocambrian				

(Courtesy of North-Holland Publishing Co.)

and Pearce and Cann, 1973).. The presence of abundant subaerial volcanics, andesitic and dacitic to rhyolitic lavas, and large trondhjemitic and gabbroic intrusive masses, have led to the interpretation that these rocks are part of a dissected island arc complex.

The basic volcanics of the Løkken, Støren and Stavfjord areas show major and trace elements characteristic of ocean-floor basalts (Cann, 1971 and Pearce and Cann, 1973). In the Løkken area there are present (actual extent not yet determined) basic pillow lavas having trace elements characteristic of low-potassium tholeiites. It is thus possible that also this area contains remnants of an island-arc sequence.

The volcanics of the Bømlo–Stord area are predominantly basaltic andesite and andesitic lavas, rhyolitic lavas and ignimbrites. These are intruded by a granodiorite–gabbro mass. Local limestone horizons attest to a relatively shallow marine depositional environment at times. They are considered to represent an island-arc complex.

Relation to a plate tectonic model

In recent years there have appeared in the literature a series of attempts to relate volcanogene massive sulphide ores of the "geosynclinal" type to the (relatively) new global-tectonics concept. These various attempts will not be discussed in the present account and for details the reader is referred to the publications of Pereira and Dixon (1971), Mitchell and Garson (1972), Sillitoe (1972), Sawkins (1972), Guild (1972) and others. As regards the Scandinavian sector of the Caledonides, a plate-tectonic model has recently been presented by Gale and Roberts (1972), while Gale and Vokes (1972) and Vokes et al. (1974) and Vokes and Gale (1975) have discussed the possible relations of the volcanogene sulphides to subduction of plates in the Palaeozoic.

The subject is patently one of great complexity and earlier ideas are being discarded as new data on the petrochemistry of the volcanics, the distribution of the different petrological types, the palaeogeography of the period in question, the areal distribution of the different ore types and the structure of the Caledonides, come to hand. The following summary is based on a more detailed account of Vokes and Gale (1975).

At the time of writing, no clear-cut ophiolite assemblages have deen distinguished in the Caledonides of southcentral Norway, although it seems possible that the islands of Hitra, Smøla and Leka in the Trøndelag region may represent slices of ophiolitic material obducted on to the continental margin. Attention is drawn to the presence of ocean-floor type basalts in the Stavfjord, Løkken and Støren areas, to low-potassium tholeiitic rocks in the Løkken and Grong areas and to the occurrence of other features characteristic of island-arc volcanic activity in the Grong area. In all these areas the volcanic rocks overlie continental clastic sediments which were deposited on a thick silicic continental margin. This fact suggests that the volcanics and related ores have been tectonically emplaced into their present positions, since the basic rocks represent material that has undergone high-level fractionation and thus cannot have been derived from a magma generated at depth and subsequently injected through a thick continental crust. Thus, the evolution of the Cambro-Ordovician volcanites may be tentatively interpreted in terms of the following plate-tectonic model (Fig. 8).

The ocean-floor basalts are envisaged as being generated during Eocambrian—Early Cambrian continental break-up at a mid-ocean ridge. During this time the trailing continental edge was receiving clastic continental sedimentation from the Baltic shield to the east. Closing of the proto-Atlantic ocean during the Ordovician generated an eastwards-dipping lithospheric plate which was subducted beneath a segment of the oceanic crust. Partial melting of the subducted plate generated tholeiitic—andesitic—dacitic magmas which were extruded on to the oceanic crust, by a mechanism similar to that which has apparently controlled the emplacement of magma in younger island-arc systems (Jakeš and White, 1969).

A back-arc sedimentary basin received clastic and volcanic debris from the island-arc complex to the west, and clastic material from the continental mass to the east during the Middle Ordovician to the earliest Silurian times. Near the endstages of the closing of the proto-ocean, slices of the island-arc complex and the oceanic crust upon which it developed, as well as the sedimentary basin behind the arc, were obducted on to the continental margin.

The massive sulphide deposits were generated either at the site of the spreading mid-ocean ridge, or during the early stages of island-arc development. At the moment, an island-arc environment is preferred, due to the association of some ore deposits with the back-arc sediments short stratigraphical distances from the volcanics.

On the other hand, several large and important sulphide deposits, such as Løkken, Gjersvik and Skorovass, are undoubtedly located intravolcanically and seem to be in a more typical oceanic crust situation. As already pointed out (pp. 98—100) the base-metal

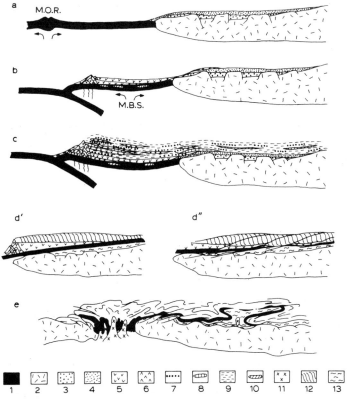

Fig. 8. Schematic sections depicting the development of the Caledonian orogen of central-southern Norway. (From Gale and Roberts, 1974.) *1* = oceanic crust; *2* = continental crust; *3* = Eocambrian arkosic psammites; *4* = lower part of Gula Group; *5* = basic volcanics; *6* = acid volcanics; *7* = polymict conglomerates; *8* = carbonates; *9* = various metasediments; *10* = gabbro bodies; *11* = acid intrusives; *12* = eugeosynclinal sequence, undifferentiated; *13* = miogeosynclinal sequence, undifferentiated; *M.O.R.,* mid-oceanic ridge; *M.B.S.,* marginal basin spreading.

a. Late Precambrian to Cambrian continental break-up and separation; Eocambrian sediments and lower part of Gula Group deposited on continental crust.

b. Early Ordovician development of island arc following subsuction, with low-potassium tholeiites erupted upon ocean floor basalts in marginal basin.

c. Middle Ordovician to Lower Silurian times. This depicts the main stage of eugeosynclinal back-arc sedimentation and volcanism above oceanic crust; miogeosynclinal regime with sedimentation of upper part of Gula Group above continental crust.

d′. The stage of obduction of the island arc/back-arc sequence (Støren Group through to Horg Group) upon the miogeosynclinal succession of the Gula Group (with Eocambrian below), in approximately early Middle Silurian time; d′ shows the idealized emplacement. d″ depicts the more likely situation wherein the obducted sequence is sliced up during its emplacement, locally with parts of the succession cut out; contiguous tectonic slices or sub-units show overriding from NW to SE.

e. The post-obduction phase of mid-Silurian orogenic deformation. Complex folding, metamorphism, intrusion and further major nappe development, with Baltic basement progressively involved towards the west and northwest. Mobile zone developed between the Baltic and North American/Greenland continental plates.

(Courtesy of North-Holland Publishing Co.)

contents of such deposits are more characteristic of sulphides supposedly generated on an ocean floor (e.g., Cyprus) than of, say, Kuroko-type deposits supposedly generated on continental crust (cf. Horikoshi, 1972).

ACKNOWLEDGMENTS

The writer is grateful to several colleagues, cited in the text, for permission to use unpublished results of their studies in progress. Drs. G.H. Gale and D. Roberts critically read the part of the text covering the results of their recent investigations.

REFERENCES

Aanerud, S., 1954. Trekk fra driften ved smelteverket i Thamshavn. *Tidsskr. Kjemi Bergves. Met.*, 6: 108–114.

Anderson, C.A. and Nash, J.T., 1972. Geology of massive sulphide deposits at Jerome, Arizona – a reinterpretation. *Econ. Geol.*, 67: 845–863.

Anger, G., 1966. Die genetischen Zusammenhänge zwischen deutschen und norwegischen Schwefelkies-Lagerstätten unter besonderer Berücksichtigung der Ergebnisse von Schwefeliso-topen-Untersuchungen. *Clausthaler Hefte Lagerstättenkd. Geol.*, 3: 115 pp.

Anger, G. and Eisbein, P., 1968. Erzmikroskopische Beobachtungen an den Schwefelkies–Kupferkies-Lagerstätten, Joma. *Bergbauwissenschaften*, 15: 93–97.

Barton Jr., P.B. and Toulmin, P., III, 1966. Phase relations involving sphalerite in the Fe–Zn–S system. *Econ. Geol.*, 61: 815–849.

Berg, G., 1928. Über das Vorkommen sogenannter porphyrischer Kieskristalle in den Lagerstätten der intrusiven Kiesgruppe. *Mitt. Abt. Gesteins-, Erz- Kohle- Salz-Unters.*, 3: 8–11.

Bergstøl, S. and Vokes, F.M., 1974. Stromeyerite and mekinstryite from the Godejord polymetallic sulphide deposit, central Norwegian Caledonides. *Miner. Deposita*, 9: 325–338.

Beyschlag, F., Krusch, P. and Vogt, J.H.L., 1910. Die Lagerstätten der nutzbaren Mineralien und Gesteine, 1. Enke, Stuttgart, 1st ed., 578 pp.

Boorman, R.S., 1967. Subsolidus studies in the ZnS–FeS–FeS$_2$ system. *Econ. Geol.*, 62: 614–631.

Boorman, R.S., Sutherland, J.K. and Chernyshev, L.V., 1971. New data on the sphalerite–pyrrhotite–pyrite solvus. *Econ. Geol.*, 66: 670–673.

Boyle, R.W., 1968. The geochemistry of silver and its deposits. *Geol. Surv. Can. Bull.*, 160: 264 pp.

Brøgger, W.C., 1901. Om dannelsen af de norske kisforekomster (kobber–svovelkisforekomster) av typen Röros–Sulitjelma. *Vidensk. Selsk. Christiania Forh. Möder.*, 25 pp.

Bugge, C., 1948. Kisene i fjellkjeden. *Nor. Geol. Tidsskr.*, 27: 97–102.

Bugge, C., 1954. The pyrite deposits in the mountain chain. In: Den Kaledonske fjellkjede i Norge. *Nor. Geol. Unders.*, 189: 73–79.

Cann, J.R., 1971. Major element variations in ocean-floor basalts. *Philos. Trans. R. Soc. Lond.*, 268: 495–505.

Carstens, C.W., 1920. Oversigt over Trondhjemsfeltets bergbygning. *K. Nor. Vidensk. Selsk. Skr.*, 1919(1): 152 pp.

Carstens, C.W., 1923. Der unterordovizische Vulkanhorizont in dem Trondheimsgebiet (mit beson-derer Berücksichtigung der in ihm auftretenden Kiesvorkommen.). *Nor. Geol. Tidsskr.*, 7: 185–269.

Carstens, C.W., 1931. Die Kiesvorkommen im Porsangergebiet. *Nor. Geol. Tidsskr.*, 12: 171–177.

Carstens, C.W., 1932. Zur Frage der Genesis der norwegischen Schwefelkiesvorkommen. *Z. Prakt. Geol.* 40 (7): 97–99.

Carstens, C.W., 1937. Zur Genese der norwegischen Schwefelkiesvorkommen. *Z. Dtsch. Geol. Ges.,* 88(4): 257–268.

Carstens, C.W., 1941. Zur Frage der Metamorphose der Schwefelkieserze. *K. Nor. Vidensk. Selsk. Forh.,* 14(3): 9–12.

Carstens, C.W., 1944. Om dannelsen av de norske svovelkisforekomster. *K. Nor. Vidensk. Selsk., Forh.,* 17: 27–54.

Carstens, H., 1955. Jernmalmene i det vestlige Trondheimsfeltet og forholdene til kisforekomstene. *Nor. Geol. Tidsskr.,* 35: 211–221.

Carstens, H., 1960. Stratigraphy and volcanism of the Trondheimsfjord area, Norway. *Int. Geol. Congr., 21st, Copenhagen, 1960, session Norden. Excursion guide A4 and C1, Norway,* 23 pp.

Chaloupsky, J., 1970. Geology of the Hølonda–Hulsjøen area, Trondheim region. *Nor. Geol. Unders.,* 266 (Årbok 1969): 277–304.

Constantinou, G. and Govett, G.J.S., 1972. Genesis of sulphide deposits, ochre and umber of Cyprus. *Trans. Inst. Min. Met.,* 81: B34–B46.

Domarev, V.C., 1956. Nekotorye geologischeskie osobennosti metamorfogennykh rudnykh mestovozhdenni. (Some geological features of metamorphic ore deposits.) *Usesoynz Nauchn. Issled. Geol. Inst. Mater. (Geol. Polezn. Iskop.),* 8: 7–41.

Duchanoy, M., 1854. Sur les gisements des minerais de cuivre et leur traîtement métallurgique dans le centre de la Norvège. *Ann. Min.,* 5: 181–243.

Durocher, M., 1849. Observations sur des gîtes métallifères de la Suède, de la Norvège et de la Finlande. *Ann. Min.,* 4: 171–443.

Foslie, S., 1925. Norges svovelkisforekomster. *Nor. Geol. Unders.,* 127: 122 pp.

Foslie, S., 1938. Betraktninger over kisforekomstenes dannelse. (Discussion of lecture by Th. Vogt.) *Nor. Geol. Tidsskr.,* 17: 214–217.

Foslie, S., 1939. Skorovass kisfelt i Grong. *Nor. Geol. Tidsskr.,* 19: 115–116.

Furnes, H., 1972. Meta-hyaloclastite breccias associated with Ordovician pillow lavas in the Solund area, west Norway. *Nor. Geol. Tidsskr.,* 52(4): 385–408.

Furnes, H. and Skjerlie, F., 1972. The significance of primary structures in the Ordovician pillow-lava sequence of western Norway in an understanding of major fold patterns. *Geol. Mag.,* 4: 315–322.

Gale, G.H. and Roberts, D., 1972. Palaeographical implications of greenstone petrochemistry in the southern Norwegian Caledonides. *Nat. Phys. Sci.,* 238(82): 60–61.

Gale, G.H. and Roberts, D., 1974. Trace element geochemistry of Norwegian Lower Palaeozoic basic volcanics and its tectonic implications. *Earth Planet. Sci. Lett.,* 22(4): 380–390.

Gale, G.H. and Vokes, F.M., 1972. Norwegian Palaeozoic volcanogene sulphide deposits in the light of a plate tectonic model. Lecture presented at Norwegian Geotraverse Meeting, Trondheim, 1972.

Geis, H.P., 1958. Die Genese norwegischer Kieslagerstätten. *Z. Erzbergbau Metallhüttenwes.,* 11: 541–543.

Geis, H.P., 1960. Frühorogene Sulfidlagerstätten. *Geol. Rundsch.,* 50: 46–52.

Geis, H.P., 1965. 100 Jahre Vigsnes Kobberverk. *Tidsskr. Kjemi Bergves. Met.,* 8–9: 194–202.

Gjelsvik, T., 1960. The Skorovass pyrite deposit, Grong area, Norway. *Int. Geol. Congr., 21st, Copenhagen, 1960, session Norden, Proc.,* XVI: 54–66.

Gjelsvik, T., 1968. Distribution of the major elements in the wall rocks and the silicate fraction of the Skorovass pyrite deposit, Grong area, Norway. *Econ. Geol.,* 63: 217–231.

Grip, E. and Frietsch, R., 1973. Malm i Sverige 2. *Norra Sverige.* Almqvist and Wiksell, Stockholm, 295 pp.

Guild, P.W., 1972. Massive sulfides vs. porphyry coppers in their global tectonic settings. In: *Joint Meeting, MMIJ–AIME, Tokyo, May 1972.* Print No. G 13: 12 pp;

Helland, A., 1873. Forekomster av kise i visse skifere i Norge. Universitetsprogram, Christiania: 97 pp.

Horikoshi, E., 1972. Geology of stratabound sulphide deposits in Japan. Stencilled report, Kyushu University, Kyushu, 104 pp.

Hutchinson, R.W. and Searle, D.L., 1971. Stratabound pyrite deposits in Cyprus and their relationship to other sulphide ores. *Soc. Min. Geol. Jap.*, 3: 198–205, Spec. issue. (*Proc. IMA–IAGOD Meetings, 1970, IAGOD vol.*).

Jakeš, P. and White, A.J.R., 1969. Structure of the Melanesian arcs and correlation with distribution of magma types. *Tectonophysics*, 8: 223–236.

Jakeš, P. and White, A.J.R., 1972. Major and trace element abundances in volcanic rocks of orogenic areas. *Geol. Soc. Am. Bull.*, 83: 29–40.

Jenks, W.F., 1971. Tectonic transport of massive sulfide deposits in submarine volcanic and sedimentary host rocks. *Econ. Geol.*, 66: 1215–1224.

Jøsang, O., 1964. En mikroskopisk undersøkelse av en del av Røros malmene. *Nor. Geol. Unders.*, 228 (Årbok 1963): 180–216.

Juve, G., 1967. Zinc and lead deposits in the Håfjell syncline, Ofoten, northern Norway. *Nor. Geol. Unders.*, 244: 54 pp.

Juve, G., 1974. Ore mineralogy and ore types of the Stekenjokk deposit, central Norwegian Caledonides, Sweden. *Sver. Geol. Unders., Ser. C.*, 706 (*Årsb.*, 18: 13): 162 pp.

Kautsky, G., 1948. Die kaledonische Sulfiderze und die palingene Prozesse. *Geol. Fören. Stockh. Förh.*, 70(2): 357–359.

Kautsky, G., 1952. Der geologische Bau des Sulitjelma–Salojaure-Gebietes in den nordskandinavischen Kaledoniden. *Sver. Geol. Unders.*, C528: 232 pp.

Kjerulf, T., 1871. Om Trondhjem Stifts geologi. *Nytt Mag. Naturvidensk.*, 18(4): 1–80.

Kjerulf, T., 1876. Om Trondhjem Stifts geologi II. *Nytt Mag. Naturvidensk.*, 21(1): 1–94.

Kleine-Hering, R., 1973. Der geologische Rahmen und die Mineralogie der Pyrit-freien, Zinkblende-führenden Kupferkies–Magnetkies-Erzvorkommen des Moskodalen, Nordreisa-Kommune, Troms, Nord-Nordwegen. Dissertation, Johannes Gutenberg Universität, Mainz, 116 pp., nicht veröffentlicht.

Krause, H., 1956. Zur Kenntnis der metamorphen Kieslagerstätten von Sulitjelma (Norwegen). *Neues Jahrb. Min. Abh.*, 89: 137–147.

Kullerud, G., 1953. The ZnS–FeS system: a geological thermometer. *Nor. Geol. Tidsskr.*, 32: 61–147.

Kullerud, G., Padget, P. and Vokes, F.M., 1955. The temperature of deposition of sphalerite-bearing ores in the Caledonides of Norway. *Nor. Geol. Tidsskr.*, 35: 121–127.

Kurek, J., 1931. Untersuchungen über die Genesis der Kieslagerstätten der Insel Stord (Norwegen). *Z. Prakt. Geol.*, 39(4): 56–58.

Lindahl, I., 1968. En undersøkelse av et mineralisert område på Kongsfjell, Korge, Nordland. Thesis, Norges Teniske høgskole, Trondheim, 127 pp., unpublished.

Ljøkjell, P., 1972. Norske sinkblende typer; kjemiske sammensetning undersøkt med mikrosonde. Delrapport I. Prosjekt: Sinkblende flotasjon. *Bergverkenes Landssammenslutnings Industrigruppe, Trondheim, Tek. Rapp.*: 22/1.

Magnusson, N.H., 1948. De svenska sulfidmalmerna och palingena processerna. En översikt. *Geol. Fören. Stockh. Förh.*, 70: 371–378.

Matsukuma, T. and Horikoshi, E., 1970. Kuroko deposits in Japan. A review. In: T. Tatsumi (Editor), *Volcanism and Ore Genesis*. Univ. Tokyo, Tokyo, pp. 153–180.

McDonald, J.A., 1967. Metamorphism and its effects on sulphide mineral assemblages. *Miner. Deposita*, 2: 200–220.

Mitchell, A.H.G. and Garson, M.S., 1972. Relationship of porphyry copper and circum-Pacific tin deposits to palaeo-Benioff zones. *Trans. Inst. Min. Met.*, 81: B10–B25.

Morton, R.D., 1972. A discussion: Sulphide mineralization and wall-rock alteration at Rødhammeren mine, Sør–Trøndelag Norway. *Nor. Geol. Tidsskr.*, 52: 313–315.

Nilsen, O., 1971. Sulphide mineralization and wall-rock alteration at Rødhammeren mine, Sør–Trøndelag, Norway. *Nor. Geol. Tidsskr.*, 51: 329–354.

Nilsen, O. and Mukherjee, A.D., 1972. Geology of the Kvikne mines with special reference to the sulphide ore mineralization. *Nor. Geol. Tidsskr.*, 52: 151–192.

Oftedahl, C., 1958. On exhalative–sedimentary ores. *Geol. Fören. Stockh. Fürh.*, 80: 1–19.

Oftedahl, C., 1961. Om dammelsen av de norske kisforekomster. *Tekn. Ukebl.*, 108: 415–419.

Oftedahl, C., 1968. Greenstone volcanoes in the central Norwegian Caledonides. *Geol. Rundsch.*, 57: 920–930.

Page, N.J., 1963. The sulphide deposit at Nordre Gjetryggen, Folldal, Norway. *Nor. Geol. Unders.*, 228: 217–269.

Peacey, J.S., 1963. Deformation in the Gangåsvann area. *Nor. Geol. Unders.*, 223 (Årbok 1962): 275–293.

Peacey, J.S., 1964. Reconnaissance of the Tömmerås anticline. *Nor. Geol. Unders.*, 227: 13–84.

Pearce, J. and Cann, J., 1973. Tectonic setting of basic volcanic rocks determined by using trace element analyses. *Earth Plan. Sci. Lett.*, 19: 290–300.

Pereira, J. and Dixon, C.J., 1971. Mineralization and plate tectonics. *Miner. Deposita*, 6: 404–405.

Ramdohr, P., 1928. Über das Mineralbestand und die Strukturen der Erze des Rammelsbergs. *Neues Jahrb. Miner. Geol. Paläontol. Abt. A*, 57(2): 1013–1068.

Ramdohr, P., 1938. Antimonreiche Paragenesen von Jakobsbakken bei Sulitjelma. *Nor. Geol. Tidsskr.*, 18: 275–289.

Ramdohr, P., 1953. Über Metamorphose und der sekundäre Mobilisierung. *Geol. Rundsch.*, 42: 11–19.

Ramdohr, P., 1960. Metamorphe Folge. In: *Die Erzmineralien und ihre Verwachsungen*. Akad. Verlag, Berlin, 3rd ed., pp. 36–76.

Ridler, R.H., 1971. Analysis of Archaean volcanic basins on the Canadian Shield using the exhalite concept. (Paper presented to: Can. Inst. Min. Metall., Ann. Western Meeting Vancouver, B.C., Oct. 1971.) Abstr., *Can. Inst. Min. Met. Bull.*, October 1971.

Ridler, R.H., 1973. Exhalite concept a new tool for exploration. *Northern Miner*, November 29th., 1973, 59–61.

Roberts, D., 1967. Structural observations from the Kopperåriksgrense area and discussion of the tectonics of Stjørdalen and the N.E. of the Trondheim region. *Nor. Geol. Unders.*, 245: 64–120.

Roberts, D., Springer, J.S. and Wolff, F.C., 1970. Evolution of the Caledonides in the northern Trondheim region, central Norway: A review. *Geol. Mag.*, 107: 133–145.

Roscoe, S.M., 1965. Geochemical and isotopic studies, Noranda and Matagami areas. *Trans. Can. Inst. Min. Met.*, LXVIII: 279–285.

Rui, I.J., 1972. Geology of the Røros district, southeastern Trondheim region, with a special study of the Kjøliskarvene–Holtsjøen area. *Nor. Geol. Tidsskr.*, 52: 1–22.

Rui, I.J., 1973a. Structural control and wall-rock alteration at Killingdal mine, central Norwegian Caledonides. *Econ. Geol.*, 68: 859–883.

Rui, I.J., 1973b. Geology and structures of the Røstvangen sulphide deposit in the Kvikne district, central Norwegian Caledonides. *Nor. Geol. Tidsskr.*, 53: 433–442.

Saager, R., 1966. Erzgeologische Untersuchungen an kaledonischen Blei-, Zink- und Kupferführenden Kieslagerstätten im Nord-Rana-Distrikt, Nord-Norwegen. Dissertation, Eidgenössische Technische Hochschule, Zürich.

Saager. R., 1967. Drei Typen von Kieslagerstätten im Mofjell-Gebiet, Nordland, und ein neuer Vorschlag zur Gliederung der kaledonischen Kieslager Norwegens. *Nor. Geol. Tidsskr.*, 47: 333–358.

Sangster, D.F., 1972. Precambrian massive sulphide ores in Canada: A review. *Geol. Surv. Can. Pap.*, 72–22: 42 pp.

Sawkins, F.J., 1972. Sulfide-ore deposits in relation to plate tectonics. *J. Geol.*, 80: 377–397.

Schermerhorn, L.J.G., 1971. Pyrite emplacement by gravity flow. *Bull. Geol. Min.*, LXXII: 304–308.

Schneiderhöhn, H., 1962. Die Lagerstätten der metamorphen Abfolge. In: *Erzlagerstätten*. Fischer, Jena, pp. 285–334.

Scott, S.D. and Barnes, H.L., 1971. Sphalerite geothermometry and geobarometry. *Econ. Geol.*, 66: 653–669.

Searle, D.L., 1972. Mode of occurrence of the cupriferous pyrite deposits of Cyprus. *Trans. Inst. Min. Met.*, 81: B189–B197.

Sen, R. and Mukherjee, A.D., 1972. A reappraisal of structural evolution and metamorphism in the Bleikvassli ore deposits, Nordland, north Norway. *Neues Jahrb. Miner. Monatsh.* 1972, (8): 375–382.

Shadlun, T., 1970. Metamorphic textures and structures of sulphide ores. *Soc. Min. Geol. Jap.*, 3: 241–250, special issue *(Proc. IMA–IAGOD Meetings 1970, IAGOD vol.)*.

Siedlecka, A., 1967. Geology of the eastern part of the Meråker area. *Nor. Geol. Unders.*, 245: 22–58.

Sillitoe, R.H., 1972. Formation of certain massive sulphide deposits at sites of sea-floor spreading. *Trans. Inst. Min. Met.*, 81: B141–B155.

Skjerlie, F.J., 1969. The pre-Devonian rocks in the Askvoll–Gaula area and adjacent districts, western Norway. *Nor. Geol. Unders.*, 258: 325–359.

Stanton, R.L., 1958. Abundance of copper, zinc and lead in some sulfide deposits. *J. Geol.*, 66(5): 484–502.

Stanton, R.L., 1959. Mineralogical features and possible mode of emplacement of the Brunswick Mining and Smelting orebodies, Gloucester County, New Brunswick. *Trans. Can. Inst. Min. Met.*, LXII: 337–349.

Stanton, R.L., 1960. General features of the conformable "pyritic" orebodies. 1. Field associations. *Trans. Can. Inst. Min. Met.*, LXIII: 22–27.

Stanton, R.L., 1962. Elemental constitution of the Black Star orebodies, Mt. Isa, Queensland and its interpretation. *Trans. Inst. Min. Met.*, 72(2): 69–124.

Stanton, R.L., 1964. Mineral interfaces in stratiform ores. *Trans. Inst. Min. Met.*, 74(2): 45–79.

Strøm, H.C., 1825. Geognostiske Bemærkninger om Værkerne i det nordenfjeldske Bergværksdistriktet. *Mag. Naturvidensk.*, 5.

Vargas Bedemar, E.R., 1819. Reisen nach dem hohen Norden durch Schweden, Norwegen und Lappland. Frankfurt am Main.

Vogt, J.H.L., 1889. Norske ertsforekomster. III. rekke. VII Foldalens Kisfelt. *Ark. Mat. Naturvidensk.*, 13: 70 pp.

Vogt, J.H.L., 1890. Salten og Ranen, med særligt hensyn til de viktigste jernmalm- og svovelkisforekomster, samt marmorlag. *Nor. Geol. Unders.*, 3: 231 pp.

Vogt, J.H.L. 1894a. Ueber die Kielagerstätten vom Typus Røros, Vigsnäs, Sulitelma in Norwegen und Rammelsberg in Deutschland. *Z. Prakt. Geol.*, II: 41–50, 117–134 and 173–181.

Vogt, J.H.L., 1894b. De norske kisforekomster av Typus Røros, Vigsnes og Sulitjelma. *Geol. Fören. Stockh. Förh.*, 16: 463–491.

Vogt, Th., 1935. Origin of the injected pyrite deposits. *K. Nor. Vidensk. Skr.*, 20: 17 pp.

Vogt, Th., 1944. Fjellkjedens flytestruckturer og malmforekomstene. I. Nord-Rana grubefelt. *K. Nor. Vidensk. Selsk. Forh.*, 17 (30): 118–121.

Vogt, Th., 1945. The geology of part of the Hølonda–Horg district, a type area in the Trondheim region. *Nor. Geol. Tidsskr.*, 25: 449–527.

Vogt, Th., 1952. Flowage structures and ore deposits of the Caledonides of Norway. *Int. Geol. Congr., 18th, 1948, Lond., Proc.*, XIII: 240–244.

Vokes, F.M., 1957. The copper deposits of the Birtavarre district, Troms, Northern Norway. *Nor. Geol. Unders.*, 199: 239 pp.

Vokes, F.M., 1962. Mineral parageneses of the massive pyritic ore bodies of the Caledonides of Norway. *Econ. Geol.*, 57: 890–903.

Vokes, F.M., 1963. Geological studies on the Caledonian pyritic zinc-lead orebody at Bleikvassli, Nordland, Norway. *Nor. Geol. Unders.*, 222: 126 pp.

Vokes, F.M., 1966. On the possible modes of origin of the Caledonian sulfide ore deposit at Bleikvassli, Nordland, Norway. *Econ. Geol.*, 61: 1130–1139.

Vokes, F.M., 1968. Regional metamorphism of the Palaeozoic geosynclinal sulphide ore deposits of Norway. *Trans. Inst. Min. Met.*, 77: B53–B59.

Vokes, F.M., 1969. A review of the metamorphism of sulphide deposits. *Earth-Sci. Rev.*, 5: 99–143.

Vokes, F.M., 1973. "Ball texture" in sulphide ores. *Geol. Fören. Förh.*, 95: 403–406.

Vokes, F.M., 1974. Structural control and wall rock alteration at Killingdal mine, central Norwegian Caledonides. (Discussion of Rui, 1973a.) *Econ. Geol.*, 69: 706–708.

Vokes, F.M. and Gale, G.H., 1975. Metallogeny related to continental drift in Scandinavia. *Geol. Assoc. Can., Spec. Vol.* (in preparation).

Vokes, F.M. and Morton, R.D., 1973. A discussion. Geology of the Kvikne mines with special reference to the sulphide ore mineralization. *Nor. Geol. Tidsskr.,* 53: 333–336.

Vokes, F.M., Roberts, D. and Gale, G.H., 1974. A subduction model for the derivation and zoning of stratiform sulphide deposits in the Norwegian Caledonides. XI Nordiska Geologiska Vintermöte, Oulu/Uleaborg, Finland, Jan. 1974. Program, Section B, Abstracts. Uleaborgs Universitet, Finland: 74 pp.

Waltham, A.C., 1968. Classification and genesis of some massive sulphide deposits in Norway. *Trans. Inst. Min. Met.,* 77: B153–B161.

Wilson, M.R., 1973. The geological setting of the Sulitjelma orebodies, central Norwegian Caledonides. *Econ. Geol.,* 68: 307–316.

Wilson, M.R., Roberts, D. and Wolff, F.C., 1973. Age determinations from the Trondheim region Caledonides, Norway: a preliminary report. *Nor. Geol. Unders.,* 288: 53–60.

Wolff, F.C., 1967. Geology of the Meråker area as a key to the eastern part of the Trondheim region. *Nor. Geol. Unders.,* 245: 123–146.

Zachrisson, E., 1969. Caledonian geology of northern Jämtland–southern Västerbotten. *Sver. Geol. Unders.,* C644: 33 pp.

Zachrisson, E., 1971. The structural setting of the Stekenjokk ore bodies, central Swedish Caledonides. *Econ. Geol.,* 66: 641–652.

Chapter 5

PRECAMBRIAN, STRATA-BOUND, MASSIVE Cu–Zn–Pb SULFIDE ORES OF NORTH AMERICA

D.F. SANGSTER and S.D. SCOTT

INTRODUCTION

Strata-bound massive sulfide ores are an important part of the mineral heritage of North America and particularly of Canada. For example, of the total Canadian metal production in 1971, 80% of the zinc, 39% of the copper, 68% of the lead, 62% of the silver, and significant amounts of gold came from this type of mineralization. In terms of metal value, Cu, Pb, Zn, Ag and Au from these deposits accounted for more than one-quarter of the value of Canada's total metal production in that year. In the United States, production from massive sulfide ores is insignificant in comparison to copper from porphyry copper deposits and lead–zinc from Mississippi Valley deposits. Nevertheless, their overall similarity to the Canadian massive sulfide ores warrants further examination.

Economic deposits of massive copper–zinc–lead sulfides have been recognized around the world in rocks ranging from Archean to Miocene in age. However, of the approximately 220 deposits of this type in North America, more than 65% occur in Precambrian rocks although those in the Precambrian of the Appalachian region of the United States may not be of Precambrian age (see p. 158). The truly Precambrian ores, which are the subject of this paper, occur in a wide variety of rock associations and types. They differ in several important aspects from their Phanerozoic analogues and, as a group, provide many excellent examples of post-ore metamorphic effects.

The ideas expressed in this paper arose from the authors' own observations together with a review of the available literature. Our starting point is Sangster's (1972a) review of the Precambrian massive Cu–Zn–Pb sulfide deposits of Canada. To this we have added deposits in the United States and have included detailed descriptions of several important districts which illustrate the principles we wish to elaborate. We have arbitrarily subdivided the massive sulfide ores into three types: (1) those which are predominantly of volcanic association; (2) those predominantly of sedimentary association and; (3) those of a mixed volcanic and sedimentary association and we offer thoughts further to Sangster (1972a) on their deformation, metal zoning and genesis.

Definition of the ore type

Strata-bound, massive, pyritic Cu—Zn—Pb sulfide deposits constitute a distinct ore type throughout the geologic column. Those in volcanic rocks are often referred to as "volcanogenic" or "exhalative" massive sulfide ores and we believe that the weight of evidence for most, but not all, deposits included in this paper indicates their formation from submarine hydrothermal emanations. Chemical sediments (including ore deposits) of predominantly volcanic—exhalative origin have been termed "exhalites" by R.H. Ridler (Ridler and Shilts, 1974, p. 1) who has also proposed a geological and chemical classification of exhalites (pp. 2, 3). For massive sulfide deposits in non-volcanic rocks, we prefer not to adopt a genetic model at the present time.

The term "massive" refers to mineralization composed of greater than 60% sulfides and carries no textural connotation. The economic minerals of these ores are predominantly chalcopyrite and/or sphalerite, but most also contain economic concentrations of silver and/or gold and, in post-Archean ores, of galena. The deposits may also have mineable concentrations of disseminated mineralization as an envelope around the massive ore or, more commonly, in a "stringer zone" in the footwall. The massive ore is strata-bound and may be stratiform but the disseminated ore is usually discordant to the stratigraphy.

The term "copper—zinc—lead" is a general one meant to imply that *as a whole* these ores are polymetallic, although individual deposits may be bi- or even mono-metallic.

Location of deposits in North America

The distribution of the Precambrian, massive, Cu—Zn—Pb sulfide ores of North America is shown in Figs. 1 and 2. Of the eighty-six Canadian deposits, all but four are in the Precambrian Shield and all but ten of them occur in the Superior and adjacent Churchill structural provinces.

Correct assignment of a Precambrian age to many of the American deposits presents a problem. The only unquestioned Precambrian deposits are those in Arizona (no. 5—10 in Fig. 2) and Balmat—Edwards, N.Y. (no. 1—4 in Fig. 2). All of the other strata-bound massive sulfide deposits, including the important ores of Ducktown, Tennessee, are in the Appalachian region of the eastern United States. These deposits are in Precambrian rocks, but the ores have traditionally been considered to be Devonian in age thus making them epigenetic. This age is indicated by K/Ar dates from minerals in schists associated with ore (Magee, 1968) and from Pb-isotope ratios in ore (Doe, in Kinkel, 1967, p. 48). On the other hand, Kinkel (1967; see also, Kinkel et al., 1965) has argued, on the basis of three K/Ar dates of greater than 1 b.y., that the ores may indeed be Precambrian and the remainder of the isotopic ages may have been updated during the Appalachian (Devonian) orogeny by argon loss and the introduction of radiogenic lead. As much as we sympathize with Kinkel's view on geologic grounds, we find it a difficult position to defend in view of

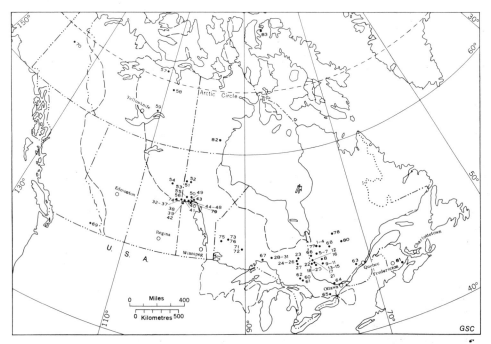

Fig. 1. Location of some Canadian Precambrian massive sulfide deposits. *1* = Mattagami Lake; *2* = New Hosco; *3* = Radiore; *4* = Orchan; *5* = Joutel; *6* = Mines de Poirier; *7* = Northern Exploration; *8* = Barraute; *9* = Manitou–Barvue; *10* = Louvem; *11* = East Sullivan; *12* = Vauze, Magusi River and Iso Copperfields; *13* = West MacDonald; *14* = Waite; *15* = Amulet; *16* = Norbec (Lake Dufault); *17* = Millenbach; *18* = Quemont; *19* = Horne; *20* = Delbridge; *21* = Mobrun; *22* = Aldermac; *23* = Texasgulf (Kidd Creek); *24* = Jameland; *25* = Kam–Kotia; *26* = Canadian Jamieson; *27* = Genex; *28* = Geco; *29* = Willecho; *30* = Willroy; *31* = Nama Creek; *32* = Flexar; *33* = Coronation; *34* = Flin Flon; *35* = Schist Lake; *36* = Mandy; *37* = Birch Lake; *38* = White Lake; *39* = Cuprus; *40* = North Star; *41* = Don Juan; *42* = Centennial; *43* Osborne Lake; *44* = Stall Lake; *45* = Rod; *46* = Anderson Lake; *47* = Ghost Lake; *48* = Chisel Lake; *49* = Dickstone; *50* = Wim; *51* = Fox Lake; *52* Ruttan Lake; *53* = Sherridon; *54* = Brabant; *55* = Bob Lake; *56* = Jungle; *57* = High Lake; *58* = Hackett River; *59* = Big Indian Mountain; *60* = Errington; *61* = Vermillion; *62* = Geneva Lake; *63* = Tetrault; *64* = New Calumet; *65* = Syngenore; *66* = Normetal; *67* = Zenmac; *68* = Coniagas; *69* = Sullivan; *70* = Hart River; *71* = Mattabi; *72* = Sturgeon Lake; *73* = South Bay (Uchi Lake); *74* = Western Nuclear; *75* = Trout Bay (Red Lake); *76* = Copper–Man; *77* = Grasset Lake; *78* = Frotet Lake; *79* = Reed Lake; *80* = Patino; *81* = Teahan and Lumsden; *82* = Spi Lake; *83* = Strathcona Sound.

the Pb-isotopic date. It is unlikely that Pb-isotopic ages of massive sulfide deposits are updated during metamorphism (Sangster, 1972b) and, rather than speculating on a Precambrian age, we have excluded the American Appalachian massive sulfides from consideration.

We are not aware of any Precambrian massive sulfide ores in Mexico or Central America with the possible exception of the La Dicha deposit in the central part of the State of Guerrero, south of Mexico City. As described by Klesse (1968), La Dicha is a

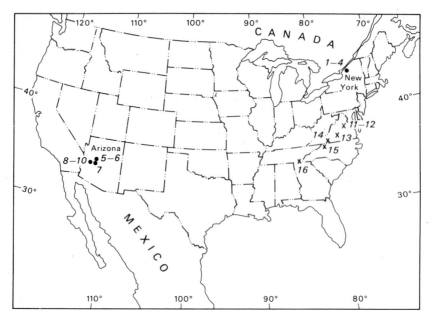

Fig. 2. Location of some massive sulfide ores in Precambrian rocks of the United States. Ages of deposits located with an X are in dispute (see text). *1* = Balmat No. 2; *2* = Balmat No. 3; *3* = Edwards; *4* = Hyatt; *5* = United Verde; *6* = United Verde Extension; *7* = Iron King; *8* = Old Dick; *9* = Copper Queen; *10* = Copper King; *11* = Sulfur Mine; *12* = Arminius; *13* = London—Virginia; *14* = Gossan Lead; *15* = Ore Knob; *16* = Ducktown.

2 m thick, stratiform bed of pyrrhotite and pyrite with minor chalcopyrite and very subordinate sphalerite. It lies between metaconglomerate and phyllite which Klesse (1970) believes are Paleozoic but, according to S.E. Kesler (personal communication, 1974) may be older.

GEOLOGY OF REPRESENTATIVE DEPOSITS

In order to provide a background for the general discussions which follow, we present in this section a description of certain key massive sulfide deposits together, in some cases, with an outline of the tectonic evolution of the district. In doing so, we have arbitrarily divided the deposits into three types: those in a predominantly volcanic environment, those in a volcano—sedimentary ("mixed") environment, and those in a predominantly sedimentary environment.[1]

---

[1] Editor's note: This approach is becoming increasingly accepted; see Chapter 4 by Gilmour, Vol. 1, on transitional types of ores, and Stanton (1972b).

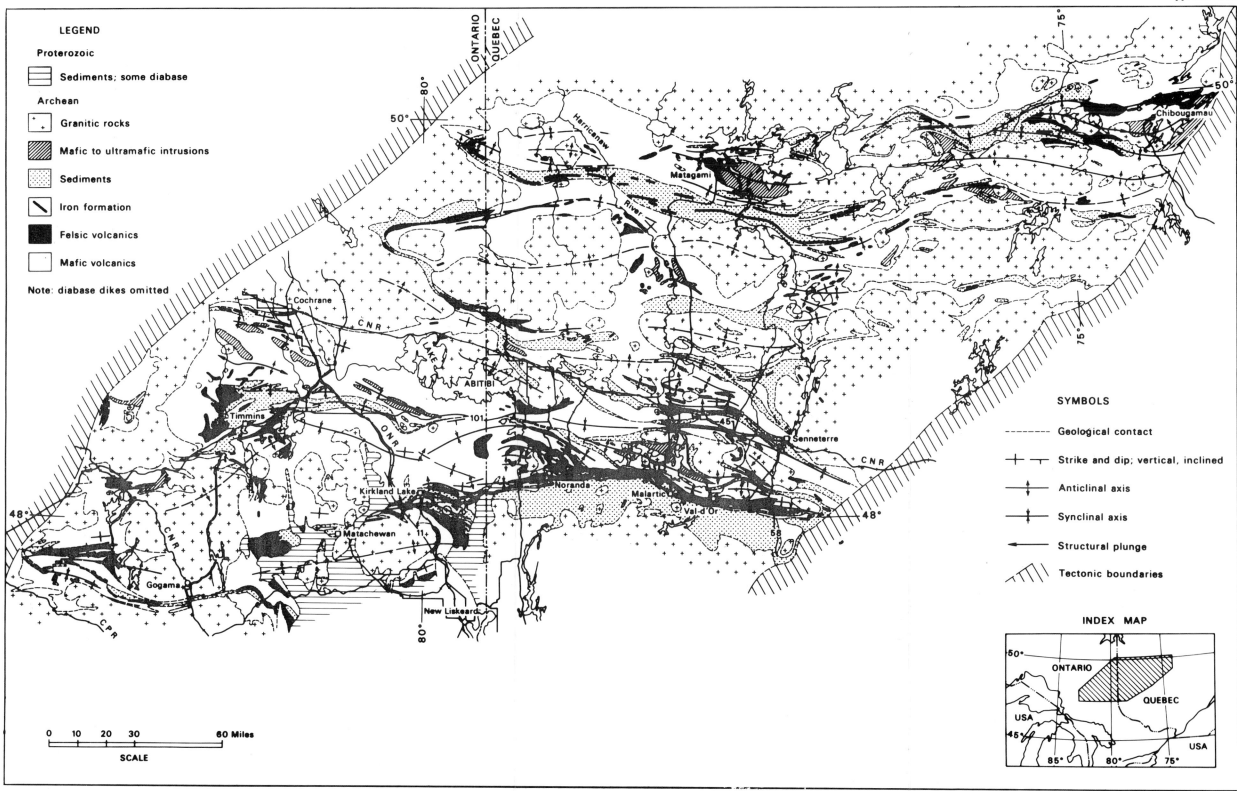

Fig. 3. Geology of the Abitibi orogen. (From Goodwin and Ridler, 1970.) A small area shown in white near New Liskeard represents Paleozoic sediments. For map of the Noranda district, see fig. 10 in Chapter 4, Vol. 1 by Gilmour. (Courtesy of Geol. Surv. Can.)

In the volcanic group we include, for example, massive sulfide ores of the Abitibi orogen (e.g., Noranda, Quebec), ores associated with the Amisk volcanics of Flin Flon and Snow Lake, Manitoba; and ores contained in the Yavapai Series volcanic rocks near Jerome, Arizona.

Many of the deposits occurring in a mixed volcano–sedimentary environment are peripheral to the volcanic centers with which the volcanic group of deposits are associated. In these "mixed"rocks, the volcanic component is clearly recognizable comprising flows and/or fine to medium-grained tuffs. Examples are the Mobrun deposit in the Noranda area (21 in Fig. 1), the Cuprus, White Lake and Centennial deposits near Flin Flon (39, 38 and 42 in Fig. 1), and the Geco orebody in the Manitouwadge area of central Ontario (28 in Fig. 1).

With a decrease in the volcanic component of the host rocks, the mixed volcano-sedimentary group of massive sulfide deposits gradually gives way to the third group, i.e. those in predominantly sedimentary rocks. In some areas, such as the Kisseynew gneiss belt north of and parallel to the Flin Flon–Snow Lake belt of volcanic rocks, the sedimentary rocks, largely greywackes, quartzites and minor carbonates, contain a small volcanic component and can be "walked out" or otherwise correlated with the adjacent volcanic pile. In other areas, the host rocks are almost or entirely sedimentary and cannot be linked to a nearby volcanic source. The best example of the latter is the siltstone and argillite-hosted Sullivan mine in the southeastern Canadian Cordillera (69 in Fig. 1). The deposit displays many characteristics in common with Mt. Isa, Hilton and MacArthur River in the Proterozoic of Australia or the Meggen and Rammelsberg deposits in Devonian clastics of Germany.

The Balmat–Edwards lithologies (1–4 in Fig. 2), dominated by carbonates and siliceous carbonates, have traditionally been considered metasediments (Lea and Dill, 1968). Similarly, the Strathcona Sound deposit in northern Baffin Island (83 in Fig. 1) is hosted in a thick dolomite between two shale units (Geldsetzer, 1973). This type of deposit is, however, uncommon on two accounts: (1) $Cu–Zn–Pb$ sulfide ores of a massive nature are uncommon in carbonate-hosted deposits of any age; and (2) carbonate-hosted ores are rare in the Precambrian relative to the abundance of carbonate in the Proterozoic.

When considering massive sulfide ores of all geological ages the wider diversity of ore types in the Phanerozoic has given rise to elaborate schemes of classification (e.g., Gilmour, 1971 and Chapter 4 by Gilmour in Vol. 1; Hutchinson, 1973; Chapter 12 by Lambert, this volume). For example, Hutchinson, in his world-wide review of massive sulfides of volcanic association, defines three fundamental types and two subtypes based on base-metal content, tectonic setting and several other parameters. Two of his fundamental types, $Zn–Cu$ pyrite and $Pb–Zn–Cu$ pyrite, are common in the Precambrian whereas his cupreous pyrite ores in ophiolites appear to be absent (cf. with Lambert's Chapter 12, this volume).

By comparison, our triple division of deposit types for the Precambrian of North America based on geological environment is simplistic and somewhat artificial. We do not intend to imply a separate origin for each. On the contrary, there is a complete gradation between the two end-member types (e.g., Gilmour, 1971, and Chapter 4 by Gilmour, Vol. 1) suggesting a common genesis which we believe to be predominantly hydrothermal solutions emanating onto the sea floor. (The carbonate-hosted deposits of Balmat–Edwards and Strathcona Sound may be exceptions to this common origin.) The differences in geological environment produce only second-order perturbations on this overall process.

Ores in predominantly volcanic rocks

Abitibi orogenic belt. The Abitibi orogenic belt is the most thoroughly studied Archean volcanic succession in North America. It occupies an area of approximately 475 × 125 miles straddling the Ontario/Quebec border (Fig. 3) and extends from near Chapleau on the west to Chibougamau on the east. The region, steeped in the romance and history of mining in Canada, encompasses the important mining centers of Timmins (Cu, Zn, Ag; Au; Ni), Kirkland Lake (Au; Fe), Noranda (Cu, Zn; Au), Val d'Or (Au; Cu, Zn; Mo, Bi, Li), Matagami (Cu, Zn) and Chibougamau (Au; Cu, Au) which have seen production from over one hundred mines in the past 60 years.

Goodwin and Ridler (1970), from whose compilation most of the following discussion on the Abitibi belt is taken, have identified eleven semi-independent volcanic–sedimentary associations which they have termed "volcanic complexes" (Fig. 4). Each complex consists of a supracrustal succession of thick mafic volcanics overlain by one or more conformable mafic to felsic cycles with intercalated volcani-clastic sediments and exhalite (Fig. 5). These are intruded by pre- to post-kinematic felsic to mafic plutons ranging up to batholithic proportions which are cut, in turn, by younger diabase dykes. Regional metamorphic grade is low, producing mineral assemblages within the lower greenschist facies. Essentially flat-lying Proterozoic and Paleozoic sediments protrude unconformably onto the margins of the volcanic belt from the south. All of the volcanic complexes have elliptical outlines whose east–west elongations are considered by Goodwin and Ridler (1970) to be a consequence of tectonic deformation. The complexes are distributed along the northern and southern margins of the orogen and are conspicuously absent from the central axis.

In the western portion of the belt the Blake River group of volcanics comprising the Noranda–Benoit complex of Fig. 4 and host to the massive Cu–Zn ores at Noranda, forms a pile up to 50,000 ft thick (Goodwin, 1973). At the base it is tholeiitic in composition but becomes increasingly calc-alkaline stratigraphically upwards. As well, the basal pillow basalts and gabbroic intrusions become increasingly intercalated upwards with andesite flows, flow breccias and minor pyroclastics. The end-stage of this upward differentiation is a concentration of felsic flows and pyroclastics at the top of the volcanic pile. The Blake River volcanics become progressively younger towards the east resulting in a greater proportion of calc-alkaline to tholeiitic basalts and of felsic to mafic

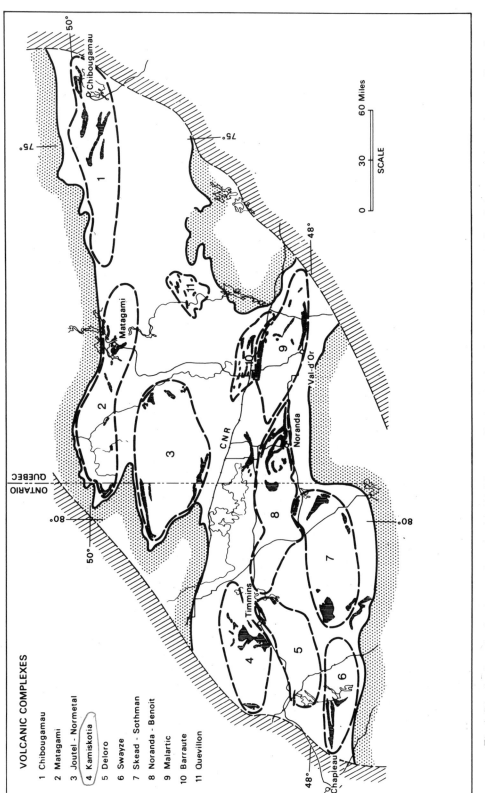

VOLCANIC COMPLEXES

1 Chibougamau
2 Matagami
3 Joutel - Normetal
4 Kamiskotia
5 Deloro
6 Swayze
7 Skead - Sothman
8 Noranda - Benoit
9 Malartic
10 Barraute
11 Quevillon

Fig. 4. Distribution of volcanic complexes in the Abitibi orogen. Felsic volcanic rocks are shown by vertically ruled pattern. Approximate outlines of the volcanic complexes including cogenetic intrusions and sediments are shown by heavy dashed lines. (From Goodwin and Ridler, 1970; courtesy of Geol. Surv. Can.)

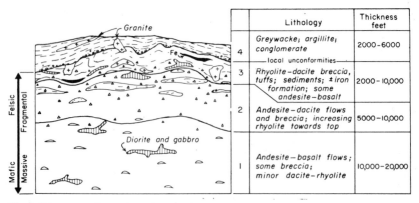

Fig. 5. Diagrammatic section of an Archean volcanic pile illustrating typical mafic to felsic succession with overlying sediments. The principal volcanic trends are from lower predominantly mafic lava flows to upper predominantly felsic fragmental rocks, the latter including both pyroclastics and flow breccias. Basalts predominate in the lower parts, andesites in middle parts and dacite plus rhyolite in upper parts. Compositional trends are transitional and local reversals are present. (From Goodwin, 1965; courtesy of *Econ. Geol.*)

volcanics towards the east near Noranda, although there are some local reversals in the tholeiitic to calc-alkaline trend in the immediate Noranda area (Spence and De Rosen-Spence, 1975).

Felsic volcanics comprise an estimated 5.5—10% of the total volume of volcanic rocks in the Abitibi belt (Goodwin and Ridler, 1970). However, they are not uniformly distributed but are concentrated within and, in part, define eruptive centers (Fig. 4) each one of which may contain up to 50% of felsic volcanic rocks. There are six principal centers representing the important mining camps at Timmins, Kirkland Lake, Noranda, Val d'Or, Matagami and Chibougamau. The majority of the felsic volcanics are of dacitic compostion, but rhyolite occurs locally at, for example, Noranda, Timmins and Matagami. According to Bennett and Rose (1973), selected samples of felsic volcanics associated with massive sulfide ores do not contain any distinctive features or trends in their *primary* major element chemistry which can be used as a guide to ore, although Descarreaux (1973) has pointed out that the massive sulfide ores are exclusively associated with calc-alkaline volcanic rocks. Both Bennett and Rose and Descarreaux agree that chemical variation resulting from alteration during mineralization, as we discuss below (p. 152), may be significant to exploration.

The lithic proportions of the volcanics vary from center to center. In most centers coarse pyroclastics are common but, in contrast, volcanism was less explosive at Noranda and such rocks are less abundant in that area. Here most of the felsic volcanics consist of intercalated andesite and rhyolite flows together with ash flows (R.S. Fiske, in Goodwin and Ridler, 1970) and agglomerate flow breccia. The reason for the quiescent volcanism at Noranda compared to violent volcanism elsewhere is not understood, but may be due to relative volatile contents of felsic magmas or to depth of water in which volcanism occurred (Rittman, 1962). It is important to note here that the *presence* of coarse pyroclastics rather than their relative *abundance* appears to be important to the formation of volcanogenic massive sulfide deposits.

Most of the sediments of the orogen are conformable with the volcanics and occupy higher stratigraphic positions. They consist mainly of clastics derived from the erosion of volcanic piles and were rapidly accumulated into nearby basins and troughs. Iron-formations are widespread within the volcani-clastic sediments and serve to outline these depositional basins. The ferruginous sediments pass laterally from oxide (magnetite) facies through narrow carbonate (siderite–ankerite–dolomite) facies to sulfide (pyrite–pyrrhotite) facies and are thought by Goodwin and Ridler (1970) to represent a shelf-to-basin transition. Typically, the massive Cu-Zn sulfide deposits are spatially associated with barren sulfide facies iron-formation and many gold ores with carbonate facies iron-formation.[1] South of Noranda and Val d'Or finely-bedded flyschoid greywackes and argillites comprise the Pontiac Group.

Deformation during the Kenoran orogeny (2560 m.y.) was about predominantly east–west, doubly-plunging, isoclinal fold axes resulting in an estimated 50% or more compression in the N–S direction (Goodwin and Ridler, 1970). In the southern portion of the belt there are two prominent east-trending shear zones, the Cadillac–Brouzon "break" running south of Kirkland Lake and Noranda to west of Val d'Or and the Porcupine–Destor "break" running from Timmins to the Noranda area. The breaks tend to follow lithofacies boundaries, including sedimentary–volcanic contacts, and are loci of gold vein mineralization. Traditionally, these breaks have been considered to be major thrust faults (Thomson, 1941; Kalliokoski, 1968), but Ridler (1970) found no evidence for large displacements nor continuity of the shear zones near Kirkland Lake. Instead, he proposed that here, and perhaps over its full extent, the Cadillac–Brouzon break simply represents shearing of very incompetent carbonate-facies iron-formation with little or no overall displacement.

Goodwin and Ridler (1970) have interpreted the Abitibi belt as an intercratonic orogen sandwiched between two sialic forelands (Fig. 6). The volcanic complexes may have been paired island arcs which shed volcani-clastic sediments into nearby basins at the same time as the Pontiac flysch was being derived from the southern foreland (Holubec, 1972). Within each "arc" were centers of effusive felsic volcanism which gave rise to the host rocks to the conformable, massive Cu–Zn sulfide ores at Timmins, Noranda and Matagami, each of which is within a different volcanic complex. Although it is premature to apply plate tectonics to the Archean (Sillitoe, 1973), Goodwin and Ridler (1970) point out that the episodic nature of the volcanic complexes, perhaps with the younger ones occurring towards the axial zone, is consistent with a progressive spreading of the

---

[1] The carbonate-facies gold ores are only one of a family of gold deposit-types recognized by Goodwin and Ridler (1970, p. 18).

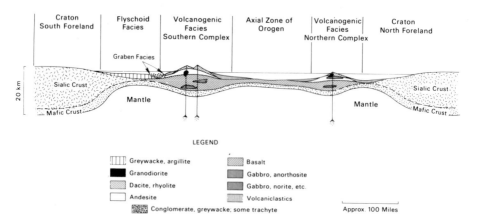

Fig. 6. Vertical cross-section of a hypothetical tectonic reconstruction of the Abitibi orogen crossing Noranda in the south and Matagami in the north. Scales are approximate. (Modified slightly by A.M. Goodwin from Goodwin and Ridler, 1970; courtesy of Geol. Surv. Can.)

sialic forelands and we might add that their subsequent closing would produce the observed structural deformation, in parallel with plate tectonic concepts. However, ophiolite complexes and suture zones have not, as yet, been recognized in the Abitibi orogen and the above chemical and structural similarities to modern plate-tectonic processes may be more apparent than real.

The Noranda district. In the Noranda area, Spence and De Rosen-Spence (1975; see also Spence, 1967) have recognized five felsic phases alternating with andesites which form distinct lithostratigraphic bands concentrically disposed about the Lake Dufault and Flavrian Lake granodiorite plutons (Fig. 7). The third of these felsic phases, the Mine Series, is host to most of the massive Cu–Zn sulfide ores, although relationships among the various rhyolite units are obscure south of the Horne Creek Fault where the important Horne deposits occur. As well, some sulfide deposits are found in the overlying Porphyritic Series (e.g., Delbridge and West Macdonald), but these deposits are quite small and are somewhat different from the "normal" Noranda-type in that they are pyritic zinc deposits without economic copper mineralization. In the fifth (Clericy) phase there is a massive pyrite body (Mobrun) containing low-grade zinc and copper. The majority of massive Cu–Zn sulfide deposits, details of which are presented below, lie on or near the contact between either the Waite rhyolite (e.g., Vauze, Norbec, East Waite, Old Waite), or the Amulet rhyolite (e.g., Amulet F, C, and A, and Millenbach) and the overlying Amulet andesite (Fig. 7). However, Spence (1967) and Spence and De Rosen-Spence (1975) believe that the Waite rhyolite, which pinches out south of the Waite mines, correlates with the Amulet rhyolite further south which, if correct, places most of the Noranda deposits at the same stratigraphic horizon. This rhyolite/andesite contact is

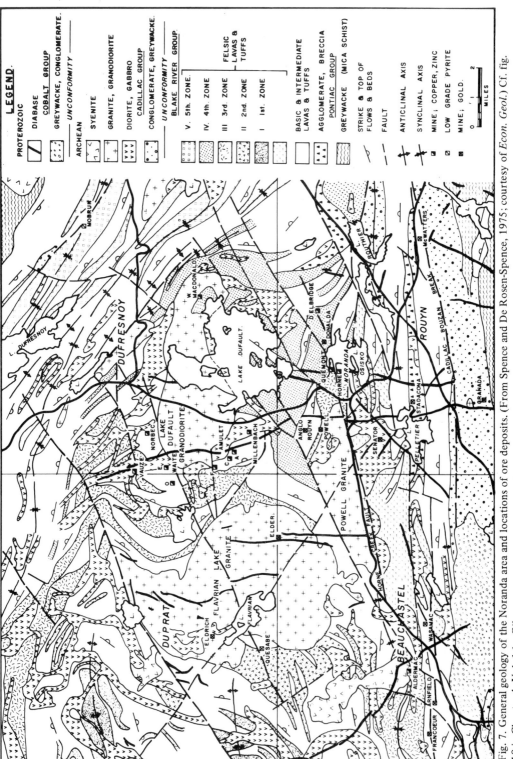

Fig. 7. General geology of the Noranda area and locations of ore deposits. (From Spence and De Rosen-Spence, 1975; courtesy of *Econ. Geol.*) Cf. fig. 10 in Chapter 4, Vol. 1 by Gilmour.

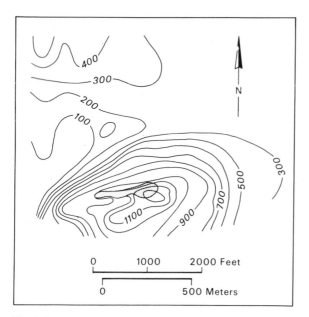

Fig. 8. Isopach map of the Waite rhyolite outlining a domal structure in the vicinity of the East Waite deposit (stippled), Noranda area. Vertical thickness is in ft. (After Spence and De Rosen-Spence, 1975; courtesy of *Econ. Geol.*) See fig. 11b in Chapter 4, Vol. 1, by Gilmour for additional information.

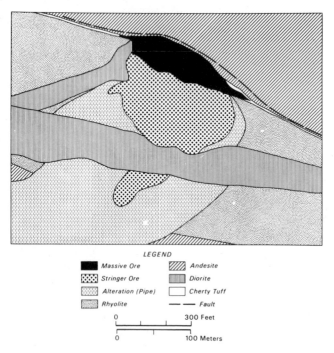

Fig. 9. Generalized east–west section, Norbec deposit, Noranda area. Stratigraphy is right-side-up as shown. (After Purdie, 1967; courtesy of Can. Inst. Min. Metall.)

sharp and, in the vicinity of the mines, is characterized by a thin but persistent cherty tuff, pyritic in part, which is used as a marker horizon in mineral exploration. This marker horizon dilates considerably near the orebodies.

The principle control on the distribution of orebodies on and near the favourable horizon, particularly in the northern part of the Noranda camp, is the formation of steep-sided domes of rhyolite and rhyolite breccia (Gilmour, 1965; Spence and De Rosen-Spence, 1975). Fig. 8, taken from Spence and De Rosen-Spence (1975), is an isopach map of the Waite rhyolite and shows clearly that the ore occurs on the flank of such a dome.

Representative sections of Noranda-area orebodies are shown in Figs. 9–13. The occurrence of massive ore at the sharp rhyolite (Mine Series)-andesite contact is evident at Norbec (Fig. 9), East Waite (Fig. 10) and Vauze (Fig. 11) but, as explained previously, Delbridge (Fig. 12) occurs stratigraphically above this favourable contact within the rhyolites of the Porphyritic Series. The position of Delbridge within a rhyolite dome is in

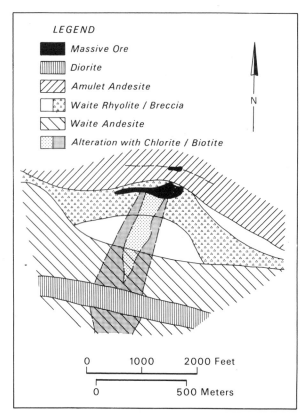

Fig. 10. Schematic east–west section of the East Waite deposit, Noranda area, showing the association of the sulfide body to a rhyolite dome and breccia. (After Spence and De Rosen-Spence, 1975; courtesy of *Econ. Geol.*)

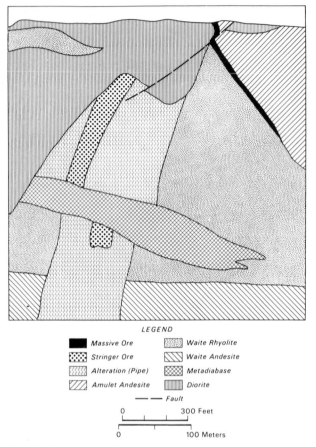

LEGEND

	Massive Ore		Waite Rhyolite
	Stringer Ore		Waite Andesite
	Alteration (Pipe)		Metadiabase
	Amulet Andesite		Diorite

— — Fault

0   300 Feet

0   100 Meters

Fig. 11. Vauze mine, Noranda area. North (left)—south section. (Slightly simplified from Spence, 1966, and Sullivan, 1968.)

parallel with that of Millenbach (Simmons and co-workers, 1973) where both the hanging and footwall are a quartz feldspar porphyry. Both deposits have the characteristic cherty tuff at the ore horizon (Figs. 12, 13) in common with Norbec and Vauze (Spence, 1975; not plotted in Fig. 11). The other common characteristic feature of these deposits is the conformable nature of the massive ore on the favourable contact underlain by a discordant cylindrical alteration "pipe" within which are disseminations and veins of sulfides ("stringer ore") which may or may not be of economic grade.

The stratiform lenses of massive ore commonly contain on the order of 80% sulfides (Van de Walle, 1972). The remainder consists of clasts up to one foot in diameter of rhyolite and highly chloritized rock and, near the top of the orebodies, cherty tuff either as layers or a pervasive matrix (Spence, 1975). Toward the hanging wall the massive ore, consisting of alternating bands of pyrite and sphalerite with lesser amounts of chalcopy-

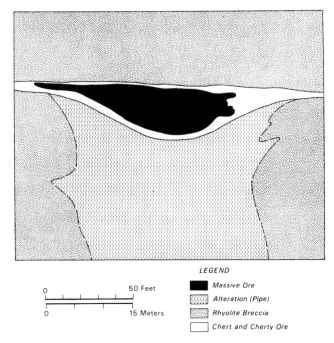

LEGEND

	Massive Ore
	Alteration (Pipe)
	Rhyolite Breccia
	Chert and Cherty Ore

Fig. 12. Delbridge deposit, Noranda area. Plan of 850-ft. level, Lens B. Stratigraphic top is to the top of the diagram. (After Boldy, 1968; courtesy of Can. Inst. Min. Metall.)

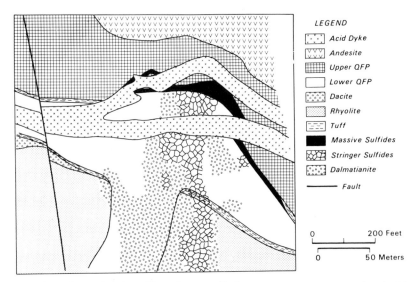

LEGEND

	Acid Dyke
	Andesite
	Upper QFP
	Lower QFP
	Dacite
	Rhyolite
	Tuff
	Massive Sulfides
	Stringer Sulfides
	Dalmatianite
——	Fault

Fig. 13. Millenbach mine, Noranda area. Dadson zone, east–west section looking north. (After Simmons et al., 1973; courtesy of Can. Inst. Min. Metall.)

rite, becomes increasingly intercalated upward and outwards with cherty tuff and is in sharp contact with the overlying andesite. This pyrite—sphalerite association grades downward to a pyrite—pyrrhotite—chalcopyrite association which is locally banded and may contain sporadic occurrences of magnetite. Thus, there is commonly a distinct stratigraphic vertical zoning in the massive sulfides from Cu-rich ore toward the footwall to Zn-rich ore toward the hanging wall. It is interesting to note, as pointed out by Spence and De Rosen-Spence (1975), that this zoning is also seen in the three different types of deposits in the 3rd (Cu-rich), 4th (Zn-rich) and 5th (predominantly pyrite) volcanic phases, reflecting a zoning with height in the succession and hence with time. In some cases, the distribution of the sulfide associations is concentric such as at the Amulet Lower A mine, Millenbach, or Norbec which contain a core of pyrrhotite—chalcopyrite ore surrounded by pyrite—sphalerite ore (Van de Walle, 1972; Simmons et al., 1973; Purdie, 1967). In others, such as the Horne and Delbridge mines, the two associations of sulfides are in separate bodies. The footwall contact of the massive ore is usually diffuse against the stringer ore in the alteration pipe, but is sharp against unaltered footwall rhyolite and rhyolite breccia.

In the footwall portion of the massive ore at the West Macdonald mine (Roscoe, 1965) and at the Vauze mine (Spence, 1975) there are zones of sulfide breccia. Large, angular blocks of sulfides of heterogeneous compositions generally resembling various ore types in the massive ore zone are chaotically mixed with rhyolite and chloritic rhyolite breccia blocks in a matrix of finely comminuted sulfides. Such breccias, found in volcanogenic massive sulfide ores of all geologic ages, have been variously interpreted as post-ore tectonic breccias, sulfide replacement of pre-ore breccias and, more recently, as products of explosive volcanism during mineralization (Hutchinson, 1965; Spence, 1975; Mannard, 1973).

The alteration pipe may extend more than 3,000 ft into the footwall rhyolites below the massive sulfide ore. The following description of the alteration pipe beneath the Vauze deposit (Fig. 11) is paraphrased from Gilmour (1965):

"The core of the alteration pipe is composed of massive chlorite containing disseminated sulphides and magnetite. The chlorite content of the rock decreases outward and the margin of the massive chlorite is gradational and poorly defined. There is an incomplete zone of massive sericite about the chloritic pipe. The chloritized and sericitized rhyolite grades into massive, siliceous rhyolite.

Within the tabular body of alteration, the disseminated sulphides occur as a shoot ... The margin of the core of intense alteration ... tends to be roughly defined as a zone which is made up of large blocks of bleached rhyolite enclosed by a matrix of chlorite and biotite."

Flin Flon and Snow Lake, Manitoba. The Flin Flon—Snow Lake region of northwestern Manitoba lies within the southern part of the Churchill Province in the Canadian Shield and comprises two, broad, parallel, east-trending belts of contrasting geologic character,

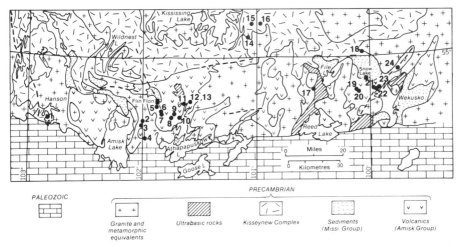

Fig. 14. General geology of the Hanson Lake–Flin Flon–Snow Lake mineral belt and locations of some massive sulfide deposits. Geology slightly simplified from Geol. Surv. Can. Map 1164A (Carrott River).

each containing distinctive massive sulfide ores (Fig. 14). To the south is the Flin Flon–Snow Lake greenstone belt of volcanic and sedimentary rocks which show a progressive increase in metamorphic grade from predominantly greenschist facies around Flin Flon in the west to lower almandine–amphibolite facies around Snow Lake 100 miles to the east. North of the greenstone belt are the middle to upper almandine–amphibolite metasedimentary rocks of the Kisseynew gneiss belt. These two belts are in faulted contact in the west, but in the vicinity of Snow Lake the contact is gradational across a zone of steep metamorphic gradients with little or no faulting (Bailes, 1971). The southern margin of the greenstone belt is covered by Paleozoic carbonates, so the true shape of the orogen is unknown.

The geology of this region has been described most recently by Bailes (1971) and Coats et al. (1972) from whose accounts the following descriptions were taken; they offer an interesting comparison with the Abitibi orogen. Bailes has devised a schematic tectonic model (Fig. 15) in which he envisages the Flin Flon–Snow Lake greenstone belt as an island arc with the adjacent Kisseynew gneiss belt as a subsiding trough which received sediments from erosion of the arc. The greenstone belt consists of a volcanic unit, predominantly flows, with intercalated volcani-clastic sediments (Amisk Group) overlain unconformably by a sedimentary unit (Missi Group). The base of the Amisk is a thick sequence of pillowed basalt flows with mafic to intermediate pyroclastic units and comagmatic intrusives. As volcanism proceeded its products became increasingly felsic, fragmental, and isolated into small extrusive centers, the two most important of which are loci of massive copper and zinc sulfide deposits at Flin Flon and Snow Lake. These centers are characterized by volcanic flows and pyroclastics of felsic composition, various

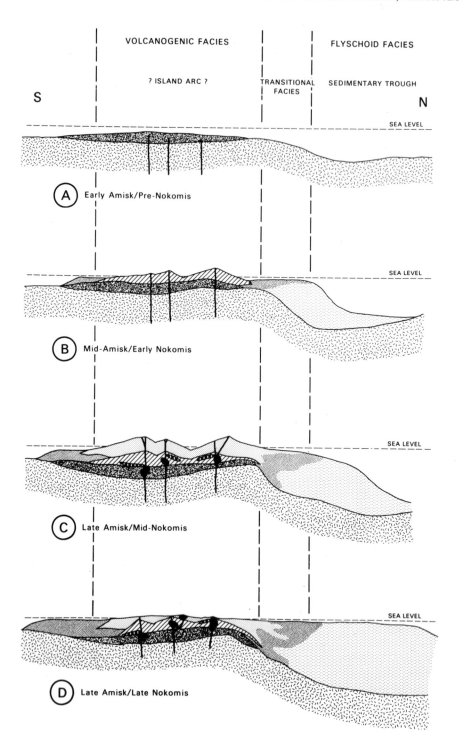

VOLCANOGENIC FACIES

? ISLAND ARC ?

FLYSCHOID FACIES

TRANSITIONAL FACIES

SEDIMENTARY TROUGH

S

N

SEA LEVEL

(A) Early Amisk/Pre-Nokomis

SEA LEVEL

(B) Mid-Amisk/Early Nokomis

SEA LEVEL

(C) Late Amisk/Mid-Nokomis

SEA LEVEL

(D) Late Amisk/Late Nokomis

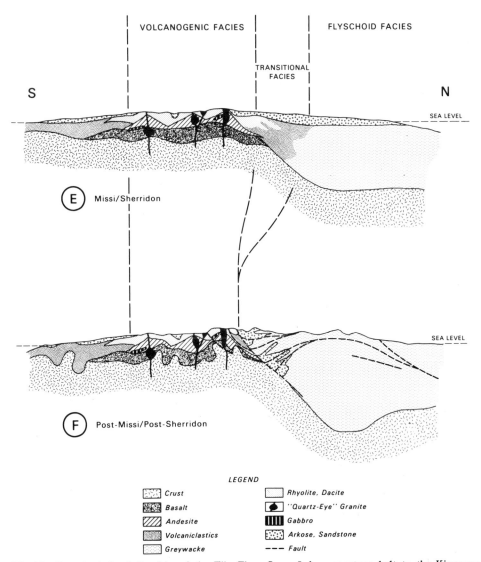

Fig. 15. Hypothetical relationship of the Flin Flon–Snow Lake greenstone belt to the Kisseynew sedimentary gneiss belt at various stages during their development. Sections are about 100 km long, with considerable vertical exaggeration. (Modified slightly from Bailes, 1971; courtesy of Manitoba Dept. of Mines, Petroleum and Resource Management.)

types of acid crystal tuffs, intrusive quartz porphyries, and siliceous, carbonate-rich tuffs. Associated with the felsic centers are thin volcani-clastic tuffaceous siltstones and grey-wackes and thicker, more widespread, turbidite greywackes and siltstones, all according to Bailes (1971) derived from the felsic pyroclastics. The Amisk is intruded by stocks and small irregular comagmatic (?) masses of "quartz-eye" porphyry and basic and ultrabasic rocks. The Missi Group, consisting of arkose, greywacke and quartzite, disconformably overlies the Amisk Group and post-Amisk intrusives and contains clasts of these units in a basal conglomerate. In turn, the Missi has been intruded by large granitic bodies up to batholithic proportions.

Traditionally, the Amisk and Missi Groups have been considered to be Archean (>2560 m.y.) and to have been deformed during both the Kenoran (2560 m.y.) and Hudsonian (1800 m.y.) orogenies. This concept is supported by their lithologic similarities with undoubted Archean rocks of the Superior Province, regional trends of major tectonic units and aeromagnetic patterns which parallel those of the Superior Province (Bailes, 1971), and Rb–Sr whole-rock analyses by Coleman (1970) from the Hanson Lake area, 40 miles west of Flin Flon, which gave ages of 2521 m.y. for Amisk-like metavolcanic rocks and 2446 m.y. for granitic rocks which intruded the metavolcanics. On the other hand, Mukherjee et al. (1971) have concluded on the basis of Rb–Sr whole-rock analyses and structural studies in the Flin Flon area that the Amisk is Aphebian (Proterozoic) in age, a view supported by Sangster (1972b) from single-stage model lead ages, and was subjected only to the Hudsonian orogeny which ended about 1800 m.y. ago.

The economic, massive sulfide ores of the Flin Flon–Snow Lake region consist predominantly of pyrite and/or pyrrhotite with varying proportions of chalcopyrite, sphalerite and galena and recoverable amounts of cadmium, silver, gold, selenium and tellurium. Galena contents are characteristically low in most orebodies; notable exceptions are the Hanson Lake and Chisel Lake deposits which are predominantly Pb–Zn in composition. The original shapes of the orebodies are difficult to ascertain. In the Snow Lake area they parallel the plunge of prominent lineations, resulting in pencil-like orebodies as, for example, described by Coats et al. (1970) for the Stall Lake (Rod) deposit. Nevertheless, where deformation is not too severe, the massive portions of the orebodies can be shown to be conformable with the enclosing rocks.

The deposits occur in various volcanic lithologies of the Amisk Group, commonly at the contact between andesite and quartz porphyry, but some such as the Flexar and Coronation mines are entirely enclosed by basic to intermediate flow rocks. The Flin Flon deposit, largest in the area, is typical and has been described by Byers et al. (1965), Coats et al. (1972), and Koo (1973).

Between 1930 when production started and 1970, the Flin Flon mine has produced 59 million tons of ore averaging 2.2% Cu and 4.4% Zn. Six lenticular orebodies are localized by subsidiary folds on a major anticline and extend en echelon for 5500 ft along the plunge of the anticline. The ores show the expected relationship to the volcanic stratigraphy, the host rock being a quartz porphyry which is bounded on the stratigraphic hanging wall by andesite flows and on the footwall by highly altered and sheared andesitic pyroclastics (Fig. 16). This extensive alteration which consists predominantly of chlorite with lesser amounts of talc, sericite and carbonate represents a typical alteration pipe beneath the massive sulfide ore. It is confined to the footwall pyroclastics and the quartz porphyry and does not penetrate the overlying andesites.

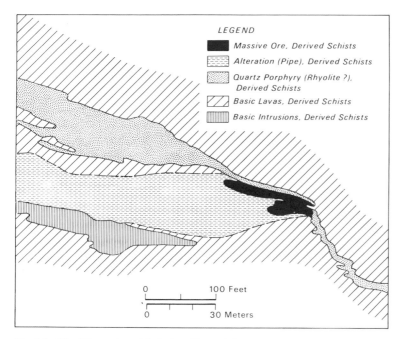

Fig. 16. Flin Flon mine, Manitoba. Plan of 3000-ft. level. Stratigraphic top is to the top of the diagram. (After Byers et al., 1965; courtesy of Saskatchewan Dept. of Mineral Resources.)

There are two types of ore: massive, constituting 70% of the total, and disseminated or stringer ore. The massive ore is predominantly fine-grained pyrite with varying proportions of chalcopyrite and sphalerite. The distribution of chalcopyrite is irregular, but it is more abundant towards the footwall. On the hanging wall, sphalerite and pyrite layers form a banded ore. Gold and silver occur largely as tellurides and selenides.

The disseminated or stringer ore forms irregular orebodies within the schists of the alteration pipe beneath the massive sulfide ore and consists of chalcopyrite in irregular veinlets and pyrite grains within the foliation planes of the schist. Gold, silver and zinc are relatively less abundant than in the overlying massive ore.

The deposits within volcanic rocks at Snow Lake, where metamorphic grades are much higher, are similar to the Flin Flon deposits in their sulfide mineralogy and geologic setting. However, they tend to be richer in copper and poorer in zinc as shown in Table III (p. 179), for example, by the Stall Lake mine and the Rod deposit No. 2, although there are exceptions such as Chisel Lake. The remainder of the differences, such as ore textures and gangue mineralogy, can be ascribed to the difference in grade of regional metamorphism to which the ores were subjected (Coats et al., 1970). The ores show distinct porphyroblastic growth of pyrite and arsenopyrite and, in places, have a well-developed cataclastic, "flow-banded" texture. Spectacular development of coarse-grained gangue silicates are not uncommon as, for example, at the Anderson Lake mine where huge blades of kyanite are found in the chlorite-sericite schists of the footwall and staurolite, garnet, and cordierite are found in the wall rocks and in the ore.

Jerome, Arizona. The United Verde, United Verde Extension and Haynes deposits near Jerome, Arizona (Fig. 2) have many similarities to the massive Cu-Zn sulfide ores of the Precambrian Shield in Canada. In fact, the similarities are so striking that they led Anderson and Nash (1972) to review an earlier comprehensive account of the Jerome deposits by Anderson and Creasy (1958) and to re-examine the critical field relations.

The massive sulfide ores occur as concordant strata-bound lenses in felsic tuffs of the Upper Precambrian Ash Creek Group. The Ash Creek Group is a eugeosynclinical accumulation in excess of 20,000 ft containing two cycles of the familiar basalt to rhyolite association in which the more mafic compositions predominate at the base and the felsic compositions at the top. The volcanic flows, tuffs and breccias are intercalated near the top of the pile with volcani-clastic sediments of probable turbidity-current deposition and with ferruginous chert beds.

The Ash Creek Group was intruded by concordant gabbros, thought by Anderson and Creasy (1958) to be older than the ore, but by Anderson and Nash (1972) to be younger, and by later quartz diorite plutons, dated by zircons at 1760 to 1770 m.y. (Anderson et al., 1971). Deformation of the area was mild producing a series of open folds and metamorphism was within the greenschist facies.

The ores are found over a narrow stratigraphic interval within a massive quartz crystal tuff (Cleopatra Member of the Deception rhyolite) and in the basal portion of an overlying, fine-grained, bedded tuff (Grapevine Gulch Formation). Zircons from the Cleopatra Member define an apparent age of 1820 m.y. (Anderson et al., 1971). Anderson and Creasy (1958) earlier interpreted the Cleopatra Member as a sill, but Anderson and Nash (1972) concluded that it is an "intertonguing complex of flow-banded porphyry, breccias and crystal tuffs" (p. 847). They envisage submarine eruption of viscous rhyolite magma, sometimes violent enough to produce coarse breccias. Volcani-clastic material moved downslope as subaqueous pyroclastic flows on the flanks of volcanoes and was redistributed in part by turbidity currents. In direct conflict with the submarine origin of the Cleopatra Member is the evidence cited by Bain (1973) for its subaerial accumulation. Certainly, the correct identification and provenance of this unit is most important in determining the genesis of its contained ores.

The ores of the United Verde mine (Anderson and Nash, 1972) consist of an elongated lens or pipe of massive sulfides underlain by a highly chloritized and mineralized alteration zone ("black schist") reminiscent of the alteration pipes at Noranda (Fig. 17). The ores are overlain by lenses of white quartz which Anderson and Nash interpret to be recrystallized chert. Pyrite with quartz and carbonates are the predominant constituents of the massive ore with minor quantities of chalcopyrite and sphalerite. Copper-rich ore shoots of chalcopyrite with minor sphalerite occupying fractures in pyrite are found in the footwall portion of the massive ore and extend with lower grades for a short distance

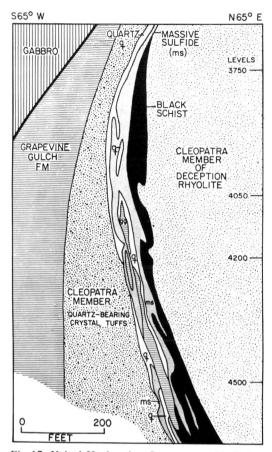

Fig. 17. United Verde mine, Jerome area. Vertical cross-section through the North orebody. Stratigraphic top is to the left. Vertical scale same as horizontal. (From Anderson and Nash, 1972; courtesy of *Econ. Geol.*)

into the black schist and weakly chloritized tuff. Sphalerite is sporadically interbanded with pyrite between the copper-rich ore and the hanging wall. The black schist represents a thorough iron—magnesium metasomatism of the quartz crystal tuff producing an almost pure chlorite rock up to 50 ft into the footwall. A zone of partial chloritization extends for an additional 1000 ft beyond the black schist. Black-schist ore consist of branching and intersecting veinlets of chalcopyrite and pyrite which merge with the higher-grade copper shoots in the footwall of the massive ore. The more weakly chloritized quartz porphyry forms small low-grade ore shoots of chalcopyrite veinlets with chlorite selvages.

Ores in volcano—sedimentary ("mixed") rocks

Manitouwadge, Ontario. The massive sulfide ores of the Manitouwadge district, north of Lake Superior (28—31 in Fig. 1), occur in a mixed sedimentary—volcanic sequence. As

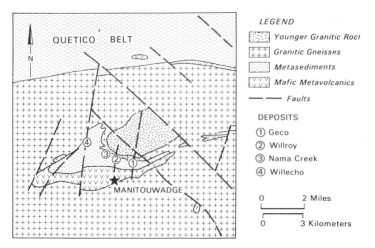

Fig. 18. Generalized regional geology of the Manitouwadge area, Ontario. (Geology simplified from Ont. Dept. of Mines Maps 1957-8 by E.G. Pye and P 494 V.G. Milne, and from Milne, 1969.)

such they display some features of the truly volcanogenic ores as well as some important differences. The geology and mineral deposits of the area have been described by Timms and Marshall (1959), Brown and Bray (1960), Suffel et al. (1971), and also by Milne (1969) from whom the following account is taken.

The deposits occur in a small (2 X 6 mile), Archean greenstone remnant within a batholith predominantly of biotite trondhjemite gneiss (Fig. 18). The metasedimentary and metavolcanic rocks have been folded into an east-northeast plunging synform cored

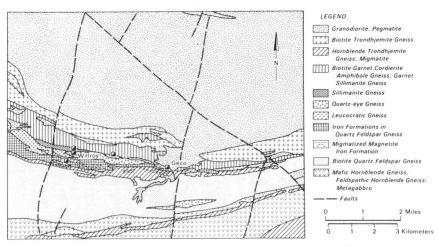

Fig. 19. Geology of the south limb of the Manitouwadge synform. (Simplified from Milne, 1969; courtesy of V.G. Milne.)

by granodiorite. Metamorphic grade is uniformly high, within the almandine–amphibolite facies.

The stratigraphic succession on the south limb of the synform, shown in Fig. 19, is relatively uncomplicated when one "looks through" the intense metamorphism to the probable original rock types. Assuming, in the absence of conventional facing indicators (see p. 164), that the synform is in fact a syncline, the succession consists essentially of a lower metabasaltic unit structurally overlain by clastic metasediments, various types of lean iron formations (or ferruginous metasediments) and extrusive (?) quartz-eye metaporphyry which are, in turn, overlain by mafic metatuffs (?) (now predominantly biotite–garnet–cordierite–amphibole gneiss). Finally, there were intrusions of biotite–trondhjemite gneisses (migmatites), granodiorites, biotite–muscovite pegmatites and later diabase dikes. The sulfide orebodies are strata-bound structurally below the biotite–garnet–cordierite–amphibole gneiss horizon. This succession is not totally unlike that of the Abitibi or Flin Flon–Snow Lake regions, the major differences at Manitouwadge being the relatively high proportion of sediments and low proportion of felsic volcanics and the position of iron-formation apparently structurally beneath the sulfide ore.

In detail, the lower metabasaltic unit contains up to 3300 ft of tholeiitic pillow basalts with mafic garnetiferous interflow tuffs, a metagabbro intrusive and a 3–4 ft thick magnetite–chert iron-formation. This metabasaltic unit is overlain by a 400-ft heterogeneous feldspathic hornblende gneiss which probably represents a mafic tuff zone. Overlying this is a 1/2-mile thick succession of fine-grained, well-bedded, clastic metasediments consisting of garnetiferous biotite gneiss and biotite–quartz–feldspar gneiss. Within the overlying zone of iron-formations there are both oxide and sulfide (pyrite) facies contained within quartz–feldspar gneisses. The gneisses are predominantly fine-grained clastic metasediments with a few beds of coarse fragmental material. For the most part, the oxide facies iron-formation is a lean, thinly-bedded, garnet–amphibolite–magnetite–quartz unit. The sulfide facies iron-formation is represented by a finely-laminated, rusty-weathering, pyrite–garnet–magnetite–hornblende–biotite–quartz–feldspar gneiss which grades laterally into oxide-facies iron-formation. Within the iron-formation zone there is a lenticular biotite–sillimanite–quartz gneiss which Milne (1969) believes was originally a kaolinitic or bentonitic chert. Structurally above the iron-formation is a white-weathering, leucocratic, laminated, magnetite–garnet–biotite–quartz feldspar gneiss which is similar to the rusty-weathering gneiss, but has a lower iron and sulfide and higher silica content. This clastic metasedimentary unit is particularly significant for it contains at two stratigraphic horizons, quartz-eye gneisses which have textural and chemical evidence (Milne, 1969) of having been extrusive or high-level intrusive quartz porphyries. These quartz-eye gneisses are closely associated with several of the Willroy orebodies (Fig. 19). The uppermost unit, structurally overlying the massive sulfide ores, is composed of lenticular masses of biotite–garnet–cordierite–anthopyllite gneiss within garnet–biotite ± sillimanite gneiss. Lenses of hornblende gneiss, similar in appearance to the lowermost metabasalts, are also present. The origin of this uppermost unit is enigmatic. Milne (personal communication, 1974) believes that it may be an altered mafic tuff with some interlensing of basalt flows and could, therefore, represent a slightly more metamorphosed equivalent of the lower metabasalts if, as suggested by Suffel et al. (1971), the sequence is overturned in a refolded syncline.

There were several major intrusive episodes in the core of the synform and in nearby volcanic and sedimentary units. Significant among these are hornblende–trondhjemite migmatite which is particularly well-developed in the gneisses of the magnetite–chert iron-formation and forms much of the "grey gneiss" wall rock south of the Geco orebody; a 1000 ft wide biotite–trondhjemite migmatitic gneiss on the north side of the biotite–garnet–cordierite–amphibolite gneiss; late or post-tectonic granodiorites occupying the core of the synform, a large protrusion of which broke through at the Geco mine (Fig. 19); "metadiorite" dikes which cut ore-bearing strata and are found as disrupted blocks in massive ore; and abundant pegmatite which cuts all units in proximity to the granodiorite except massive sulfide ore.

There are four mines in the Manitouwadge district, two of which (Willroy and Geco) are within the map area of Fig. 19. Willroy consists of six separate massive or disseminated orebodies whereas Geco is one continuous unit.

Geco orebody. The Geco orebody is a pinching and swelling, vertically-dipping body of massive sulfides partially enclosed by an envelope of disseminated sulfides. Representative vertical and horizontal sections are shown in Figs. 20 and 21. Most of the orebody lies within a zone of muscovite–quartz schist between migmatized iron-formations ("grey gneiss") to the south and biotite–garnet ± cordierite ± anthophyllite gneisses to the north although, at one locality in the upper part of the mine, massive ore transgresses the muscovitized zone into iron-formation (Fig. 20). Milne (1969) has interpreted the muscovite–quartz schist as a major fault zone which extends across the entire map area of Fig. 19. Although not denying that some movement took place in this zone, we disagree with the fault interpretation and believe, instead, that the zone represents a large-scale boudin, a view that we will develop further in the section on "Metamorphic effects" (p. 187).

The massive ore is composed principally of pyrite, sphalerite, pyrrhotite and chalcopyrite in order of abundance (see Table I). Inclusions of silicate country rocks composed of "metadiorite", gneiss, schist, and pegmatite (Suffel et al., 1971) ranging in size from large blocks to small fragments, and centimeter-sized quartz eyes constitute about 25% of the massive orebody. In the thicker parts of the orebody, the sulfides are very coarse grained ranging in texture from tightly packed, coarse pyrite grains with interstitial black sphalerite and peppered with blobs and wispy stringers of pyrrhotite and chalcopyrite to coarse-grained, black sphalerite containing scattered grains of coarse, angular pyrite. Pyrite–sphalerite banding is well developed in places. Where the massive ore pinches down to one or two inches, it is Cu-rich ("chalcopyrite vein" in local mine terminology) and consists of very fine-grained pyrrhotite with a few tiny rounded pyrite grains and cut by wispy chalcopyrite stringers.

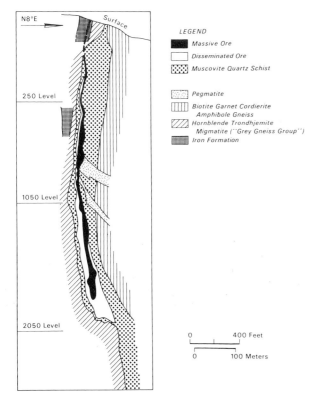

Fig. 20. Geco mine, Manitouwadge. Vertical section. (After Milne, 1969; courtesy of V.G. Milne.)

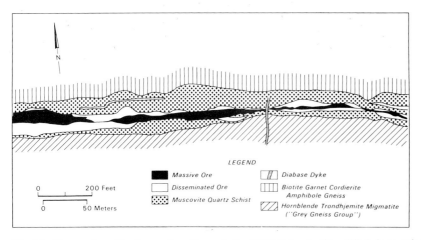

Fig. 21. Geco mine, Manitouwadge. Plan of the 1200-ft level. (Simplified from mine plans of R. Weeks, Chief Geologist Geco mine, and staff.)

TABLE I

Major sulfide mineral content of selected Precambrian massive Cu—Zn sulfide ores (as % of total sulfides)

	Pyrite	Pyr-rhotite	Sphalerite	Chalco-pyrite	Reference
Geco	50	17	20	10	Milne (1969)
Lake Dufault	39	14	26	20	Purdie (1967)
Little Stall Lake (Rod)	30	12	15	40	Coats et al. (1970)
Coronation	25	13	—	62	Whitmore (1969)
Kidd Creek	56	2	31	10	Matulich et al. (1974)

The disseminated ore does not enclose the entire massive orebody and often occurs in schistose zones separated by several feet from the massive ore. The disseminated ore minerals to the north, structurally above the massive sulfide ore, and to the south in the deep "keel" area, are predominantly chalcopyrite and pyrrhotite with pyrite and particularly sphalerite much less important. On the other hand, south of the massive ore near the muscovite schist—"grey gneiss" contact there is an intermittent disseminated zinc zone (Ross Weeks, chief geologist Geco mine, personal communication, 1974).

Although sphalerite-rich and chalcopyrite-rich areas occur in the massive ore, their distribution does not unfold a simple pattern of metal zoning. However, the disseminated ore displays a crude copper-to-zinc zoning from what is thought to be hanging wall to footwall, the reverse of that commonly found at Noranda, Flin Flon and elsewhere. On the basis of this apparently anomalous zoning, coupled with the occurrence to the south of the iron-formation (which usually *overlies* massive sulfides) and the north-to-south succession of mafic to felsic gneisses at Geco and Willroy, Suffel et al. (1971) have speculated that the commonly accepted stratigraphy presented by Milne (1969) is upside-down and that the Manitouwadge synform is overturned, a concept we strongly endorse.

Ores in predominantly sedimentary rocks

Sherridon and Bob Lake mines, Manitoba. The Sherridon and Bob Lake massive sulfide deposits, located 40 miles northeast of Flin Flon (14 and 15 in Fig. 14), lie within the Kisseynew gneiss belt whose rocks were derived mainly by regional metamorphism and granitization of sedimentary rocks.

Relationships within the Kisseynew gneiss belt are difficult to sort out because of the complex deformation and the intense regional metamorphism which has transformed large proportions of the paragneisses into migmatites and granitoids. In spite of these difficulties, Bailes (1971) was able to subdivide the Kisseynew into the Nokomis and

Sherridon groups and a "basement group" of granitized gneisses, which is now thought to be part of the Sherridon Group (Bailes, personal communication, 1974). The Nokomis is a thick monotonous sequence of fine-grained plagioclase + quartz + biotite + garnet ± staurolite ± sillimanite ± cordierite ± graphite gneisses derived from argillites and grey-wackes. The Sherridon is another monotonous sequence, this time of fine- to medium-grained quartz + plagioclase + potash feldspar + biotite ± hornblende ± sillimanite ± magnetite ± garnet gneisses derived from arkoses and quartzites. The two groups represent contrasting environments of sedimentary deposition (Robertson, 1953; Bailes, 1971). Turbidites of the Nokomis Group are indicative of a deep-water environment with an adjacent elevated land mass. The fine laminations of the Sherridon Group are indicative of a shallow-water and/or fluvial-deltaic environment. The economically important unit within the Kisseynew is a hornblende–garnet–plagioclase gneiss which occurs as a con-formable, thin (commonly less than 50 ft thick; Bailes, 1971) layer at the Nokomis–Sherridon contact and is host to the Sherridon and nearby Bob Lake orebodies. Bateman (1944) interpreted this unit to be a basic flow, but more recent field (Robertson, 1953) and geochemical evidence (Orville, 1969; Bailes, 1971) strongly suggest that these gneisses are the product of metamorphic reaction between calcareous and pelitic sedimentary horizons. The sedimentary environment envisaged by Robertson (1953) for the precur-sors to the hornblende–garnet–plagioclase gneiss is shallow water with limited influx of detrital material and was stable for a significant time interval prior to deposition of the Sherridon Group.

A post-Sherridon tectonic event produced the intense metamorphism of the Nokomis and Sherridon groups and emplacement of syntectonic granites and other intrusives.

The Sherridon and Bob Lake deposits have been described by Farley (1949). They consist of massive pyrite, pyrrhotite, sphalerite, chalcopyrite ore in folded clastic meta-sedimentary rocks intruded by felsic plutons and pegmatites. The footwall of the Sher-ridon deposit is comprised of metamorphic derivatives of quartzite, arkose and grey-wacke. The hanging-wall rock is the basic hornblende–garnet (–plagioclase) gneiss con-sidered, above, to be a metamorphic equivalent of calcareous pelitic layers.

The Sherridon deposit consists of two orebodies, the east and west zones, both of which are conformable with the enclosing rocks. The east zone has an average width of 15 ft, a strike length of 4300 ft and a maximum down-dip extent of 250 ft. Similarly, the west-zone dimensions were 15 X 7900 X 1500 ft. Average grade of the east zone was 2.45% Cu, 2.97% Zn, 0.018 oz/ton Au and 0.58 oz/ton Ag (Farley, 1949).

Compared with deposits of "purely" volcanic affiliation, the Sherridon orebodies are thinner, stratiform, and more blanket-like in their morphology. In this regard they mimic the layered, laterally-continuous nature of their host rock metasediments. However, the metal ratios of the Sherridon and Bob Lake deposits are so unlike those hosted in argillites and carbonates (see Fig. 26C) and are so similar to those of volcanic or volcano-sedimentary environments that they may be reflecting their spatial linkage to the adjacent Flin Flon–Snow Lake volcanic belt to the south. In other words, even though the host rocks to the Sherridon and Bob Lake deposits are essentially sedimentary, the loci of deposition may not have been sufficiently distant from the volcanism to the south to establish the typical Pb- and Zn-rich, Cu-poor type of deposit which is typical of sedimentary environments not linked to an adjacent volcanic belt. The Sherridon and Bob Lake deposits are examples illustrating the complete gradation between end-members of the division of deposit types that we have adopted (p. 129). In other words, these deposits appear to be of volcanic composition occurring in sedimentary rocks adjacent to a contemporaneous volcanic belt.

Sullivan mine, British Columbia[1]. The Sullivan mine located at Kimberley in southern British Columbia has produced more than 100 million tons of lead, zinc and silver ore making it historically one of the largest lead—zinc deposits in the world. The deposit occurs on the east side of the Purcell anticlinorium in fine-grained clastic sediments of the Middle Proterozoic Purcell Group, a 35,000 ft thick sequence interpreted by Price (1964) as a westward prograding continental terrace wedge.

The lowermost Purcell, the Aldridge Formation, is at least 15,000 ft thick and has been subdivided by Leech (1957) into three members, the ore occurring at the transition between the Lower and Middle Aldridge (Morris, 1972). Extensive sills and dykes of Moyie intrusions (Purcell diorite) intrude the entire Aldridge (Freeze, 1966) and have been dated by Hunt (1961) at 1500—1100 m.y. by K-Ar methods. Volcanic rocks are distinctly absent from the Lower Aldridge. Regional metamorphism was low, within the greenschist facies, but nevertheless was intense enough to recrystallize completely the finest grain sizes in all rocks and mask some interrelations among sedimentation, mineralization and alteration. The coarse fraction does not appear to have been affected, so primary sedimentary structures are preserved.

The deposit has been described by several authors (Swanson and Gunning, 1945; Carswell, 1961; Leech and Wanless, 1962; Freeze, 1966) but the following brief description is taken largely from Morris (1972). Fig. 22 is a generalized cross-section of the deposit. The orebody occurs in a single generally conformable zone between 200 and 300 ft thick. The footwall rocks comprise a rhythmic succession of grey-green, rusty weathering, thin, graded beds ranging from fine-grained, impure quartzite to argillite. The immediate footwall is either these regularly bedded sediments or a large lens of intraformational conglomerate which is widespread in the mine area. Other lenses of similar conglomerate are interbedded with ore in one part of the mine and are known in the immediate hanging wall. The rocks above the ore zone (Middle Aldridge) are very similar to the Lower Aldridge, but are more thickly bedded and more arenaceous. Interbedded siltstone and argillites give way to the upper quartzites 100 to 150 ft above the ore zone.

The sulfide mineralogy is principally pyrrhotite, sphalerite, galena and pyrite with

---

[1] See also Chapter 2 by Thomson and Panteleyev (Vol. 5) for their discussion on the origin of the Sullivan orebody.

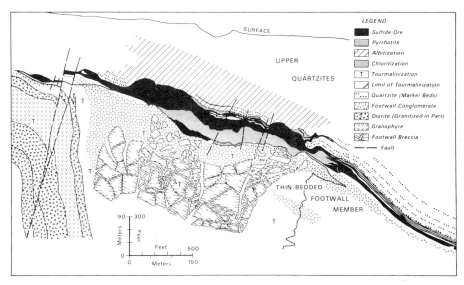

Fig. 22. Sullivan mine, British Columbia. Geological cross-section. (After Morris, 1972.)

minor chalcopyrite and arsenopyrite. Magnetite is common in places. Cassiterite is wide-spread in small amounts and is recovered commercially. Nonmetallic gangue minerals are common and constitute the typical mineralogy of a metamorphosed arenaceous sedimen-tary rock. Quartz and sericite are prominent together with substantial amounts of albite, biotite, spessartite, chlorite and calcite and local concentration of tremolite, clinozoisite, sphene and microcline.

The orebody has been arbitrarily divided by company geologists on the basis of dis-tinctive lithologies of ore into an upper mine in the central and western portions and a lower mine to the east. In the lower mine, the ore is composed of fine-grained sulfides in distinct, continuous layers from less than an inch to more than a foot thick. The stratigra-phy is extremely regular and consists of interbanded sulfides and metasediments within both of which are remarkably preserved, primary sedimentary structures. The upper mine consists of recrystallized coarse-grained, massive ore within which the delicate banding characteristic of the lower mine ore is preserved only as small remnant patches. Large zones of essentially pure iron sulfides (pyrrhotite and pyrite) with little or no galena or sphalerite, the "iron zone", form a core to these ores. According to Leech and Wanless (1962), there is a roughly symmetrical zonal distribution of metals about the central iron zone. They cite as an example the decrease in the Pb/Zn ratio from the central iron core towards the periphery of the deposit. Within the upper mine is found the strongest alteration as well as thin sulfide zones which occur in narrow shears extending as far as 100 ft both above and below the main ore zone. One such transgressive structure is enriched in cassiterite.

Beneath the central part of the orebody there are two distinctive, overlapping features which are highly discordant to the footwall stratigraphy and form a roughly funnel-shaped structure at right angles to the massive ore. These are zones of brecciation and tourmalinization which together constitute a feature being similar in its morphology and its relation to the massive ore to the alteration pipes beneath the massive sulfide ores of volcanic association. Both are confined to the footwall and are known to extend in depth for several hundred feet. The breccia is in irregular discontinuous bodies. In places, the matrix has been heavily mineralized by pyrrhotite and, in rare instances, by galena and sphalerite. Tourmalinization is spread over a wider area than the breccia. Alteration was so complete that the original argillites are now an anastomosing network of tourmaline needles which penetrated even the margins of relict quartz grains. The tourmalinized rocks have sharp contacts against sulfide ore, but interfinger and have diffuse contacts with unaltered footwall sediments. Besides tourmalinization there is a pervasive chloritic alteration around the sulfide zone in both the footwall and hanging wall as well as adjacent to the footwall intrusive diorite, and an albitic alteration which is confined to the hanging wall. In places, the albitic alteration is so intense that all sedimentary features and marker horizons are lost.

GENERAL GEOLOGICAL FEATURES

Nature and depositional environment of host rocks

A majority of the Precambrian massive sulfide ores occur within a volcanic complex several tens of thousands of feet thick comprised of three main volcanic lithologies (Goodwin, 1968). Typically, the first and lowermost portion of the complex is a thick series of pillowed and vesiculated flows characteristically basaltic. This is followed upward by flows, flow breccias and tuffs of mainly andesitic composition. The third and uppermost portion of the complex contains abundant — and is sometimes predominantly composed of — volcanics of dacitic to rhyolitic composition which may be either massive, textureless flows and/or pyroclastics of various size ranges. The acidic phase usually marks the end of a single volcanic cycle after which another cycle may begin, in which case the succeeding rocks are the typical basal basalts, or the volcanism may cease. In the latter case, the complex becomes surrounded or mantled in a thick sequence of grey-wacke-type sediments derived partly or entirely from the volcanics themselves. In some instances, the waning phases of volcanism appear to be contemporaneous with sedimentation and, in such cases, the rocks may grade imperceptibly, both laterally and vertically, from "true" tuffs and fine pyroclastics, through reworked tuffs and volcanic sediments, into greywackes containing various amounts of volcanic component, and finally into "true" sediments in which the volcanic component is lacking or minimal.

It is also characteristic of the ore-containing volcanic piles that they are intruded by rocks of diverse composition such that, as a whole, they exhibit as much differentiation as the extrusives themselves. Hence, within many regions of massive sulfide deposits it is not uncommon to find major intrusions of ultramafics, gabbro, diorite, granodiorite, and even granite. Felsic intrusions are usually found as large stocks at or near the centre of the volcanic pile. As such, they may represent part of the magmatic hearth which originally spawned the volcanism and later moved upward to intrude its own daughter products. The Matagami area (no. 1–4 in Fig. 1) is somewhat anomalous in that, instead of an acidic pluton in the volcanic core, it is centred around an elongate intrusion of layered gabbroic rocks (Sharpe, 1968).

The most common host rocks in the immediate vicinity of most orebodies are the acidic, usually clastic, phases of volcanism. Agglomerates, coarse tuffs, or occasionally, massive dacitic to rhyodacitic flows constitute the ore horizon of many massive sulfide bodies. Ore may occur entirely within the acidic portion, or at the contact between dacite (rhyolite) and andesite, or, less commonly, entirely within the andesitic portion of the pile. The volcanic pile (and volcanogenic sediments) is diagrammatically illustrated by Fig. 23. In multicyclic volcanic accumulations, ore formation often occurs in association with only one of the acidic phases; similar rocks in other phases of the same pile will be barren. Even within one volcanic cycle, the acidic portion sometimes consists of several rhyolite— andesite phases, only one of which contains important ore as, for example, at Noranda where Spence and De Rosen-Spence (1975) have recognized five acidic phases of volcanism but significant sulfide ores are associated with only one of these.

The close spatial association between acid agglomerate (or coarse pyroclastics) and massive sulfide ores is a characteristic feature of most established mining camps in the Precambrian. Commonly more than one-third of the volume of pyroclastics consists of block-size fragments or larger (greater than 64 mm), so the rock is properly classified as tuff-breccia or pyroclastic breccia in Fisher's (1966) scheme for pyroclastic rocks.

However, the term that we prefer to use is Sangster's (1972a) nonscientific "mill-

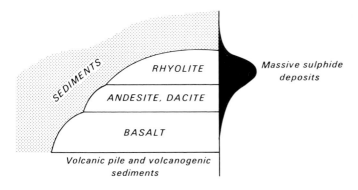

Fig. 23. Relative abundance of massive sulfide deposits in the volcanic and sedimentary components of a schematic Archean volcanic pile. (Modified from Goodwin, 1968.)

Fig. 24. "Mill-rock" in the Mine Series rhyolites, Noranda area.

rock" (Fig. 24), so named because he could usually hear a mine mill whenever he stood on such an outcrop! "Mill-rock" is generally found in, or close to (commonly stratigraphically above), the volcanic unit in which the massive sulfides occur. With diminution of fragment size, "mill-rock" grades laterally into lapilli-tuff or tuff. Although the method of fragmentation to produce "mill-rock" and its equivalents is reasonably clear (i.e., explosive volcanism), the method of deposition is not. The extremely coarse fragments may have been deposited directly after the explosion but the finer particles could have been air-borne, water-borne or borne by pyroclastic flows. The latter, generally referred to as ignimbrites when they occur sub-aerially (Marshall, 1935) have recently been suggested as occurring under water as well (Fiske, 1963). In our experience, "mill-rock" (the coarsest fragments), whatever its method of deposition, can be found within one half-mile of most Precambrian massive sulfide deposits. In addition to volcanic rock fragments, it is not uncommon to find fragments of sulfides also occurring in "mill-rock" (Rokachev, 1965). The sulfide clasts are either angular or subround (or both) and are generally composed predominantly of pyrite, although lesser amounts of other sulfides have also been observed (e.g., Sinclair, 1971; Spence, 1975). In some of the larger fragments sulfide banding can be discerned. The maximum size of the sulfide fragments in any exposure of "mill-rock" is generally slightly less than the largest fragments of volcanic rock. The pyroclastic origin of the larger sulfide fragments is reasonably clear, but with diminishing fragment size in the pyroclastic unit, sulfide *fragments* become increasingly difficult to distinguish from sulfide *grains* which may or may not be of pyroclastic origin.

In addition to "mill-rock", quartz porphyry is a common host to, or close associate of, these ores and in many areas the geologist has difficulty in deciding whether the porphyry is extrusive or intrusive. In the Flin Flon area, for example, massive lenses of quartz porphyry originally considered by Stockwell (1960) as extrusive, can now, with new exposures available, be traced along strike into pyroclastic units of the same lithology. It is likely that, near the top of a volcanic pile, shallow intrusions could "break through" and form local extrusive accumulations and result in both intrusive and extrusive equivalents of the same lithology.

The abundance of pillows in the basalts and andesites of the complex, the graded bedding in greywackes, and the frequent association with cherty iron-formation implies a submarine depositional environment for the Precambrian volcanics and sediments. Phanerozoic rocks, containing similar massive sulfide ores, commonly contain marine fossils in addition to the above features. The Precambrian host rocks are considered to have been deposited under submarine conditions, although a more complete appreciation of the depositional environment of Precambrian host rocks awaits comprehensive interpretation of their volcanic and sedimentary textures.

Compositionally, the volcanic host rocks appear to have been derived by differentiation of a tholeiitic parent magma and chemically resemble standard tholeiitic and calc-alkalic rocks of the basalt–andesite–rhyolite association (Goodwin, 1968, p. 75; Baragar and Goodwin, 1969; Irvine and Baragar, 1971). Phanerozoic rocks of this suite are typically identified with the island arc or eugeosynclinal environment. In such an environment, extrusion of calc-alkaline volcanics is frequently closely followed by orogenesis, thereby accounting for the almost universally metamorphosed nature of most Phanerozoic volcanogenic massive sulfide ores. Perhaps Precambrian ores were formed in a similar environment. In comparison to the Precambrian massive sulfide ores of volcanic association, those in truly sedimentary piles are numerically insignificant, but individually may be of large tonnage; e.g., the Sullivan orebody contains in excess of 175 million tons of ore. According to Stanton (1960), King (1965) and Mannard (1973), the most common host rocks of massive sulfide ores in sediments on a world-wide basis are shales and siltstones, commonly limy or dolomitic, of shallow to moderately-deep marine environment. The ores of the Kisseynew metasedimentary gneiss belt in Manitoba and the Sullivan deposit are compatible with this generalization, but the only other significant Precambrian massive sulfide deposits in North America in a truly sedimentary environment, Balmat and Strathcona Sound, are not compatible. The Sullivan orebody is in a sedimentary sequence which Bishop et al. (1970) have interpreted to be the initial infilling of a geosynclinal trough. The Balmat ores (no. 1—4 in Fig. 2), on the other hand, occur in pure and silicated dolomitic marbles, the precursors of which according to Solomon (1963; see also Lea and Dill, 1968) formed in shallow carbonate-precipitating basins of localized fetid bottom depressions. In fact, the carbonate host rocks and mineralogy of the Balmat ore (predominantly sphalerite and pyrite with minor galena, pyrrhotite and chalcopyrite) coupled with evidence that the ores took part in the Grenville metamorphism (Solomon, 1963; Brown, 1965) have suggested to some (e.g., Wedow et al., 1973, p. 704) that the Balmat ores are a metamorphosed Mississippi Valley type. The Strathcona Sound deposit occurs in a thick dolomite. In summary, unlike the Precambrian massive sulfide ores in volcanic environments, those few North American representatives in metasedimentary rocks do not conform to a single geological environment but, rather, are characterized by their diversity.

Rocks which are transitional between the "pure" volcanics and "pure" sediments, and variously referred to as volcanic sediments, sedimented volcanics, slightly reworked tuffs, "tuffwackes", etc., constitute the next most common host rock of massive sulfide ores after the acid volcanics. These rocks, as a rule, exhibit better, and laterally more persistent, layering than do the true volcanic rocks. They contain deposits which generally show more pronounced layering (bedding?) than do those in the massive volcanic rocks and have a tendency to be more blanket-like in form rather than ovoid. In all other respects, however, (size, grade, mineralogy, composition, etc.) the deposits are identical with those occurring in "pure" volcanic rocks. The layered host rocks, exhibiting more sedimentary features than volcanic, are best developed in the waning stages of volcanism and hence are usually found on top of, and peripheral to, the main volcanic pile. In spite of difficulties in correlating Precambrian volcanics and sediments, there is some evidence to show that sediments, correlative with an ore-bearing volcanic cycle (or pile), will themselves contain massive sulfide ores adjacent to the volcanic centre. A correlation of this type has been suggested by Bailes (1971) for the Flin Flon—Sherridon area of Manitoba.

Such a mixture of the two end-member environments is represented by the deposits of the Manitouwadge area (Fig. 19). The relationship to ores in volcanic rocks is enhanced by the observations that lean oxide and sulfide facies iron-formation possibly overlies (see p. 164) the Geco ore; several of the Willroy orebodies are in close proximity to quartz-eye gneiss which is thought to have been originally a quartz-eye porphyry; a former mafic volcanic unit underlies (?) many of the orebodies; and the bulk composition and mineralogy of the ores are similar to those in purely volcanic environments. On the other hand, felsic metavolcanics are absent at Geco; the orebodies (particularly Geco) are very much elongated or blanket-like as is Sullivan, although at Geco some of this elongation may be a consequence of structural deformation; and a 1/2 mile thick sequence of well-bedded clastic metasediments occurs within a few hundred feet of the ore.

Sulfide deposits

By far the majority of North American massive sulfide orebodies occur in volcanic or mixed volcanic–sedimentary rocks where the volcanic component is high. In their metal ratio, grade, habitat, and internal characteristics (including metal zoning), ores of the "mixed" environment more closely resemble those of the volcanic environment than those with "pure" sedimentary host rocks. Because of the review nature of this paper and the predominance of deposits in volcanic and "mixed" terranes, the general comments discussed in the following pages pertain to these deposits rather than the relatively few sediment-hosted ores.

Distribution. Examination of Figs. 1 and 2 shows that most of the deposits occur in clusters, particularly in the Superior and Churchill Provinces of the Canadian Shield, and that over half of the more than one hundred deposits shown are in six or seven main centres. The ores in volcanic environments, occurring as deposits of various sizes, are found clustered within roughly circular areas usually 10–20 miles in diameter. The oval or circular nature of these areas may be in part tectonic but, because the distribution of ores frequently coincides with that of felsic volcanic rocks, it may also be in part an original depositional feature. These circular areas, characterized by massive sulfide orebodies within or peripheral to a central area of acidic volcanism, are considered to be only a small part of larger volcanic complexes. For example, Goodwin and Ridler (1970) have proposed eleven volcanic complexes, each with its felsic volcanic component, within the Abitibi orogen of the eastern Superior Province (Fig. 4). Similarly, the Flin Flon and Snow Lake mineral areas of Manitoba (Fig. 14) are marked by local areas of acidic volcanic rocks with closely associated sulfide orebodies. The tendency to occur in swarms or clusters is so characteristic of this type of sulfide deposit that it is sometimes a major factor leading to re-exploration of volcanic belts containing only one or two known massive sulfide orebodies.

The near-universal post-ore metamorphism mentioned previously frequently precludes preservation of certain primary features of the sulfide masses. However, in a few areas such as Noranda and Matagami, metamorphism is minimal and a few observations on the original nature and distribution of the ores are possible. For example, it is often possible to demonstrate that deposits in any one area tend to occur within a fairly narrow stratigraphic interval or even along a single horizon. In the Noranda area, for example, significant ore occurs close to or at both the upper and lower contacts of only one of five rhyolite formations in the area (Spence, 1967; Spence and De Rosen-Spence, 1975). These acidic phases are separated by andesitic units and the major ore-producing contact between rhyolite and andesite is referred to as the "favourable horizon" (Sharpe, 1968). Considering the rapidity with which a lava pile is accumulated, it is apparent that the Noranda area massive sulfide ores formed virtually simultaneously within a near-circular area approximately 10 miles in diameter. Similarly, in the Matagami area, most of 13 sulfide orebodies are distributed along a distinctive and extensive stratigraphic zone which is "marked by concomitant accumulations of pyroclastic deposits, siliceous chemical sediments, and probably iron formation" (Sharpe, 1968, pp. 106–107). The "favourable horizon" in this area is traceable for 24 miles along strike around the periphery of a mafic complex which has intruded the volcanic pile. In the Kamiskotia area near Timmins, Pyke and Middleton (1971, p. 161) state " ... the known sulfide deposits occur mainly in felsic pyroclastic rocks of brecciated zones confined to a particular stratigraphic unit ...". In this case the favourable unit is traceable over a strike length of about ten miles.

Allied to the concept of a "favourable horizon" of ore deposition, is the relationship between massive sulfide deposits and the other forms of "exhalite". The most common forms of exhalite are chert and the four facies of iron-formation, namely oxide, carbonate, sulfide and silicate. The oxide facies of the iron-formation is not of universal occurrence being absent, for example, at Flin Flon and Snow Lake. In the Noranda and Matagami areas mentioned above, exhalite, in the form of chert and lean sulfide iron-formation, is the regionally stratigraphic equivalent of the economic base-metal sulfide bodies and as such constitutes a valuable marker in the search for favourable horizons. The three major forms of exhalite iron-formation, i.e., oxide, carbonate, and sulfide, are considered to represent a change of facies corresponding to a possible increase in water depth and change in Eh, i.e., there does appear to be a tendency for economic sulfide accumulations to occur toward the centre (deeper-water) portions of the basin as defined by the various facies of iron-formation (Goodwin and Ridler, 1970; Ridler, 1971a; Hutchinson et al., 1971). This is largely an empirical observation and independent of whether or not the various iron-formation facies are precise stratigraphic equivalents of the massive sulfides or not.

In addition to this demonstrable stratigraphic control, studies in relatively undeformed areas have suggested a possible tectonic control as well. Just as thermal springs and volcanoes are frequently aligned along, or parallel to, a major fault, rift, or other major linear feature, so it can be demonstrated, in some areas, that massive sulfide orebodies, considered to be exhalative or hot-spring deposits (see section on "Genesis", pp. 199–208), can be aligned parallel to prominent local linears. The linear concept of ore control was proposed as early as 1904 by Hobbs and has lately been developed by Kutina (Kutina et al., 1967; Kutina, 1968, 1969, 1971, 1972; Kutina and Fabbri, 1972). The relationship between linears and orebodies is best developed in the Noranda area, probably the least deformed of the major Precambrian massive sulfide camps (Fig. 7). The major lineaments of this mineral area are northeast-trending faults. Examination of the sulfide ore distribution reveals that three orebodies (Old Waite, East Waite, and Norbec) also lie along a northeast-trending line parallel to nearby faults. If lines are then drawn through the other orebodies of the area, parallel to the Waite–Norbec trend, then two features become apparent: (1) one such line passes through two widely-separated sulfide bodies, the Amulet F and Mobrun; and (2) there is a certain regularity in the spacing of the lines. The nature of these lines passing through the sulfide ores, parallel as they are to recognized geological features such as faults, may, as suggested by Kutina, be indicative of fundamental tectonic processes which have expressed themselves over long periods of time. Because the rocks of the Noranda area, particularly those in the northwestern part, are only slightly deformed, the possibility exists that the faults are merely the more obvious manifestations of a pervasive structural grain which began very early in the geological history of the area and continued to affect the rocks long after they were deposited. The structural stress may have been active during actual deposition of the volcanics as subtle zones of weakness along which the ore-bearing solutions rose to precipitate sulfides at or near the existing ocean floor. Thus, only when these postulated lines of weakness intersected the ocean-floor interface during a particular phase of volcanic activity (i.e., the favourable horizon) did sulfide accumulations take place. In other words, ore formation occurred at the intersection of a structural plane with a sedimentary horizon. Further investigations of the distribution of massive sulfide deposits adjacent to volcanic centres may reveal that the ore clusters are not as random as they might appear at first sight.

Ore types. Many massive sulfide bodies comprise two main ore-types, massive ore and stringer ore, distinguished largely by the relative proportions of sulfide and silicates. Of the more than one hundred deposits shown in Figs. 1 and 2, half are known to contain both massive and stringer-ore types. The remainder, with one exception, are comprised of massive ore only. The one exception, Louvem (no. 10 in Fig. 1) described by Guha and Darling (1972), appears to be a conformable, but disseminated, chalcopyrite–pyrite deposit in tuffaceous rocks.

Massive ore. Although "massive" is a relative term and rather wide variations in total sulfide content exist between one deposit and the next, in our experience the term would normally apply to ore consisting of at least 50% sulfides by volume (corresponding to approximately 60% by weight). Schermerhorn (1970) has coined the term "pyrite" to refer to massive pyritic ore.

The two longest dimensions of massive sulfide bodies and layering within the mass are, in relatively undeformed ores, parallel to bedding planes or flow contacts in the host rocks. Sharp contacts of the sulfide-body hanging wall with country rock are the rule while footwall contacts are generally much more diffuse. Several orebodies are known where unaltered country rock on the hanging wall passes directly into high-grade ore (20–25%, or higher, combined metals), whereas the footwall contact can only be determined by assay.

Stratigraphic thickness commonly range up to 70 ft with a few (e.g., Kidd Creek, Horne) over 100 ft; deposits of greater dimension are known, but in most of these there is usually evidence of structural thickening of ore such as in the hinge zone of folds. The other two dimensions of the massive ores are several times greater than their thickness. Unmetamorphosed deposits, with some exceptions, are roughly circular in plan, particularly those occurring in massive volcanic rocks.

Mention has been made previously of the relation, in the case of undeformed deposits, between shape of the massive ores and the nature of the country rock. Briefly, sulfide bodies in layered rocks (sediments, tuffs, etc.) are blanket-shaped with the two dimensions parallel to the layering greatly exceeding the thickness. Orebodies in massive volcanic rocks, on the other hand, tend to be much more ovoid or lenticular in cross-section.

Stringer ore. Also referred to as "disseminated" or "vein" ore, stringer ore, when present, is always in the stratigraphic footwall of the massive sulfide ore lens (Figs. 9, 10 and 13). The total sulfide content of stringer ore is considerably less, and much more variable, than massive ore and seldom exceeds 25%. In undeformed deposits, it consists of anastomosing sulfide veinlets, veins, and irregular replacements extending downward, for a few inches, a few feet, or a few hundreds of feet, roughly perpendicular to the main massive sulfide orebody. In doing so, the stringer-ore zone cuts across stratigraphy, in contrast to the massive sulfide lens, which roughly parallels stratigraphy. An example of stringer ore is shown in Fig. 25.

Stringer ore is not found with all massive sulfide deposits. Where present, it constitutes a distinct, mappable unit which, when followed upward, frequently grades into massive ore by coalescence of the veins and small replacement bodies which comprise stringer ore. In other deposits, stringer ore changes into massive ore rather abruptly, within 3—5 ft. We have noted, however, that ore-grade material in the stringer zone is almost invariably in direct contact with massive ore and only in some orebodies does it constitute ore more than 200 ft stratigraphically below the massive lens. Barren sulfides, however, are known

0   5   10 cm

Fig. 25. Copper-rich stringer ore, Kidd Creek mine.

to continue downward for thousands of feet, but in diminishing abundance until background sulfide content of the host rock is reached.

The shape of the stringer-ore zone, again in undeformed ores, is generally cylindrical or funnel-shaped with the massive ore lying in the widest part of the "funnel" at the top. The upward dilation of the stringer-ore zone ends at the massive sulfide orebody and in only one or two instances are we aware of it extending beyond the massive ore, either laterally or upward. The Amulet "A" orebodies in the Noranda area are an example of several massive sulfide deposits "stacked" one on top of the other within stringer ore in the intervening area (Dresser and Denis, 1949, p. 378).

Mineralogy and composition of ore. The mineralogy of massive sulfide bodies (including stringer-type ore) is remarkably simple considering their abundance and diversity in size, geological setting, and post-ore metamorphism. The iron sulfides, particularly pyrite and pyrrhotite, characteristically comprise at least half the total sulfide content. Sphalerite, chalcopyrite, and galena, in widely varying proportions, constitute most of the remaining sulfides.

The contents of major sulfide species in several selected deposits are shown in Table I. These values represent averages for the entire orebody; in detail, the proportions of various sulfides vary considerably within the deposit.

Detailed mineralogical studies of massive sulfide ores have revealed a wide diversity of minerals to be present in very minor amounts. Table II presents some examples of this varied assemblage.

TABLE II

Minor metallic minerals in selected massive sulfide ores

Chisel Lake	Lake Dufault (Norbec)	Texasgulf (Kidd Creek)	Coronation	Zenmac	Sullivan
Arsenopyrite	argenite	cassiterite	arsenopyrite	cassiterite	chalcopyrite
Bournonite	silver	stannite	cubanite	spinel	arsenopyrite
Native gold	(antimonial)	rutile	marcasite	violarite	boulangerite
Altaite	chalcocite	covellite	native gold	magnetite	jamesonite
Hessite	cubanite	digenite	magnetite	ilmenite	tetrahedrite
Tennantite	dyscrasite	bornite	ilmenite		magnetite
Geocronite	galena	chalcocite	hematite		cassiterite
Arsenic	mackinawite	marcasite			pyromorphite
Tetrahedrite	stannite	arsenopyrite			
Boulangerite	magnetite	acanthite			
Pyrargyrite-		stromeyerite			
proustite		native silver			
		galena			

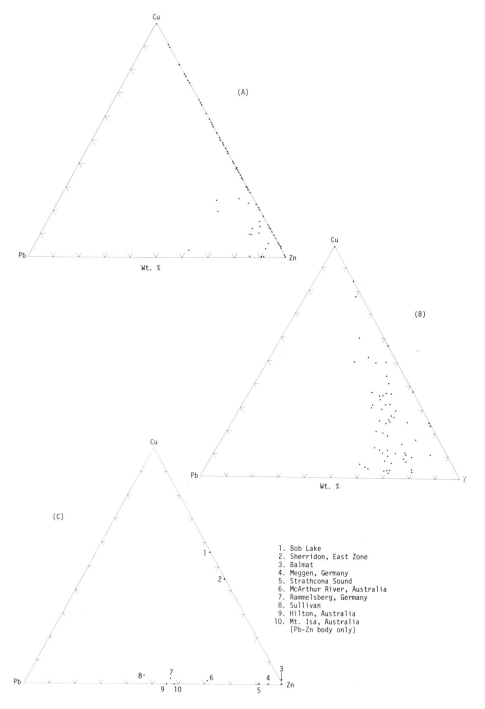

Fig. 26. Weight ratios of Cu, Pb and Zn in: A. North American Precambrian massive sulfide deposits in volcanic or volcano-sedimentary host rocks; B. Canadian and Japanese Phanerozoic massive sulfide deposits in volcanic or volcano–sedimentary rocks; C. massive sulfide deposits in sedimentary rocks. (Because this type is relatively uncommon in North America, a few selected deposits in similar rocks elsewhere in the world are shown for comparison; compare also with fig. 12 in Chapter 7, Vol. 6, on the Kupferschiefer.)

When the bulk composition of Precambrian massive sulfide ores is expressed in terms of the three main ore elements (Cu—Zn—Pb), two features immediately become apparent (Fig. 26): first, the ores are remarkably deficient in lead relative to copper plus zinc and are considerably lower in lead than similar deposits in Phanerozoic rocks; and secondly, more than 70% of the deposits contain more zinc than copper.

Major byproducts won from massive sulfide ores are cadmium, silver and gold with lesser amounts of selenium, tellurium, bismuth, and tin. Recoverable silver contents generally fall in the range 1—3 oz/ton, seldom exceeding 4 oz/ton. Recoverable gold values range from 0.01 to 0.2 oz/ton.

In Table III are metal grades and tonnages of several Cu—Zn massive sulfide ores, taken in part from the compilation by Hutchinson (1973).

Zoning. An extremely important and characteristic feature of massive sulfide deposits is that the vast majority of them are zoned, in one form or another, and that the zonal arrangement shows a consistent pattern relative to host rock stratigraphy. The zoning in the deposits is expressed in four ways: morphology, mineralogy, texture, and composition.

Morphology. Morphological zoning has already been alluded to in the discussion of ore types. Where both massive and stringer ore coexist in a deposit, the stringer ore occurs on the stratigraphic footwall of the massive orebody (Figs. 9, 10 and 13).

TABLE III

Base- and precious metal content of selected Precambrian massive Cu—Zn—Pb sulfide deposits (partly after Hutchinson, 1973)

Name, district	%Pb	%Zn	%Cu	oz./ton Ag	oz./ton Au	Remarks, reference
Horne mine; Noranda, Que. (Can.)	n.s.	n.s	2.19	n.s	0.152	52.6 million tons production to 1966 (Spence, 1967)
Quemont; Noranda, Que. (Can.)	n.s	1.86	1.27	0.523	0.129	13.5 m.t. production to 1966 (Spence, 1967)
Lake Dufault Norbec; Noranda, Que. (Can.)	n.s.	7.2	4.0	2.2	0.03	initial ore reserve estimate 2.3 m.t. (Purdie, 1967, p. 53)
	n.s.	6.88	5.05	1.81	0.032	1.08 m.t. production to 1966 (Spence, 1967)
Waite; Noranda, Que. (Can.)	n.s.	4.47	7.11	0.70	0.03	total production 1.1 m.t. (Price and Bancroft, 1948, p.748)
Amulet Lower A; Noranda, Que. (Can.)	n.s.	5.69	5.13	1.41	0.041	Millhead ore assays, 2.1 m.t. (Suffel, 1948, p.757)

TABLE III (*continued*)

Name, district	%Pb	%Zn	%Cu	oz./ton Ag	oz./ton Au	Remarks, reference
Amulet C; Noranda, Que. (Can.)	n.s.	14.30	2.62	4.19	0.020	Millhead ore assays, 240,000 t. (Suffel, 1948, p.757)
Amulet F; Noranda, Que. (Can.)	n.s.	10.88	4.28	1.48	0.017	Millhead ore assays, 150,000 t. (Suffel, 1948, p.757)
Vauze; mine, Noranda, Que. (Can.)	n.s.	0.94	2.90	0.69	0.019	0.39 m.t. production to 1965 (Spence, 1967)
West MacDonald; Noranda, Que. (Can.)	n.s.	2.91	0.012	0.0051	0.0019	1.03 m.t. estimated production 1955–59 (Spence, 1967)
Magusi River deposit; Noranda, Que. (Can.)	n.s.	4.18	1.32	0.92	0.024	3.6 m.t. reserves 1973 (Jones, 1973)
Mattagami Lake; Matagami, Que. (Can.)	n.s.	10.4	0.69	1.13	0.014	18.7 m.t. ore reserves 1966 (Can. Mines Handbook 1967–68 p.213)
Manitou Barvue; Val d'Or, Que. (Can.)	0.38	4.74	0.03	2.57	0.034	production from 6.1 m.t. zinc–lead ore (Ramsay and Swail, 1967, p.19)
	n.s.	n.s.	0.9	0.068	0.012	production from 3.1 m.t. copper ore (Ramsay and Swail, 1967, p.19)
Mattabi; Sturgeon Lake, Ont. (Can)	0.84	7.6	0.91	3.13	0.007	12.9 m.t. ore reserves, 1973; recoverable lead (Franklin et al., 1973)
Flin Flon; Flin Flon, Man. (Can.)	n.s.	4.24	2.99	1.25	0.089	26 m.t. ore reserves, 1964 (Geol. Staff, Hudson Bay Mining and Smelting and Stockwell, 1948, p.295)
	n.s.	4.4	2.2	n.s	n.s	59 m.t. production to 1970 (Coats et al., 1972)
Cuprus mine; Flin Flon, Man. (Can.)	n.s.	6.4	3.75	0.84	0.038	avg. grade 510,000 t. ore mined (Geol. Staff, Hudson Bay Mining and Smelting, 1957, p.253)
Stall Lake mine; Snow Lake, Man. (Can.)	n.s.	0.5	5.2	0.31	0.044	2.1 m.t. reserves, 1968 (Coats et al., 1970, p.971)
Rod Deposit, No. 2; Snow Lake, Man. (Can.)	n.s.	2.88	7.10	0.47	0.05	0.3 m.t. reserves, 1968 (Coats et al., 1970, p.971)
Chisel Lake mine; Snow Lake, Man. (Can.)	0.7	12.00	0.39	1.23	0.049	3.2 m.t. reserves, 1968 (Coats et al., 1970, p.971)

TABLE III (*continued*)

Name, district	%Pb	%Zn	%Cu	oz./ton Ag	oz./ton Au	Remarks, references
Ghost Lake mine; Snow Lake, Man. Can.)	0.7	11.6	1.42	1.14	0.013	0.26 m.t. reserves, 1968 (Coats et al., 1970, p.971)
Osborne Lake mine; Snow Lake, Man. (Can.)	n.s.	1.4	4.17	n.s.	n.s.	2.7 m.t. reserves, 1968 (Coats et al., 1970, p.971)
Fox Mine; Lynn Lake, Man. (Can.)	n.s.	2.70	1.84	n.s.	n.s.	13.1 m.t. reserves, 1970; (Coats et al., 1972)
Jerome District; Ariz. (U.S.A.)	n.s.	?	8.5	2.50	0.068	avg. of 5 different mines, 19.2 m.t. produced 1908–51 (Anderson and Creasey, 1958, p.100)
United Verde Extension mine; Jerome, Ariz. (U.S.A.)	n.s.	n.s.	?	<2.0	0.04	avg. grade 3 m.t. ore (Anderson and Creasey, 1958, p.136)
Iron King; Prescott, Ariz. (U.S.A.)	2.50	7.34	0.19	3.69	0.123	based on metal recovered from 5 m.t. production (Gilmour and Still 1968, p.1241)
Errington mine; Sudbury Basin (S.A.)	0.75	3.24	1.02	1.49	0.017	ore reserves 7.5 m.t. after 15% dilution (Martin, 1957, p.365)
Vermilion Lake mine; Sudbury Basin (S.A.)	1.10	4.56	1.10	1.78	0.020	ore reserves 2.8 m.t. after 20% dilution (Martin, 1957, p.365)
Balmat No. 2 mine; Balmat, N.Y. (U.S.A.)	0.73	9.65	n.s.	n.s.	n.s.	10.6 m.t. production to 1960 (Lea and Dill, 1968)
Balmat-Edwards district, N.Y. (U.S.A.)		10.11				17.4 m.t. production from all 4 mines (Lea and Dill 1968)
Sullivan mine; Kimberley, B.C. (Can.)	6.4	5.5	n.s.	2.1	n.s.	>175 m.t. production and reserves, 1900–1974 (B.C. Dept. Mines Ann. Rep. and Cominco Ann. Rep.)
Kidd Creek; Timmins, Ont. (Can.)	0.40	9.75	1.52	4.30	n.s.	25 m.t. production to December 1973; significant Sn and Cd recovered; 95 m.t. reserves remaining above 2800′ level with higher Cu and lower Zn grade; massive ore extends below 4000′ level (Matulich et al., 1974)

n.s. = not stated; ? = data unknown but metal present.

Mineralogy. Mineralogical zoning is, in part, a function of morphological zoning and manifests itself in the varying proportions of the five main sulfide species: pyrite, pyrrhotite, sphalerite, chalcopyrite, and galena. Hence, in an ideal deposit, sphalerite would be considerably more abundant in the massive ore than in the stringer ore. Similarly, the ratio pyrite/pyrrhotite is commonly higher in the massive than in the stringer ore. Galena, where it occurs, is found only in the massive ore. Consequently, an ideal orebody would consist of a pyrite–sphalerite (–chalcopyrite–galena) massive orebody underlain by pyrrhotite–chalcopyrite stringer ore. Furthermore, within the massive sulfide ore, galena and sphalerite are more abundant in the upper half of the mass (with galena increasing toward the hanging wall), whereas chalcopyrite increases toward the footwall and grades downward into chalcopyrite stringer ore.

This ideal zoning pattern is not present in all deposits and is, of course, best developed in ores having the full complement of ore minerals. As the number of mineral phases decreases, so the zonation tends to become obscure and may not be in evidence at all in some orebodies. For example, the Flexar mine in Saskatchewan, a massive lens consisting mainly of chalcopyrite and pyrrhotite without a stringer zone, shows no evidence of zonation whatsoever. In some deposits, the zoning is concentric, such as at the Amulet Lower A and Millenbach mines of Noranda, where a core of pyrrhotite–chalcopyrite ore is surrounded by pyrite–sphalerite; or at Sullivan, where the galena/sphalerite ratio decreases peripherally from an essentially barren pyrite–pyrrhotite core. In other deposits, sphalerite-rich ores are in separate orebodies, such as at the Horne and Delbridge mines, Noranda; or their distribution in massive ore may be random, such as at Geco. However, at Geco the disseminated ore is of two distinct spatially-separated types, namely sphalerite-rich and chalcopyrite-rich.

Texture. Textural zoning is concomitant with mineralogical zonation in that the more sphalerite-rich portions of the ore are generally banded, expressed as monomineralic layers of pyrite and sphalerite (Fig. 27), whereas the chalcopyrite-rich portions seldom show banding even though the total content of the two portions may be similar.

Composition. Compositional zoning, of course, parallels the distribution of the three major ore sulfides – sphalerite, chalcopyrite, galena – but has the advantage of being more quantitative and can usually be detected in routine assay plots. Sphalerite, because of its wide variation in colour, is sometimes difficult to recognize and a weak or poorly-developed zonation could easily be missed on the basis of mineralogical examination alone. Compositional zoning, whereby zinc and lead are most abundant toward the hanging-wall side of the massive ore with copper increasing toward the footwall and passing into Cu-rich stringer ore, results in the exceptionally rich "copper keels" of some orebodies. In most ores, however, the compositional zoning is due more to a downward decrease in zinc rather than an increase in copper. For example, a typical deposit may contain 12% Zn and 1% Cu in the upper part of the massive ore decreasing to 2% Zn and

Fig. 27. Banded sphalerite (dark)—pyrite (light) ore, Mattabi mine. Head of hammer gives scale.

3% Cu toward the footwall, i.e., the zinc has decreased six-fold whereas the copper has increased only three-fold. In the United Verde orebody of Arizona (Anderson and Creasey, 1958), the Cu/(Cu + Zn) ratio increases from 0.14 in the upper part of massive ore to 0.87 in the stringer ore. In some orebodies there is a sharp transition between Cu-rich ore and Zn-rich ore, whereas in others the transition is more gradual. The Mattabi orebody is an exceptional case of repetitive Cu—Zn zoning with two portions of massive ore separated by a 10 ft thick bed of felsic tuff and chert (Franklin et al., 1973).

The ideal zonal pattern from Cu-rich footwall to Zn-rich hanging wall, although not universal in massive sulfide ores, is nevertheless common enough that Hutchinson (1973, p. 1224) believes it to be a reliable indicator of stratigraphic tops in deformed areas. This concept is undoubtedly valid in a majority of cases. However, we urge caution in its uncritical application in view of the observed concentric zoning in some deposits, cited above.

The origin of the compositional zoning is a problem that continues to plague geologists and geochemists and has not been adequately explained. Any discussion of this problem must ultimately include the origin of the massive Cu—Zn—Pb sulfide ores, so we will reserve our further comments for the section on "Genesis" (pp. 199—208).

Ore textures and structures. [1] Most of the textures now present in massive sulfide ores are the result of post-depositional processes, such as diagenesis and metamorphism. Metamorphic textures are by far the most common and will be dealt with later in a discussion of metamorphism of ores (see p. 187).

In studies of younger sediments, great care is generally taken to distinguish between depositional processes (and textures) and diagenetic processes (and textures). In the comparatively young sediments the distinction is important and made possible because of the very short total history of the material being studied. In Precambrian ores and sediments, however, so much of earth history has passed, and so many other processes have been operative since the original deposits were formed, that to distinguish between primary depositional and diagenetic textures, is, in most instances, decidedly futile.

Recent experiments have indicated that, of the common sulfides, pyrite alone remains brittle over a wide range of temperature, confining pressure, and differential stress (Graf and Skinner, 1970; Stanton and Willey, 1970; Clark and Kelly, 1973; Salmon et al., 1974). These studies would, therefore, suggest that primary textures of sulfides would be best preserved in pyrite, and indeed, most of the textures described in younger pyrite deposits have also been found in Precambrian ores. Features, such as colloform and framboidal pyrite, are common, even in Archean orebodies, particularly those with a high carbon (graphite) content. Some of the best botryoidal pyrite that we have observed have come from deposits in the Timmins area, where graphitic bands are a common occurrence in even the massive ore (Fig. 28). Similarly, the Cuprus deposit in Manitoba has both a high carbon content and well-developed botryoidal or colloform pyrite.

Extremely thin layering of fine-grained pyrite can also, by analogy with modern sulfide sediments, be attributed to primary deposition. Thin layering can also result from comminution of pyrite during deformation, but this can generally be recognized by the abundance of fractured pyrite grains, whereas primary layering will contain, albeit on a microscopic scale, undisturbed colloform rosettes of pyrite.

Besides texture, several primary structures are also recognizable in certain well-preserved deposits. Mention has been made previously of the common occurrence of sulfide clasts in silicic volcanic agglomerates (i.e., "mill-rock") coeval with the orebodies. By a similar token, angular blocks and fragments of volcanic host rocks are frequently found in the massive sulfide portion of several deposits (Fig. 29). The low metamorphic grade and lack of schistosity in the clasts bespeaks of a primary origin and distinguishes it from the "Durchbewegung" fabric described by Vokes (1969, p. 129) for the flowage of a sulfide mass around detached silicate fragments during metamorphism.

In a very few deposits, we have observed such soft sediment structures as slump-folds, and flame-structures, or load-casts. The former are recognized by their confinement to a single sulfide layer or a small number of adjacent layers; beds above and below the structure are undisturbed. The best-developed load-casts were observed in a thin-bedded

---

[1] For additional relevant discussions refer to Chapter 5 by Dimroth in Vol. 7, and to Chapter 5 by Mookherjee in Vol. 4.

Fig. 28. Botryoidal pyrite from a graphite band, Kidd Creek mine.

Fig. 29. Rhyolite block (R) in massive sulfide ore (M), Mattabi mine.

alternating series of pyrite and graphitic shale, part of a more massive pyrite deposit. The load-casts were developed by slumpage of the pyrite into the underlying sediment resulting in "flames" of shale which disrupted the pyritic layers. Soft-sediment deformation has also been recognized in layered sulfides of the Mattagami mine (Roberts, 1966).

Excellent examples of graded bedding in monomineralic sulfide layers, again mainly of pyrite, have been observed in the Kidd Creek deposit in the Timmins area, and by Jones (1973) in the Magusi River deposit. The layers are generally only a few inches thick and are composed of angular pyrite grains in a carbonaceous matrix. The structure is entirely similar to that commonly found in greywackes and results from differential settling of sulfide fragments. A related structure, consisting of a relatively large fragment of sulfide (or sulfides), surrounded by finer grained sulfide layers abutting the fragments on both sides, but passing above and below it, is frequently observed in some deposits. In appearance, the structure is analogous to the "drop-stones" of glacial origin and has on occasion been referred to as "volcanic drop-stones" in reference to the assumed pyroclastic origin of the sulfide fragments.

Alteration. Wall-rock alteration adjacent to sulfide ores has been a topic of investigation by economic geologists and geochemists for decades. Even in areas of lowest-grade metamorphism, host rocks to sulfide ores have undergone at least some change in their mineralogical and chemical composition. The alteration associated with massive sulfide deposits is easily recognized and invariably occurs as a well-defined zone on the footwall side of the massive ore (e.g., Dugas, 1966, pp. 49—50; Gilmour, 1965; Sharpe, 1968, pp. 107—109). In some cases it has been traced for 3000 ft stratigraphically below the main ore mass. In undeformed deposits, the alteration zone is pipe-like in shape and contains within it, usually as a shoot toward the centre, the pyrrhotite—chalcopyrite stringer-type ore (Figs. 9, 10 and 13). The diameter of the alteration pipe increases upward until it is coincident with that of the massive ore. Only rarely does the alteration pipe extend into the hanging wall, or laterally beyond the massive sulfide lens, as it does at the Amulet "A" deposit.

The pipe-like shape of the alteration zone is perhaps the most spectacular form of this diagnostic feature of these ores in volcanic environments. In many deposits, however, the altered zone is expressed merely as a chloritization of the footwall rocks, which is lacking in the hanging wall. The abundance of disseminated sulfides in the alteration zone (and the contained stringer ore) leads to the common occurrence of a diffuse footwall contact of the orebody, in contrast with the hanging wall which is normally very abrupt.

Development of chlorite is the main mineralogical expression of the footwall alteration followed, in decreasing order of frequency, by sericitization, silicification, and carbonation. Chloritization and sericitization, the two most common types of alteration, result in a desilication of the host rocks, particularly where they are developed in rhyolite. In deposits where chemical studies have been made, the most obvious chemical changes in the alteration pipe are a relative increase of Fe, Mg, and S and a decrease of Si, K, and Na

(Riddell, 1952; Lickus, 1965; Roberts, 1966; Sakrison, 1966; Descarreaux, 1973). Magnesium metasomatism in particular, appears to have been the dominant process in the formation of the footwall alteration pipes and zones.

Of the North American massive sulfide ores in truly sedimentary environments, only Sullivan has an alteration pipe which is so characteristic of the volcanic associated ores. The Sullivan pipe is a brecciated and tourmalinized zone beneath the massive ore (Fig. 22) and contains considerable pyrrhotite mineralization with rare galena and sphalerite. Similar alteration has not been identified beneath the deposits at Balmat nor those in the Kisseynew gneiss. The cordierite–anthophyllite gneisses north of the Manitouwadge orebodies may possibly represent a metamorphosed alteration pipe (see p. 198, dalmatianite), but such an assertion must remain speculative until detailed geochemical and structural information is available.

METAMORPHIC EFFECTS

The changes brought about in massive sulfide ores and their host rocks as a result of metamorphic processes is not merely of academic interest. Sulfides, because of their wide stability range, are capable of surviving even the highest metamorphic grades. A massive sulfide deposit, after undergoing even mild metamorphism, is often radically different in many ways than it was before deformation; metamorphic changes can directly influence the exploration for, and development of, these ores.

For example, during metamorphism the textures and structures of the host rocks are often destroyed and their mineralogy radically altered making it difficult for the exploration geologist to recognize, and to trace, favourable host rocks. Similarly, changes are also wrought in the sulfide ores resulting in a different mineralogy, grain size, and form of the ore deposit which can affect its geophysical, and even geochemical, response. Mineral textures affect the beneficiation of ores, and the shape of the deposit influences the techniques and efficiency of the mining process.

Furthermore, some sulfide assemblages, such as those of the Zn–Fe–S system, may be highly useful in delimiting the pressure–temperature–sulfur fugacity relations attending metamorphism of the deposits.

Elsewhere in this multi-volume publication, Mookherjee (Chapter 5, Vol. 4) describes in greater detail the effects of metamorphism on sulfides. Here we present those aspects which are pertinent to the form and genesis of massive Cu–Zn–Pb sulfide ores.

Effects on host rocks

Metamorphism of volcanic, volcani-clastic and sedimentary rocks is a subject treated at length in many standard petrology texts and will not be discussed in detail here. The ability to "see through" metamorphism and to deduce the original nature of the rock is a

TABLE IV

Possible metamorphic equivalents of primary rock types commonly associated with Precambrian massive sulfide deposits (modified slightly from Hutchinson, 1970)

Primary or low-grade metamorphic rock	Medium-grade metamorphism	High-grade metamorphism
Chert	siliceous schist	quartzite
Pyritic, cherty iron-formation	pyrite—pyrrhotite—magnetite mica schist	pyrrhotite—magnetite mica quartzite
Rhyolite		
Rhyolite tuff		
Rhyolite breccia	quartz—feldspar—sericite gneiss	quartz—feldspar gneiss
Rhyolite agglomerate		
Andesitic tuff (chlorite-schist)	biotite—chlorite—quartz schist	biotite—quartz gneiss
Andesite (chlorite-schist)	epidote—plagioclase—amphibolite	hornblende—plagioclase—amphibolite gneiss
Basalt (chlorite-schist)	epidote amphibolite	amphibolite (gneiss)

valuable asset to any geologist, particularly when working in the Precambrian. Table IV has been compiled from the literature as an aid in the interpretation of "what it was" from "what it is".

Because most massive Cu—Zn—Pb sulfide ores are closely associated with the acid phases of volcanism, the distinction between metamorphosed silicic volcanic rocks and metaquartzites becomes of paramount importance. Similarly, the recognition of a cherty tuff layer can be of assistance as a marker horizon and aid in the structural interpretation of volcanic terrains, which are characteristically complex because of rapid lithologic and facies changes.

The host rocks often supply the best or only quantitative indicators of strain around deformed massive sulfide ores. Quartz-eye and feldspar porphyries, which are of common occurrence in many mining camps, are particularly useful in this respect inasmuch as changes in shape and orientation of the phenocrysts can accurately reflect the intensity and attitude of physical deformation (Stauffer, 1967). For example, at the South Bay Mine, where such a porphyry forms one contact with ore, Berezowski (1972) was able to document pronounced increasing elongation of the quartz phenocrysts, indicating increasing intensity of deformation, approaching an ore contact and a slightly different pattern of elongation of the phenocrysts approaching a barren tuff contact (Fig. 30). Presumably,

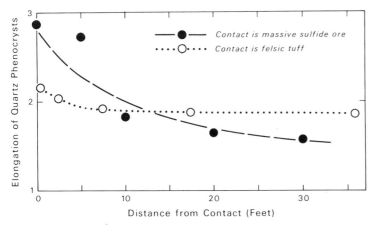

Fig. 30. Average elongation of quartz phenocrysts in quartz-eye porphyry at the South Bay mine, northwestern Ontario, relative to distance from a contact with massive ore and with felsic tuff. Elongation is expressed as the ratio of the maximum to minimum dimensions of the quartz grains. Plots of median and maximum elongations show similar trends. (Data from Berezowski, 1972.)

these differing effects are a consequence of the relative differences in competency of porphyry vs. ore and porphyry vs. tuff. These same strain indicators are being used at South Bay to map regional lineations in an attempt to sort out the seemingly complicated fold patterns near ore as an aid to locating extensions of ore zones (Asbury, 1975; Asbury and Scott, 1975).

Effect on sulfide deposits

Form. Judging from examples in undeformed areas, the original form of most massive sulfide bodies was probably roughly circular or oval in plan with dimensions parallel to bedding, several times greater than thickness. Beginning with this assumption, certain preliminary deductions may be drawn concerning the changes of form or morphology which can occur in sulfide masses during metamorphism.

One of the most common of these is the development of linear or blade-like orebodies aligned parallel to lineations, such as fold axes, hornblende crystals, or corrugations in the foliation planes, in the enclosing rocks. Examples of bodies of this shape (and origin) would be those of Balmat, N.Y. (Lea and Dill, 1968), Chisel Lake, Manitoba (Martin, 1966) and Stall Lake Mines Ltd., Manitoba (Coats et al., 1970). Rod-shaped orebodies, such as these develop in areas of medium to high metamorphic grade which show penetrative lineation. In rocks of lower metamorphic grade, where deformation is expressed mainly by shearing and folding, without well-developed lineation, the deposits tend to be flattened parallel to schistosity, perhaps as a result of transportation along the shear planes. Frequently, orebodies deformed predominantly by shearing give the appearance of having been sliced into several en-echelon ore lenses separated by narrow widths of highly schistose and altered wallrock. The effects of these two types of deformation, i.e., shear and penetrative lineation, on the shapes of massive sulfide orebodies, has been described by Howkins and Martin (1970) for deposits in the Flin Flon and Snow Lake areas. In the former area, where shears are the predominant structural element, ore lenses have plunge to strike-length ratios of about 3 : 1. In the Snow Lake area, where lineation in the high metamorphic grade rocks is well developed, orebodies have plunge to strike-length ratios up to 10 : 1. Howkins and Martin also note that "pencil-like structures of the Snow Lake deposits have a continuity along plunge which results in more accurate predictions of the ore locations at depth".

The pinching and swelling of the Geco orebody (Fig. 21), as well as the distribution of ore lenses in the Little Stall Lake (Rod) deposit (Coats et al., 1970 p. 983), are reminiscent of boudinage in deformed rocks with the massive sulfide ore (predominantly pyrite) behaving as a relatively competent and, hence, disrupted unit. Although we have not made extensive structural measurements, we have observed features at Geco which are consistent with and, in some cases, demanded by boudinage:

(1) All of the massive ore is in one continuous stratigraphic horizon.

(2) Pinch and swell occurs both along strike (Fig. 21) and down dip (Fig. 20).

(3) "Necking" is abrupt; e.g., 30 ft thick massive ore pinches down to a 1-inch "vein" within 75 ft.

(4) The ore in the neck zones, where failure occurred, is characterized by very fine grain size. In many places pyrite grains have a cataclastic texture.

(5) Chalcopyrite, a mineral which migrates into fracture zones of deformed ore deposits, appears to be more concentrated in the neck zones of the "boudins" than in the thicker portions.

(6) The massive ore is enclosed within a heavily muscovitized schist whose schistosity appears to parallel the massive ore contact suggesting differential movement between the two units during deformation.

(7) The thickness of the muscovite schist increases substantially relative to thickness of massive ore in the pinch zones.

(8) The schist in the pinch zones is characterized by small buckle (?) folds as might be expected if it had flowed.

These qualitative observations do not prove that boudinage was operative at Geco, but they do suggest that it may be the mechanism of altering the original shape of the massive ore. We have seen such features in several massive sulfide deposits so the boudinage concept, if proven correct, may be important in tracing extensions of orebodies.

Areas of intense, polyphase regional deformation, or local areas adjacent to large intrusions, commonly contain orebodies which have been so distorted that their form is best described as amoeba-like. Such deposits are extremely difficult to exploit because of their irregular and unpredictable shapes. Successful exploitation of these ores generally depends on studies of detailed structures in the host rocks. In cases where this has been done, the irregular nature of the ore zones has generally been found to be controlled by the interference pattern produced by intersecting deformation trends, such as two periods of folding with widely divergent axial planes. The highly distorted zones adjacent to large intrusions are particularly difficult to interpret because of the irregular nature of the strain on country rocks resulting from the forceful emplacement of the intrusive body. Examples of deposits with amoeba-like shapes are Joutel and Mines de Poirier.

Mineralogy. As Vokes (1969) has quite correctly pointed out, the abundant mineralogical changes brought about in silicate rocks during metamorphism do not find their counterparts in sulfide assemblages, including massive sulfide ores. The wide stability range of sulfides, the relatively few phases (and components) present in most ores, and the ease with which high-temperature sulfide phases revert to low-temperature phases on cooling, all contribute to ensure that the metamorphism of sulfide masses is generally not reflected in their mineralogical assemblages.

Nevertheless, in spite of these restrictions, certain mineralogical changes do occur in sulfide masses which have undergone metamorphism. Most of these changes are caused by thermal events and most occur in the iron sulfide species. The formation of pyrrhotite from primary pyrite is perhaps the most common change, and perhaps most of the pyrrhotite presently found in Precambrian ores has formed in this manner, particularly pyrrhotite occurring with sphalerite in the massive sulfide portion of the ores; stringer-type ore can, and commonly does, contain what appears to be primary pyrrhotite as the iron sulfide phase. The breakdown of pyrite to magnetite can also occur as well as the generation of "new" pyrite from pyrrhotite or ferromagnesian silicates. The pyrrhotite in massive sulfide ores is usually an intergrowth of hexagonal (or ordered pseudohexagonal) and monoclinic polymorphs. These can be seen best by treating a polished section with a magnetic colloid which is attracted to the monoclinic pyrrhotite phase (Fig. 31). Inasmuch as monoclinic pyrrhotite is not stable above $253\,^{\circ}C$ (Kissin, 1974; see also Scott and Kissin, 1973) this intergrowth must have developed upon cooling from the higher metamorphic temperatures.

Exsolution of chalcopyrite in sphalerite is also a common phenomenon in metamorphosed ores and could result in a reduced Cu-content in the Zn-concentrate, if ground finely enough. The sphalerite itself commonly emerges from the metamorphism considerably enriched in iron relative to its original composition, because of the general increase in FeS activity brought about by conversion of some of the original pyrite to pyrrhotite. This can be deleterious in that it depresses the Zn-content of sphalerite concentrate since sphalerite commonly contains 10–20 mol% FeS in solid solution. On the other hand, the FeS content of sphalerite coexisting with pyrite and pyrrhotite is a reliable measure of the pressure during metamorphism of the ore (Scott and Barnes, 1971; Scott, 1973). For example, Bristol (1974) and Scott (1974a) have applied the sphalerite geobarometer to the metamorphosed massive sulfide ores of the Flin Flon and Snow Lake region, where they found that the variation in indicated pressure paralleled the metamorphic grade. In another application (Scott, 1974b), the sphalerite geobarometer indicates a pressure of 6

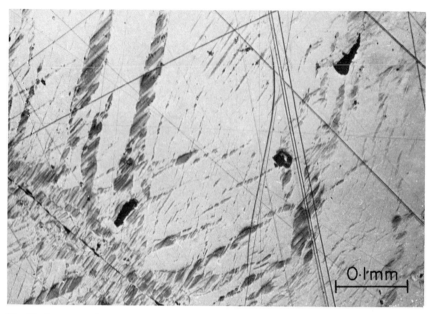

Fig. 31. Intergrowth of monoclinic pyrrhotite (dark) and hexagonal pyrrhotite (light) as seen in polished section after application of magnetic colloid. Colloid is preferentially attracted to the ferrimagnetic monoclinic pyrrhotite. (From Kissin, 1974.)

kbar at Balmat during metamorphism, a figure that is compatible with geological estimates.

Small amounts of gahnite ($ZnAl_2O_4$) found in some metamorphosed zinc ores probably formed by reaction between sphalerite and aluminous silicates.

McDonald (1967) has suggested that increased metamorphism of sulfide assemblages results in ore of increasingly complex mineralogy, but admits that these mineralogical variations may be due to original metal contents. We have also noted this trend among Precambrian sulfide ores, but would suggest that the increased complexity is more apparent than real due to the increased grain size of the more highly metamorphosed ores, thereby making the accessory minerals more readily seen.

Texture. Increase in grain size is probably the most common textural change wrought in sulfide masses during metamorphism. Although there is a general correspondence between grain size and metamorphic grade, a recent attempt to quantify this relationship yielded only fair correlation (Templeman-Kluit, 1970).

In ores of high metamorphic grade, recrystallization and grain-growth produce a sulfide assemblage which is best described as a "sulfide pegmatite"; large porphyroblasts of pyrite and sphalerite are surrounded by a crystalline matrix of pyrrhotite and chalco-

Fig. 32. Pyrite porphyroblasts in a matrix of sphalerite and pyrrhotite, Geco mine.

pyrite (Fig. 32). Original banding in the sphalerite portion of the ore is almost totally destroyed. The Geco orebody is a good example of a massive sulfide which has undergone high-grade regional metamorphism (Suffel et al., 1971).

Although Precambrian massive sulfide ores have undergone at least some post-ore deformation, evidence of strain in the ore is not as common as one would expect. For example, Asbury's (1975) careful examination of the South Bay deposit, which lies in complexly deformed metavolcanics of greenschist grade, has failed to reveal well-developed strain indicators within the predominately sphalerite + pyrite massive ore itself. The reason for this is not clear, but the relative timing of maximum stress with respect to maximum thermal effects may be an important factor. Sulfides lacking strain texture may indicate that thermal annealing took place after the release of stress. When the reverse is the case, i.e., stress in the absence of appreciable heat, such as deformation in rocks of low metamorphic grade, or the continuation of deformation *after* the thermal "peak", ores may exhibit such textures as comminution of pyrite and sphalerite, deformation twinning in sphalerite, "steely" galena, and "streaky" chalcopyrite. Preferred *morphological* orientation of sulfide grains, particularly galena and sphalerite, is a commonly observed phenomenon, but preferred crystallographic orientation is much more difficult to document. Through the use of an X-ray technique, Roberts (1966) was able to demonstrate that the (0001) plane in pyrrhotite of the Mattagami Lake orebody lay in the axial plane of foliation. Similarly, Kanehira (1969) through a study of the optical extinction positions of pyrrhotite in the Coronation Mine, showed that there existed a preferred orientation of the c-axes perpendicular to foliation. Wang (1973) has found X-ray pole figures of sphalerite to be a reliable indicator of foliation in the Mattagami Lake ores.

Annealing of sulfide grains, if allowed to proceed to completion, will result in textural equilibrium among the sulfide phases. Equilibrium of this type is best recognized by a study of the interfacial angles between adjacent sulfide grains; thermodynamic stability is indicated when the interfacial angles approach $120°$ (Stanton, 1964). Certain Pb–Zn ores in the Grenville Province as well as highly metamorphosed Cu–Zn ores contain sulfides approaching textural equilibrium (Fig. 33). The increase in sulfide grain size brought about by regional metamorphism is of more than academic interest because the coarser grained ores, during benefication, generally yield cleaner concentrates with less grinding than do the fine-grained ones. Texturally-equilibrated ores, in particular, are more economical to mill and concentrate because of minimal interfacial contacts between the grains, as described by Stanton (1964).

Massive sulfides which have undergone intense deformation sometimes have the appearance of having flowed plastically. The aspect of flowage is enhanced by the common occurrence of fragments of schistose wall rock within the sulfide mass and the "flow-lines" in the sulfides curve around these silicate fragments (Figs. 34, 35). Spurr (1923) noting this phenomenon in the Mandy deposit of Manitoba, concluded that the sulfides had been introduced as a magma and that the "flow-lines" he observed were akin to flow-banding in rhyolite and other viscous magmas. The silicate intrusions were regarded by Spurr as having been detached from wall rock during the forceful intrusion of the

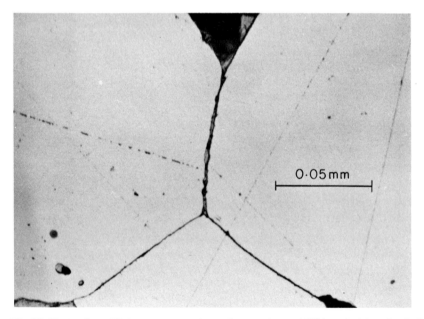

Fig. 33. Textural equilibrium among pyrite grains meeting at $120°$ interfacial angles. Polished section.

Fig. 34. "Flow lines" of banded sulfides around an elongated rhyolite fragment (white), Mattabi mine.

Fig. 35. Host-rock schlieren in ore.

sulfide magma. More recently, the occurrence of schistose fragments in flow-banded sulfides has been termed "Durchbewegung" fabric by Vokes (1969) who attributes the banding to post-ore deformation and the silicate inclusions as fragments of wall rock which have been detached as a consequence of plastic flow of the sulfides during metamorphism. Kalliokoski (1965) and Shadlun (1971) have presented pictorial catalogues of textural changes in massive sulfides attending increasing intensity of regional metamorphism.

Selective melting of sulfide assemblages, at geologically reasonable temperatures, to produce myrmekitic textures has been experimentally demonstrated by Brett and Kullerud (1967) in the Fe—Pb—S system and by Craig and Kullerud (1968, 1969) in the Cu—Pb—S and Cu—Zn—S systems, respectively. There is some evidence that certain sulfide ores may have melted, at least to some degree. For example, Mookherjee and Dutta (1970) describe vermicular intergrowths of pyrrhotite—chalcopyrite—sphalerite in small veins projecting into a post-ore diabase dyke from the main sulfide mass at the Geco mine. This, and other textures in the veins, was interpreted as evidence of incipient melting of the sulfides adjacent to the diabase/ore contact.

Composition of ore. Except for minor changes in Fe-, Zn- and S-content due to interaction between sulfides and silicates, desulfurization of pyrite to pyrrhotite and oxidation of some pyrrhotite to magnetite, there is little evidence that the bulk composition of massive ores is significantly altered during metamorphism. Internally, however, there is much evidence of a redistribution of elements which, in some instances, tends to disrupt, modify, or otherwise obscure the compositional zoning previously described as a characteristic feature of massive Cu—Zn sulfide ores.

The rearrangement of elements in metamorphosed ores is due almost entirely to the relative mobilities of the ore-forming sulfides, either by solid state flow or by hydrothermal dissolution and redeposition in dilatant zones. Chalcopyrite and galena are much more mobile than sphalerite and pyrite and respond plastically to deformative and thermal processes. For example, metamorphism of Archean sulfide deposits, which contain only minor galena, will mobilize the small amounts of galena out of its original position in the uppermost portion of the massive ore and redeposit it in small gash veins in the stratigraphic hanging wall of the deposit. Sulfosalts, such as tetrahedrite—tennantite, frequently accompany galena in these veins which, if large enough, can constitute important silver-rich concentrations. In the massive ore, gash veins of all sizes, consisting predominantly of chalcopyrite and quartz, respectively, the most mobile of the sulfides and gangue silicates, are a common occurrence. Characteristically, these Cu-rich gash veins are aligned roughly perpendicular to banding or the long dimension of the orebody and are usually contained within the main sulfide mass. In some deposits, chalcopyrite veins, with or without sphalerite, penetrate into the host rock for short distances often beyond the ends of the massive ore. These veins can be distinguished from stringer-type ore because they are less abundant and not accompanied by heavy chloritic alteration. Also, remobilized chalcopyrite veins commonly cross-cut schistosity, whereas metamorphosed stringer ore is essentially parallel to metamorphic foliation.

Intrusion of sulfide bodies by later igneous bodies can result in re-mobilization of the sulfides particularly chalcopyrite, by plastic flowage "back into" the intrusion. This phenomenon has been documented by, for example, Mookherjee and Suffel (1968) for diabase dykes and has been observed by the senior author of the present paper at several granite/sulfide contacts.

On a large scale, intense deformation is capable of moving the entire sulfide mass by transposition along the metamorphic foliation planes. This may result in a detachment of the massive ore from the stringer ore, such as may have occurred at the Manitou—Barvue deposit (Ramsay and Swail, 1967). Where the deposit has been folded, particularly by isoclinal folding, mobilization of the sulfide mass can be brought about by plastic flowage over short distances of the more ductile sulfides (chalcopyrite, galena, and to a lesser extent, sphalerite) and comminution of the brittle ones (pyrite). In well-developed instances, this differential mobility during isoclinal folding will result in a noticeable increase in grade (particularly copper) and thickness toward the nose of the fold, whereas the limbs will tend to be thinner and more pyritic.

In highly metamorphosed ores, there may be a slight lowering of over-all grade due to the inclusion of numerous contorted siliceous fragments, which have been forcefully detached from the wall rock by plastic flowage of the sulfides and incorporated into the ore mass.

Effects on the footwall alteration zone

Mention has been made earlier of the flattening effect produced on massive sulfide ores by transposition along shear or schistosity planes. If the first schistosity developed in the rock undergoing metamorphism is parallel to depositional layering, transposition along these planes will effectively stretch the massive sulfide lens in a plane parallel to its original stratigraphic dimensions. The same effect may be achieved by sliding a pack of cards to form a cross-sectional parallelogram from the original rectangle. If the process is continued, it will produce a body several times longer than the original, but which retains its original position relative to stratigraphy, i.e., parallel to stratigraphic layering. If an alteration pipe occurs vertically beneath the massive sulfide lens, its original position will be nearly perpendicular to regional stratigraphy and incipient schistosity. Movement along the schistosity planes will rotate or transpose the pipe into a position nearly parallel to both schistosity and the massive sulfide lens (Fig. 36). The final position of the pipe will be en echelon to the main sulfide body rather than perpendicular to it, as it was originally. Transposition also has the effect of decreasing the apparent diameter of the pipe due to the stretching action during transposition. The stringer-zone ore, under these circumstances, assumes a final position near one end, but still in the footwall, of the massive ore. The appearance of this metamorphosed complex has been described in the

older literature as a chloritic zone, within which the massive sulfide ore has been emplaced. Excellent examples of this type of deformation are found in the Flin Flon (see
Fig. 16), Stall Lake and Normetal orebodies.

Metamorphism can also have a profound effect on the mineralogical constitution of
the alteration zones. The original chloritic alteration is commonly altered to an assemblage consisting of various proportions of hornblende, biotite, cordierite and anthophyllite (Froese, 1969). If sufficient calcium was present in the original alteration zone,
tremolite-actinolite, garnet, and Ca-pyroxenes may also form. Chlorites are unstable
above 590°C (at P_{H_2O} = 2 kbar) in the presence of quartz and will re-equilibrate into the
mineral assemblages noted above (Fleming and Fawcett, 1974). In the absence of quartz,
Mg-rich chlorites [Mg/(Mg + Fe) > 0.4] can coexist with cordierite + talc (± magnetite ±
orthorhombic amphibole) to higher temperatures (J.J. Fawcett, personal communication,
1974) which may account for the preservation of chlorite in some alteration pipes in very
high-grade metamorphic rocks.

Cordierite–anthophyllite assemblages in the alteration zone are usually the result of
regional metamorphism. Under the proper conditions, however, simple thermal metamorphism may also produce cordierite–anthophyllite by isochemical metamorphism of alteration pipes. In the Noranda area, "dalmatianite", i.e., cordierite-bearing pipes (Walker,
1930; Dugas, 1966, p. 49), is only found in deposits occurring within the thermal aureole
of the Dufault granodiorite stock (De Rosen-Spence, 1969). Deposits such as Vauze,

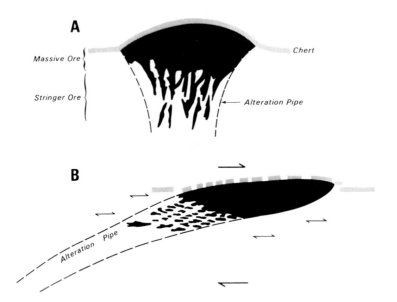

Fig. 36. A. Schematic diagram illustrating main geological features of an undeformed massive sulfide
deposit. B. Same deposit as A but modified by shearing in the sense shown by the heavy arrows above
and below the orebody. Compare with Fig. 16 of the Flin Flon mine.

Quemont, Mobrun, Horne, etc., occurring at some distance from the intrusive margin, do not contain cordierite in their alteration pipes.

GENESIS

Conclusions regarding ore genesis of virtually all types of mineral deposits have had a long history of pendulum-like swings[1] and the massive sulfide ores of North America are no exception. For example, Spurr (1923) in commenting on the Mandy mine in Manitoba, concluded "that the ore could have been introduced in no other way than in plastic form, as an intrusive mass ... the viscosity was like that of tar and is analogous to that of those rhyolites and obsidians which show a fine flow-banded structure".

By 1933, Lindgren's text *Mineral Deposits* was in its fourth printing and the hydrothermal theory of ore genesis predominated in North America, as elsewhere, for several decades. This was evident in opinions expressed in two separate volumes (1948 and 1957) on the *Structural Geology of Canadian Ore Deposits*, published by the Canadian Institute of Mining and Metallurgy. In these volumes, the authors were unanimous in ascribing a hydrothermal origin to massive sulfide ores of the type described in the present paper.

In 1957, however, Knight proposed the "Source bed concept" and suggested that a great number of sulfide orebodies had formed contemporaneously with their host rocks.[2]

This was followed by Oftedahl's (1958) popularization (in English) of the exhalative—sedimentary theory of ore formation previously expressed by a number of European authors (e.g., Schneiderhöhn, 1944; Ramdohr, 1953), in which base-metal accumulations were thought to have been produced by sea-floor fumaroles of volcanic origin. Thus, when the Canadian Institute of Mining and Metallurgy held its 1959 symposium on *The Occurrence of Massive Sulphide Deposits in Canada* (C.I.M. Bull., Feb. 1960), although most authors tenaciously clung to the hydrothermal replacement theory, several advocated a close genetic relationship between massive sulfide ores and volcanism. Stanton (1959), in particular, strongly supported a volcanic—exhalative and biologic origin for ores in the Bathurst area of New Brunswick. He later (Stanton, 1960) expanded this concept for all orebodies of the "conformable pyritic" type.

By 1960, P. Gilmour had applied Oftedahl's and Stanton's concepts to the Noranda area, but did not publish his conclusions until 1965 at which time he remarked (Gilmour, 1965): "It is suggested that the massive sulphide deposits were formed on or near the surface through the agency of fumarolic emanations." Similar views were expressed by Goodwin (1965) with reference to base-metal deposits in the entire volcanic belt from

---

[1] For discussions on theories of ore genesis, including their historic developments, see the Chapters by King and Ridge, in Vol. 1, and Vokes in this volume.

[2] A general review of the influence of compaction of sediments (including pyroclastics) and ore genesis has been offered by Wolf (1976).

Noranda to Timmins and by Roscoe (1965) to the Noranda—Matagami Lake areas. The volcanic—exhalative concept gained ready acceptance among economic geologists and its general tenets were subsequently applied to sulfide ores of the Matagami Lake and Val d'Or districts of Quebec (Latulippe, 1966), and recently to the Arizona massive sulfides (Anderson and Nash, 1972).

In general terms, the volcanic—exhalative theory advocates that massive sulfide ores in predominantly volcanic environments are an integral part of, and coeval with, the volcanic complex in which the deposits occur. The common occurrence of several orebodies at or near the same stratigraphic horizon over large areas, the consistent relation between stratigraphy and zoning of the ores, the widespread association with silicic volcanism, the many occurrences of coarse pyroclastics ("mill-rock") containing massive sulfide clasts, the general absence of alteration in the hanging wall in contrast to that of the footwall, the sharp contact between hanging-wall rocks and massive ore, and the commonly associated chemical sediments (iron-formations) all combine to make a deep-seated, epigenetic site of ore formation untenable. Furthermore, volcanic—exhalation is demonstrated by a present-day example of sulfide deposit forming from fumaroles related to an 1890 volcanic eruption on the island of Vulcano (Honnorez et al., 1973).

Proponents of the epigenetic or replacement origin of the massive sulfide deposits (e.g., Miller, 1973) advance as evidence:

(1) Not all massive sulfides are stratabound; some, such as Geco, cross-cut stratigraphy to a degree and, in some deposits, chalcopyrite or galena veins cut the host rocks.

(2) At Noranda many of the deposits are peripheral to the Lake Dufault granodiorite (Fig. 7), a possible source of mineralizing fluids.

(3) Many deposits are related to major faults and to folds which are thought to be favourable loci for distribution and entrapment, respectively, of fluids.

(4) Alteration pipes underlying most deposits represent undoubted hydrothermal replacement of pre-existing rocks.

(5) In a few cases, there is alteration of the hanging-wall rocks, such a the albitization and chloritization at Sullivan (Fig. 22).

(6) Gold mineralization at the Horne Mine in Noranda is spatially associated with felsic dykes suggesting a hydrothermal emplacement of the gold.

(7) At the Horne mine, massive sulfides occur in the youngest diabase dykes which cut the volcanic rocks, whereas the older diabase dykes near massive sulfides are intensely altered.

(8) Rounded pyrite grains are found within massive pyrrhotite suggesting a replacement texture.

In response to these views, many of the "epigenetic" features, such as 1, 3, 6 and 8 above, can be ascribed to metamorphic overprinting and deformation slightly rearranging the constituents of the massive ore, as we have discussed previously. Mookherjee and Suffel (1968) and Mookherjee (1970) have adequately explained the ore-dyke relationships of points 6 and 7 as being due to thermal contact effects or subsequent regional metamorphic effects. The location of deposits at Noranda (point 2) is controlled primarily by the volcanic stratigraphy (Spence and De Rosen-Spence, 1975), the distribution of which, peripheral to the granodiorite, was, in part, structurally-controlled by the forceful emplacement of the intrusive. The mere presence of an intrusive stock is not evidence of its having spawned replacement deposits any more than nearby faults are necessarily the conduits for the ore-forming fluids (point 3). In fact, the Lake Dufault granodiorite is decidedly post-ore as evidenced by the formation of "dalmatianite" (cordierite-bearing rocks) from pre-existing chloritic alteration pipes only in those deposits which were close enough to be affected by its thermal aureole (De Rosen-Spence, 1969) and by cross-cutting relationships of granodiorite and dalmatianite pipes. On the question of the discordant footwall alteration pipes (point 4), there is no argument; they were the feeders of the mineralizing fluids and hence are epigenetic. However, pipes do not underlie all massive orebodies and their mineralization is totally unlike that of the massive ore suggesting either that the massive ore had a different, albeit related, origin to the stringer ore or that its depositional environment was substantially different. The hanging-wall alteration of the Sullivan deposit is a natural consequence of waning fumarolic activity as elaborated below.

Recently, Hutchinson et al. (1971) advanced the views of Goodwin and Ridler (1970) on the Abitibi orogeny by presenting a generalized model relating metallogeny to lithic elements within a typical Archean volcanic—sedimentary basin. A slightly modified version of their model is given in Fig. 37. This model ties together the massive sulfide ores in volcanic rocks with those in the nearby volcani-clastic sediments (i.e., "mixed facies") and relates the different facies of iron-formation to these ore-forming processes and to the paleogeography of the sedimentary basin. Various aspects of this model as well as relationships to massive sulfide ores in truly sedimentary successions (e.g., Sullivan) are considered below.

The exhalative concept does not differ markedly from the classical hydrothermal replacement theory in terms of process. In both instances, ore deposition was from hydrothermal solutions originating within the crust of the earth and rising along fractures or other zones of weakness. The passageway through which the fluids rose is now preserved as the alteration pipe stratigraphically beneath many orebodies. Extensive leaching, metasomatism and replacement by ore sulfides undoubtedly took place within the pipe, leaving us clues as to the original chemistry of the ore fluids. The main point of departure between the exhalative and replacement theories is the site or, more precisely, the *timing* of massive ore deposition relative to the host rocks. In the exhalative theory, ore deposition takes place at or near the volcanic rock/seawater interface between successive volcanic episodes, whereas the advocates of replacement normally regard ore deposition as a secondary feature imposed on the host rocks a considerable time after their formation and lithification. Metal-sulfide precipitation in both theories, however, was probably influenced by the long-held concepts of temperature and pressure gradients, dilution of ore fluids, pH change, etc.

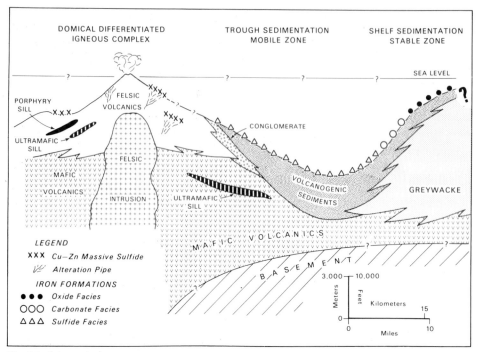

Fig. 37. Schematic tectonic and stratigraphic relations of an Archean volcano—sedimentary complex. (Modified from Hutchinson et al., 1971, and Ridler, 1973.)

Most likely the ore fluids were hot saline brines, such as those of the present-day Salton Sea (Skinner et al., 1967), Cheleken, U.S.S.R. (Lebedev and Nikitina, 1968) or the Red Sea deeps (Degens and Ross, 1969; see also their Chapter 4, Vol. 4). Brines are capable of carrying enough metal (>10 p.p.m.) as chloride complexes together with reduced sulfur (H_2S, HS^-) (Helgeson, 1969) to produce massive orebodies, so long as they are neutral to weakly acid and their temperatures are greater than $100°C$ (for sufficient galena and sphalerite solubility in a 3-molal NaCl brine; Anderson, 1973) to $200°C$ (for sufficient pyrite and chalcopyrite solubility in a 3-molal NaCl brine; Helgeson, 1969, Fig. 21) or $250°C$ (for sufficient chalcopyrite solubility in a 1-molal NaCl brine; Crerar and Barnes, 1974). These salinities and temperatures are in reasonable agreement with the sparse fluid inclusion data from the alteration pipes and massive portions of the "Kuroko" volcanogenic massive Cu—Pb—Zn sulfide ores of Japan (Sato, 1972; Tokunaga and Honma, 1974). As well, weakly acid solutions are required to produce the sericitic altera-tion by hydrolysis of potash feldspar (Hemley and Jones, 1964; Meyer and Hemley, 1967; Shade, 1974) and the chloritic alteration (Hemley and Montoya, 1971; Beane, 1974, Fig. 4) in the alteration pipes beneath many Precambrian massive sulfide ores. At lower temperatures and significantly lower salinities, or in alkaline solutions, sulfide solubilities are too low to produce a massive deposit within a reasonable length of time.

The actual locus of sulfide precipitation from metal-bearing saline exhalations, whether at the fumarolic vent or further away, has been shown by Sato (1972) to be dependent largely on the original salinity and temperature of the ascending fluid which govern its mixing behaviour with sea water. Fig. 38, taken from Sato, considers four theoretical paths of temperature and density that would be followed by brines mixing with sea water of 20°C and 0.5 molal NaCl assuming that hydrologic equilibrium is established. In type I mixing results in a steady decline in temperature and density of the brine. In type IIa the ore brine is originally more dense than sea water and upon mixing rises to a maximum density before declining. In type IIb, the original brine is initially less dense than sea water, but upon mixing becomes more dense than sea water rising to a maximum before declining as in type IIa. In type III, mixing results in a steady increase in density. The consequences of these four models are illustrated in Fig. 39 as follows:

Type I. Because of its high density, the ore fluid will flow down-slope from the vent. It will not readily mix with sea water because fresh ascending fluid will always form the bottom layer. The brine will collect in a depression and will retain its density layering for a considerable time precipitating metals from the upper portion, where slow mixing occurs. The resulting ore deposit may be considerably removed from the vent and proba-

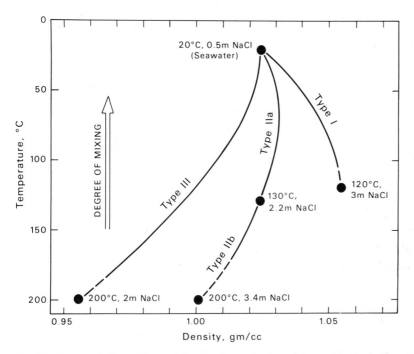

Fig. 38. Representative patterns of density change during mixing various hydrothermal brines with sea water. (After Sato, 1972; courtesy of *Min. Geol.*)

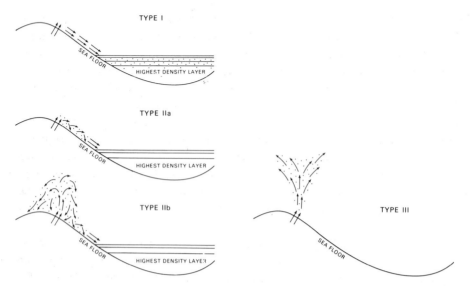

Fig. 39. Probable behaviours of ascending ore-forming solutions welling up onto the sea floor, according to the four basic types of temperature—salinity relations in Fig. 38. Dotted areas are sites of mineral precipitation, assuming that minerals are precipitated during early stages of mixing with sea water. (After Sato, 1972; courtesy of *Min. Geol.*)

bly relatively thin and wide-spread, its distribution controlled by the shape of the depression.

Type IIa. As in type I the dense solution will flow down the slope, but upon mixing with sea water will become more dense and sink beneath freshly ascended brine. Consequently, fresh brine is always in contact with sea water, resulting in the entire mass ultimately achieving maximum density. Because of the efficient mixing, there will be a rapid precipitation of sulfides close to, but downhill from, the vent, resulting in a thick ore-mass with little topographic control.

Type IIb. The low-density brine will float up from the vent rapidly, mixing with sea water until the density maximum is reached, at which point the brine will sink following the same course as IIa. The resulting deposit would be similar to that produced in IIa, with the exception that sulfides may be deposited anywhere in the vicinity of the vent.

Type III. The ascending solution, being less dense than sea water, will float up and disperse rapidly, depositing a thin layer of sulfides over a large area remote from the vent.

    The majority of massive sulfides in volcanic rocks occur above their vents (alteration pipes), particularly in the Precambrian and to a lesser extent in the Phanerozoic, and their shape is accurately predicted by cases IIa or b. In most deposits, it is apparent that precipitation took place almost immediately upon emanation of the ore fluid from the fumarolic vent aided by rapid cooling and/or dilution. The sulfide was probably precipitated either as a heavy gelatinous mud, as evidenced by soft sediment slump features in some ores, or else as a sinter, as suggested by Spence (1975). In either case, it must have rapidly become a hard, brittle solid in order to remain on a 30° slope at Vauze and at Millenbach (Spence, 1975) and to produce clasts in "mill-rock".

The sulfide clasts which are found either in "mill-rock" and/or in the massive sulfide body itself, may actually have formed by any of several processes. Some may be the result of post-depositional slumpage and brecciation of an original bedded sulfide layer, such as described by Kajiwara (1970a), to explain the brecciated nature of ore in the Shakanai mine in Japan. A variation of this method would be brecciation of a sulfide layer by a later volcanic explosion erupting underneath it. Clark (1971, p. 213) has suggested a third method involving original precipitation of sulfides in open fissure veins and breccia pipes. Continued deposition of ore minerals in these vertical channels would result in blockage of the passageways, leading to an increase in pressure, and climaxing in one or more minor explosions, which free the sulfide accumulations from the channel walls and distribute the sulfide breccia fragments around the volcanic orifice(s) as a volcani-clastic.

Massive sulfide ores occurring in the volcanically-derived sedimentary apron surrounding and/or mantling the volcanic complexes seldom include the stringer-ore portion or alteration pipe, so common in the ores within the volcanic pile. If the alteration pipe is regarded as the source orifice for these ores, then deposits in the mantling sediments probably formed from solutions emanating from similar orifices remote from the site of sulfide deposition. The metal-rich solutions are pictured as moving away from the source, possibly downslope on the ocean floor, to then collect, and later precipitate sulfides in a slight hollow or other depression (Gilmour, 1971) as in Sato's type I, or else they may have been more widely dispersed as in Sato's type III. The ores of the Manitouwadge district might well be an example of type I. As pointed out previously (p. 186), some footwall alteration zones (e.g. White Lake) are blanket-like and conformable, rather than pipe-like and disconformable, with the footwall contact. This alteration type may have been produced by "auto-alteration" of footwall rocks after the dense, metalliferous brine (Sato's type II) came to rest in a slight sea-floor depression. Alternatively, the lack of an alteration pipe beneath some massive sulfide ores could be explained by early, submarine transport such as sliding down the volcanic slope, as suggested by Jenks (1971) and Schermerhorn (1970), or stripping by turbidity currents (Suffel, 1965). Post-ore faulting could also account for the lack of appreciable alteration beneath the Horne mine (Hodge, 1967).

Of the massive sulfides in a truly sedimentary environment, the Sullivan mine appears to be an unusual case of a deposit *with* an alteration pipe. Beneath the central part of the orebody, and extending for more than 1500 ft below the footwall of the ore zone, is a roughly funnel-shaped zone of tourmalinized and brecciated wall rock (see Fig. 22).

Alteration in the hanging wall has affected considerably less rock and has resulted in chloritization and albitization of the metasediments more-or-less directly above the "iron zone" in the ore and the central part of the tourmaline alteration pipe. Consideration of the geological relationships described previously, and shown in Fig. 22, suggests that formation of the Sullivan orebody may be similar to that of the volcanic–exhalative type of Stanton (1960) and that ore deposition was essentially contemporaneous with sedimentation of the host rock argillite. In this genetic model, the tourmaline-rich pipe would represent the feeder zone through which the ore-bearing solutions penetrated the footwall sediments to reach the sea floor, at which point deposition of sulfides took place. The annular, zonal arrangement of elements in the orebody supports the idea of a central feeder zone. The thicker part of the orebody, the abundance of pyrrhotite, and the higher tin content are all coincident immediately above the central part of the tourmaline pipe, i.e., the feeder zone source vent of the hot mineralizing solutions. Chlorite and albite in the hanging-wall sediments above the central zone could represent small amounts of post-ore "leakage" of the (by then) barren solutions. These solutions, which produced tourmaline $[(Na, Ca) (Al, Fe, Mg)_3 B_3 Al_6 Si_6 O_{27} (OH)_4]$ in the hotter environment of the feeder zone, would, when passing into the sedimentary–diagenetic environment above the ore horizon, be reconstituted into chlorite $[(Mg, Fe)_5 (Al, Fe)_2 Si_3 O_{10} (OH)_8]$ and albite $[(Na, Ca) AlSi_3 O_8]$; minerals which are stable under these less intense conditions of diagenesis and sedimentation. Boron, a natural constituent of seawater, would escape into the Proterozoic seas.

Besides the Sullivan, at least three other small orebodies also occur in the Aldridge metasediments and bear many similarities to the main Sullivan ore. These ores, as well as the ubiquitous iron sulfide disseminated throughout the Lower Aldridge and which gives it its characteristic rusty weathering appearance, are considered to have reached their present position in the Aldridge in a manner analogous to that of the Sullivan, i.e., through deep fractures to the sea floor, whereupon the metals were deposited contemporaneously with sedimentation. A similar origin for the Sullivan deposit has been proposed by Gilmour (1971).

The tourmaline zone beneath the orebody (analogous to the chlorite alteration pipe beneath massive sulfide ores in volcanic rocks), the metal zoning, the relative lack of alteration on the hanging wall, and the massive disrupted nature of the ore in the central part of the body (over the postulated feeder pipe), giving way laterally to increased banding and continuity of layers on the fringe of the ore (quieter depositional environment?), all combine to make an exhalative theory of origin reasonably plausible.

The major difference between the Sullivan orebody and the "normal" volcanogenic massive sulfide bodies is, in our opinion, the depositional environment, which Price (1964) has suggested was mainly deltaic. Recently, Kanasewich (1968) postulated, on the basis of deep seismic reflections, that a Precambrian rift valley may exist beneath southern Alberta and British Columbia. The position of the postulated rift is drawn by Kanasewich in such a way that the Sullivan orebody would be situated just within the north edge. Sedimentation trends, however, in the Proterozoic rocks in the Kimberley area are north—south and small-scale sedimentary features can be followed across the trace of Kanasewich's proposed rift (G.B. Leech, personal communication), thereby casting considerable doubt on its existence in that area. Nevertheless, Kanasewich speculates that the deposit was formed in an ancient rift in a manner similar to that occurring at present for the Red Sea hot metalliferous brines (Degens and Ross, 1969, and their Chapter 4, Vol. 4), i.e., that metalliferous solutions rose along deep-seated fractures in a rift, or ocean-floor spreading, tectonic environment and deposited metals directly on the ocean floor. The process is entirely analogous to that postulated for the volcanogenic massive sulfides, the only difference being that, for the latter, the process was operative in an active, eugeosynclinal, volcanic, possibly island-arc environment. The similarity in ore-forming processes, however, has resulted in deposits with many parallel internal features, as discussed above, and for these reasons Sullivan is regarded as being of exhalative origin but in a sedimentary, rather than volcanic environment, regardless of whether Kanasewich's rift exists or not.

The pronounced zoning of massive sulfide ores from Cu-rich footwall to Zn-rich hanging wall must be an inheritance of their origin, rather than a feature superimposed by metamorphism as it is generally best displayed in the least metamorphosed deposits (compare Noranda and Geco). For those ores deposited away from the exhalative orifice (e.g., type I of Fig. 39) precipitation could have been slow enough from a static brine reservoir to allow separation of sulfides in reverse order of their solubilities. The expected order of precipitation would be chalcopyrite followed by galena and sphalerite, in agreement with the observed zonal sequence. However, for large deposits such a mechanism requires a large reservoir of metalliferous brine over the site of deposition (more than 2 cubic miles of brine containing 10 p.p.m. metal per million tons of ore). Such a mechanism clearly is not possible for those deposits which have accumulated rapidly over the orifice. Zoning in these deposits, representing over half of the Precambrian massive ores in North America, requires that the chemistry of the ore fluid must have changed with time becoming increasingly Zn-rich. Kajiwara (1970b) has attempted to quantify the thermochemistry of these changes in the ore fluids as reflected by the mineral assemblages in the Kuroko ores.

Several models, not all of which are independent, come to mind to explain the progressive change with time in the chemistry of the ore fluid from predominantly Cu-rich to predominantly Zn-rich: (1) variation in the temperature of the heat source driving the fumarolic activity; (2) changes in the salinity and/or pH of the ore fluid affecting the solubility of the metal sulfides; (3) selective removal by the ore fluids of metals from the source area (either a magma, or leached volcanic rocks); and, (4) reaction with wall rocks in the alteration pipe.

Mention was made earlier in this paper, in a discussion of a favourable horizon for sulfide bodies, that the exact relationship between exhalite (generally chert and/or iron formation of the oxide, carbonate, or sulfide facies) and economic massive sulfide bodies is not clear. Although many orebodies are overlain by, or grade laterally into, chert with or without variable amounts of iron oxide or sulfide, it has yet to be demonstrated, with the exception of Stanton's (1972a) studies at Broken Hill in Australia, that these occurrences are the stratigraphic correlatives of any of the widespread oxide, carbonate, or sulfide facies of iron formation, which are so characteristic of Precambrian volcano—sedimentary basins. In spite of the great number of massive sulfide deposits and iron formations in the Precambrian (the Superior Province is a good example), examples of where it is possible to trace oxide iron formation through to a massive sulfide body, containing economic amounts of base metals, are rare. Oxide facies can, however, be traced into sulfide facies barren of base metals and, as Goodwin and Ridler (1970) and Ridler (1971a) have pointed out, this zonation may be indicative of a shelf-to-basin transition. As such, the barren sulfide facies may, by its presence, be a valuable indication of a depositional environment favourable for sulfide accumulation, irrespective of whether or not the barren sulfides constitute lateral equivalents of the economic sulfide bodies. The empirical observation relating the spatial correlation between the two differing forms of sulfide deposits (metal-rich and metal-poor) has led Ridler (1971a) to summarize as follows: "Wherever a volcanic complex is within the sulfide or sulfidic carbonate facies and exhaling base or precious metals, significant mineralization may occur." If, in fact, the barren sulfides are the stratigraphic, as well as the genetic, equivalents of the metal-rich sulfide deposits, then, as Ridler (1973) quite rightly points out, geochemical investigation of the regional exhalite may prove a rewarding exploration exercise. Since any single volcanic pile may be comprised of one, two, or even three volcanic cycles, each containing one or more exhalite horizons, demonstration that the "favourable horizon" exhalite equivalent is geochemically distinctive from the other (barren) exhalites, would constitute an exciting breakthrough in the exhalative concept of metallogenesis. Lickus (1965) and Sakrison (1966) have shown that the chemistry of a cherty tuffaceous layer, stratigraphically equivalent to the Vauze and Lake Dufault orebodies in the Noranda area, significantly changes its character as the ore zones are approached. This is a valuable observation because it at once renders the exhalite ("favourable horizon") amenable to exploration for "blind" orebodies.

COMPARISON WITH THE JAPANESE KUROKO DEPOSITS

With the recent publication of the books *Volcanism and Ore Genesis* (Tatsumi, 1970) and *Geology of Kuroko Deposits* (Ishihara, 1974), Japanese geologists have revealed the extent of their excellent studies and descriptions of volcanically-related ore deposits in Japan. Although several types of mineralization are included in the former book, of interest to the present study are the comments relating to "Kuroko-type" deposits.

Briefly, a Kuroko deposit is "a strata-bound polymetallic mineral deposit genetically related to submarine acid volcanic activity of Neogene Tertiary in Japan" (Matsukuma and Horikoshi, 1970, p. 153). The unmetamorphosed nature of these ores, together with their excellent documentation by Japanese geologists, has naturally led to the unofficial acceptance of these as the "type-deposits" for volcanogenic sulfide ores. It is, therefore, of interest to compare (and contrast) similar ores in the Precambrian with those of the type locality.

All the Kuroko deposits of Japan occur in association with Miocene volcanic rocks and fossiliferous sediments deposited on the eastern margin of a major geosynclinal basin. The depositional environment of the host rocks, on the basis of paleontological studies, is considered to be a warm inland sea. Paleontology has also shown that Kuroko mineralization occurred within a relatively short period of time (Middle Miocene) over a strike-length of approximately 500 miles, the length of the so-called Green Tuff region of Japan. Within this "greenstone belt" (to apply Precambrian terminology) more than a hundred Kuroko-type occurrences are known, but most are clustered into eight or nine districts (Aoki et al., 1970).

Host rocks to the Kuroko deposits are acidic pyroclastic flows considered to have been deposited from turbidity currents accompanying submarine eruption (Horikoshi, 1969). Much of the Kuroko mineralization shows close spatial association with lava domes or masses, particularly those showing evidence of explosive activity (see Fig. 40; compare with Fig. 8).

Within the deposit, Kuroko ores show a consistent stratigraphic succession of ore and rock types and an idealized, fully-represented Kuroko deposit (Fig. 41) would contain the following types in descending stratigraphic succession (Matsukuma and Horikoshi, 1970, p. 163):

hanging wall: upper volcanic and sedimentary formation;
ferruginous quartz zone: composed chiefly of hematite, quartz and minor pyrite;
barite ore zone: massive barite ore;
Kuroko zone: sphalerite—galena—barite;
Oko zone: cupriferous pyrite ores;
Sekkoko zone: anhydrite—gypsum pyrite ore;
Keiko zone: copper-bearing, siliceous, disseminated
and/or stockwork ore;
footwall: silicified rhyolite and pyroclastic rocks,
disseminated and veined by sulfides.

Fine laminations and compositional layering parallel to associated tuff beds, graded bedding of sulfide fragments, colloform textures, and penecontemporaneous brecciation of sulfide layers are characteristic features of the Kuroko and Oko ores.

Compositionally, the ore deposits are zoned roughly in the following sequence (from top to bottom): barium, lead, zinc and copper. In the upper part of Kuroko ore, the predominant copper-bearing mineral is tetrahedrite—tennantite while in the Oko and Keiko ores chalcopyrite predominates.

Alteration directly related to mineralization generally consists of silicification and argillization (montmorillonite—zeolite and sericite—chlorite). Argillization is usually more

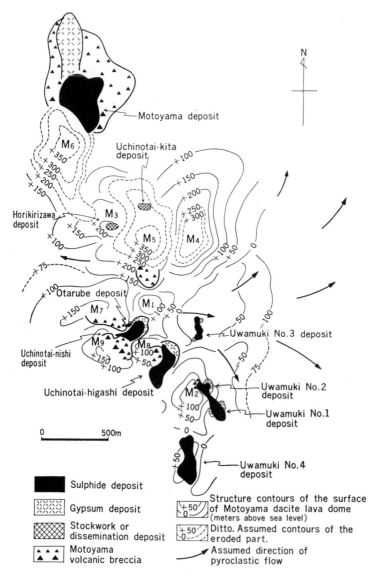

Fig. 40. Distribution of dacite lava domes and Kuroko-type mineral deposits, Kosaka district, Japan. (From Horikoshi and Sato, 1970; courtesy of University of Tokyo Press.)

extensive in the rocks overlying the deposits and is considered by Japanese geologists to represent post-ore "leakage" of mineralizing solutions. Silicification of host rock is most common in the footwall to the orebodies and is an integral part of Keiko (siliceous copper) ore. Silica in the ferruginous chert horizon overlying the sulfide ore may be the exhalative equivalent of the footwall silicification.

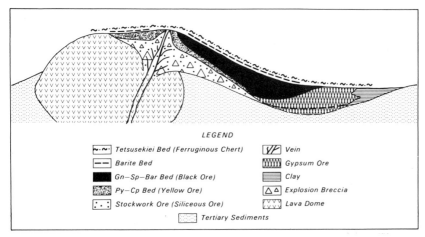

LEGEND

~·~	Tetsusekiei Bed (Ferruginous Chert)	⫽	Vein
— —	Barite Bed	▩	Gypsum Ore
■	Gn—Sp—Bar Bed (Black Ore)	▤	Clay
▨	Py—Cp Bed (Yellow Ore)	△△	Explosion Breccia
⁚ ⁚	Stockwork Ore (Siliceous Ore)	⋁⋁	Lava Dome
	⁚⁚⁚ Tertiary Sediments		

Fig. 41. Schematic section of a Kuroko deposit. Hanging-wall rocks (mainly felsic tuffs) are omitted. (After Sato, 1972; courtesy of *Min. Geol.*)

To best explain all the geological features of Kuroko mineralization, Tatsumi and Watanabe (1970) proposed three main ore-forming processes:

(1) Fissure-filling, dissemination or replacement by ascending ore-forming fluids in pre-existent rocks or sediments (Keiko orebodies).

(2) Chemical precipitation from ore-forming fluids emanating onto a sea floor of pre-existing sediments or volcanic rocks (Kuroko and Oko orebodies).

(3) Mechanical sedimentation of ore fragments formed by crushing of early orebodies during precipitation or by later explosion.

In some deposits, all three processes appear to have contributed to ore deposition, in others only one or two appear to have been operative.

Even this extremely brief outline of some of the main characteristics of Kuroko deposits reveals several features in common with Precambrian volcanogenic ores, namely:

(1) Both occur with calc-alkaline, submarine, volcanic rocks.

(2) Both show a tendency to occur in clusters or districts related to centres of volcanic activity.

(3) Both show a strong spatial correlation with the acidic, explosive phase of volcanism.

(4) Both consist of two main ore types, massive ore (Kuroko and Oko) and stringer ore (Keiko). The massive ore in both cases is essentially conformable with surrounding rocks, whereas stringer ore clearly cross-cuts stratigraphy. The massive ore in both instances is banded.

(5) Both are frequently capped by a layer of ferruginous chert (hematite for the Japanese ores, magnetite for the Precambrian) which may extend beyond the limits of the orebody and serves as a marker horizon.

(6) Both show a compositional zoning relative to stratigraphy with lead—zinc decreasing and copper increasing downward.

(7) Both are underlain by a zone of alteration enclosing the stringer-type ore.

In view of these many points of similarity between Kuroko and Precambrian massive sulfide deposits, there seems ample justification for comparing the two and, by inference, ascribing to them a similar mode of origin. There are, however, several consistent, and therefore perhaps important, differences between the two which may impose certain restrictions on a straight "one-to-one" correlation between Kuroko and Precambrian ores:

(1) Footwall alteration in Kuroko deposits is predominantly silicification ($>90\%$ SiO_2) of pre-existing felsic rocks, whereas most Precambrian (mainly Archean) footwall alteration is characterized by magnesian metasomatism (Riddell, 1952; Lickus, 1965; Sakrison, 1966; Roberts, 1966) accompanied by a marked *decrease* in silica relative to the unaltered host rock. No explanation for this difference in alteration chemistry is offered but, because rocks and ores in the two instances are separated by about 3 b.y. of earth history, it may be related to fundamental and long-term changes in earth chemistry, whereby volcanic-related hydrothermal solutions may have changed with time in terms of major-element chemistry.

(2) Post-ore alteration, which can be directly related to mineralization, of hanging-wall rocks is rare (or largely unrecognized) in the Precambrian but common in the Kuroko. The difference, however, may be more apparent than real because mineralogical alteration of hanging-wall rocks may conceivably be more obvious in the younger rocks where direct comparison with fresh, unaltered rocks is possible. The pervasive mineralogical breakdown in Precambrian volcanic rocks may have effectively masked any subtle alteration of rocks immediately overlying the orebodies. At the present state of knowledge, however, with the exception of the Sullivan mine which is not in a volcanic environment, hanging-wall alteration related to mineralization is not a recognized feature of Precambrian massive sulfide ores. Hanging-wall chemical alteration of both the Vauze and Norbec orebodies has been documented by Lickus (1965) and Sakrison (1966), but it is very sporadic and confined to only a few feet above the ore contact.

(3) Assuming, for the present, that chalcopyrite—pyrite—quartz (Keiko) stringer ore is the equivalent of the Precambrian chalcopyrite—pyrrhotite—chlorite stringer ore, then all the major ore types in Kuroko deposits (Kuroko, Oko, Keiko) can be matched with their counterparts in the Precambrian. However, the Japanese Sekkoko (or Sekko) ore type, consisting of bedded anhydrite and gypsum, does not have an equivalent in the older sulfide masses. In fact, bedded sulfates of any kind are unknown in the Archean which contains the bulk of the known North American Precambrian massive sulfide ores, although they are known in the Archean Pilbara block of Western Australia in association with massive sulfide deposits (Allchurch et al., 1975). Because sulfur-isotope studies of Sekkoko ore (Sakai et al., 1970; Tatsumi, 1965; Sasaki, 1974) suggest that the sulfate is derived from sea water, the paucity of sulfates in the Archean may indicate a corresponding lack of sulfate in Archean oceans possibly resulting from a low oxygen content in the

atmosphere at that time. Hence, the absence of Sekkoko-type bodies associated with Precambrian volcanogenic ores may be attributed more to the chemical evolution of the earth's atmospheric and oceanic compositions than to major differences in ore-forming processes in Tertiary and Precambrian deposits.

(4) Although the major ore types (with the exception of Sekkoko) are common to both, the relative abundances of certain minerals differ between the North American and Japanese ores. For example, bornite and tetrahedrite—tennantite are commonly major constituents of Kuroko deposits, but are rarely more than accessory minerals in the Precambrian ores. Similarly galena and barite, the former a minor, and the latter a seldom, constituent of Precambrian massive sulfide ores, comprise the major portion of the massive Kuroko ore. The reason for the paucity of bornite and tetrahedrite—tennantite in Precambrian ores is not clear but absence of barite may, as with anhydrite and gypsum, reflect a lack of sulfate in Early Precambrian seas. Similarly, the lack of lead in the ancient ores may be a function of long-term geochemical evolution, because the crust (and/or mantle) would not have had time to generate and accumulate lead produced by radioactive decay of uranium and thorium. Because the earth's over-all "lead budget" is continually increasing by this process, the paucity of lead in the Early Precambrian, like the paucity of sulfate, can probably be attributed to over-all changes in earth chemistry. If this premise is accepted then, in spite of these major differences in mineralogy, comparison of the Tertiary and Precambrian ores would still be justified.

In contrast to the above trend of a more varied assemblage in the Tertiary ores, pyrrhotite, a very abundant mineral in most Precambrian ores, is a rare constituent in the Japanese ones. Primary pyrrhotite does not occur in Kuroko ores; minor occurrences of secondary pyrrhotite have been reported as the products of thermal alteration of pyrite by nearby post-ore dykes. This absence of primary pyrrhotite in Kuroko ores may suggest that pyrrhotite in Precambrian deposits is largely or entirely the result of regional metamorphism, even in deposits in areas of low-grade metamorphism such as Vauze and Matagami Lake. Conversely, the Precambrian pyrrhotite could, in large part, be primary and indicative of a slightly different depositional environment relative to the Kuroko.

In summary, although the Kuroko deposits are similar to their Precambrian massive sulfide counterparts in most of their major features, the Precambrian ores, particularly those in the Archean, possess unique properties, which must be taken into consideration before rigorously equating the two. The relative lack of lead and sulfates and the predominantly Mg- instead of Si-metasomation in the footwall rocks are characteristic of the older Precambrian ores. In our opinion, these chemical characteristics are "restraints" imposed on the bulk compositions of Archean ores by the primitive geochemical evolution of the earth, as recently suggested by Hutchinson (1973), and evolution of its atmosphere at that time.

If, as proposed earlier, volcanogenic massive sulfide ores have, in fact, formed from submarine exhalations, perhaps the chemical differences between Precambrian and Kuroko ores are manifestations of the geochemical evolution of the earth's lithosphere, hydrosphere, and even atmosphere. In fact, Hutchinson (1973) has developed this thesis in demonstrating a clear evolutionary trend in massive sulfide ores with time.[1]

ACKNOWLEDGEMENTS

Much of what we have learned about massive sulfide deposits is from our many colleagues in industry, government and universities who have passed on their special knowledge to us in discussions and have spent valuable time with us in the field. It would be futile to name all of our benefactors. In extending thanks, we hope that they recognize their contributions and we hasten to absolve them of any misrepresentations which might appear in our text.

Many people have, as well, contributed directly to the preparation of the manuscript by sending us unpublished reports, preprints, figures and samples. These include A.H. Bailes, S.N. Charteris, J.M. Franklin, A.M. Goodwin, R.W. Hutchinson, Hugh Jones, R.R. Large, R.J.M. Miller, V.G. Milne, H.C. Morris, J.T. Nash, Andre Spence, C.D. Spence and Ross Weeks. Fred Jurgeneit did most of the drafting and Brian O'Donovan the photography. R.H. Ridler is to be commended for agreeing to the formidable task of reviewing the manuscript and for the constructive comments arising therefrom. Similarly we are grateful to G.B. Leech for pointing out certain irregularities in the text.

Scott gratefully acknowledges the support of his research on massive sulfides by the National Research Council of Canada (Grant A7069).

REFERENCES

Allchurch, P.D., Brook, W.A., Marshall, A.E. and Reynolds, D.G., 1975. Volcanogenic copper–zinc deposits in the Pilbara and Yilgarn Archean blocks. In: *Economic Geology of Australia and Papua New Guinea, Metalliferous Volume.* Australasian Institute of Mining and Metallurgy, Melbourne, in press.
Anderson, C.A. and Creasey, S.C., 1958. Geology and ore deposits of the Jerome area, Yavapai County, Arizona. *U.S. Geol. Surv. Prof. Pap.*, 308: 185 pp.
Anderson, C.A. and Nash, J.T., 1972. Geology of the massive sulfide deposits at Jerome, Arizona–a reinterpretation. *Econ. Geol.*, 67: 845–863.
Anderson, C.A., Blacet, P.M., Silver, L.T. and Stern, T.W., 1971. Revision of the Precambrian stratigraphy in the Prescott–Jerome area, Yavapai County, Arizona. *U.S. Geol. Surv. Bull.*, 1324-C: 16 pp.
Anderson, G.M., 1973. The hydrothermal transport and deposition of galena and sphalerite near 100°C. *Econ. Geol.*, 68: 480–492.
Aoki, K., Sato, K., Takeuchi, T. and Tatsumi, T., 1970. Kuroko deposits and Towada and Hakkoda volcanoes. *IAGOD Guidebook, 3: Japan.*

---

[1] Editor's note: Cf. Chapter 1 by Veizer, Vol. 3, for a detailed treatment of the geochemical evolution of the earth as related to ore genesis.

Asbury, B.C., 1975. *Sulfide and Wallrock Deformation at the South Bay Mine, Confederation Lake, Northwestern Ontario.* Unpublished M. Sc. Thesis, Univ. of Toronto, Canada.

Asbury, B.C. and Scott, S.D., 1975. Sulfide and wallrock deformation at the South Bay Mine, Northwestern Ontario. *Geol. Soc. Am. (North Central Sect.) – Geol. Assoc. Can. – Miner. Assoc. Can., Ann. Meet., Abstr. Progr.,* 7: 712.

Bailes, A.H., 1971. Preliminary compilation of the geology of the Snow Lake–Flin Flon–Sherridon area. *Manitoba Dep. Mines Nat. Resour., Geol. Pap.,* 1–71.

Bain, G.W., 1973. Discussion of geology of the massive sulfide deposits at Jerome, Arizona reinterpretation. *Econ. Geol.,* 68: 709–714.

Baragar, W.R.A. and Goodwin, A.M., 1969. Andesites and Archean volcanism of the Canadian Shield. *Oreg. Dep. Geol. Min. Ind., Bull.,* 65 (Proc. of the Andesite Conf.): 121–142.

Bateman, J.D., 1944. Sherrit Gordon mine area. *Geol. Surv. Can., Pap.,* 44-4.

Beane, R.E., 1974. Biotite stability in the porphyry copper environment. *Econ. Geol.,* 69: 241–256.

Bennett, R.A. and Rose Jr., W.I., 1973. Some compositional changes in Archean felsic volcanic rocks related to massive sulfide mineralization. *Econ. Geol.,* 68: 886–891.

Berezowski, M., 1972. *A Study of the 2A Orebody at South Bay Mines, NW Ontario.* Thesis, Univ. Toronto, unpublished.

Bishop, D.W., Morris, H.C. and Edmunds, F.R., 1970. Turbidites and depositional features in the lower Belt–Purcell supergroup. *Geol. Soc. Am. Ann. Meet., Abstr. Progr.,* 2 (7): 497.

Boldy, J., 1968. Geological observations on the Delbridge massive sulphide deposit. *Can. Inst. Min. Metall. Trans.,* 71: 247–256.

Brett, R. and Kullerud, G., 1967. The Fe–Pb–S system. *Econ. Geol.,* 62: 354–369.

Bristol, C.C., 1974. Sphalerite geobarometry of some metamorphosed orebodies in the Flin Flon and Snow Lake districts, Manitoba. *Can. Mineral.,* 12 (5): 308–315.

Brown, J.S., 1965. Oceanic lead isotopes and ore genesis. *Econ. Geol.,* 60: 47–68.

Brown, W.L. and Bray, R.C.E., 1960. The geology of the Geco mine. *Can. Min. Metall. Bull.,* 53: 3–11.

Byers, A.R., Kirkland, S.S.T. and Pearson, W.J., 1965. Geology and mineral deposits of the Flin Flon area, Saskatchewan. *Sask. Dep. Miner. Resour., Rep.,* 62.

Carswell, H.T., 1961. *Origin of the Sullivan Lead, Zinc, Silver Deposit, British Columbia.* Thesis, Queen's University, unpublished.

Clark, B.R. and Kelly, W.C., 1973. Sulfide deformation studies, 1. Experimental deformation of pyrrhotite and sphalerite to 2000 bars and 500°C. *Econ. Geol.* 68: 332–352.

Clark, L.A., 1971. Volcanogenic ores: comparison of cupriferous pyrite deposits of Cyprus and Japanese Kuroko deposits. In: Y. Takeuchi (Editor), *IAGOD Volume: IMA–IAGOD Meetings '70,–Soc. Min. Geol. Japan,* Spec. Issue 3: 206–215.

Coats, C.J.A., Clark, L.A., Buchan, R. and Brummer, J.J., 1970. Geology of the copper–zinc deposits of Stall Lake Mines Ltd., Snow Lake area, northern Manitoba. *Econ. Geol.,* 65: 970–984.

Coats, C.J.A., Quirke, Jr., T.T., Bell, C.K., Cranstone, D.A. and Campbell, F.H.A., 1972. Geology and mineral deposits of Flin Flon, Lynn Lake and Thompson areas, Manitoba, and the Churchill–Superior front of the Western Precambrian Shield. *Int. Geol. Congr., 24th, Guide to Excursions A31 and C31,* 96 pp.

Coleman, L.C., 1970. Rb/Sr isochrons for some Precambrian rocks in the Hanson Lake area, Saskatchewan. *Can. J. Earth Sci.,* 7: 338–345.

Craig, J.R. and Kullerud, G., 1968. Phase relations and mineral assemblages in the copper–lead–sulfur system. *Am. Miner.,* 53: 145–161.

Craig, J.R. and Kullerud, G., 1969. The Cu–Zn–S system. *Carnegie Inst. Wash. Yearbook,* 67: 177–179.

Crerar, D.A. and Barnes, H.L., 1974. Reactions of chalcopyrite-rich sulfides in hydrothermal solutions. *Trans. Am. Geophys. Union,* 55: 483 (abstract).

Degens, E.T. and Ross, D.A. (Editors), 1969. *Hot Brines and Recent Heavy-Metal Deposits of the Red Sea.* Springer, New York, N.Y., 600 pp.

De Rosen-Spence, A., 1969. Genèse des roches à cordiérite—anthophyllite des gisements cuprozin-cifères de la région de Rouyn-Noranda, Québec, Canada. *Can. J. Earth Sci.*, 6: 1339—1345.

Descarreaux, J., 1973. A petrochemical study of the Abitibi volcanic belt and its bearing on the occurrences of massive sulphide ores. *Can. Min. Metall. Bull.*, 66 (2): 61—69.

Dresser, J.A. and Denis, T.C., 1949. Geology of Quebec, III. Economic geology. *Que. Dep. Mines Geol. Rep.*, 20: 371—383.

Dugas, J., 1966. The relationship of mineralization to Precambrian stratigraphy in the Rouyn—Noranda area, Quebec. *Geol. Assoc. Can., Spec. Pap.*, 3: 43—56.

Farley, W.J., 1949. Geology of the Sherritt Gordon orebody. *Can. Min. Metall. Bull.*, 42: 25—30.

Fisher, R.V., 1966. Rocks composed of volcanic fragments and their classification. *Earth-Sci. Rev.*, 1: 287—298.

Fiske, R.S., 1963. Subaqueous pyroclastic flows in the Ohanapecosh Formation, Washington. *Bull. Geol. Soc. Am.*, 74: 391—406.

Fleming, P.D. and Fawcett, J.J., 1974. Upper stability of chlorite + quartz assemblages in the Ni/NiO-buffered system $MgO—FeO—Al_2O_3—SiO_2—H_2O$ at 2 kbar water pressure (abstr.). *Trans. Am. Geophys. Union*, 55: 479.

Franklin, J.M., Gibb, W., Kerr, B. and Tammam, A.O., 1973. *Guide to the Mattabi Massive Sulphide Deposit. Can. Inst. Min. Metall. Field Excursion, September 1973.*

Freeze, A.C., 1966. On the origin of the Sullivan orebody, Kimberley, B.C. In: *Tectonic History and Mineral Deposits of the Western Cordillera. Can. Inst. Min. Metall.*, Spec. Vol. 8: 263—294.

Froese, E., 1969. Metamorphic rocks from the Coronation Mine and surrounding area. In: A.R. Byers (Editor), *Symposium on the Geology of the Coronation mine, Saskatchewan. Geol. Surv. Can., Pap.*, 68-5: 55—77.

Geldsetzer, H., 1973. Syngenetic dolomitization and sulphide mineralization. In: G.C. Amstutz and A.J. Bernard (Editors), *Ores in Sediments*. Springer, Berlin, pp. 115—127.

Geological Staff, Hudson Bay Mining and Smelting Company, Ltd., 1957. Cuprus mine. In: *Structural Geology of Canadian Ore Deposits, 2*. Can. Inst. Min. Metall., Montreal, Que., pp. 253—258.

Geological Staff, Hudson Bay Mining and Smelting Company. Ltd. and Stockwell, C.H., 1948. Flin Flon mine. In: *Structural Geology of Canadian Ore Deposits, 1*. Can. Inst. Min. Metall., Montreal, Que., pp. 295—301.

Gilmour, P., 1965. The origin of the massive sulphide mineralization in the Noranda district, north-western Quebec. *Geol. Assoc. Can. Proc.*, 16: 63—81.

Gilmour, P.C., 1971. Strata-bound massive pyritic sulphide deposits—a review. *Econ. Geol.*, 66: 1239—1244.

Gilmour, P. and Still, A.R., 1968. The geology of the Iron King mine. In: J.D. Ridge (Editor), *Ore Deposits of the United States, 1933—1967. Am. Inst. Min. Eng.*, 2: 1239—1257.

Goodwin, A.M., 1965. Mineralized volcanic complexes in the Porcupine—Kirkland Lake—Noranda region, Canada. *Econ. Geol.*, 60: 955—971.

Goodwin, A.M., 1968. Archean protocontinental growth and early crustal history of the Canadian Shield. *Int. Geol. Congr., 23rd*, 1: 69—89.

Goodwin, A.M., 1973. Petrochemical trends in Archean volcanic assemblages, Abitibi belt, Ontario and Quebec, Canada. *Geol. Soc. Am. Ann. Meet. 1973, Abstr. Progr.*, 5 (7): 641—642.

Goodwin, A.M. and Ridler, R.H., 1970. The Abitibi orogenic belt. In: A.J. Baer (Editor), *Symposium on Basins and Geosynclines of the Canadian Shield. Geol. Surv. Can., Pap.*, 70-40: 1—30.

Graf, J.L., Jr. and Skinner, B.J., 1970. Strength and deformation of pyrite and pyrrhotite. *Econ. Geol.*, 65: 206—215.

Guha, J. and Darling, R., 1972. Ore mineralogy of the Louvem copper deposit, Val d'Or, Quebec. *Can. J. Earth Sci.*, 9: 1596—1611.

Helgeson, H.C., 1969. Thermodynamics of hydrothermal systems at elevated temperatures and pressures. *Am. J. Sci.*, 267: 729—804.

Hemley, J.J., 1959. Some mineralogical equilibria in the system $K_2O—Al_2O_3—SiO_2—H_2O$. *Am. J. Sci.*, 257: 241—270.

Hemley, J.J. and Jones, W.R., 1964. Chemical aspects of hydrothermal alteration with emphasis on hydrogen metasomatism. *Econ. Geol.*, 59: 538–569.

Hemley, J.J. and Montoya, J.W., 1971. Some mineral equilibria in the system $K_2O-MgO-Al_2O_3-SiO_2-H_2O$ (abstr.). *Geol. Soc. Am., Ann. Meet., Abstr. Progr.*, 3: 597.

Hobbs, W.H., 1904. Lineaments of the Atlantic border region. *Geol. Soc. Am. Bull.*, 15: 123–176.

Hodge, H.J., 1967. Horne Mine, Noranda Mines Ltd. In: M.K. Abel (Editor), *C.I.M. Centennial Field Excursion, Northwestern Quebec and Northern Ontario.* Can. Inst. Min. Metall., Montreal, Que. pp. 41–46.

Holubec, J., 1972. Lithostratigraphy, structure and deep crustal relations of Archean rocks of the Canadian Shield. *Krystalinikum*, 9: 63–88.

Honnorez, J., Honnorez-Guerstein, B., Valette, J. and Wauschkuhn, A., 1973. Present-day formation of an exhalative sulfide deposit at Vulcano (Tyrrhenian Sea), II. Active crystallization of fumarolic sulfides in the volcanic sediments of the Baia di Levante. In: G.C. Amstutz and A.J. Bernard (Editors), *Ores in Sediments. Int. Union. Geol. Sci., Ser. A*, 3 – Springer, Berlin, pp. 139–166.

Horikoshi, E., 1969. Volcanic activity related to the formation of the Kuroko-type deposits in the Kosaka district, Japan, *Miner. Deposita*, 4: 321–345.

Horikoshi, E. and Sato, T., 1970. Volcanic activity and ore deposition in the Kosaka mine. In: T. Tatsumi (Editor), *Volcanism and Ore Genesis.* Univ. Tokyo Press, Tokyo, pp. 181–195.

Howkins, J.B. and Martin, P.L., 1970. A comparison between the Flin Flon and Snow Lake orebodies of Hudson Bay Mining and Smelting Co. Ltd. *Can. Inst. Min. Metall. Ann. Meet., April, 1970* (presented paper).

Hunt, G.H., 1961. *The Purcell Eruptive Rocks.* Thesis, Univ. Alberta, unpublished.

Hutchinson, R.W., 1965. Genesis of Canadian massive sulphides reconsidered by comparison to Cyprus deposits. *Trans. Can. Inst. Min. Metall.*, 68: 286–300.

Hutchinson, R.W., 1970. Mineral potential in greenstone belts of northwestern Ontario. *Lake Superior Inst. Conf., Thunder Bay, Ont. May, 1970* (presented paper).

Hutchinson, R.W., 1973. Volcanogenic sulfide deposits and their metallogenic significance. *Econ. Geol.*, 68: 1223–1246.

Hutchinson, R.W., Ridler, R.H. and Suffel, G.G., 1971. Metallogenic relationships in the Abitibi belt, Canada: a model for Archean metallogeny. *Trans. Can. Inst. Min. Metall.*, 74: 106–115.

Irvine, T.N. and Baragar, W.R.A., 1971. A guide to the chemical classification of the common volcanic rocks. *Can. J. Earth Sci.*, 8: 523–548.

Ishihara, S. (Editor), 1974. *Geology of Kuroko Deposits. Soc. Min. Geol. Japan*, Spec. Issue, 6: 435 pp.

Jenks, W.F., 1971. Tectonic transport of massive sulfide deposits in submarine volcanic and sedimentary host rocks. *Econ. Geol.*, 66: 1215–1224.

Jones, H., 1973. The Copperfields–Iso, Magusi River deposit. *Ann. Meet. Prospectors Developers Assoc., 41st, Toronto, March 1973* (presented paper).

Kajiwara, Y., 1970a. Syngenetic features of the Kuroko ore from the Shakanai mine. In: T. Tatsumi (Editor), *Volcanism and Ore Genesis.* Univ. Tokyo Press, Tokyo, pp. 197–206.

Kajiwara, Y., 1970b. Some limitations on the physico-chemical environment of deposition of the Kuroko ore. In: T. Tatsumi (Editor), *Volcanism and Ore Genesis*, Univ. Tokyo Press, Tokyo, pp. 367–380.

Kalliokoski, J., 1965. Metamorphic features of North American massive sulfide deposits. *Econ. Geol.*, 60: 485–505.

Kalliokoski, J., 1968. Structural features and some metallogenic patterns in the southern part of the Superior Province, Canada. *Can. J. Earth Sci.*, 5: 1199–1208.

Kanasewich, E.R., 1968. Precambrian rift: genesis of strata-bound ore deposits. *Science*, 161: 1002–1005.

Kanehira, K., 1969. Sulphide ores from the Coronation mine. In: A.R. Byers (Editor), *Symposium on the Geology of the Coronation Mine, Saskatchewan. Geol. Surv. Can., Pap.*, 68-5: 79–136.

King, H.F., 1965. The sedimentary concept in mineral exploration. *Commonw. Min. Metall. Congr., 8th, Australia and New Zealand — Explor. Min. Geol.*, 2: 25—33.

Kinkel, A.R., Jr., 1967. The Ore Knob copper deposit, North Carolina, and other massive sulfide deposits of the Appalachians. *U.S. Geol. Surv., Prof. Pap.*, 558: 58 pp.

Kinkel, A.R., Jr., Thomas, H.H., Marvin, R.F. and Walthall, F.G., 1965. Age and metamorphism of some massive sulfide deposits in Virginia, North Carolina and Tennessee. *Geochim. Cosmochim. Acta*, 29: 717—724.

Kissin, S.A., 1974. *Phase Relations in a Portion of the Fe—S System.* Univ. Toronto, 294 pp., unpublished.

Klesse, E., 1968. Geology of the El Ocotito—Ixcuinatoyac region and of La Dicha stratiform sulphide deposit, State of Guerrero. *Bol. Soc. Geol. Mex.*, 31: 107—140.

Knight, C.L., 1957. Ore genesis — the source bed concept. *Econ. Geol.*, 52: 808—818.

Koo, J., 1973. *Origin and Metamorphism of the Flin Flon Cu—Zn Deposit, Northern Saskatchewan and Manitoba, Canada.* Thesis, Univ. Saskatchewan, 154 pp., unpublished.

Kutina, J., 1968. On the application of the principle of equidistances in the search for ore veins. *Int. Geol. Congr., 23rd*, 7: 99—110.

Kutina, J., 1969. Hydrothermal ore deposits in the western United States: a new concept of structural control of distribution. *Science*, 165: 1113—1119.

Kutina, J., 1971. A contribution to the correlation of structural control of ore deposition between North America and Western Europe. In: Y. Takeuchi (Editor), *IAGOD Volume: IMA—IAGOD Meetings '70. Soc. Min. Geol. Japan*, Spec. Issue 3: 70—75.

Kutina, J., 1972. Regularities in the distribution of hypogene mineralization along rift structures. *Int. Geol. Congr., 24th, Sect. 4, Miner. Deposits*, pp. 65—73.

Kutina, J. and Fabbri, A., 1972. Relationships of structural lineaments and mineral occurrences in the Abitibi area of the Canadian Shield. *Geol. Surv. Can., Pap.*, 71-9: 36 pp.

Kutina, J., Pokorný, J. and Veselá, M., 1967. Empirical prospecting set based on the regularity distribution of ore veins with application to the Jihlara mining district, Czechoslovakia. *Econ. Geol.*, 62: 390—405.

Latulippe, M., 1966. The relationship of mineralization to Precambrian stratigraphy in the Matagami Lake and Val d'Or districts of Quebec. In: A.M. Goodwin (Editor), *Precambrian Symp., Geol. Assoc. Can., Spec. Pap.*, 3: 21—42.

Lea, E.R. and Dill, D.B., Jr., 1968. Zinc deposits of the Balmat—Edwards district, New York. In: J.D. Ridge (Editor), *Ore Deposits of the United States, 1933—1967.* A.I.M.E. 1: 20—48.

Lebedev, L.M. and Nikitina, I.B., 1968. Chemical properties and ore content of hydrothermal solutions at Cheleken. *Dokl. Akad. Nauk. S.S.S.R. (Earth Sci. Sect.)*, 183: 180—182.

Leech, G.B., 1957. *St. Mary's Lake Map Sheet 15, 1957, with Marginal Notes.* Geol. Surv. Can.

Leech, G.B. and Wanless, R.K., 1962. Lead-isotope and potassium-argon studies in the East Kootenay district. *Geol. Soc. Am., Buddington Vol.*, pp. 241—279.

Lickus, R.J., 1965. *Geology and Geochemistry of the Ore Deposits at the Vauze Mine, Noranda District, Quebec.* Thesis, McGill Univ., unpublished.

Magee, M., 1968. Geology and ore-deposits of the Ducktown district, Tennessee. In: J.D.Ridge (Editor), *Ore Deposits of the United States, 1933—1967.* Am. Inst. Min. Metall. Pet. Eng., 1: 207—241.

Mannard, G.W., 1973. The syngenetic massive sulphide deposits. *A.I.M.E. Ann. Meet., Chicago, Ill.* (preprint of presented paper).

Marshall, P., 1935. Acid rocks of the Taupo—Rotorua district. *Trans. R. Soc. N. Z.*, 64: 323—366.

Martin, P.L., 1966. Structural analysis of the Chisel Lake orebody. *Trans. Can. Inst. Min. Metall.*, 69: 323—366.

Martin, W.C., 1957. Errington and Vermillion Lake mines. In: *Structural Geology of Canadian Ore Deposits, 2.* Can. Inst. Min. Metall., Montreal, Que., pp. 363—376.

Matsukuma, T. and Horikoshi, E., 1970. Kuroko deposits in Japan, a review. In: T. Tatsumi (Editor), *Volcanism and Ore Genesis.* Univ. Tokyo Press, Tokyo, pp. 153—180.

Matulich, A., Amos, A.C., Walker, R.R. and Watkins, J.J., 1974. The geology department. In: *The Ecstall Story. Can. Min. Metall. Bull.*, 67: 56–63.

McDonald, J.A., 1967. Metamorphism and its effect on sulphide assemblages. *Miner. Deposita*, 2: 200–220.

Meyer, C. and Hemley, J.J., 1967. Wall rock alteration. In: H.L. Barnes (Editor), *Geochemistry of Hydrothermal Ore Deposits*. Holt, Rinehart and Winston, New York, N.Y., pp. 166–235.

Miller, R.J.M., 1973. The morphology of some Canadian massive sulfide deposits. *A.I.M.E. Ann. Meet. Chicago, Ill.* (preprint of presented paper).

Milne, V.G., 1969. Progress report on a field study of the Manitouwadge area ore deposits. *Can. Inst. Min. Metall. Ann. Meet., April, 1969* (presented paper).

Mookherjee, A., 1970. Dykes, sulphide deposits, and regional metamorphism: criteria for determining their time relationship. *Miner. Deposita*, 5: 120–144.

Mookherjee, A. and Dutta, N.K., 1970. Evidence of incipient melting of sulfides along a dike contact, Geco mine, Manitouwadge, Ontario. *Econ. Geol.*, 65: 706–713.

Mookherjee, A. and Suffel, G.G., 1968. Massive sulfide/late diabase relationships, Horne Mine, Quebec: genetic and chronological relationships. *Can. J. Earth Sci.*, 5: 421–432.

Morris, H.C., 1972. An outline of the geology of the Sullivan mine, Kimberley, British Columbia. In: W.T. Irvine et al. (Editors), *Major Lead–Zinc Deposits of Western Canada. Int. Geol. Congr., 24th, Guide Excursions A24 and C24*, pp. 26–34.

Mukherjee, A.C., Stauffer, M.R. and Baadsgaard, H., 1971. The Hudsonian orogeny near Flin Flon, Manitoba: a tentative interpretation of Rb/Sr and K/Ar ages. *Can. J. Earth Sci.*, 8: 939–946.

Oftedahl, C.A., 1958. A theory of exhalative–sedimentary ores. *Geol. Fören. Stockh. Förh.*, 80: 1–19.

Orville, P.M., 1969. A model for metamorphic differentiation origin of thin-layered amphibolites. *Am. J. Sci.*, 267: 64–86.

Price, P. and Bancroft, W.L., 1948. Waite-Amulet mine. In: *Structural Geology of Canadian Ore Deposits, 1*, Can. Inst. Min. Metall., Montreal, Que., pp. 748–756.

Price, R.A., 1964. The Precambrian Purcell system in the Rocky Mountains of southern Alberta and British Columbia. *Bull. Can. Pet. Geol.*, 12: 399–426.

Purdie, J.J., 1967. Lake Dufault Mines Ltd. In: M.K. Abel (Editor), *C.I.M. Centennial Field Excursion, Northwestern Quebec and Northern Ontario*. Can. Inst. Min. Metall., Montreal, Que., pp. 52–57.

Pyke, D.R. and Middleton, R.S., 1971. Distribution and characteristics of the sulphide ores of the Timmins area. *Trans. Can. Inst. Min. Metall.*, 74: 157–168.

Ramdohr, P., 1953. Mineralbestand, Strukturen und Genesis der Rammelsberg-Lagerstätte. *Geol. Jb.*, 67: 367–494.

Ramsay, B.A. and Swail, E.E., 1967. Manitou–Barvue Mines Ltd. In: M.K. Abel (Editor), *C.I.M. Centennial Field Excursion, Northwestern Quebec and Northern Ontario*. Can. Inst. Min. Metall., Montreal, Que., pp. 36–39.

Riddell, J.E., 1952. *Wall-rock Alteration around Base-metal Sulphide Deposits of Northwestern Quebec*. Thesis, McGill Univ., unpublished.

Ridler, R.H., 1970. Relationship of mineralization to volcanic stratigraphy in the Kirkland–Larder Lakes area, Ontario. *Proc. Geol. Assoc. Can.*, 21: 33–42.

Ridler, R.H., 1971a. Relationships of mineralization to stratigraphy in the Archean Rankin Inlet–Ennadai Belt. *Can. Min. J.*, 92 (4): 50–53.

Ridler, R.H., 1971b. Analysis of Archean volcanic basins in the Canadian Shield using the exhalite concept. *Can. Min. Metall. Bull.*, 64 (714): 20 (abstract).

Ridler, R.H., 1973. Exhalite concept, a new tool for exploration. *Northern Miner*, Nov. 29, pp. 59–61.

Ridler, R.H. and Shilts, W.W., 1974. Exploration for Archean polymetallic sulphide deposits in permafrost terrains: an integrated geological/geochemical technique, Kaminak Lake area, District of Keewatin. *Geol. Surv. Can., Pap.*, 73-34.

Rittman, A., 1962. *Volcanoes and Their Activity*. Wiley, New York, N.Y., 305 pp.

Roberts, R.G., 1966. *The Geology of the Mattagami Lake Mine, Galinee Township, Quebec*. Thesis, McGill Univ., unpublished.

Robertson, D.S., 1953. Batty Lake map-area, Manitoba. *Geol. Surv. Can., Mem.*, 271.

Rokachev, S.A., 1965. More facts about fragmental sulfide concentrations in roof rocks of the Urals Sibay ore deposit. *Dokl. Acad. Sci. U.S.S.R., A.G.I. Transl.*, 162: 85–87.

Roscoe, S.M., 1965. Geochemical and isotopic studies, Noranda and Matagami areas. *Trans. Can. Inst. Min. Metall.*, 68: 297–285.

Sakai, H., Osaki, S. and Tsukagishi, M., 1970. Sulphur and oxygen isotopic geochemistry of sulphate in the black ore deposits of Japan. *Geochem. J.*, 4: 27–39.

Sakrison, H.C., 1966. *Chemical Studies of the Host Rocks of the Lake Dufault Mine, Quebec*. Thesis, McGill Univ., unpublished.

Salmon, B.C., Clark, B.R. and Kelly, W.C., 1974. Sulfide deformation studies, II. Experimental deformation of galena to 2000 bars and 400°C. *Econ. Geol.*, 69: 1–16.

Sangster, D.F., 1972a. Precambrian volcanogenic massive sulphide deposits in Canada: a review. *Geol. Surv. Can. Pap.*, 72-22: 44 pp.

Sangster, D.F., 1972b. Isotopic studies of ore-leads in the Hanson Lake-Flin Flon-Snow Lake mineral belt Saskatchewan, Manitoba. *Can. J. Earth Sci.*, 9: 500–513.

Sasaki, A., 1974. Isotopic data of Kuroko deposits. In: S. Ishihara (Editor), *Geology of Kuroko Deposits. Soc. Min. Geol. Japan*, Spec. Issue, 6: 389–397.

Sato, T., 1972. Behaviours of ore-forming solutions in seawater. *Min. Geol.*, 22: 31–42.

Schermerhorn, L.J.G., 1970. The deposition of volcanics and pyritite in the Iberian pyrite belt. *Miner. Deposita*, 5: 273–279.

Schneiderhöhn, H., 1944. *Erzlagerstätten*. Fischer, Jena, 371 pp.

Scott, S.D., 1973. Experimental calibration of the sphalerite geobarometer. *Econ. Geol.*, 68: 466–474.

Scott, S.D., 1974a. Sphalerite composition in the Zn–Fe–S system. In: *Proceedings of the Ninth Meeting on Experimental and Engineering Mineralogy and Petrography, Irkutsk, U.S.S.R., 1973* (in press; in Russian).

Scott, S.D., 1974b. Sphalerite geobarometry of regionally metamorphosed terrains. *Geol. Soc. Am. Ann. Meet. Abstr. Progr.*, 6 (7): 946–947.

Scott, S.D. and Barnes, H.L., 1971. Sphalerite geothermometry and geobarometry. *Econ. Geol.*, 66: 653–669.

Scott, S.D. and Kissin, S.A., 1973. Sphalerite composition in the Zn–Fe–S system below 300°C. *Econ. Geol.*, 68: 475–479.

Shade, J.W., 1974. Hydrolysis reactions in the SiO_2-excess portion of the system $K_2O–Al_2O_3–SiO_2–H_2O$ in chloride fluids at magmatic conditions. *Econ. Geol.*, 69: 218–228.

Shadlun, T.N., 1971. Metamorphic textures and structures of sulphide ores. In: Y. Takeuchi (Editor), *IAGOD Volume: IMA–IAGOD Meet. '70. Soc. Min. Geol. Japan*, Spec. Issue 3: 241–250.

Sharpe, J.I., 1968. Geology and sulphide deposits of the Mattagami area. *Que. Dep. Nat. Resour., Geol. Rep.*, 137.

Sillitoe, R.H., 1973. Environments of formation of volcanogenic massive sulfide deposits. *Econ. Geol.*, 68: 1321–1325.

Simmons, B.D. and co-workers, 1973. Geology of the Millenbach massive sulphide deposit, Noranda, Quebec. *Can. Min. Metall. Bull.*, 66: 67–78.

Sinclair, W.D., 1971. A volcanic origin for the No. 5 zone of the Horne mine, Noranda, Quebec. *Econ. Geol.*, 66: 1225–1231.

Skinner, B.J., White, D.E., Rose, H.J. and Mays, R.E., 1967. Sulfides associated with the Salton Sea geothermal brine. *Econ. Geol.*, 62: 316–330.

Solomon, P.J., 1963. *Sulfur Isotopic and Textural Studies of the Ores at Balmat, New York and Mount Isa, Queensland*. Thesis, Harvard Univ., 162 pp., unpublished.

Spence, C.D., 1966. Volcanogenetic settings of the Vauze base metal deposit, Noranda district, Quebec. *C.I.M. Ann. Meet., April, 1966* (presented paper).

Spence, C.D., 1967. The Noranda area. In: M.K. Abel (Editor), *C.I.M. Centennial Field Excursion, Northwestern Quebec and Northern Ontario*. Can. Inst. Min. Metall. Montreal, Que., pp. 36–39.

Spence, C.D., 1975. Volcanogenic features of the Vauze sulfide deposit, Noranda, Quebec. *Econ. Geol.*, 70: 102–114.

Spence, C.D. and De Rosen-Spence, A.F., 1975. The place of mineralisation in the volcanic sequence at Noranda, Quebec. *Econ. Geol.*, 70: 90–101.

Spurr, J.E., 1923. *The Ore Magmas*. McGraw-Hill, New York, N.Y., 915 pp.

Stanton, R.L., 1959. Mineralogical features and possible mode of emplacement of the Brunswick Mining and Smelting orebodies, Gloucester County, New Brunswick. *Can. Min. Metall. Bull.*, 52 (570): 631–643.

Stanton, R.L., 1960. General features of the conformable "pyritic" orebodies. I. Field association. II. Mineralogy. *Trans. Can. Inst. Min. Metall.*, 63: 22–27; 28–36.

Stanton, R.L., 1964. Mineral interfaces in stratiform ores. *Bull. Inst. Min. Metall. Aust.*, 696: 45–79.

Stanton, R.L., 1972a. A preliminary account of chemical relationships between sulfide lode and "banded iron-formation" at Broken Hill, New South Wales. *Econ. Geol.*, 67: 1128–1145.

Stanton, R.L., 1972b. *Ore Petrology*. McGraw-Hill, New York, N.Y., 713 pp.

Stanton, R.L. and Willey, H.G., 1970. Natural work–hardening in galena, and its experimental reduction. *Econ. Geol.*, 65: 182–194.

Stauffer, M.R., 1967. Tectonic strain in some volcanic, sedimentary, and intrusive rocks near Canberra, Australia: a comparative study of deformation fabrics. *N.Z. J. Geol. Geophys.*, 10: 1079–1108.

Stockwell, C.H., 1960. Flin Flon–Mandy. *Geol. Surv. Can.*, Map 1078A.

Suffel, G.G., 1948. Waite Amulet mine, Amulet section. In: *Structural Geology of Canadian Ore Deposits, 1*. Can. Inst. Min. Metall., Montreal, Que., pp. 757–763.

Suffel, G.G., 1965. Remarks on some sulphide deposits in volcanic extrusives. *Can. Min. Metall. Bull.*, 58: 1057–1063.

Suffel, G.G., Hutchinson, RW. and Ridler, R.H., 1971. Metamorphism of massive sulphides at Manitouwadge, Ontario, Canada. In: Y. Takeuchi (Editor), *IAGOD Volume IMA–IAGOD Meetings '70. Soc. Min. Geologists Japan* Spec. Issue 3: 235–240.

Sullivan, C.J., 1968. Geological concepts and the search for ore, 1930–1967. In: E.R.W. Neale (Editor), *The Earth Sciences in Canada, A Centennial Appraisal and Forecast. R. Soc. Can. Spec. Publ.*, 11: 82–99.

Swanson, C.O. and Gunning, H.C., 1945. Geology of the Sullivan mine. *Trans. Can. Inst. Min. Metall.*, 48: 645–667.

Tatsumi, T., 1965. Sulfur isotope fractionation between coexisting sulfide minerals from some Japanese ore deposits. *Econ. Geol.*, 60: 1645–1659.

Tatsumi, T. (Editor), 1970. *Volcanism and Ore Genesis*. Univ. Tokyo Press, Tokyo, 448 pp.

Tatsumi, T. and Watanabe, T., 1970. Geological environment of formation of the Kuroko-type deposit. In: Y. Takeuchi (Editor), *IAGOD Volume: IMA–IAGOD Meetings '70, Soc. Min. Geol. Japan*, Spec. Issue, 3: 216–220.

Templeman-Kluit, D.J., 1970. The relationship between sulfide grain size and metamorphic grade of host rocks in some strata-bound pyritic ores. *Can. J. Earth Sci.*, 7: 1339–1345.

Thomson, J.E., 1941. Geology of McGarry and McVittie townships. *Ont. Dep. Mines, Ann. Rep.*, 50 (7): 1–94.

Timms, P.D. and Marshall, D., 1959. The geology of the Willroy mines base-metal deposits. *Proc. Geol. Assoc. Can.*, 11: 55–65.

Tokunaga, M. and Honma, H., 1974. Fluid inclusions in the minerals from some Kuroko deposits. In: S. Ishihara (Editor), *Geology of Kuroko Deposits. Soc. Min. Geol. Japan*, Spec. Issue, 6: 385–388.

Van de Walle, M., 1972. The Rouyn–Noranda area. In: G.O. Allard, G. Duquette, M. Latulippe and M. van de Walle, *Precambrian Geology and Mineral Deposits of the Noranda–Val d'Or and Matagami–Chibougamau Greenstone Belts, Quebec. Int. Geol. Congr.*, 24th, Guide to Excursions *A41* and *C41*, pp. 41–51.

Vokes, F.M., 1969. A review of the metamorphism of sulphide deposits. *Earth Sci. Rev.*, 5: 99—143.
Walker, T.L., 1930. Dalmatianite, the spotted greenstone from the Amulet mine, Noranda, Quebec. *Univ. Toronto Stud., Geol. Ser.*, 29: 9—12.
Wang, S., 1973. *Sphalerite Pole Figure Analysis and Metamorphic Textures, Matagami Lake Mine, Quebec*. Thesis, Columbia Univ., 194 pp., unpublished.
Wedow, H., Jr., Kiilsgaard, T.H., Heyl, A.V. and Hall, R.B., 1973. Zinc. In: D.A. Brobst and W.P. Pratt (Editors), *United States Mineral Resources. U.S. Geol. Surv. Prof. Pap.*, 820: 697—711.
Whitmore, D.R.E., 1969. Geology of the Coronation copper deposit. In: A.R. Byers (Editor), *Symposium on the Geology of the Coronation Mine, Saskatchewan. Geol. Surv. Can., Pap.*, 68-5: 37—54.
Wolf, K.H., 1976. Ore genesis influenced by compaction. In: G.V. Chilingarian and K.H. Wolf (Editors), *Compaction of Coarse-Grained Sediments*, II. Elsevier, Amsterdam, in press.

Chapter 6

GEOLOGY OF THE ZAMBIAN COPPERBELT

V.D. FLEISCHER, W.G. GARLICK and R. HALDANE

INTRODUCTION (V.D.F.)

The well-known Zambian Copperbelt is one of the great metallogenic provinces in the world. Its eight operating mines produced in 1974 about 750,000 metric tonnes of copper, to place the country in the fourth position on the world production table.

Several mines that outcropped were discovered near the turn of the century, others, including those covered with deep overburden, were found later in the 1920's.

Credit must be given to the "ancients" who mined several deposits down to the water-table, using primitive techniques. The early prospectors were quick to locate these deposits, being led to them by the local inhabitants. Subsequently, further ore shoots have been discovered by diamond drilling on strike near established mines and it is an optimistic belief that more success stories will continue to be forthcoming — after all, there is the old prospector's adage that "the best place to look for elephants is in elephant country"!

A useful plant indicator for discovery of the shale-type orebodies or in areas with a thin soil cover is the copper flower (*Ocimum homblei* De Wild). It has been found growing as a dense carpet in clearings over many outcrops and sub-outcorps of ore shale beyond the Copperbelt as well as far afield as Shaba in Zaire to the north, and in the Lusaka district to the south. The more porous and deeply leached sandstone (arenite) deposits show insignificant geochemical copper values in overlying soils and there the copper flower is absent.

The geology of the Copperbelt is truly fascinating and the chapter that follows relates some of the more interesting features and geological thoughts recorded over the last 45 years of active mining and prospection.

The Geology of the Northern Rhodesian Copperbelt, edited by F. Mendelsohn and published in 1961 by Macdonald, London, is mainly out of date in its title, since the country on attaining independence, changed its name to Zambia. The facts and data presented in that book, are still as valid 14 years later, but mining has removed many millions of tons of ore, more ore has been proved by drilling, and new orebodies have been discovered. A greatly increased geological staff has also discovered further evidence confirming the syngenetic origin of the ore. The above book is no longer readily available, not even in its Russian translation, and has become a collector's treasure. This chapter will serve as a digest of the data presented in Mendelsohn's book and provides much of the new data available from the continued exploration of this rich and intriguing mineral field.

The sections on each mine or orebody comprise a brief description of the Basement Complex and of the footwall formations, a more detailed description of the ore formation, with sulfide distribution, zoning, and especially the relation to lithological facies, shore lines, bioherms, and barren gaps. The hanging-wall formations are briefly mentioned to complete the stratigraphic section, and the later history of folding, faulting, and metamorphism, erosion and weathering completes the picture of the ore and orebody as

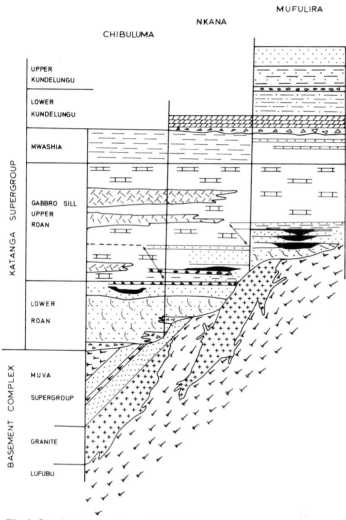

Fig. 1. Stratigraphic section and facies diagram.

revealed by mining. All recent publications were consulted in attempting to bring this latest contribution up to date and all mine geologists are thanked for criticism and corrections of the original manuscripts.

There have been many changes of company names, and with township names different to those of the adjacent mines, only the mine names have been used in these reports, thus enabling ready comparison with Mendelsohn's book.

REGIONAL SETTING

The Copperbelt (Fig. 3, pp. 229–230) is situated on the southern slope of the flat watershed between the Zaire and Zambezi drainage systems, about 1300 m above sea level. Only 13° south of the equator, it has tropical climate ameliorated by the altitude, with heavy summer rains, mostly as thunderstorms, and a long dry winter with rare frost in open meadows at stream heads. Interfluve areas have deep soils with a well developed "stone line" and nodular laterite profile at the base of the "A" zone. Sporadic outcrops are, therefore, confined to some stream beds and scattered hills of more resistant rocks.

Mapping in the field is difficult and inferential deduction must be applied to some extent for the production of worthwhile geological maps. For example, soil colour and texture and the associated vegetation, especially tree species, can be most useful indicators of the main rock type hidden underneath. Broad geology can be mapped this way, especially with the aid of aerial photographs.

The detailed geology of the Copperbelt proper has been obtained by manual pitting along structure lines down to hard rock or the watertable (3–30 m deep), systematic diamond drilling, and eventually from mine openings. Aerial and ground geophysical surveys continue to assist in deciphering broad structures and stratigraphic correlation.

The stratigraphic diagram (Fig. 1) shows the general relations of the basal arenaceous facies, termed the Lower Roan, of the Katanga Supergroup (Sequence) to the underlying Basement Complex, consisting of the Lufubu and Muva Supergroups, the former intruded by many and varied, usually potassic, Granites.

SEDIMENTARY BASINS

Figure 2 shows the all important sedimentary basins which contain the stratiform copper–cobalt deposits of the Copperbelt region, and it can be seen that there are in fact only two basins, the one fairly simple and confined, the other complex and widespread.

The former is the Roan-Muliashi basin at Luanshya which contains the Roan, Muliashi and Baluba orebodies and spreads over a small area in comparison with the sinuous sub-outcrop of the Lower Roan which demarcates the outer rim of the larger one. This basin is actually an aggregate of several interconnecting synclines. The Lower Roan is seen to

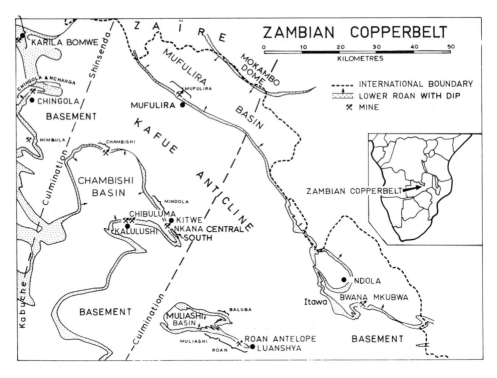

Fig. 2. Copperbelt location map.

stretch continuously along the periphery from south of Kalulushi through the mines at Chibuluma, Nkana, Chambishi, round to Mimbula, Nchanga, Karila Bomwe (Bancroft) then back from Zaire through Mufulira, Ndola, Bwana Mkubwa and returning into Zaire.

It is considered that the present deep basins represent only the remaining roots of a very extensive spread of Katanga sediments in the geological past. Several cycles of erosion to present levels must have removed the top parts of existing orebodies (those that outcropped) and more than likely several other discrete orebodies in the process as well.

It is estimated that some synclinal structures may extend beyond 10,000 m below surface. The base configuration is now known to be highly irregular in most of the synclines, and not simple as previously inferred. This feature has obvious economic attractions, as the potential Lower Roan rocks may be at minable depths in the more central regions of the synclines, and, even more important, the ore formation will be found to boast of greater back (dip) lengths due to folding in many localities within the basins.

BASEMENT COMPLEX

The Lufubu schists and gneisses have a dominant northeasterly grain and the intervening granite intrusions, dated at 1975 million years (m.y.) (Cahen et al., 1970), have the same trend, controlling local depositional basins and cross-folds on the later Lufilian folding of the Katanga sediments. Thus, the minor axis of the Chambishi-Nkana basin is directed northeast over a belt of Lufubu schists.

The Muva quartzites and schists have an east-northeast trend well south of the Copperbelt, but appear haphazardly with many irregularities, in the vicinity of the mines. The Muva contains a thick band of hard cross-bedded quartzite that outcrops more abundantly than any other Copperbelt formation. The distribution of the Muva is, therefore, well mapped and is found to be highly erratic due to intense episodes of folding and major faulting. Belts of the hard quartzite member occur in sinuous disposition with no preferential trends. In the mines themselves, no Muva is intersected in underground workings, hence detail study of this group is lacking. The Muva rocks are younger than most of the granites, but the Nchanga red microline granite is younger. It is not in contact with the Muva, which may be coeval with the Kibaran of Zaire, about 1300 m.y. in age. The Kibaran post-tectonic tin granites have been dated at 900 m.y., giving an earliest date for Katanga sedimentation. Several small mineral occurrences have been recorded in diverse rock types in the basement. No economic deposits have been found, perhaps because of lack of detailed prospection. Gold, manganese, iron and copper mineralization is known and recently a small emerald mine has opened.

KATANGA SUPERGROUP

The Katanga sediments overlie an extremely irregular unconformity, with hills and ridges of granite and of Muva quartzite rising several hundred metres above adjacent valleys. In addition, the land surface had an overall gentle tilt to the southwest, so that the earliest arenaceous sediments occur to the southwest of the Copperbelt, and similar sediments, still labelled Lower Roan, are of progressively younger age to the northeast. The first sediments were terrestrial talus screes, valley boulder conglomerates, and aeolian[1] sands. These were followed by aqueous sediments deposited near the shore of a north-eastward transgressing sea, consisting of beach gravels, beach or deltaic sands, locally algal bioherms, passing seaward into sands and muds. In both of the latter anaerobic bacteria were active in producing hydrogen sulfide. The copper—cobalt sulfide deposits are restricted to these near-shore sediments. Most of the Lower Roan arenites and argillites contain carbonate and sulfate. Both seaward and upward in the succession, Lower Roan

---

[1] Recently, Van Eden (1974) has re-interpreted the aeolian deposits as of marine origin but the present writer believes that a good case can be made to support the older concept.

sediments give way to dolomites and dolomitic argillites, with beds of anhydrite. These latter are called Upper Roan and the contact is taken at the base of the first significant dolomite bed at the top of the uppermost arenite, or the top of the argillite overlying the latter; significantly the division between Upper and Lower Roan is arbitrary and may be at higher time-stratigraphic horizons to the northeast.

The Upper Roan is overlain by up to 600 m thickness of dolomite and shale, the latter generally carbonaceous and pyritic, of the Mwashia. In the Mufulira syncline three feldspathic arenites occur near the top of the Mwashia.

A tillite or fluvio-glacial conglomerate, up to 150 m thick of granite, quartz, quartzite, dolomite, and shale fragments in a massive argillaceous matrix, forms the base of the Lower Kundelungu. It is overlain by the Kakontwe limestone and dolomite, which contain minor disseminations of sphalerite, galena, and purple fluorite of probable sedimentary origin. This association of zinc, lead, and fluorite in strata 500 m above the Lower Roan is in significant contrast to the syngenetic copper, cobalt, and iron sulfides associated with mauve anhydrite in the latter. The Kakontwe is overlain by about 500 m of pyritic shales grading up into massive sandy argillites.

The base of the Middle Kundelungu is marked by a thinner tillite, followed by limestone of the Calcaire Rose and then argillite and purplish feldspathic sandstone. This section is little known.

NOMENCLATURE

The metamorphosed arenites of the Copperbelt have been labelled "quartzite" and arkose since the initial geological investigations of 1926 to 1930. Only the Muva quartzite has more than 90% quartz, the lower limit as defined by Pettijohn (1948). Some of the feldspathic "quartzites" of aeolian origin contain up to 85% quartz, but most come under the category of arkose. One of the aeolian "quartzite" beds at Mufulira has the following composition: 50% quartz, 15% feldspar, 35% anhydrite. Practically all of the Katanga aqueous arenites are arkose, sub-graywacke or graywacke. As the term quartzite is entrenched in the literature and used for local formational names, it is here perpetrated with a capital Q or inverted commas.

Many of the mineralized "quartzites" contain over 20% feldspar, 10% or more sericite, and over 20% sulfides. Where the sulfide content exceeds 33% the term "sulfidite" has been coined. Thus, the Chibuluma orebody consists of mineralized arenites, thin shales, and thin to massive sulfidite layers, whilst the Mufulira orebodies consist of mineralized grits and arkoses, thin mineralized argillites, slump breccias derived from these sediments, and carbonaceous graywacke, with numerous thin sulfidite layers (Garlick, 1967).

In common with mining geology nomenclature in mining regions the world over, local names given to certain rocks early in the development of the orebodies for stoping, persist. For example, "far-water dolomite", "near-water dolomite" and "ultra far-water dolomite"

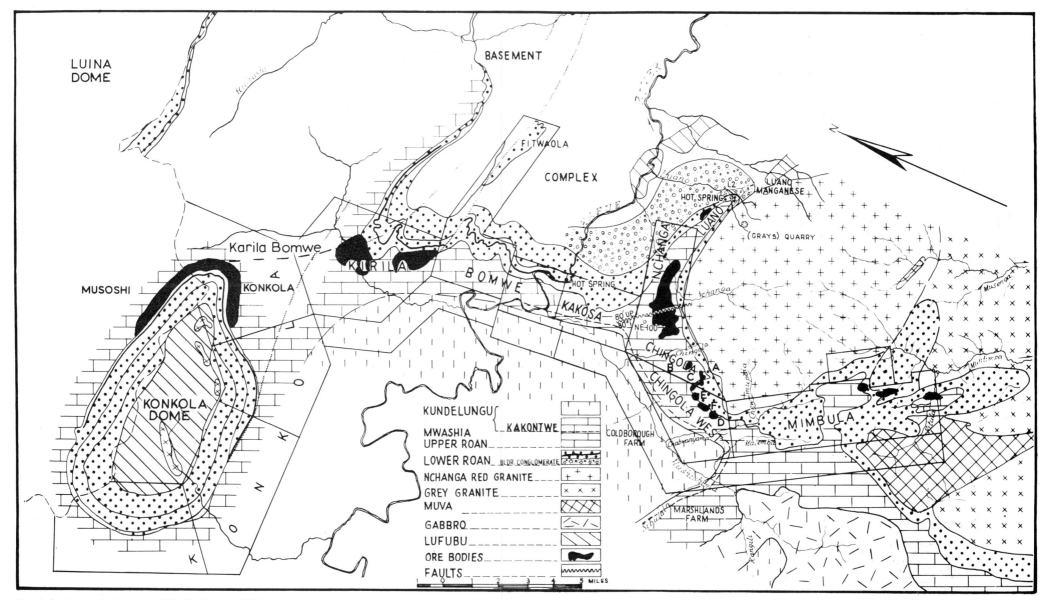

Fig. 4. Surface geological map of Karila Bomwe—Nchanga area.

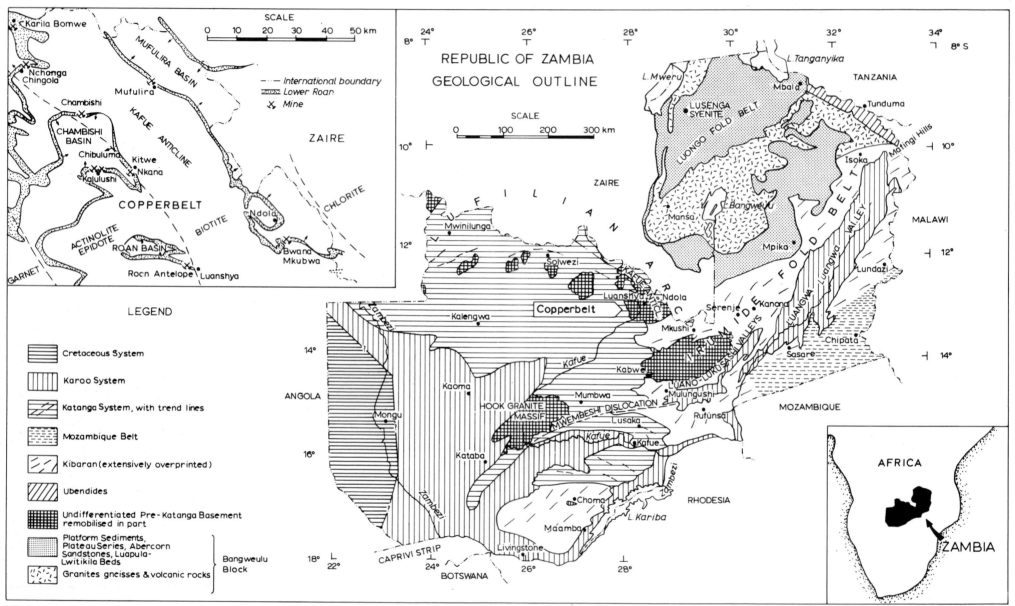

Fig. 3. Geological outline for Zambia.

of Nkana, the "Chingola dolomite" and the "cherty dolomite" of Chambishi, the "Mud-seam" of Mufulira typify local terminology. The "Mudseam" at Mufulira obtained this name at shallow depths where a thin dolomitic siltstone was weathered into a black manganiferous clay. In present mine workings it is fresh. Further examples can be found in the stratigraphic column for Mufulira (see Fig. 28).

ACKNOWLEDGEMENTS

Permission from the Management of Roan Consolidated Mines Limited and Nchanga Consolidated Copper Mines Limited for publication of this chapter is gratefully acknowledged. All draughting office personnel who worked on the illustrations and several typists involved in final preparation of the texts are warmly thanked.

STRUCTURES OF THE ZAMBIAN COPPERBELT (W.G.G.)

THE LUFILIAN ARC

The Zambian Copperbelt occupies 160 km at the southeast end of the 800 km long fold arc, which stretches from Angola, through Mwinilunga, Kolwezi, Fungurume, Lubumbashi (Elisabethville) in Zaire, then into Zambia just south of Musoshi and to the end of the Congo pedicle. The fold arc was formed by a northward push of Katanga sediments between the Kibaran massif in the west and the Bangweulu massif in the east (Fig. 3). At Kolwezi and Fungurume, nappes of Upper Roan sediments were thrust north over middle Kundelungu, but northward the folds die out rapidly into the Lufira plain, which gave the name to the fold arc and the orogeny (Van Doorninck, 1928). The uppermost Kundelungu may post-date the orogeny.

Through the Shaba province of Zaire the copper–cobalt deposits occur in diapyric anticlines and overthrust the Upper Roan shale and dolomite associated with stromatolite reefs (Roche siliceuse cellulaire), but near the Zambian border the first domes of the basement appear, with the Kinsenda deposit in Lower Roan graywacke and arenite on the south flank of the Luina dome, and Musoshi in shale on the north flank of the Konkola dome (Fig. 4).

THE KAFUE ANTICLINE

From the border just south of Kinsenda, the major Kafue anticlinal axis emerges from under an east–west syncline of Katanga sediments and for 50 km is followed by the Kafue river, exposing Lufubu schists, gneiss, and intrusive granite, and one infold of

Muva quartzite. This arch separates by 20 km the deep Mufulira syncline from the shallower Chambishi-Nkana and Muliashi basins.

The ore-shale deposits are confined to the west flank of the Kafue anticline and there is
strong evidence, detailed under the descriptions of the mines, to show that this arch was
initiated along an axis roughly parallel to the old shoreline of the copper sea and the
margin of the geosyncline.

EARLY RECUMBENT FOLDS

At Chingola, Nchanga West, and at 13 km south of the Mufulira mine, there are folds
with flat or diving axial planes pushed over from the southwest. They are evidently an
extension of the nappe and fold structures of Shaba. Figure 5 shows a recumbent fold
deep in the later Nchanga syncline.

EN-ECHELON SYNCLINES

West-northwesterly plunging synclines lie along the west side of the Kafue anticline
and are from north to south Fitwaola, Nchanga, Nkana, Baluba and Roan. They are
generally asymmetric, with axial planes dipping northeast, with northeast limbs vertical or
locally overturned, indicating a thrust from that side. This may indicate that, after the

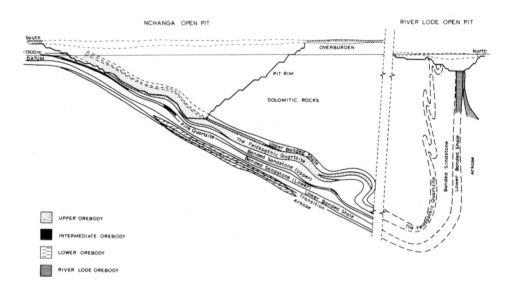

Fig. 5. Geological section across Nchanga syncline.

initial northeastward directed push piled the recumbent folds towards the Bangweulu massif, the latter overrode the thickened pile of sediments in front of it; the increased pressure developed a cleavage or foliation parallel to axial planes.

CROSS-FOLDING

Cross-folding developed in a north-northeast direction parallel to major structures in the basement (Fig. 2). This resulted in the Kabuche—Kinsenda culmination: an alignment of granite domes; the depression containing the wide part of the Mufulira syncline and the Chambishi—Nkana basin opening west of Chibuluma into a wide irregular syncline striking west of south; the next culmination from the Mokambo dome to the tail of the Nkana syncline and west of Muliashi; and, finally, the Muliashi—Itawa depression oriented nearly east—west due probably to the influence of the Walamba granite south of Roan Antelope stretching across the Kafue anticline. Basins of Katanga and depressions of fold axes tend to coincide with areas of basement schist.

DOMES

In this fold syntaxis, the Katanga has been updomed at the crossing of anticlinal axes and in particular the Kabuche—Kinsenda culmination appears due to the coalescence of aligned domes, by stripping of much of the sedimentary cover. With less erosion some basement areas, mostly granite, have retained a complete or nearly complete ring of outward dipping Lower Roan sediments. The best example is the Konkola dome between Musoshi and Karila Bomwe; less perfect are the Ndola and the Mokambo domes. Originally, these domes were presumed to be due to upward punch of granitic magma intrusive into the Katanga system, but careful mapping has shown that they contain only pre-Katanga granite, gneiss, and, in places, Lufubu schist; none even approach Eskola's (1948) definition of a mantled gneiss dome. Study of the Nchanga red granite has shown that it acted as a rigid buttress during folding of the overlying sediments, but that mild reheating during the orogeny caused homogenization of the strontium isotopes to give an apparent age of 570 ± 40 m.y. (Snelling et al., 1964). Cahen et al. (1970) obtained a rubidium—strontium age of about 1320 m.y. for the granite at Roan Antelope, but zircons indicate an age over 1750 m.y. and zircons from the Mufulira granite give 1975 ± 20 m.y.

DRAGFOLDS

The dragfolds are described in the detailed structures of individual mines.

FAULTS

The east–west Phantom fault crosses the Mufulira syncline diagonally with Mwashia and Kakontwe thrust south over Kundelungu, but causes only two sharp flexures in the Lower Roan on the east flank of the Kafue anticline, where the fracture is marked by basic dikes, and further west results in Muliashi porphyritic gneiss being thrust over the north limb of the Fitwaola syncline.

There is little faulting of mining importance, with the exception of a fault occupied by a basic dike under the Mindola stream at Nkana, the main fault in the Nchanga open pit, and the minor fault occupied by the lamprophyre at River Lode, Nchanga. Elsewhere, it is remarkable how efficiently the sediments have been moulded on the irregularities of the basement surface and have stretched, flowed and bent around both open and tight folds, without appreciable faulting, although joint systems are well developed.

METAMORPHISM (W.G.G.)

Regional metamorphism accompanied the folding of the Katanga sediments, with increase in grade from the lower greenschist facies in the extreme east, to high greenschist facies over most of the mine areas, and to lower epidote–amphibolite facies in the southwest. Immediately west of the Copperbelt, almandine garnet appears, and around basement domes, 40–200 km to the west, kyanite is abundant in Lower Roan argillaceous sediments.

This metamorphism is important in having increased the grain size of the sulfides as well as that of the other constituent minerals of the sediments and thus aiding the beneficiation of the ores. However, even at kyanite grade, there has rarely been appreciable migration of the sulfides from their original host; thus kyanite-bearing beds may contain copper sulfides whilst adjacent garnetiferous layers are generally barren (McGregor, 1964).

BASEMENT COMPLEX

The Basement was previously subjected locally to much higher temperatures and pressures than the overlying Katanga system, yet evidence for higher grades of metamorphism are rare or subtle. Sillimanite and cordierite have not been found, whilst kyanite and garnet are rare. In a recent study of the Nchanga granite (Garlick, 1973), it was a surprise to discover that garnet and perthitic microcline were preserved throughout except in shear zones, whilst in the overlying arkose detrital garnet is absent and microcline is non-perthitic.

Mendelsohn (1961) recognized that the Basement had undergone a long-lasting and pervasive retrogressive metamorphism during the folding of the cover. Biotite and sericite are thus characteristic of both the Lufubu and Muva argillaceous rocks, whilst amphibole,

garnet, and kyanite, are erratically preserved. Chlorite sericite schists occur in the Muva and may indicate late zones of intense shearing. Much research has yet to be done on Basement rocks.

KATANGA SYSTEM

Sandstone and arkose have been completely recrystallized to quartzitic rocks, with occasional preservation of original grain outline in the new quartz and feldspar. Perthitic microcline has changed to non-perthitic microcline and plagioclase has recrystallized as albite or albite—oligoclase. West of Nkana, much of the microcline has been albitized, showing step twinning. Micas and clay mineral were converted to biotite and sericite. Some tourmaline, apatite, and most of the zircon show original rounded outlines, but tourmaline, like quartz and feldspar, commonly shows considerable overgrowths. Bedding, cross-bedding, and ripple marks are generally well preserved, although the iron of the heavy mineral streaks is recrystallized to specularite or maghemite.

Argillaceous rocks are recrystallized to biotite sericite rocks with usually considerable quartz and feldspar from the coarser detritus, and although some have developed axial plane cleavage, the more massive argillites still preserve scour and fill features, ripple marks, and preconsolidation slump structures.

Carbonate rocks have been recrystallized to dolomitic marbles and impure varieties have, in the western part of the Copperbelt, developed tremolite or actinolite, the former usually developing in the presence of copper sulfides. Talc is abundant in the Mufulira syncline.

Scapolite is abundant in the argillites, including the ore-shale, at Roan Extension and less so at Nchanga. Mendelsohn (1961) has shown that it grew under directed stress and this is conspicuous where it causes "spotting" of the middle stretched limbs of certain dragfolds at Roan, whilst the syncline, anticline, and outer limbs are relatively free from scapolite.

Epidote occurs in tightly folded calcareous argillaceous arenites at Muliashi, the southern part of Nkana, Chibuluma, and Mimbula and, with the actinolite of these southwestern areas, indicates the epidote—amphibolite facies.

Anhydrite and dolomite occur in most Lower and Upper Roan rocks.

BASIC INTRUSIVES

Amphibole and scapolite are ubiquitous in the gabbro sills (basement highs) and relict labradorite and pyroxene are rare. Olivine has been recorded. Restriction of chlorite to shear zones and the abundant amphibole suggest that the gabbros have been metamorphosed to higher grade than the enclosing sediments, but replacement of pyroxene and feldspar by uralite and scapolite may be auto-metasomatic effects whilst the sills retained some of their original heat. Tourmaline is locally conspicuous in the sediments adjacent to the sills.

METAMORPHIC VEINS

During folding, the more competent beds fractured and the openings were filled by minerals crystallizing from connate and metamorphic liquids squeezed from the immediately adjacent wall rock, exemplified at Roan by *bc* gash veins in thin arkose beds between argillites. These gash veins, up to 20 cm thick, thin abruptly and terminate against the adjacent argillites, and the arkose shows boudin structure. The vein filling is specularite, microcline, anhydrite and quartz in that order of crystallization. Rubidium—strontium age determination on the microcline gives 840 ± 40 m.y., but the feldspar of the adjacent wall rock has not been determined (Cahen et al., 1970).

Larger veins are usually in the *ac* direction, but contain the same minerals in the same order. At Roan, several of these extend for over 100 m into the Basement, where their mauve anhydrite content immediately distinguishes them from magmatic veins and pegmatites. At Mufulira, they usually contain only anhydrite, dolomite and quartz in the footwall arenites, indicative of a considerable anhydrite content of these beds and lack of migration of feldspar at the lower metamorphic grade. Only a little albite has been observed in the Mufulira veins, and specularite is not common. At Nkana North orebody, vein feldspar has the shape of adularia, but X-rays indicate low microcline. At Roan, Chambishi, and Nchanga, the feldspar is mostly microcline, but without perthitic structure as occurs in the microcline of the Basement.

Veins in dolomitic beds are dominantly dolomite, but usually with a little quartz.

Veins in the argillite beds may be conformable or discordant; the former are often corrugated or folded and show pinch and swell structures. Microcline (or adularia at Nkana) is as abundant in these veins as in those of the arkose; anhydrite is frequent, but dolomite and quartz are ubiquitous. In barren argillite, there may be a little specularite, but in sulfide-bearing argillite there is no specularite and instead copper, cobalt and iron sulfides succeed quartz in the order of crystallization.

There is variation in the mineral content of the veins according to metamorphic grade. At Mufulira, feldspar is practically absent and specularite is rare. Albite crystals have been recorded in less than a dozen veins. Mauve anhydrite, dolomite and quartz are ubiquitous in all veins, and bornite, chalcopyrite, and, less commonly, pyrite occur in veins cutting the orebodies. At Nkana, specularite, adularia and anhydrite are common outside the orebody and anhydrite, pink manganiferous dolomite, quartz and sulfide beautify the ore. At Nchanga and Chambishi, flesh-coloured microcline and white dolomite with the usual specularite or sulfides occur. At Roan, flesh-coloured, rarely pale green, microcline, with mauve anhydrite, dolomite, specularite or sulfides are accompanied by muscovite and less commonly tourmaline. Tremolite and scapolite rarely appear in the veins and then only in Roan Extension, Muliashi and Chibuluma.

TYPES OF OREBODIES (V.D.F.)

Copperbelt orebodies are tabular, they boast of long strike lengths and are relatively thick, extending from near-surface to depths with ore back lengths of 2000 m and more. A few completely blind orebodies have been discovered, some by grid drilling.

There are two main varieties or types of ore-bearing rock: shale and arenite/arkose. About 60% of the reserves are of the shale-type.

A remarkable line-up of the shale-type of copper—cobalt orebodies can be seen in Fig. 6 through Luanshya—Nkana—Chambishi—Chingola—Karila Bomwe—Musoshi. In each case, carbonaceous shale with pyrite gives way eastward to low-carbon cupriferous shale. The deposits in this category are considered to be aligned along an ancient shoreline, the pyritic fraction representing the deeper water facies. Algal reefs associated with the orebodies suggest a relatively shallower environment about the time of ore deposition. The

TABLE I

Ore-reserve statistics (1974)

Mine	Ore reserves				Copper metal (in tonnes)	
	gross tonnes	copper grade (%)	net tonnes	copper grade (%)	gross	net (remaining)
1 Nchanga	374,000,000	4.11	254,000,000	3.45	15,371,000	8,763,000
2 Mufulira	282,000,000	3.47	141,000,000	3.22	9,786,000	4,530,000
3 Luanshya (Baluba)	318,000,000	2.82	142,000,000	2.58	8,957,000	3,664,000
4 Rokana (Nkana)	312,000,000	2.81	122,000,000	2.51	8,767,000	3,062,000
5 Konkola (Bancroft)	161,000,000	3.65	126,000,000	3.56	5,877,000	4,486,000
6 Chambishi	55,000,000	2.92	43,000,000	2.93	1,594,000	1,265,000
7 Chibuluma	20,000,000	4.84	8,000,000	4,74	983,000	387,000
8 Bwana Mkubwa	8,000,000	3.40	4,000,000	3.62	138,000	265,000
					Total 26,422,000	

1, 4, 5 and 8 are 49% owned by Anglo American Corporation (Central Africa) Limited.
2, 3, 6 and 7 are 49% owned by Amax Zambia, Inc. (formely Roan Selection Trust) and others.
Note: (1) Since 1970 the Government of Zambia acquired 51% of the copper mining company shares.
(2) Prospection continues on all properties by way of diamond drilling. Several areas of significant mineralization in the mining leases are being evaluated with a view to mining them; these are excluded from the above tabulation at this time.

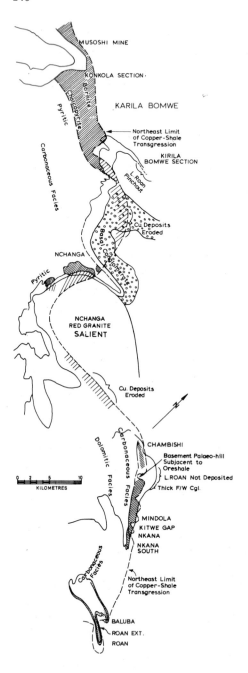

Fig. 6. Shoreline ore deposits along northeastern edge of carbonaceous shale facies.

arenite/arkose (wacke) type of orebodies are well developed to the northeast of the Kafue anticline (Mufulira and Bwana Mkubwa) but are not entirely confined to this strike. Similar orebodies of varying dimensions, usually rich in grade, occur both above (Nchanga) and below (Nchanga, Chambishi, Chibuluma and other new discoveries) the recognized ore shale horizon.

A common structural feature of the ore shale is a display of intense and tight drag folding in contrast to large-amplitude, open-type folds in the more competent arenite orebodies.

Ore-reserve statistics for each producing mine are given in Table I.

MINERALIZATION

The valuable sulfide mineralization in the Copperbelt is now regarded by most geologists working on the mines to have been emplaced contemporaneously with the enclosing sediments. This classifies these stratiform deposits as syngenetic or synsedimentary. Recently, the importance of gypsum/anhydrite has been recognized. The calcium-sulfate minerals constitute a significant proportion of the critical geological column at each property and its overabundance testifies to saturated chemical conditions. This would imply that ample sulfur was available for the formation of any of the sulfides possibly through the action of sulfate-reducing bacteria. The widespread action of sulfate-reducing bacteria is now given more credence and the only conjectural part of ore genesis seems to have settled on the somewhat irrelevant question directed at the original source of the metal fluxes. The author doubts if satisfactory explanations for the latter will ever be forthcoming. The matter is considered purely academic and would have little practical application. However, two major sources can be postulated. The older Basement rocks contain many occurrences and deposits of copper. For example, $1-3$ m thick veins of quartz-carbonate with pyrite–chalcopyrite–bornite assemblages in the Basement are occasionally exposed in the maze of footwall tunnels in the mines. This could have been one primary source of the metals derived through denudation processes that acted on the provenance of the sediments themselves. The other source may have been more distant, originating from exhalative solutions rich in metal leading to a concentration such as in the brines of the Red Sea (see Chapter 4 by Degens and Ross, Vol. 4).

Whatever the source, it is considered that the various metals were transported into marine waters and eventually concentrated through chemico-bacterial processes in favourably confined basins at river mouths or gulfs under anaerobic conditions during specific stages in the subsequent sedimentation. Carbon occurs to varying degrees in all the deposits and it is acknowledged that organic carbon provided some of the nutrients for bacterial activity.[1]

---

[1] Cf. Chapter 6 by Trudinger, Vol. 2, and Chapter 5 by Saxby, Vol. 2, for bacterial processes and influence of organic matter, respectively.

At the start of each sulfide epoch (at the base of an orebody), there is usually evidence of an interface and change from prevailing oxidizing to reducing conditions. Examples from waste to ore are: siderite to sulfide, barren brown shale to bluish shale with sulfide, red limonitic arenite to grey arenite with sulfide. Of prime significance also is a notable reduction of the anhydrite content within the orebody itself compared with some of the adjacent sediments. Interstitial dolomite and sulfides appear to be antipathetic to anhydrite and may testify to pH and physicochemical variations in the sedimentary basins at critical times. There is evidence also that sedimentary reworking took place in the basal parts of the arenaceous orebodies and that some of the sulfides were finally emplaced in a detrital fashion – similar to the more common mechanically concentrated iron and titanium oxides ("ores") found in cross-bedded sandstone units.

The primary mineralization at each mine is essentially sulfide with a cap of secondary oxidation due to weathering, usually to a depth range between 30–70 m below surface, depending upon the permeability of host rocks. Deep leaching down thin porous beds to 500 m or more is common. The remarkable feature of the mineralization, apart from its disseminated nature, is that cobalt (carrolite) occurs as an economic accessory (up to 0.25% Co) in the mines situated west of the Kafue anticline (in shales and arenites).

Bornite is the most abundant sulfide followed by chalcocite (partly secondary near surface) and chalcopyrite. Various admixtures exist, but it is uncommon to have end members of the series together, e.g., pyrite and bornite. Grain sizes of the sulfides, where relatively free of metamorphism, are directly related to the grain size of the host rock. For this reason, fine disseminations of bornite and/or chalcocite in dark argillaceous beds are more difficult to evaluate megascopically than mineralization in coarser clastics (e.g., wackes of Mufulira and Bwana Mkubwa).

Lateral secretion veins occur abundantly in all orebodies. A diagnostic feature of these veins is that they traverse ore and waste, yet copper mineralization is emplaced within the limits of the orebodies exclusively, which mitigates strongly against the theory that they were feeder veins.

DISTRIBUTION OF ORE

As orebodies are confined to certain shales and arenites of the Lower Roan, all the mines sub-outcrop in the narrow meandering strip of Lower Roan overlying the unconformity on the Basement. The latter occupies nearly two-thirds of the area of the Copperbelt proper, but in the areas west to the Angolan border and in Zaire, the Basement appears in only a few isolated domes, suggesting that there is much deep-lying Lower Roan to be explored.

In the isolated Roan–Muliashi basin, Fig. 2, footwall arenite ore occurs in Muliashi, and shale-type ore through Muliashi, Roan Extension, Roan, and Baluba, the latter with useful cobalt values. West of the Kafue anticline, on the south flank of the Nkana syncline are the Chibuluma West and Chibuluma deposits in arenites and on the northeast flank of the same syncline the Nkana South, Nkana North, and Mindola orebodies in shale. On the north flank of the Chambishi—Nkana basin are the Chambishi Main and West orebodies in shale. The Lower Roan with pyritic carbonaceous shale trends west and then south around the Chambishi basin and eventually connects with the highly folded east flank of the Nchanga—Karila Bomwe basin. From south to north under pyritic Lower Banded Shale are the four ore lenses of Mimbula (Fitula) in arkose, the Chingola D, F, E, and C orebodies also in arkose, Chingola A in shale, Chingola B in arkoses above and below the Lower Banded Shale; at Nchanga West the lower orebody is in arkose and shale and overlapping this the upper orebody in arkose at the top of the Lower Roan. On the other side of the Nchanga syncline is the River Lode in shale and after a 12-km gap the Bancroft orebodies of Kirila Bomwe and Konkola in shale. The Musoshi shale deposit on the north flank of the Konkola dome is in Zaire. From Kirila Bomwe, the Lower Roan swings east and thins over the plunging Kafue anticline. Lower Roan arenites flank the northeast limb of the anticline, containing the Mufulira orebodies and 72 km southeast the Bwana Mkubwa orebodies, also in arenaceous hosts.

Copperbelt orebodies generally boast of long strike lengths and have handsome thicknesses extending from near surface to depths of 1600 m and more. However, a feature not always appreciated is the total thickness of pay and non-pay mineralization through the mineralized formation. It is estimated that had nature condensed all the copper occurring in these sediments more efficiently to 2% grades or better, then the mining gross ore reserves would be trebled.

That sporadic copper mineralization continued over such an extended time and in diverse rock types ranging, as they do in the Mufulira and Chingola deposits for example, from fine shales to coarse grits, suggests that sulfide mineralization continued intermittently during sedimentation of low through to high energy levels.

OREBODY DESCRIPTIONS

Most of the orebodies mentioned above are described briefly in the following text. Emphasis is on the syngenetic features, the footwall formations leading up to deposition of the ore-bearing sediments and the distribution of various ore minerals in relation to transgressive conglomerates, to algal bioherms, and to water depth of the original sedimentary environment. Hanging-wall formations are mentioned briefly to complete the stratigraphic picture. Structure and metamorphism with veining are summarized to give the present shape and condition of the orebody.

Under the heading "Syngenetic explanation", a model for formation of the orebodies is presented, amplified as geological histories under "Genesis of oreshale deposits" and "Genesis of arenaceous ore deposits". These genetic considerations are supplemented by a section entitled "Syngenesis versus Epigenesis".

KONKOLA AND MUSOSHI (W.G.G.)

The Konkola mine consists of the South and North Kirila Bomwe deposits dipping west off the Kafue anticline and the Konkola deposit dipping southeast off the Konkola dome, separated only by the international boundary from the Musoshi deposit on the northern flank of the same dome (Fig. 2).

Whereas the other Copperbelt deposits are associated with a thin development of Lower Roan and, hence, are influenced by the paleotopography of the Basement, the Karila Bomwe—Musoshi deposits are underlain by great thicknesses of Lower Roan meta-arenites and conglomerates, forming prominent hills and ridges on the mid-Tertiary peneplain.

The better known Kirila Bomwe deposits are described in detail, but Konkola is now being mined and is, with Musoshi, mentioned only briefly to complete the picture of the north end of the ore-shale line (see Fig. 6).

BASEMENT COMPLEX

Biotite gneiss and schist are intruded by the Muliashi granitic "porphyry", which has large rounded microcline porphyroblasts, simulating pebbles, in a gneissic matrix of oligoclase, quartz, biotite, and epidote.

KATANGA SYSTEM OR SUPERGROUP

Basal Conglomerate (0–300 m thick)

The Basal Conglomerate is well exposed in the Kafue valley, where it may be twice the above indicated thickness, but duplication by folding or thrusting is suspected. It consists of slightly deformed quartz and quartzite boulders in a matrix of flattened boulders of sericitic granite and gneiss as well as layers of micaceous arenite.

Footwall quartzite formation (300 m)

The conglomerate is overlain by light and dark-grey meta-arenite, gritty arkose with red feldspars and conglomerate layers. Cross-bedding indicates currents from the north-east. In the mine area, a 2 m thick aeolian arenite occurs in the thick aqueous arenite, but to the south, near the Kafue, thick aeolian beds form outcrops with spectacular cross-bedding marked by banding due to slight concentrations of iron oxides.

Schwellnus (1961) notes that the pebbles and boulders in this footwall formation are dominantly quartzite and grey granite, whilst the overlying formation is highly feldspathic. A grey argillaceous arenite about 20 m thick occurs at the top of the footwall formation.

Porous Conglomerate—sandstone formation (15—50 m)

The Porous Conglomerate is 9—21 m thick. With a zone of boulders at the base grading up into pebble conglomerate in an arkosic matrix with dolomite and anhydrite cement (the latter leached to considerable depth) it is, with the overlying beds, very similar to the Lower Conglomerate and overlying beds at Nkana and Chambishi.

Dark-grey sandy argillites, grey argillaceous arenites, and some beds of pink arkose overlie the Porous Conglomerate. They are 30 m in the south, but only 3 m thick in the north, suggestive of transgressive overlap by the Footwall Conglomerate. Cross-bedding is mainly from the northeast.

Footwall Conglomerate (0 in south —3.5 m in north)

This is included in the above formation by mine geologists, but as it represents the transgressive beach gravel of the ore-shale formation, it should logically be included with the latter. Over large areas it is a highly porous arkose with slightly rounded red feldspar fragments. On soft weathered and kaolinized surfaces, the somewhat rare quartz pebbles stand out conspicuously. As the lower contact of the Footwall Conglomerate is especially wet and weathered, evidence for transgressive features are hard to see. The bed is not recognized in the southern 600 m of the south orebody.

Banded Shale and Sandstone formation (60 m)

A cyclic sequence of siltstone, shale, and feldspathic arenite overlies the Footwall Conglomerate. The lower 9 m contains the only economic copper mineralization and is sub-divided into six horizons with distinctive lithology and usually distinctive grade of mineralization:

Unit A (0.6—1 m). A thinly bedded calcareous sandy siltstone, with erratic mineralization commonly about 1.5% copper as oxide or chalcocite. In the south where it rests on foot-wall sandstone, it is a fresh, laminated grey shale with pink sandstone; over the Footwall Conglomerate in the north, it is decomposed in upper levels to a sandy clay with manganese. It is generally crenulated by movement on the competent footwall beds and compares to the schistose ore of Nkana.

Unit B (1.4—2 m). This consists of a thickly bedded to massive grey siltstone fairly uniformly mineralized with about 5.5% copper finely disseminated and compares with the low-grade argillite of Nkana. It is more resistant to weathering than adjacent units.

Unit C (0.9—2 m). This consists of alternating 7—12 cm layers of grey siltstone and pink calcareous sandstone (or cherty dolomite?), with wavy or wrinkled bedding and copper between 3 and 6% grade largely concentrated along bedding planes, both as sulfidite layers and as metamorphic veins. This unit may be mineralized where adjacent layers are barren of copper and compares to the Banded Ore of Nkana.

Unit D (1—1.8 m). There is a gradational contact of the previous unit into a grey siliceous siltstone, thinly laminated, with erratic mineralization commonly in small lenses and concentrated along bedding planes. It is intermediate in appearance between the Cherty Ore of Nkana and the Roan Antelope ore-shale and is in places the richest bed at Karila Bomwe with up to 8% copper.

Unit E (0.6—1.5 m). This is a thinly bedded grey to reddish-brown sandstone with fine mica flakes along the bedding. Mineralization is highly erratic and generally up to 1.3% as minor oxide minerals in the upper levels. It compares with the leached shallow parts of the Nkana "Porous Sandstone".

Unit F. Forms the hanging wall of the ore formation and is a feldspathic sandstone commonly kaolinized, with no cross-bedding and no detrital iron concentrations.

Barren "gap"

Between the north and south orebodies is a gap nearly 1.5 km wide with no mineralization in any of the units. Approaching the gap from the north orebody, ore is restricted to the C unit and lenticles of sandy dolomite increase in frequency and the shale layers between the thickening dolomites are crumpled and folded, partly due to pre-consolidation slump. Cherty quartz lenses become more abundant in the dolomite bed and in the weathered formation form "quartz rubble" in a soft dirty micaceous formation. These facies changes are reminiscent of those approaching the Irwin Shaft gap at Roan and it may be suspected that an algal bioherm occupies part of the unexplored barren gap at Karila Bomwe.

Eastern margin

Approaching the eastern margin of the Kirila Bomwe North orebody, the ore is confined to the top of the B unit and to the C unit. The assay hanging-wall drops as the assay footwall rises and the calcareous sandy laminae become more sandy and increase in thickness. A dolomitic argillite "marker" in the C unit grades eastwards into a calcareous sandstone. Arkose lenses in the E unit show load casts and other beds show pre-consolidation slumps directed to the west. After all ore has terminated, the ore-shale grades into a feldspathic arenite, which cannot be differentiated from footwall beds and the latter diminish in thickness rapidly along the northeast limb of the Karila Bomwe anticline, with eventual pinchout of the Lower Roan arenites.

Konkola orebody

The ore-shale here is a dark grey siltstone, more siliceous than at Kirila Bomwe, but westwards it grades into a carbonaceous shale.

Mineral zoning

Zoning of the sulfides is the same as in the other shale-type deposits and is best preserved in the more impermeable B unit, with pyrite in the west at considerable depth, chalcopyrite at intermediate depths, and bornite in the east. In the Kirila Bomwe South orebody above the 580-m level, chalcocite (partly primary) is predominant in the B unit and above the 420-m level malachite comes in and is the predominant mineral to the sub-outcrop.

At both Konkola and Musoshi, the bornite zone is in the east and passes westwards into a chalcopyrite zone and then into pyritic carbonaceous shale. Malachite and supergene chalcocite zones truncate the uptilted and weathered portions of the bornite and chalcopyrite zones in the sub-outcrop of the ore-shale, both south and north of the Konkola dome.

As shown in Fig. 6, the mineral zones are probably continuous at great depth in the saddle between the east end of the Konkola dome and the North orebody on the nose of the Karila Bomwe anticline, allowing for possible complications such as gaps due to algal bioherms. Mineralized ore-shale in the overthrust syncline at Fitwaola, 5 km southeast of the South orebody, suggests that further exploration and development may require modification of Fig. 6, possibly as a deep eastward embayment of the ore-shale shoreline.

Footwall mineralization

Below the Kirila Bomwe South orebody, mineralization in places extends down to a parting in the Footwall Conglomerate or for 10 ft into the Footwall Sandstone. Although consisting largely of supergene minerals, it probably represents oxidation of sulfides in situ, comparable to the sulfide mineralization in footwall beds at Muliashi and Nkana South.

Hanging-wall formations

Arkose (10.7 m). This overlies the 45–60 m thickness of Banded Shale and Sandstone, and is probably equivalent to the "Feldspathic Quartzite" of Nchanga and the Upper quartzite of Chambishi.

Feldspathic sandstone, grits, and shales (58 m). These form a gradation into the Upper Roan, but are here included in Lower Roan.

Shale with grit (82 m).

Upper Roan (220 m)

Interbedded dolomite, dolomitic sandstone, and dolomitic shale form this group and are deeply weathered.

Mwashia (200 m)

Conglomerate and shale, 45 m thick, are overlain by 152 m of pyritic shales, in part carbonaceous.

Lower Kundelungu

Tillite, Kakontwe limestone, and thick shales overlie the Mwashia in the syncline north of Kirila Bomwe.

FOLDING, FAULTING, AND METAMORPHISM

Although the Konkola dome and the Karila Bomwe anticline probably developed originally on a northwestward striking flexure, later pressure from a little east of north has resulted in en-echelon anticlines plunging in opposite directions. The southern anticline plunges about 8° in direction north 70° west and has five bench-like parasitic folds on its southwest limb plunging at low angles in the same direction, corrugating the south orebody. These folds are, however, not as intense as the dragfolds of Chambishi and Nkana mines. The west limbs are generally steep to nearly vertical, whilst the middle limb has a moderate dip to north and only in the major folds show a component of dip to east. Both crests and troughs are gently rounded, with no sign of carinate troughs. Supergene oxidation of sulfides is generally intense on crests and troughs, but is much less on vertical limbs. Lower Roan footwall arenites form Kamenza Hill at the crest and along the north limb of the Karila Bomwe anticline. The northern orebody lies on the north limb of this anticline, dipping gently north at first and then steepening to 60° below the 300-m level. The north orebody peters out to the east against the old shoreline, whilst to the south the "Barren gap", lying diagonally across the nose of the anticline, interrupts the ore before it can sub-outcrop. Hence, this orebody is "blind".

Some faults, somewhat sinuous but generally perpendicular to fold axes, are marked by intense kaolinization of the beds. Some have differential slip of fold axes on either side. South of the South orebody is the major fault striking east-southeast into the Phantom fault of the Mufulira syncline. It results in the Karila Bomwe anticline of gneiss being thrust south over the Fitwaola syncline of Lower Roan containing mineralized ore-shale.

Metamorphism at Karila Bomwe appears intermediate between that of Chambishi and Mufulira. Biotite is extensively developed in argillaceous beds. Considerable slippage or shearing occurs along bedding planes and, along some steep footwall formations, sericite schist with lenticular quartz feldspar veins had formed.

In the footwall formations, quartz-specularite veinlets are common, some with feldspar, carbonate, and rare anhydrite. In the ore formation, veins are less common than at Chambishi and contain mainly carbonate and sulfides.

Sulfide concentrations are controlled very strictly by stratigraphy, bedding, and lamination. However, on a small scale, grains of sulfides and also malachite pseudomorphous after sulfide are commonly oriented parallel to axial planes, especially where folding is obvious.

CHAMBISHI (W.G.G.)

Situated midway between Nkana and Nchanga (Fig. 2), the Chambishi Main and Western deposits have many features similar to the former and a few to the latter. Except for a few outcrops of granite and of barren ore-shale in the Chambishi stream, all the formations are buried under laterite and soil, so that geology was interpreted from prospecting pits into weathered formations and from drill holes. An open pit in the Main Orebody has provided magnificent exposures of all formations up to the gabbro with text book illustrations of the fold structures in contrasting arkose and argillite beds (Fig. 7). Two mineral clearings with the copper flower, *Ocimum homblei* De Wild, one with a small copper stained outcrop, indicated the presence of the two orebodies.

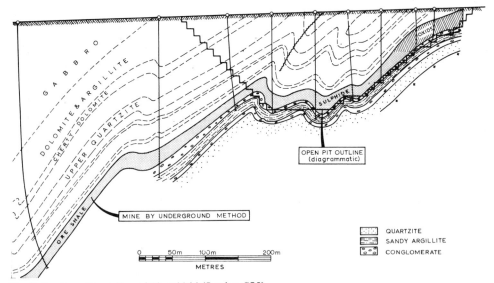

Fig. 7. Stratigraphic section of Chambishi (Section CS5).

BASEMENT COMPLEX

Grey, microcline-biotite-granite with opalescent quartz contains a few xenoliths of Lufubu schist and is cut by numerous aplite dikes. It is overlain by Muva conglomerate, quartz schist, and quartzite striking northeast across the Kafue anticline. This formation has been intersected in only one drill hole east of the mine, where thick quartz-sericite-schist contains a sparse dissemination of pyrite and chalcopyrite.

KATANGA SYSTEM OR SUPERGROUP

Lower Roan Group

Basal Conglomerate (0–67 m). This is developed only under the Western orebody.

Aeolian Quartzite (0–33 m). This overlaps the basal conglomerate and overlies the flanks of the granite paleo-ridges. It is a white even-grained feldspathic arenite, more quartzitic than the overlying arkoses, with neither silty, argillaceous, gritty, nor pebbly layers. Smooth parallel truncation planes are widely spaced. Cross-bedding is of large scale and foresets, making an angle of up to 30° with the truncation planes, indicate a wind from about N40°W, similar to directions in the aeolianites of Chibuluma and Nkana. One layer contains anhydrite lenses flattened parallel to fold axial planes. Anhydrite quartz veins, perpendicular to fold axes, are barren of copper.

Footwall Transition Arenites (0–18 m). Arkose, grading up into argillaceous arenites and thin argillites, overlies the aeolian beds. In the north wall of the pit a few metres above the aeolian, there is disseminated chalcopyrite and pyrite in an arkose bed, too thin and localized to make ore. It is cut by a few *ac* feldspar, quartz, and sulfide veins.

Cobble Conglomerate (0–18 m). Granite and quartzite cobbles up to 20 cm in diameter occur in a carbonate- and anhydrite-rich sandy matrix leached to porous manganiferous material near surface. Schist and some granite fragments are flattened parallel to axial plane in folded areas. At the east end of the Western orebody, disseminated bornite and chalcopyrite make ore with underlying Transition Arenites, where banked against a granite ridge.

Arkose and argillite (0–12; –23 m under Western orebody). Poorly bedded white to pink arkoses, with argillite beds increasing to west.

Footwall Conglomerate (0.3–9 m). This is variable in composition and thickness, being composed of pebbles and cobbles of granite, schist, and quartz adjacent to paleo-ridges of granite and thinning to less than a metre of arkose with sparse pebbles down dip and to the west. Both east and west of the Main orebody, there are promontories of granite projecting south into the ore-shale, which near surface thins to one fifth its usual thickness. Yet, there is always a thin blanket of arkose or pebbly arkose between shale and granite, due to facies change from shale to arkose and to conglomerate approaching the paleo-ridges. Thus, the Footwall Conglomerate is a blanket formed by overlapping sandy and pebbly beaches at the transgressive margin of the ore-shale sea.

Ore-shale (0–30 m). This is a fine-grained biotite quartz argillite with up to 8% potash and only 0.2% soda. Distinct banding is due to varying dolomite content, but, unlike Nkana, there are few dolomite layers — a banded dolomite-argillite bed at the base, usually contorted and schistose, and a brown dolomite bed less than 1 m thick at about 2 m below the top.

Over the eastern granite ridge, it is barren and the upper half contains numerous arkose layers; westwards and down dip this thins and chalcocite disseminations appear in the lower third of the ore-shale. The only stratigraphic marker is a sandy layer above the ore showing pre-consolidation slump and westward the top of the ore rises above this marker.

Chalcocite gives way to bornite and in the basal dolomitic schist chalcopyrite also appears with much bornite; both sulfides are richly segregated along the bedding and cross-cutting quartz dolomite veins occur in the east wall of the pit. The basal dolomitic schist thickens westward and down dip and eventually forms an ore layer 3 m thick with up to 10% copper. This high-grade layer makes up for the diminishing grade in the overlying argillite, where bornite gradually gives way to chalcopyrite. Down dip and westward chalcopyrite in the upper layers of the ore-shale gives way to pyrite. The bornite zone is again developed against the western granite ridge over which the ore-shale thins to 6 m and is barren for 460 m along strike. The Main orebody thus occupies a bay between the two granite headlands and has a strike of 800 m increasing to 1500 m at 300 m below surface, with a thickness of up to 30 m (see Fig. 8).

West of the western granite ridge, as the ore-shale thickens, only a little bornite appears and the Western orebody is dominantly chalcopyrite dissemination with pyrite above and in basal beds. Consequently, this orebody with a strike of 1800 m is low-grade, containing a little over 2% copper. Of special interest is the development of a small orebody in arkose below the Cobble Conglomerate, where these beds abut the western flank of the ridge.

Except for the repetition of zones around the western granite ridge, Chambishi provides a magnificent east–west slightly oblique cross-section of the marginal sediments of the sea in which the ore-shale was deposited. East of the Main orebody, the ore-shale is only 9–6 m thick for 2 km, beyond which it becomes very sandy and cannot be distinguished from overlying arkoses. These beds eventually overstep the basement. By contrast, from the Western orebody to the west flank of the Chambishi–Nkana basin the ore-shale is thick, carbonaceous, and pyritic and the Footwall formations are thicker than at Chambishi.

PLAN
CHAMBISHI

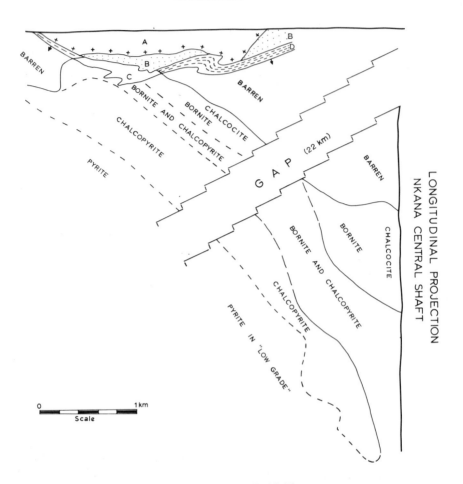

Fig. 8. Plan of mineral zones at Chambishi compared with Nkana.

Stromatolites. No algal structures have been found at Chambishi. Midway between this
and the Mindola deposits drilling revealed a domal structure under the Mwambashi stream,
where the ore-shale is draped over a granite hill truncated at the base of the shale and
much of the thickness of the latter is replaced by a barren dolomitic facies in which
stromatolites are preserved. This algal bioherm, nearly a kilometre in diameter, is flanked
on the east by a narrow zone of disseminated bornite and then by chalcopyrite, but on
the west the bioherm projects into the chalcopyrite zone, which after 700 m passes into
disseminated pyrite and pyrrhotite in carbonaceous shale.

Hanging-wall Quartzite (3–12 m). This is a white arkose with cross-bedding marked by detrital iron oxides with silty and argillaceous layers. Currents were from north or north-east.

Interbedded quartzite and argillite (30–37 m). Two, 3 m thick, arenites are underlain and separated by similar thicknesses of argillite and overlain by about 15 m of schistose argillite and dolomite. The top arenite is a gritty to pebbly pink arkose equivalent to the Pink Quartzite of Nchanga.

Upper Quartzite (12–24 m). A coarse white arkose with up to 25% microcline and a few pebbles is an excellent marker at the top of the Lower Roan. Concentrations of iron-oxide mark parts of foresets and curving bottomsets of cross-bedded units. Direction of currents was from north and northeast.

Eastwards the formation thickens, especially as underlying and overlying argillite beds become sandy, and it merges with underlying arenites and the Hanging-wall Quartzite, and overlaps the ore-shale; eventually at about 3 km east it thins out against the unconformity and is overlapped by Upper Roan argillite and sandstone. This complete pinchout of the Lower Roan on the Basement is repeated at the north end of the Mindola section of the Rokana mine (Fig. 9).

Upper Roan Group

Interbedded schist and quartzite (23–27 m). Thick beds of biotite schist and schistose sandy argillite contain interbeds less than 1 m thick of cross-bedded arkose indicating currents from the northeast.

Cherty dolomite (12–24 m). A white to pink cherty dolomite weathers to black manganiferous talcose clay with chert boulders. Four kilometres to the east, where the Upper Roan overlaps the Upper Quartzite, it contains a meagre dissemination of chalcopyrite.

Sandy talc schist (76–90 m). A greenish grey chloritic biotite schist with a little talc contains numerous lenses of arkose up to 5 cm thick and at increasingly close intervals upwards are 7–30 cm layers of dolomite.

Dolomite with gabbro sills (400 m). White to pink dolomite with anhydrite in some layers. Some cherty layers occur with talc and pyrite is disseminated in shale layers.

About 100 m above the base, meta-gabbro sills totalling 170 m of scapolite amphibolite plus 52 m of granophyre invade the section. Chlorite amphibole shear zones with a little dolomite and sparse pyrite, pyrrhotite, and meagre chalcopyrite occur in the gabbro.

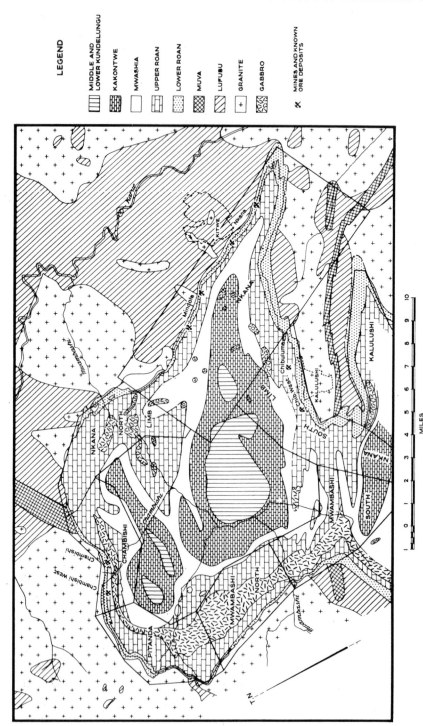

Fig. 9. Geological map of Chambishi–Nkana basin.

Mwashia Group (550 m)

In the east–west syncline between Chambishi and Mindola, the formations above the Upper Roan have been explored by prospecting pits and a few drill holes. The Mwashia consists of 300 m of grey argillites overlain by 240 m of black carbonaceous shales with considerable disseminated pyrite and very meagre chalcopyrite.

Lower Kundelungu Group

Tillite (30 m). Grey gritty argillites contain scattered pebbles and boulders of quartz, granite, and Upper Roan dolomite. It is of fluvio-glacial origin (Van Eden and Binda, 1972).

Kakontwe Formation (426 m). Limestone and dolomite, 152 m thick, is overlain by 122 m of pyritic shales and then by another 152 m of dolomite.

Kundelungu Shales (over 460 m). These shales are covered usually with thick laterite, on which the trees are stunted, with termitaries covered by bamboo and trees making prominent features. They occupy central shallow synclines in the Nkana–Chambishi basin.

FOLDING AND METAMORPHISM

The pinchout of the Lower Roan east of Chambishi and north of Mindola, appears to have controlled the position of the west flank of the Kafue anticline (Mendelsohn, 1961, p. 94), and this eastward pinchout of the ore-shale and of the Lower Roan follows the whole length of this anticline.

A monoclinal feature parallel to the Kafue anticline surfaces near the western granite promontory, but east–west drag-folds east of this granite ridge are more conspicuous. Axial planes and cleavage in argillites dip $55-60°$ south steepening into footwall formations. On the overturned limbs the Upper Quartzite is commonly overthrust over the adjacent syncline. The folds plunge west at $10°$, but over the west granite ridge the plunge steepens and the folds die out on the steep limb of the monocline. Over the eastern granite ridge, the fold axes flatten and then plunge eastward. Although the arenites fold concentrically, the assemblage of arenaceous and argillaceous beds form similar-type folds and may be due to drag of the alternating hard and weaker strata between the rigid granite below and the massive meta-gabbro and granophyre sills in the Upper Roan above (see Fig. 7).

Below the competent aeolinite the Old Granite has been sheared for one or two metres to a sericitic orthogneiss.

Metamorphism is similar to that at Nkana, except that tremolite and actinolite are developed only in the thin metamorphic aureoles of the gabbro sills. Specularite, feldspar, anhydrite, dolomite—quartz veins are common in the arenites, and feldspar, anhydrite, dolomite, quartz and sulfide veins occur in the orebodies. The feldspar is microcline pseudomorphous after adularia. Anhydrite is commonly leached to depths of 300 m, leaving cavities in the quartz and dolomite.

A late event in the folding and metamorphic history was the formation of a breccia zone — trending N65°E across the Main orebody and dipping south 70°. It varies from 3 to 13 m thick and the adjacent ore-shale is bleached and leached of its copper sulfides for a thickness of up to 1 m and veined by albite quartz veins. The breccia is generally barren and albitized with quartz and phlogopite and these minerals, with dolomite, minor anhydrite and chlorite and trace epidote, barite, and molybdenite, occur in veins in the white breccia.

NCHANGA (W.G.G. and R.H.)

The Nchanga group of deposits consist of the extensive Lower or Nchanga West orebody in shale and underlying arkose; a similarly extensive Upper or Nchanga orebody in the overlying 'Feldspathic Quartzite', both on the south limb of the Nchanga syncline; the small River Lode in shale on the north limb; the Chingola A and C orebodies mostly in shale to the southwest; and the Chingola E, F and D orebodies successively to the south in footwall arkose; and, finally, well to the southeast in the Mimbula syncline, four overlapping orebodies in footwall arenites (Fig. 4).

With at least 12 orebodies along a strike of 24 km, description of the first two will be stressed and only significant variations in facies and structure noted for the other ten. However, a new orebody, the Luano deposit, was discovered 2.5 km east of the River Lode to which it is, as known at present (end of 1974), equivalent in size. For this reason some specific data is given on the Luano deposit first.

LUANO DEPOSIT

Location

The Luano deposit lies on the north limb in Lower Banded Shale and Footwall Arkose of the Nchanga syncline (Figs. 10 and 11) approximately 2500 m east of the River Lode open pit.

Stratigraphy

The stratigraphic succession is essentially the same as the type Nchanga—Chingola succession as established in the main underground mine and the Nchanga and Chingola open pits.

pp. 259–260

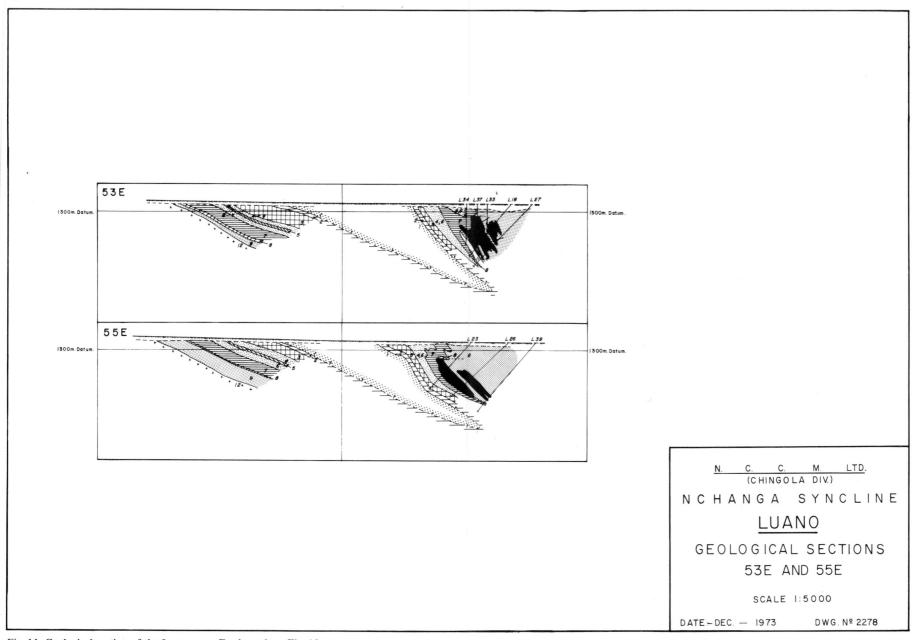

Fig. 11. Geological section of the Luano area. For legend see Fig. 10.

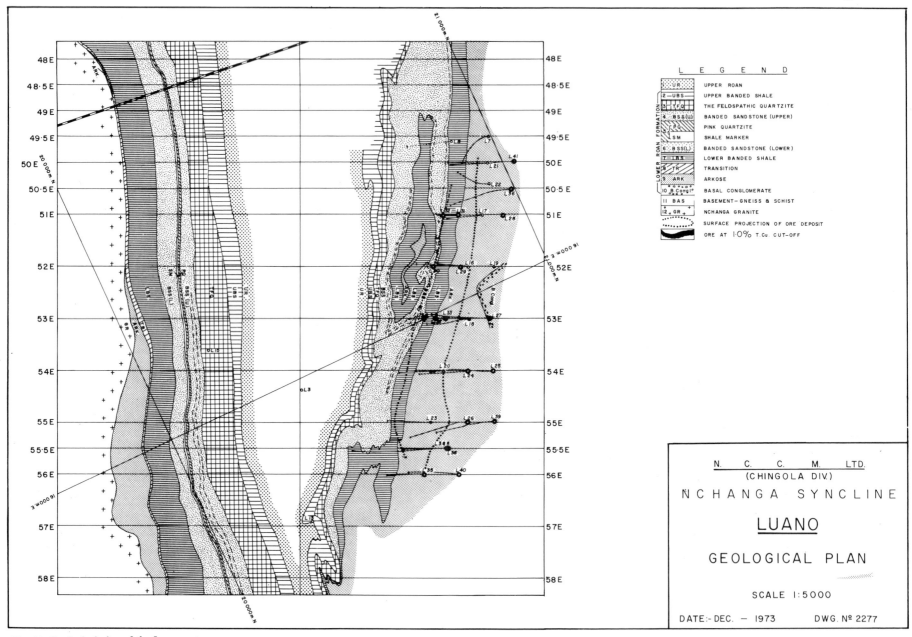

N. C. C. M. LTD.
(CHINGOLA DIV.)

NCHANGA SYNCLINE

LUANO

GEOLOGICAL PLAN

SCALE 1:5000

DATE:- DEC. — 1973   DWG. Nº 2277

LEGEND

1 UR	UPPER ROAN
2 UBS	UPPER BANDED SHALE
3 TFQ	THE FELDSPATHIC QUARTZITE
4 BSS(U)	BANDED SANDSTONE (UPPER)
5 PQ	PINK QUARTZITE
5 SM	SHALE MARKER
6 BSS(L)	BANDED SANDSTONE (LOWER)
7 LBS	LOWER BANDED SHALE
8 TR	TRANSITION
9 ARK	ARKOSE
10 B Cong lº	BASAL CONGLOMERATE
11 BAS	BASEMENT—GNEISS & SCHIST
12 GR	NCHANGA GRANITE
	SURFACE PROJECTION OF ORE DEPOSIT
	ORE AT 1·0% T.Cu. CUT-OFF

LOWER ROAN FORMATION

Fig. 10. Geological plan of the Luano area.

Basement granite gneiss. The granite gneiss is typically a pinkish white and grey, medium to coarse-grained crystalline rock consisting of about equal amounts of quartz, weathered feldspar and flakes or clusters of flakes of greenish to brownish mica. The gneiss is locally intermixed with mica schists on various scales. Typical mica schists are yellowish to greenish to brownish block rocks, almost entirely composed of mica.

Arkose. This formation is a heterogeneous assemblage of poorly-sorted, medium- to coarse-grained feldspathic sandstones to conglomerates. In places a thin (up to 2 m thick) conglomerate is found at the top of the arkose. However, this conglomerate is much patchier and more inconsistent than the Footwall Conglomerate at the River Lode. The thickness of the arkose is highly variable and is controlled by the pre-Katanga topography (see attached sections in Fig. 11).

Lower Banded Shales. The L.B.S. are generally similar to those found on the south limb, a bluish grey to black well-banded shale. Locally, yellowish-brown to pinkish white shales are interbedded and in places greyish fine-grained quartzites are present. The thickness is highly variable ranging from zero to 25 m.

Banded Sandstones. Again this member is similar in lithology to the Banded Sandstone Shales (B.S.S.) of the south limb, i.e., a series of pink to brown feldspathic and micaceous, fine- to medium-grained sandstones to sandy shales. Although numerous thin, lenticular quartzites are present, there is no consistent marker quartzite like the Pink Quartzite of the south limb. The thickness of the B.S.S. ranges from 12–40 m.

The Feldspathic Quartzite (T.F.Q.). This formation forms a consistent marker along the entire north limb and comprises a sequence of white to pink to yellowish-brown, medium- to coarse-grained, well-sorted feldspathic quartzites. The thickness ranges from 15 to 20 m.

Upper Banded Shales. Rocks similar to the Upper Banded Shales of the south limb overlie the Feldspathic Quartzite (T.F.Q.) along the entire north limb. The contact with the dolomitic schists of the Upper Roan is gradational.

Structure

The north limb of the Nchanga syncline in this area is formed by a tightly folded overturned succession of Lower Roan metasediments. The structure is characterized by a number of synclines causing a repetition of the sub-outcrop of the Arkose, L.B.S. and B.S.S. The beds in general dip northeast at 50–60°. Like the strata, the axial planes of the folds are overturned to the southwest. The fold axes run roughly parallel to the general strike and normally plunge to the west. There is a fairly persistent anticline which forms a distinct axial culmination on section 52E. Drilling has proved that this anticline is underlain by a NW—SE trending pre-Katanga basement ridge.

As W.G. Garlick (personal communication, 1974) stated: "The most interesting feature of this deposit is its association with a ridge of Basement gneiss, projecting through the (?) 1000 m of Footwall arkoses and conglomerate, and the way in which subsequent folding has concentinated a relatively narrow orebody into a compact mass, amenable to open-pit working."

Ore distribution

Ore grade copper mineralization in the form of mixed oxides and sulphides occurs in the upper part of the Arkose and the L.B.S. The ore is mainly associated with the syncline formed on the northern flank of the pre-Katanga ridge mentioned above. Drilling to date has indicated potential ore over a strike length of 750 m at depths between 25 and 140 m below surface. However, the deposit is still open-ended along strike and down dip, although in the deepest holes (100—140 m below surface) there are indications of a reduction in copper grades and the ore splits into a number of thin (2—3 m thick) lenses. The copper mineralization in the Lower Banded Shale gives way — west, south, and east — to pyritic dissemination in this carbonaceous shale.

BASEMENT COMPLEX

Lufubu garnetiferous schist and paragneiss are intruded by grey granite. South of Mimbula, Muva orthoquartzite and mica schist strike northwest. An oval mass of Nchanga red granite, 10 km by 14 km is a late pluton intruding the Basement complex and may be post-Muva in age. The red granite contains large pink phenocrysts of microcline microperthite, oligoclase, opalescent quartz, biotite, and accessory zircon with notably fluorite and garnet. This red granite grades locally into a similar grey granite and both are cut by many aplite dikes, but relatively few pegmatites. Drilling has shown that the red granite extends for another 4 km to the northwest under Katanga metasediments, giving a total area of over 120 km^2.

KATANGA SYSTEM OR SUPERGROUP

Lower Roan Group

Basal Boulder Conglomerate (0—1200 m). This is well developed only south of Mimbula and on the north limb of the Nchanga syncline, where it is sheared and poorly-sorted with subangular to rounded boulders of granite and of quartz—sericite—schist derived from granite, with Muva quartzite boulders in the south. Quartz pebbles are ubiquitous.

On the south limb of the Nchanga syncline, boulders up to 1 m in diameter and insolated slabs 3 m long of red granite occur at the base of the arkose, but buolders of aplite are more abundant.

Arkose (0–150 m thick on south limb of Nchanga syncline). Roughly equal amounts of pink non-perthitic microcline and opalescent quartz occur as coarse grains and pebbles in a matrix of quartz, feldspar, sericite, and biotite. The rock is more quartzose than the underlying granite. It is well-bedded with silty layers and is commonly cross-bedded.

The uppermost 27 m, generally adjacent to red granite hills where arkose is of minimal thickness, is the host for the "Arkose ore" of the Nchanga West orebody and at Chingola B, C, D, and E orebodies. At Chingola F the mineralization lies between 15 and 40 m below the Lower Banded Shale. Still farther south at Mimbula, this formation is represented by porous argillaceous feldspathic arenites, which are host to four ore lenses distributed over a stratigraphic thickness of 75 m, the uppermost lens being 30 m below the Lower Banded Shale. The lowermost lens has intercalated conglomerates up to 2 m thick.

Transition Sandstone (0–10 m). This consists generally of weathered sandstone of sub-rounded quartz grains and sericite flakes in a clay matrix, commonly iron-stained and mineralized with copper oxides and carbonates in the orebodies.

Transition Quartzite (0–6 m). A finely laminated silicified banded dolomite, mineralized with chalcocite and copper carbonates and oxides in the orebodies. It is equivalent to the schistose ore of Chambishi and Nkana.

Lower Banded Shale (12–27 m). A grey to black laminated shale of quartz, sericite, carbon, feldspar, and accessory zircon and tourmaline is ubiquitously peppered with pyrite (or limonite blebs) beyond the orebodies. At Nchanga West, Chingola A, and River Lode, chalcocite, bornite, and chalcopyrite take the place of the pyrite as disseminations and concentrations along laminations and coarser aggregates in bedding veinlets. Commonly, the sulfides are oxidized to depths of over 300 m.

Even over the orebodies in the Lower Banded Shale the upper part of the shale is pyritic or, where leached, with limonite nodules. An horizon of geodes with quartz and pyrite occurs near the top of the shale.

Brown Chert (0–6 m). This consists of fine crystalline silica with sericite flakes and may be due to silicification of a thin dolomite as recognized at the top of the ore-shale at Chambishi.

Lower Banded Sandstone (12–30 m). This consists of weathered clayey rocks with much sericite, brown mica, and detrital quartz, commonly mineralized with cupriferous vermiculite, and some malachite and cuprite in the vicinity of orebodies.

Shale Marker (1–1.5 m). This laminated shaly sandstone is a persistent marker on the south limb of the Nchanga syncline. Locally with the Pink Quartzite it forms the Intermediate orebody.

Pink Quartzite (4.5–6 m). An arkose with abundant microcline and quartz in sericitic matrix.

Upper Banded Sandstone (15–30 m). This is similar to the Lower Banded Sandstone, with cupriferous vermiculite near orebodies.

The Feldspathic Quartzite (T.F.Q.) (15–37 m). This is a thick competent arkose with well-sorted, medium-sized grains of quartz and microcline in a matrix of same minerals with sericite and dolomite. Feldspar varies from 5 to 50%. It shows tabular cross-stratification, predominantly from northeast, mudcracks, and current ripples.

It is the host for copper sulfides and oxides in the Nchanga orebody, which is fairly exactly superimposed over the Nchanga West orebody, but extends further east along the south limb of the Nchanga syncline and is mined mostly by open pit. At Chingola B, the Upper orebody is again in this formation.

Upper Banded Shale (15–37 m). This unit is composed of grey to dark-grey rock of quartz–microcline laminae alternating with micaceous laminae, which becomes interbedded with dolomitic schist and dolomite upwards. The basal 3 m contain copper sulfides or oxides over large areas where the T.F.Q. is mineralized, but the upper part and the more dolomitic schist contain fairly extensive cupriferous vermiculite. In the other mines this formation is included in the Upper Roan.

Upper Roan Group (300–600 m thick)

Above the Upper Banded Shale is 46 m of dolomitic schist followed by 60 m of crystalline dolomite, and then further alternations of schist and beds of crystalline dolomite, with considerable anhydrite. The carbonate formations are usually totally leached to depths of 300 m and the shales and schists weathered to red clay.

Mwashia Group

This is a dolomitic facies here and not readily distinguishable from the Upper Roan.

Lower Kundelungu Group

The schistose equivalent of the basal tillite is recognized in a drill-hole west of Nchanga.

Kakontwe Limestone. The Kakontwe is well-developed and exposed in the same drill-hole. Weathering of this formation provides distinctive drainage patterns, soils, and vegetation.

The above formations are overlain westwards by calcareous shaly sandstones of considerable thickness. They must have extended east across the Kafue anticline, so that the Nchanga granite and adjacent Basement Complex were originally buried under 1500 m of Katanga sediments, thickened considerably by the folding observed above the Lower Banded Shales.

POST-KATANGA INTRUSIVES

Gabbroid sills occur in the Upper Roan west and southwest of Nchanga, but apparently do not intrude the Upper Roan in the Nchanga syncline. However, a so-called "lamprophyre", or biotite—scapolite—schist, possibly a metamorphosed gabbro, forms a dike striking west-southwest across the basement north of the Nchanga syncline and penetrating the footwall formations up to the Lower Banded Shale at the River Lode. It contains copper sulfides and a little cupriferous vermiculite and could be a feeder for the River Lode orebody. Alternatively, copper from this orebody may have contaminated the intrusive, considered a more likely explanation as copper in the lamprophyre tails off rapidly northwards from the orebody (R. Haldane, personal communication).

STRUCTURES

SHINSENDA

Kabuche—Kinsenda Culmination

Erosion of the Katanga sediments over this high, extending from the Muliashi porphyry, over the Nchanga red granite, the Mwambashi dome, to the Kabuche Basement anticline far to the south, has separated the Chambishi—Nkana basin from the Mimbula—Nchanga basin. Strong orogenic pressure from the southwest developed a series of folds which plunge northwestward off this culmination. These are, from south to north, the nearly symmetrical Mimbula syncline, the recumbent Chingola folds pushed over the southern flank of the Nchanga granite, the asymmetrical Nchanga syncline pushed in the opposite direction over the north flank of the Nchanga granite, the Fitwaola syncline with Muliashi porphyry thrusted southward over the north limb, and finally the Karila Bomwe anticlinorium at Karila Bomwe mine. Within 30–50 m of surface on the south limb of the Nchanga syncline, there is a Chingola-type recumbent fold in the Lower Banded Shale splitting into two digitations in T.F.Q. and Upper Banded Shale. As this is in the opposite sense for a drag-fold on the south limb of the asymmetric Nchanga syncline, it is evident that the Chingola recumbent folds must be older than the latter. Figure 5 shows a similar recumbent fold at depth.

The *Nchanga red granite* behaved as a rigid buttress during the early folding, protecting the footwall arkoses immediately above from involvement in the recumbent folds affecting the Lower Banded Shale and overlying strata. However, the southern margin was sheared and small thrusts developed. Later, the whole pluton with its blanket of sediments over the bevelled surface was tilted northward and probably pushed under the thick pile of sediments to the north, forming the Nchanga syncline. On the south limb thereof, the granite surface and overlying sediments dip very evenly at 27° north, but near the nose of the syncline, beds on the north limb are overturned to dip as low as 45°, gradually steepening to vertical westwards. At the same time or later, the surface of the granite was slightly arched, but not nearly sufficiently to bring it into the category of a mantled gneiss dome, as claimed by McKinnon in Mendelsohn (1961) and quoted by Davidson (1962). Rupture of the granite in the "Main Fault" is apparently associated with northward tilt of the granite surface and sediments into the Nchanga syncline; its 24 m upthrow on the north side results in an offset in the south wall of the Nchanga open-pit. Whereas the Main Fault strikes N40°W, the fault occupied by the lamprophyre dike in the River Lode footwall arkoses does so N45°E.

The Kabuche–Kinsenda culmination developed over varied granitic and gneissic rocks, flanked on the east by Lufubu schist pendants striking north-northeast and partly covered by the Chambishi–Nkana basin. It developed probably at a late stage in the tectonism as an alignment of granitic domes, some separated by synclines of Katanga, others coalescing by erosion of the Katanga sediments in synclinal saddles.

METAMORPHISM

The Katanga sediments at Nchanga have by deep burial and tectonism achieved the higher greenschist facies of regional metamorphism. Sericite and biotite are the typical minerals in the argillaceous rocks, quartz grains show overgrowths, and feldspar is usually microcline. Scapolite occurs rarely in the Lower and Upper Banded Shales, but is abundant in the dike rock below the River Lode. Southwards through Chingola, scapolite increases in the argillites and at Mimbula epidote, actinolite, and tremolite occur.

Retrograde metamorphism of Basement

The Basement had attained varying degrees of metamorphism before Katanga sedimentation, generally of amphibolite facies, but the tectonism and lower-grade metamorphism of the Katanga system caused widespread retrograde metamorphism of most of the Basement. In a few localities the higher-grade minerals were preserved, such as garnet in Lufubu schist in the Kafue River and kyanite in Muva quartzite south of Mimbula. Notably, the Nchanga red granite, which behaved as a rigid buttress during the Lufilian folding, has preserved its tiny red garnets, its fluorite, and the microcline microperthite phenocrysts, except where it has been sheared along the south boundary and the Main Fault. The overlying arkose contains the same pink to red grains, but they are no longer perthitic, and garnet is practically absent, although typical Nchanga granite zircons are abundant (Binda, 1972). These relations are stressed, as the rubidium–strontium whole-rock age of the Nchanga granite of 570 ± 40 m.y., must therefore refer to a mild reheating of the granite and a homogenization of the strontium during the Lufilian orogeny and not to the date of intrusion of the granite (Garlick, 1973).

Migration of soda

The destruction of the perthitic structure of the microcline derived from the Nchanga and other granites, during the regional metamorphism of wet sediments, liberated soda and caused net enrichment in potash of both arkose and argillite. Similarly, little of the oligoclase of the granite survived in the sediment, thus explaining the high potash content of arkose and argillite. Some of the expelled soda has formed albite veins in the Main Fault.

SUPERGENE PROCESSES

Part of the high grade of the Lower (Nchanga West) orebody is due to surface enrichment and most of the Nchanga mineralization contain nearly as much oxide copper minerals as they do sulfide copper. The Lower Banded Shale is near the surface leached to a light grey or white clayey rock, due to removal of all the copper and iron sulfides and oxidation of the carbon. Much of the shale, the arkose ore and T.F.Q. are iron-stained, with streaks and patches of malachite, azurite, and chrysocolla, commonly with grains and blebs of secondary chalcocite. In the central part of the arkose ore of Nchanga West, there are considerable areas of primary sulfides, bornite and chalcopyrite, some layers being extremely rich and warranting the term sulfidite. Thus, the Nchanga West orebody was probably rich by Copperbelt standards before surface leaching and enrichment.

With this prevalent secondary migration, local geologists have so far failed to recognize the primary sulfide zoning, which is a feature of the other Copperbelt deposits. As mining reaches greater depths at Nchanga West and in the T.F.Q. of the Nchanga orebody, the primary sulfide zoning will probably be revealed.

A stratigraphic zoning is seen in places at Nchanga West, where a metre of barren shale overlies pyritic Lower Banded Shales, in turn overlying chalcopyrite and bornite in the lower part of the shales, commonly chalcocite in the Transition Quartzite, and mixed bornite—chalcopyrite in the arkose.

PALEOTOPOGRAPHY

Many of the lower orebodies at Nchanga are closely associated, if not actually con-trolled, by the topography of the surface eroded on Basement granite, gneiss, and schist during deposition of the first sediments of the Lower Roan.

Near the outcrop at Nchanga West, a hill of Nchanga red granite projects through most of the Arkose; ore in the latter and in the Lower Banded Shale flanks both sides of the hill symmetrically. On the 970-ft level another hill of red granite forms the west margin of Arkose ore and here, as the Arkose wedges out, the mineralization extends for 6 m into the granite, partly in fractures and partly disseminated in metamorphosed paleosol, in which the microcline has lost its perthitic texture. The hill is bevelled at the level of the base of the Lower Banded Shale, which over the hill is pyritic, but copper occurs in the shale for a small distance southwest, giving way to pyrite again as the barren arkose thickens southwards (Fig. 12).

The Chingola A orebody in Lower Banded Shale partly overlies a hill of red granite buried under arkose. Chingola C and its extension flank a hill of Basement gneiss pro-jecting through the Arkose into Lower Banded Shale, all inverted in the overturned limb

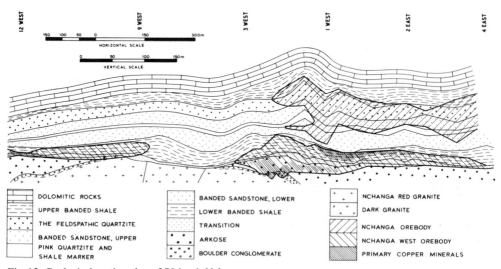

Fig. 12. Geological section along 970-level, Nchanga.

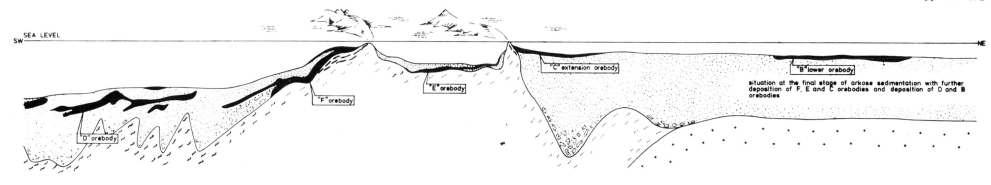

SEA LEVEL

SW   NE

"B" lower orebody

"C" extension orebody

"E" orebody

"F" orebody

"D" orebody

situation at the final stage of arkose sedimentation with further deposition of F, E and C orebodies and deposition of D and B orebodies

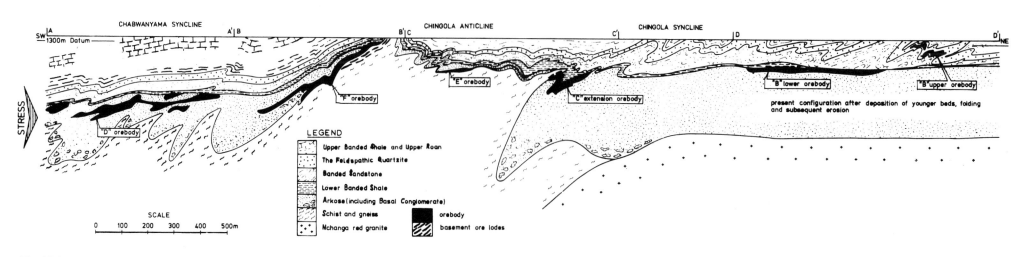

CHABWANYAMA SYNCLINE   CHINGOLA ANTICLINE   CHINGOLA SYNCLINE

SW A   A' B   B' C   C'   D   D' NE

1300m Datum

STRESS

"E" orebody

"F" orebody

"C" extension orebody

"B" lower orebody

"B" upper orebody

"D" orebody

present configuration after deposition of younger beds, folding and subsequent erosion

LEGEND

Upper Banded Shale and Upper Roan

The Feldspathic Quartzite

Banded Sandstone

Lower Banded Shale

Arkose (including Basal Conglomerate)

Schist and gneiss   orebody

Nchanga red granite   basement ore lodes

SCALE

0   100   200   300   400   500m

Fig. 13. Diagrammatic sections of the Chingola area.

of the Chingola recumbent syncline. Westwards down the plunge of the fold, copper in Lower Banded Shale gives way to pyrite, but ore in the arkose persists and some mineralization occurs in the adjacent Basement gneiss (Fig. 13).

South of the above mentioned hill of Basement gneiss, Chingola E orebody occurs in arkose in a local basin, in the central part of which the ore does not reach up to the pyritic Lower Banded Shale; in part the ore is draped over the Basement and here mineralization again occurs in the upper part, probably an old paleosol, of the Basement.

Chingola F occurs on the south flank of the next Basement ridge and extends from 4 m in the Basement through the arkose to the pyritic Lower Banded Shale contact, but southwards the upper part of the arkose becomes barren and the ore follows the arkose along the unconformity, with mineralization up to 4% extending into the Basement. As the arkose thickens southward the mineralization follows a stratum 15–40 m below the Lower Banded Shale (Fig. 13).

The Chingola D orebody is further south in arkose overlain by 20–30 m of barren arkose and resting on pebbly arkose occupying valleys between rugged steep ridges up to 300 m high, but just failing to reach the base of the ore. In places a lens of ore occurs at top of the arkose.

Finally, at Mimbula there is a change of facies, the Arkose being represented by argillaceous feldspathic arenites. The uppermost lens corresponds to the Chingola D orebody, with the lower three lenses occupying a vertical thickness of 75 m. The lowermost lens contains conglomerate layers.

Thus, from Mimbula to Nchanga West, a distance of 17 km, the ore is consistently close to the unconformity and steps up gradually across the strata, from 105 m below the pyritic Lower Banded Shale at Mimbula to within 3 m of the top of the shale at Nchanga West. This transgression across the strata is consistent with the northeastward transgression of the Lower Roan beds up the tilted flank of the geosyncline, as recorded at Nkana, Chambishi, and Bancroft.

The Upper or Nchanga orebody in The Feldspathic Quartzite is superimposed over that part of the Lower (Nchanga West) orebody lying east of the granite hills projecting through the arkose, but it extends further east than the ore in the Lower Banded Shale. The Nchanga orebody may have formed in a local basin ascribed to compaction of the lower beds, whilst lesser compaction of the thinner sediments over the granite ridge, caused shallower waters in the T.F.Q. above and, hence, non-deposition of copper. To the southwest of the ridge, over thicker sediments there is ore again in the T.F.Q. at Chingola B. Study of cross-bedding in the T.F.Q. should be rewarding.

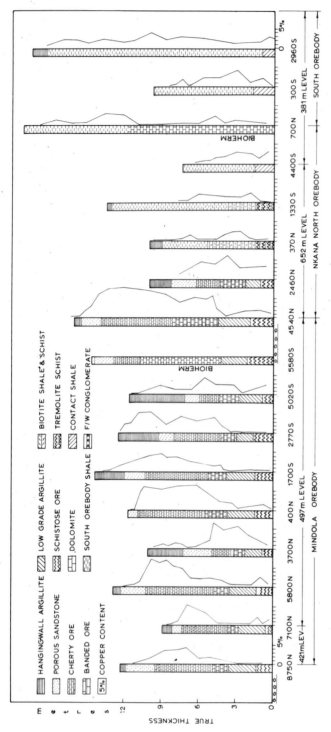

Fig. 14. Lithology and copper grades of the Nkana area.

NKANA (W.G.G.)

The Nkana South, Nkana North, and Mindola copper–cobalt deposits of Rokana Division occur for 14 km along the east limb of a northwesterly plunging synclinorium forming part of the Chambishi–Nkana basin (Figs. 9 and 14).

BASEMENT COMPLEX

The oldest rocks are Lufubu schist, with minor chloritic quartzite, and gneiss, intruded by grey to pink biotite–muscovite–granite, and by small pegmatite dikes. The grain of the basement is northeasterly, at right angles to the unconformity with the overlying Katanga sediments. Near the unconformity a foliation parallel to that of the Katanga is superimposed.

KATANGA SYSTEM OR SUPERGROUP

Lower Roan Group (200 m thick)

Basal Conglomerate (0–24 m). A cobble to boulder conglomerate occurs in hollows of the old landsurface of Lufubu schist and on the flanks of Basement gneiss ridges. Angular to rounded boulders of granite, gneiss, and quartzite, and cobbles of quartz occur in sandy matrix and grade down into flattened granitic boulders and less distorted quartz pebbles in a sheared matrix formed by extreme distortion of schist fragments at the unconformity.

Footwall Quartzite (up to 152 m). Thin grit and pebbly beds overlie the Basal Conglomerate or rest directly on the unconformity and may attain 6 m thickness. They are overlain by white to light-grey feldspathic quartzite with characteristic large-scale cross-bedding, foresets banded by only minor changes of mineral content — quartz, feldspar, and iron oxide — rigidly parallel and curving down gently into bottomsets with finer banding. There are, however, no silty or muddy layers and a complete absence of pebbles or grit grains larger than 2 mm. The angle between truncation plane and foresets attains 34° and this with the above features was recognized in 1940 as being characteristic of aeolian sands, but planar smooth parallel truncation planes, seen also in the Coconino sandstone of Arizona, were explained only by Stokes (1968).

Midway in the aeolian quartzite section is 2 m of a breccia of ill-sorted angular fragments of granite, quartzite, and quartz in dark muddy, gritty matrix, probably formed by sheet flood. It is parallel to the multiple parallel truncation bedding planes of the adjacent aeolian beds.

Footwall Transition Formation (20 m). Thick beds of pink to grey sandstone, arkose, and thin argillite layers overlie the aeolian conformably, but at Nkana North are represented by very porous dirty micaceous sand due to near-surface leaching of an anhydrite-rich evaporite. Rectangular cavities are conspicuous in cross-cutting quartz veins. Directly overlying the basement schist at Nkana South, this formation is the host of the footwall orebody.

Footwall Sandstone Formation (30 m). (a) The Lower Conglomerate, 10—14 m thick, consists of cobbles of granite, quartzite, and quartz in a porous argillaceous sandy matrix, from which anhydrite has been leached. Near the top, the matrix is coarsely arkosic with pebbles. Recrystallization has given a granitic appearance and in drill cores was mistaken for intrusive granite, but pebbles and cross-bedding (as iron oxide streaks) proves to be a metasediment. (b) Pink cross-bedded arkose, 6 m thick, overlies the Lower Conglomerate. Current directions are persistently from northeast. (c) A massive sandy argillite with minor lenses of arkose is usually 6 m thick and is overlain by (d) 5—7 m of cross-bedded pink to grey arkose with thin argillite beds. Current direction is consistently from northeast.

 These three distinctive beds between the Lower Conglomerate and the Footwall Conglomerate were useful markers in guiding the footwall haulages. At Nkana South the three-fold division becomes indistinct and southwards the Lower Conglomerate thins out. Attenuation of the beds on the limbs of the tight folds can reduce the thickness of these beds to a quarter.

Ore Formation (24 m). This consists of Footwall Conglomerate, Ore-shale and its variants, and the Hanging-wall argillite. Ore is generally confined to the central member.

Footwall Conglomerate (1 m to rarely 40 m). Although usually relegated to the Footwall formation, this conglomerate is an essential part of the Ore formation and heralds the transgression of the sea in which the ore-shale was deposited. It is usually barren, but the upper 20 cm where less pebbly and if "muddy" may carry copper sulfides. Its base is mildly transgressive and scours the underlying unit, but channels are rarely seen in the underlying formations. Below the cupriferous shale it is about 1 m thick, but north of Mindola, under barren thin shale, it reaches 40 m in thickness. Critical exposures, usually near paleo-hills of the Basement, reveal that the apparent blanket consists of an overlapping series of shoestring pebble beds elongated northwest—southeast, representing pebbly beaches formed during marine transgression. Westward, each shoestring deposit grades into dolomitic biotite schists or crumpled finely laminated dolomite and mud, usually mineralized. The conglomerate consists of rounded pebbles of quartz, quartzite, granite, and microcline in a feldspathic sand matrix. Where the conglomerate is thicker, the pebbles are larger, up to cobble in size, commonly with one or more facets.

Ore shale (21—23 m). Through Nkana North and Mindola this can be sub-divided as follows:

stratigraphic unit	average thickness	mineralization			
Hanging-wall argillite	11 m	py.	py.	ore	barren
Porous sandstone (dolomite)	1.5 m	py.	ore	ore	barren
Cherty ore (argillite)	3 m	py.	ore	ore	ore
Banded ore (dolomite + argillite)	2 m	ore	ore	ore	ore
Low-grade argillite	2 m	ore	ore	ore	barren
Schistose ore	1.5 m	py.	ore	ore	barren

mineralization at deeper levels ⟶

mineralization at medium depths ⟶

mineralization at medium depths, maximum thickness ⟶

mineralization at shallower depths ⟶

The lower half of the ore-shale, about 10 m, contains excellent marker beds that allow the above division—a geologists' paradise for mapping in three dimensions. Recognition in 1940 that the characteristic grade of each sub-division remained constant through both the attenuated beds of the squeezed limbs and the expanded section in the axial regions of the dragfolds provided the first proof that mineralization was pre-folding (Fig. 15). The restriction of the ore to the shale—dolomite formation, with richest ore in the most impermeable bed, and the zonal arrangement of the sulfides uncontrolled by either folding or the attitude of the beds, but closely related to shorelines and sedimentary facies, were explainable only by the mineralization being contemporaneous with sedimentation.

On the right of the stratigraphic column, ore sections are shown for various depths in the mine. These are controlled by the mineral zones, which are generally inclined north-ward. Thus the shallow-depth ore, confined to the Cherty and Banded ore, occurs through Mindola, but only in the uppermost levels of Nkana North adjacent to the Kitwe Gap. The deep ore section of Nkana North appears at 600 m depth opposite Central Shaft, with ore only in the Banded ore and Low-grade argillite between pyrite above and below, but does not appear at Mindola until below 1400 m depth.

In each stratigraphic sub-division, from surface downdip, the following mineral zones occur: barren (chalcocite); bornite; bornite with chalcopyrite and over 0.1% cobalt as carrollite; chalcopyrite and cobalt; chalcopyrite with minor pyrite; pyrite with chalco-pyrite; and, finally, pyrite with only traces of chalcopyrite (Fig. 8).

The 14 km of mineralized ore-shale along the east limb of the Nkana syncline is inter-rupted by "barren gaps" of up to 600 m width, such as the Kitwe Gap between Nkana North and Mindola orebodies. Here the Footwall formations are thinner over the Kitwe granite paleo-ridge, but the Ore formation is thicker by 50%, massive siliceous dolomite taking the place of the cupriferous shales and thin-bedded dolomites and extending down dip for more than 800 m. This feature, later recognized south of Chambishi and at Irwin

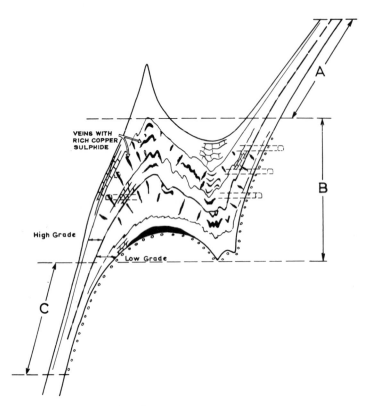

VEINS WITH
RICH COPPER
SULPHIDE

A

B

C

High Grade

Low Grade

Fig. 15. Major drag-fold at Nkana.

Shaft at Roan Extension, is now interpreted as an algal bioherm. Similar barren dolomitic sections occur between Nkana North and South orebodies and a spectacular thickness of dolomite occurs at the Central section of Nkana South. These barren bioherms occur over paleo-hills of Basement, but a 600 m wide gap at No. 3 shaft in the northern half of Mindola extends to only 200 m below surface and has no biohermal dolomite; instead the Footwall formations lose their identity and thin to a wedge of unsorted terrigenous debris over a granite hill, the Footwall Conglomerate steps up through the ore formation to the base of the Banded ore, which with the Cherty ore, changes to barren very sandy argillite (Fig. 16).

Except in the barren gaps, the above sub-divisions can be followed from the north end of Mindola 9.5 km to the south end of the Nkana North orebody, but below 600 m in the latter and throughout the Nkana South orebody, the ore-shale passes into a deeper-water facies of black carbonaceous shale with chalcopyrite, carrollite (catterite), and pyrite. At the base, there is commonly tremolitic schist, but this grades laterally into banded dolomite and quartzitic argillite, the shearing now being taken up by the fissile overlying

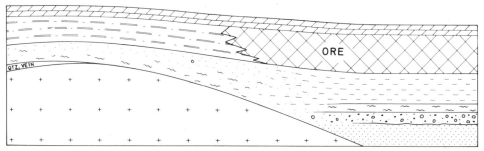

Fig. 16. Granite protuberance into the Lower Roan under the Mindola stream.

carbonaceous shale, usually intensely veined with most of the chalcopyrite expressed into the quartz-carbonate veins.

The Hanging-wall argillite, about 11 m thick, blankets the ore for the whole 14 km strike of the east limb and the lower half thereof is mineralized with finely disseminated pyrite. However, near surface in the north part of Nkana North and in Mindola, the Hanging-wall argillite and locally the Porous sandstone are barren. Between the up-dip barren and the down-dip pyritic zones, bornite and chalcopyrite disseminations occur in the Hanging-wall argillite. Where contiguous with mineralization in the Porous sandstone, these cupriferous zones have been mined with the usual ore-shale, as at 2000–5000 ft south on 1630-ft level, Mindola.

Dolomite and anhydrite lenticles become increasingly abundant down dip and form marker horizons in and above the pyritic argillite.

Hanging-wall sandstone formation (45 m thick). This consists of brown and grey feld-spathic sandstone, grey to greenish argillite, red dolomite, and at the top a conspicuous marker of 10–15 m of white feldspathic quartzite.

Upper Roan Group (366 m thick)

About 20 m of grey to green pyritic calcareous argillite overlies the feldspathic quartzite and is followed by 10 m of white crystalline dolomite, succeeded by numerous inter-beds of argillite and pink to red dolomite. Higher in the section are some interbeds of dolomitic feldspathic sandstone and a few grit and conglomerate layers.

Mwashia and Lower Kundelungu Groups

These occupy the core of the northwest plunging syncline and are recognized from their vegetational patterns and soil types, confirmed by deep pitting. They have not been intersected by drill holes; the description in Mendelsohn (1961) was of sedimentary breccias and red limestone later recognized as Upper Roan. A brief description of these higher sediments is given under Chambishi, from pitting data between that mine and Nkana.

POST-KATANGA INTRUSIVES

Gabbroic intrusives into Upper Roan and doubtfully into Mwashia are marked at surface by float boulders, forming a crude "V" in the plunging syncline. The scapolitized gabbro probably forms irregular sills in the dolomites as elsewhere in the Chambishi basin. A possible feeder to these sills occurs as a much altered dike-like body in a fault zone under the Mindola stream. From its micaceous nature it has been called a lamprophyre. It has sharp contacts with the Footwall formations, but in the Ore formation it is wider and packed with xenoliths, usually on the one side, mostly of barren arkose and argillite, but including some ore-shale with characteristic sulfide dissemination. There is slight thermal metamorphism of the wall-rock, partly masked by later regional metamorphism. Adjacent to the dike there is slight impoverishment of copper values in the ore-shale and there are a few carbonate-sulfide veinlets.

FOLDING AND METAMORPHISM

The Nkana synclinorium plunges northwest at 20–30° and the nose shows much nearly isoclinal folding. Both limbs have numerous parasitic folds, involving Basement, Lower, and Upper Roan. In the ore-shale and above, the axial planes of dragfolds dip 65–70° southwest, but curve into the more competent Footwall formations to dip 60° northeast (Fig. 17). The axes of these dragfolds plunge at about 20° northwest, but approaching granitic paleo-hills, the plunge steepens to 40° and the fold dies out into a bench with numerous minor folds in the higher beds. In the reverse direction, approaching granite up the plunge, the axes flatten and the drag-fold again degenerates to a bench and may die out completely on the footwall. Thus, in longitudinal projection, the axes of these drag-folds have the plan of breakers entering a bay and being retarded against the headlands either side.

In the Mindola section, there is a bench-like structure with southward plunge over the No. 4 Shaft gap adjacent to the Kitwe gap, but elsewhere the westward dip is not so steep and drag-folds are practically absent. In the absence of later fold structures, the eastward embayment of the ore horizon into the areas of Basement schist between the westward projecting granites is a feature.

The attenuated isoclinal folds in the incompetent carbonaceous ore-shale of Nkana South dip at 65° southwest and the axial planes, passing into the more competent footwall, steepen to vertical as seen in Fig. 18. Here, a squeezed anticline of Lufubu schist forms a narrow screen (formally a paleo-hill) between the rounded trough of Footwall orebody and an isoclinal keel of ore-shale only 4 m west.

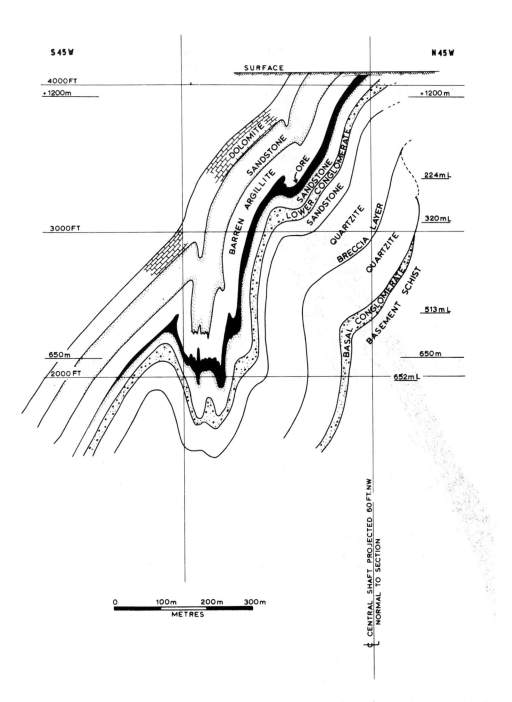

Fig. 17. Central shaft section at Nkana.

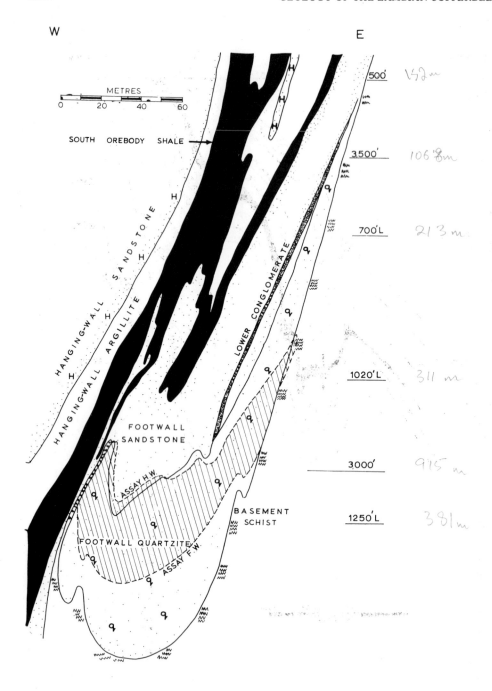

Fig. 18. Cross-section of the South Orebody, Nkana. The Footwall Orebody lies in the upper part of the Footwall quartzite.

METAMORPHISM

Biotite is the characteristic regional metamorphic mineral of the argillaceous rocks throughout Mindola and Nkana North, but is not particularly oriented parallel to axial planes and, hence, slaty cleavage is not well developed. In the bench-type fold at the southern end of Mindola, tremolite occurs in the schistose dolomitic argillites of both the Schistose ore and the Banded ore. In the argillites quartz, microcline, and albite constitute nearly 60% of the rock, the rest being a fine decussate aggregate of biotite, sericite, minor carbonate, anhydrite, and chlorite and, where mineralized, a fine peppering of bornite, chalcopyrite, or pyrite. Coarser aggregates of sulfides lie along most laminations and, in the Cherty ore particularly, there are numerous cross-cutting veinlets with microcline, albite, dolomite, anhydrite, quartz, and the same sulfide as characterizes the adjacent wall-rock. The walls of many of these veinlets are bleached to a flesh colour by migration of the disseminated sulfides into the vein, this evidence of lateral secretion being conspicuous for 20—50 mm from the vein wall. In the drag-folds of the Nkana North orebody the Cherty ore approaches a crackle breccia, due to numerous veins and veinlets in all directions, with flesh-coloured margins against grey disseminated sulfide ore. Sulfides in the veins are far more conspicuous than the fine disseminations and, hence, the ore in the folds was originally considered richer than that on the limbs. The dolomitic layers show the same minerals, but with more dolomite and generally coarser grained sulfides. Argillite layers between dolomite are intensely drag-folded, often forming a concertina pattern, especially in the Schistose ore above anticlines of competent Footwall Conglomerate.

Approaching the Kitwe gap from the north, the Banded ore loses its copper and thickens, at the expense of the Cherty ore, to a massive tremolitic siliceous dolomite, originally a bioherm. South of the Kitwe gap below the 500-m level, tremolite appears again in the crumpled banded dolomite and schist in anticlines over Footwall Conglomerate.

Footwall Orebody

This small orebody, 370 m strike by 350 m along the dip, occurs below the Nkana South Orebody in the equivalent of the Footwall Transition formation. It is immediately below the Lower Conglomerate, which is here barely 3 m thick and thins to an arkose with scattered cobbles down dip approaching a paleo-hill of Lufubu schist. Unlike the anhydrite-rich micaceous arenites at Nkana North, the Transition beds, filling a local basin eroded in Lufubu schist, are much cleaner with little mica and less anhydrite. They still have over 15% feldspar and, therefore, should be called arkoses rather than the local name "Footwall or Basal Quartzite". Their recrystallization to an "arkosite" has given them a competency which controlled the later folding, forming a rounded syncline in the bottom of the original basin, whilst the Lufubu schist hill to the west was squeezed up as an anticlinal screen only 3 m thick between the cupriferous carbonaceous ore-shale and the rounded trough of cupriferous quartzite to the east. Several isoclinal keels of carbonaceous ore-shale overlie this trough (see Fig. 18).

Stratigraphically this orebody is at the same horizon as Chibuluma, whilst structurally it resembles Chibuluma West. Bornite dissemination is dominant in the short vertical west limb and in the trough, but chalcopyrite is dominant up the long 200 m east limb. Here, the orebody thins on the east flank of the basin and, especially to the north, the mineralization gives way to pyrite. The sulfides show concentrations along the bedding and clots are conspicuously flattened parallel to the axial plane of the syncline, as are the phlogopite and pennine chlorite. Mineralization extends in places as disseminations and veinlets for up to 8 m into the Lufubu schist.

Although Richard (1964) claims that the orebody was formed by migration of copper from the squeezed and contorted nearby ore-shale, with replacement of aeolian quartzite in the pressure shadow of the Lufubu schist hill, much of the orebody shows parallel bedding due to quiet aqueous sedimentation in the centre of this local basin, and the uncommon cross-bedding shows dihedral angles of 20° or less contrasting with the 30–34° of windblown sands. Moreover, this formation immediately below the Lower Conglomerate is normally up to 20 m thick and overlies the aeolinite at Nkana North and at Chibuluma.

Richard supplies in Table II the variations of mineral content approaching and in the Footwall Orebody, and in arkose and dolomite below the Footwall Conglomerate (Table II). The increase of sulfide, dolomite, and albite into the orebody is similar to that at Chibuluma, where there was no overlying shale orebody to supply copper to the footwall. The albite is evidence of deposition in the saline playa lake.

TABLE II

Mineral content (in %) near and in the Footwall Orebody, and in arkose and dolomite below the Footwall Conglomerate

Mineral	18 m from FW Orebody	9 m from FW orebody	In FW Orebody	Mineralized arkose	Dolomite
Quartz	75–80	65–70	55	35	5
Microcline	15	18	18	20	–
Albite	nil	5	14	15	–
Muscovite	3	3	3	–	–
Phlogopite	2	1	0	15	5
Chlorite	0	1	2	–	–
Dolomite	1–2	6	8	10	45
Tremolite					45
Sulfide	0	pyrite	chalcopyrite and bornite	2–5	5 with much Co.

The grade is fairly even up to 5% copper with negligible cobalt, unlike Chibuluma with its sulfidite layers, which assay up to 10% copper and 5% cobalt.

At Nkana South, under the Footwall Conglomerate, reduced to a pebbly grit, the Footwall Sandstone formation is commonly rich in anhydrite, locally up to 70%, and in places grades into a very dolomitic rock with much tremolite. For short distances these beds are well mineralized with bornite, chalcopyrite, and carrollite and the mineral content of these are shown in Table II, fourth and fifth columns. This again demonstrates the association of sulphides with soda (albite), mica, and dolomite in probably local basins in these arenaceous formations.

Mindola uranium

Recognition in 1940 of a syngenetic origin for the copper and cobalt sulfides as well as by assuming that ore-mineral zonation was the result of the geochemical properties of the metals involved, stimulated the search for other metals, leading to the discovery in 1945 of uranium with a little molybdenite at the southern end of Mindola. It was mined out by 1960. Near the No. 4 Shaft, bornite in the Cherty and Banded Ore horizons of the ore-shale give way to barren dolomitic sandstone with some pebbly arkose at the top. The Low-grade argillite is represented by pebbly arkose, indistinguishable from Footwall Conglomerate. These barren sediments were interpreted as a sand spit diagonal to the Kitwe Gap, towards which it plunges south at $25°$, enclosing a lagoon with richly cupriferous sediments between it and the Kitwe Gap bioherm. The uranium occurs on the southeast margin of the spit, adjacent to cupriferous Cherty and Banded Ore.

Pitchblende occurs as fine disseminations up to 0.5 mm and as blebs, with minor uranite, brannerite, coffinite, molybdenite, and vanadinite. The uranium grade was inverse to the copper grade. Coarse pitchblende occurred also in cross-cutting anhydrite dolomite quartz veins. One vein carried botryoidal pitchblende along its walls. An age determination gave 520 m.y.

ROAN ANTELOPE AND BALUBA (W.G.G.)

These copper deposits occur as infolds across the grain of the Basement and consist of: the Roan Basin, which is a canoe-shaped syncline in Lufubu schist; the Roan Extension composed of a tightly squeezed syncline in a granite opening out into a synclinorium overlying Basement gneiss and schist and, which joins at Muliashi the main basin formed by the en-echelon westward plunging Baluba synclines. Muva quartzite and schist form the western frame of the Muliashi–Baluba basin (Fig. 19, pp. 289–292). The Roan Antelope deposit is described first.

BASEMENT COMPLEX

Lufubu System

The system contains quartz–biotite schist, foliated parallel to the adjacent Katanga; the schist is well exposed in the Roan basin, especially in the Storke Shaft cross-cuts. A few minor beds of dolomite with actinolite strike northeasterly.

Granite and gneiss

Westwards, pink to grey granite intrudes the Lufubu; the axis of the syncline rises and the Roan syncline is pinched as between the jaws of a vice and is only 400 m across at its narrowest.

Pink pegmatites with perthitic microcline intrude the granite and adjacent Lufubu, but truncation of these by the Katanga Basal Conglomerate first proved the pre-Katanga age of the granite (Garlick and Brummer, 1951). Some pegmatites contain red microcline up to 30 cm long, coarse quartz, muscovite up to 20 cm and often bunches of radiating black tourmaline (schorl).

Farther west in Roan Extension and in Muliashi, gneissic sills in Lufubu schist show boudin structures and both are intruded by a complex of granite and pegmatite.

Muva System

Muva schist and quartzite form the western and part of the northern boundary of the Muliashi-Baluba basin. Kyanite occurs in some of the schist and magnetite streaks preserve cross-bedding in the quartzite. West of the Kafue River, good-grade emeralds are mined from a pegmatite intruding a metamorphosed ultrabasic sill.

KATANGA SYSTEM

Lower Roan Group

RL7 Footwall Formation (0–240 m). 100 m of argillaceous feldspathic arenites, with a lenticular conglomerate, fill the lower part of a hollow eroded in Lufubu schist, destined to become the Roan basin. At the top of the section is 18 m of clean, well-sorted, white to grey feldspathic arenite of aeolian origin. This is overlain by pink to white aqueous arkose and grey argillite and then by the persistent "Second Conglomerate", consisting of sub-rounded pebbles and cobbles of quartz, quartzite, granite, and gneiss in a micaceous sandy matrix.

Above are 15 m of argillaceous arenite and argillite with preconsolidation slump structures. Hills of Basement granite project through these lower beds.

Footwall Conglomerate (0–2 m). Usually included as the top bed of the RL7 formation, this gravel blanket of 2–10 cm quartz and quartzite pebbles marks the transgression of the copper-shale marine environment across both the RL7 formation and the Basement hills projecting through the footwall, but truncated at the level of the Footwall Conglomerate. Local gravel-filled scours and channels transect underlying argillite and arkose. Significantly, the superface of the conglomerate is marked by overlaps as basal dolomitic schist, with copper sulfides, grades eastwards into barren arenite and further on into pebbly arkose – which is then included in the Footwall Conglomerate. Thus, the latter blanket is actually a series of transgressive overlapping "shoestring" gravelly beach deposits.

In the Roan Extension, the pebbles gradually become smaller and sparser and in Muliashi it grades into an argillaceous sandstone. Here it carries, as do subjacent beds, disseminated copper sulfides, generally chalcocite, contiguous with bornite and chalcopyrite in the overlying ore-shale.

RL6 ore-shale (17–55 m). With a transgressive footwall and regressive hanging wall, the ore-shale is 55 m thick in Muliashi, mostly pyritic and with many dolomitic laminae, and gradually thins to 17 m near the limit of the transgression at the east end of the Roan basin, where it contains chalcocite in a silty massive argillite.

Originally it was a mud, but with a bottom layer 1 m in the east, thickening to 10 m in the west, with numerous dolomite layers. This bottom layer was a zone of shearing against the competent footwall formations and was severely crumpled grading westwards into a dolomitic schist. In the east, the muds were indurated to a laminated argillite of hornfelsic texture, westward developing a poor axial-plane cleavage, and then grading into a biotite–quartz–feldspar schist with some carbonate. However, in tight folds in the Roan Extension, the upper layers with bornite, chalcocite, or overlying barren shale retain some of the character of the massive argillite of the east end. At this latter locality of the Roan basin, the lowermost two-thirds of the ore-shale contain disseminated chalcocite and bornite with a sharp cutoff at the top against argillite.

Westwards, chalcopyrite appears in the basal dolomitic schist and, in turn gives way to pyrite. The stratigraphic section is thus: footwall conglomerate–pyrite–chalcopyrite–bornite–chalcocite (sharp cutoff)–barren argillite. The sharp hanging wall cutoff steps upwards across the laminae of the argillite in the westwards direction, so that barren argillite becomes chalcocite-bearing, whilst the chalcocite of the subjacent layer gives way to bornite and, in turn, to chalcopyrite.

In the centre and west half of the Roan basin, the lower half of the ore-shale contains disseminated pyrite, much as sparkling cubes, which thus underlies the "Upper Orebody", in turn overlain by barren argillite.

West of a granite hill, probably due to a "step" down of the Footwall Conglomerate, a narrow zone of chalcopyrite dissemination appears in the basal dolomite schist and this thickens westward through Roan Extension to form the Lower Orebody, separated from the Upper Orebody by a variable thickness of pyritic shales. As the Upper Orebody gradually steps up higher to the top of the ore-shale, it becomes thinner and eventually uneconomic. But the Lower Orebody, although not so consistent, continues far into Muliashi.

A most interesting and significant complication to the above mineral zoning occurs on the south limb of Roan Extension. Here, opposite the Irwin Shaft, a barren talcose siliceous dolomite occupies all the lower section of the ore-shale. It represents an algal bioherm built up in shallow water over a Basement ridge. Flanking the bioherm to the east, north, and west is an arcuate zone nearly 100 m wide of richly mineralized ore-shale with both bornite and chalcopyrite. Away from the bioherm, bornite diminishes in the lower half and soon pyrite appears near the footwall, forming the usual stratigraphic zonal sequence for the Upper Orebody of pyrite—chalcopyrite—bornite—barren argillite. Westwards from the bioherm, chalcopyrite persists on the footwall to form the Lower Orebody.

The Upper Orebody with its perfect regressive zoning of sulfides extends for 8 km along the Roan syncline; for 2 km it overlaps, with intervening pyritic layer, the Lower Orebody with its transgressive mineral zoning, which extends for 6 km to the west into Muliashi. Beyond that, the whole thickness of the ore-shale is pyritic and carbonaceous for another 10 km. For most of the eastern 10 km the Footwall Conglomerate is well developed, but west of that it is either patchy or a dirty arkose and the mineralization of the Lower Orebody extends down through it and for 3—6 m into the sandy dolomitic argillites of the RL7 formation. Chalcocite and bornite, commonly with malachite and black oxides, occur in the Footwall formation and grade into chalcopyrite dissemination in the dolomitic ore-shale, and then up into 40 m of pyritic shale.

Near the top of the Upper Orebody, there are thin sandy layers in the mineralized argillite; they are usually barren of sulfides except for occasional concentrations of chalcocite or bornite on the lower scoured surface. They demonstrate that waxing currents were able to scour immediately preceding cupriferous muds and bring in barren sands from nearby oxygenated environment.

RL5 Hanging-wall quartzite (60—120 m). A few small lenses of arkose occur in the barren argillite above the Upper Orebody, but the base of the RL5 is taken arbitrarily at the base of the first bed of arkose. With the regression at the top of the ore-shale, this means that the base of the RL5 steps up across the strata westwards. The formation consists of a sharply defined alternation of pink to grey cross-bedded arkose and dark-grey argillite, the latter becoming thinner upwards. Westwards, the formation becomes more dolomitic and a 7 m thick bed of dolomite is taken as the base. Cross-bedding is due to currents dominantly from north east.

RL4 dolomite and green shale (12—30 m). At the base is a white dolomite, cherty in the west, overlain by soft green-grey shale.

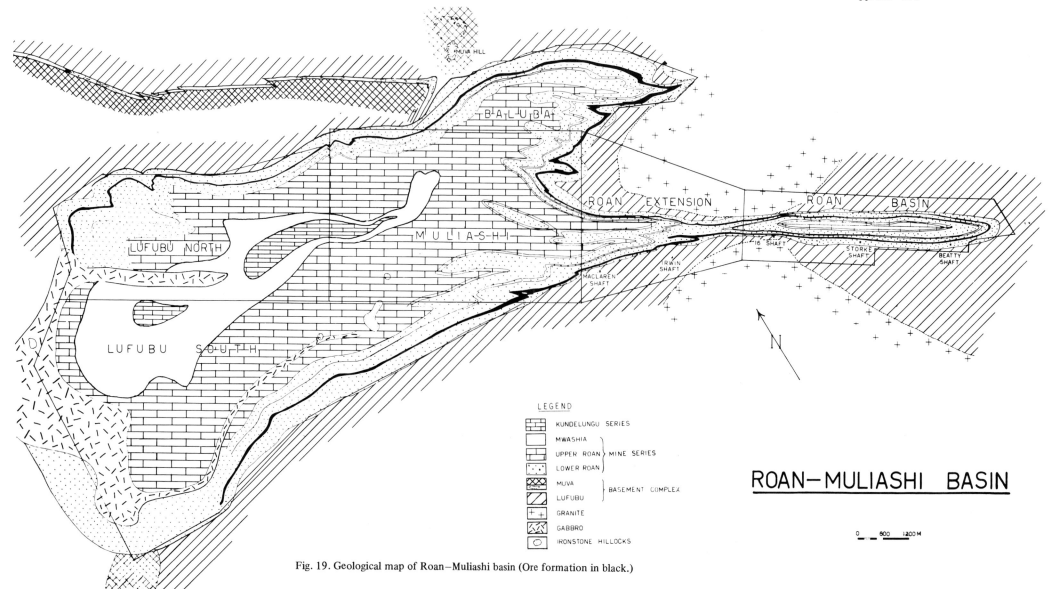

LEGEND

	KUNDELUNGU SERIES
	MWASHIA
	UPPER ROAN } MINE SERIES
	LOWER ROAN
	MUVA } BASEMENT COMPLEX
	LUFUBU
	GRANITE
	GABBRO
	IRONSTONE HILLOCKS

ROAN—MULIASHI BASIN

0   600   1200 M

Fig. 19. Geological map of Roan—Muliashi basin (Ore formation in black.)

RL3 Arkose formation (105–130 m). This thick sequence of medium- to coarse-grained pebbly arkoses develops some argillite beds westwards, and is the most competent formation of both Lower and Upper Roan.

Upper Roan Group

RU2 Shale formation (40–50 m). This consists of pyritic argillites with interbedded dolomite and thin arenites.

RU1 Dolomites (460–610 m). Mottled pale green to pink dolomite occurs in beds up to 30 m and is interbedded with nearly equal amounts of dolomitic pale grey argillites. Both are usually pyritic. Gabbro sills intrude these dolomites west of Muliashi.

Mwashia Group (100 m)

Black, thin bedded carbonaceous shales with disseminated pyrite occur in and west of Muliashi.

Lower Kundelungu Group

Tillite. A grey to brownish argillite with scattered irregular pebbles and boulders of quartzite and feldspathic quartzite overlies the Mwashia.

BALUBA

The lithology and the thicknesses of the formations are so similar at Roan and at Baluba that it is unnecessary to repeat the descriptions for the latter; it is preferable to stress the differences.

RL7 Footwall Formation

The basal conglomerate is better developed and contains more numerous boulders of white quartzite approaching the Muva System to the west.

Footwall Conglomerate

This is, generally, a little thicker and better developed than at Roan with pebbles and cobbles up to 15 cm diameter of quartz and quartzite. Yet there are abrupt changes between drill holes and the bed may be represented by a gritty argillite, in places mineralized, with red oxides and cupriferous vermiculite. The matrix is usually calcareous.

RL6 Ore-shale (55 m)

Cobalt minerals are lacking in the Upper Orebody at Roan, but low cobalt values of no economic value occur in the Lower Orebody of Roan Extension and Muliashi. The Baluba Orebody has a strike length of 4600 m and a dip extension measured around the folds of 1500 m, with an average thickness of 8 m. It is developed immediately above the Footwall Conglomerate in a light green tremolitic dolomite or dolomitic schist and consists of a dissemination of chalcocite at the base, grading up into bornite with carrollite, and then into chalcopyrite with some carrollite. The chalcopyrite in places extends into the over-lying grey argillites, overlain by nearly 50 m of intensely pyritic argillites. This orebody is thus equivalent to the Lower Orebody at Muliashi, but differs by carrying an average of 0.16% cobalt in a basal development of the ore-shale.

At the top of the pyritic argillite there is a thin zone of chalcopyrite mineralization. Near the east end of the Baluba syncline, the upper half of the pyritic argillite grades into disseminated chalcopyrite and chalcocite (in large part supergene), and then into barren argillite with arkose lenses and beds. This part thus makes a parallel to the Upper Orebody of Roan, but it is of limited strike. Being relatively shallow and duplicated in the limbs of an overturned fold, it may be mined by open pit.

Basement mineralization

The Lufubu schist and the intrusive granite at Roan have erratic uneconomic mineraliza-tion in the form of chalcopyrite, pyrite, and pyrrhotite in quartz veins and pegmatites or more rarely, sparsely disseminated.

West of the Irwin Shaft in Roan Extension a paleo-hill of Basement schist with granite sills projects through the RL7 Footwall formations and is truncated at the base of the ore shale. For 100 m along the strike, Footwall Conglomerate is absent and mineralization contiguous with the Lower Orebody persists as chalcopyrite dissemination down into the Basement schist for 10 m. Mineralization in the granite sills is poor or negligible. Where Footwall Conglomerate intervenes between the ore-shale and the Basement, mineraliza-tion in the latter terminates laterally. Evidently copper, either detritally as sulfide or in solution, penetrated vertically down cracks from the weathered and truncated surface of the hill exposed to the waters of the ore-shale.

FOLDING AND METAMORPHISM

The Roan syncline in its envelope of Lufubu schist is isoclinal, striking $300°$, and is overturned towards the southwest with axial plane dipping $65°$ northerly. The trough is rounded and plunges $15°$ westerly to a depth of 800 m opposite Storke Shaft (Fig. 20). Here the axis is nearly horizontal and an en-echelon synclinal axes appears and then, as

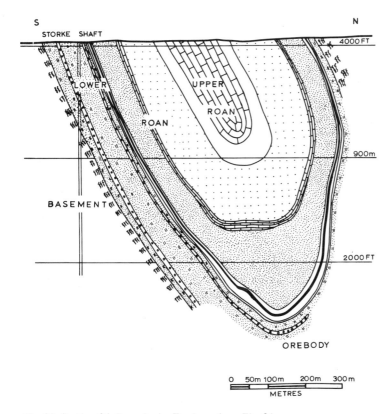

Fig. 20. Section 24, Roan basin. For legend see Fig. 21.

the axes rise to the west, the first syncline dies out. As the bottom of the syncline reaches to within 450 m of surface, it becomes pinched between the Basement granite, as between the jaws of a vice, and the incompetent ore-shale is squeezed into numerous isoclinal drag-folds (Fig. 21). Westwards into Roan Extension numerous en-echelon drag-folds involve both hanging and footwall formations as the synclinorium plunges west again over mixed Basement schist, gneiss and granite (Fig. 22).

In the Roan basin, perhaps due to the cushioning effect of the Basement schist, the muds of the ore-shale were metamorphosed to a hornfels-like aggregate of biotite, quartz, microline, and carbonate with, in the orebody, generally fine-grained sulfides. Westward, the axial-plane cleavage becomes increasingly evident and dolomitic beds are schistose. In the tightly squeezed trough between the granites, dynamic effects are most obvious with formation of dolomitic biotite schist and a gneissic appearance in banded dolomite and argillite. Sulfides are coarser and elongated parallel to axial planes. Microcline, dolomite, quartz, sulfide veins parallel to bedding and perpendicular to fold axes are abundant here. Tremolite and actinolite appear in the dolomitic rocks.

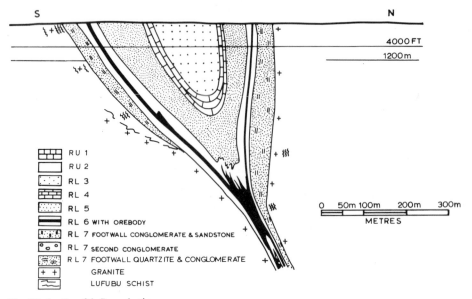

Fig. 21. Section 34, Roan basin.

In the Roan Extension, tremolite and actinolite become more prominent in the dolomitic schists and scapolite appears in the less dolomitic argillites. The scapolite forms white ovoid spots, which increase in size and abundance westward and notably in the

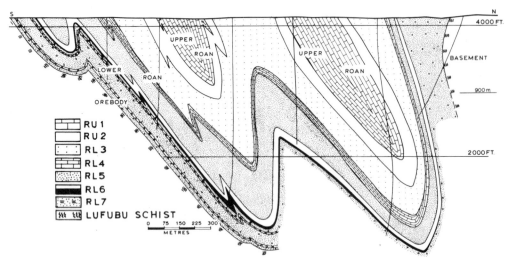

Fig. 22. Section 50, Roan extension.

stretched overturned limbs of drag-folds. This mineral tends to clear itself of the sulfide dissemination in the rock, resulting in a marginal concentration of sulfide.

In Muliashi, all the above metamorphic minerals are even more conspicuous in the ore-shale, and actinolite and epidote appear in the dolomitic arenites of the footwall.

Tourmaline shows increasing amounts of overgrowths on the original detrital grains (Mendelsohn, 1957) westwards and in the Roan Extension argillite in a closed anticlinal structure in the ore-shale is strongly tourmalinized.

At Baluba the folds are more open than at Roan and drag-folding is much less. The syncline at Baluba east is overturned to the south with steep to overturned north limb, but becomes nearly upright in the central area, where it is joined to a more southerly syncline through a mutual anticline, all axes plunging west-northwest. Farther west, the northern syncline again becomes overturned to the south and a bench-like fold develops on its north limb (Fig. 23). The synclinal axes eventually rise and the folds die out en-echelon against a steep homocline striking west-northwest.

The metamorphism at Baluba is of upper greenschist facies, just entering the epidote amphibolite facies westwards. In the dolomitic schist, the host for the richer mineralization, there is considerable development of tremolite, whilst in the overlying argillite there is often conspicuous scapolite and a little epidote and clinozoisite. Tourmaline is minor, mainly as overgrowths on detrital grains of tourmaline. Epidote and actinolite occur in some of the more dolomitic impure arenites of the footwall, especially in the west.

A conspicuous feature of the dolomitic schist and of the immediately overlying cupriferous argillite is the presence of white veins along the bedding, in places cross-cutting, consisting of mainly calcite showing fine columnar growth perpendicular to the walls of the veins. In addition the veins carry some quartz and copper sulfides. In appearance they are reminiscent of asbestos veins; similar veins occur in the Nkana South Ore-body in carbonaceous shales.

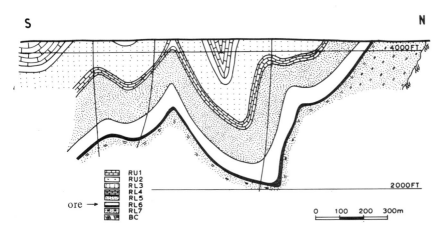

S   N

RU1
RU2
RL3
RL4
RL5
ore →   RL6
RL7
BC

4000FT.

2000FT

0   100   200   300m

Fig. 23. Section 52, Baluba.

CHIBULUMA AND CHIBULUMA WEST (W.G.G.)

These copper—cobalt deposits lie on the southern flank of the Chambishi-Nkana basin at 13 and 15 km west of Nkana. The similarities in stratigraphic sequence require only one general description, with emphasis on any differences.

BASEMENT COMPLEX

At Chibuluma a granite gneiss hill is buried under aeolian arenites and had little effect on the overlying orebody, but at Chibuluma West Basement gneisses projected above the ore horizon as a ridge and controlled the shape and location of the orebody. The rock here is a quartz—microcline—biotite—gneiss, but where the ridge has been squeezed up into a knife-edged screen 1—3 m thick in the crests of folds, it has been converted to quartz—biotite—schist.

KATANGA SYSTEM OR SUPERGROUP

Lower Roan

Basal Conglomerate (0—5 m thick). Angular boulders of granite and quartz—biotite—gneiss in a sandy matrix flank Basement hills and at Chibuluma West fill erosional gullies between hills.

Aeolian "Quartzite" (0—130 m). Well-banded feldspathic arenites have large-scale cross-bedded cosets up to 5 m thick and angles of over 30° between foresets and widely spaced parallel truncation surfaces. Quartz, albite, minor biotite, iron oxide, and rutile with variable carbonate and anhydrite cement are the usual constituents. Quartz—anhydrite—dolomite veins are common. Direction of cross-bedding is generally up dip, indicating wind from the north.

At Chibuluma West, the Basement gneiss projected as an isolated ridge above the dunes and the north wind excavated a shallow depression on the north, and east sides and the rear eddy excavated a deep narrow depression in the dune field on the south flank.

Aqueous arkose (3—15 m). Overlying an undulating and in places deeply potholed surface on the aeolianites, are feldspathic grits with pebbles up to 7 cm diameter, cross-bedded arkose, and minor shale beds. Greater mud and carbonate content in some beds allowed formation of biotite and tremolite. Small-scale cross-bedding, ripple marks, pebbles, gritty and muddy layers testify to an aqueous origin. Currents were mainly from the north.

Ore Formation (0–23 m). A layer of arkose with scattered quartz and quartzite pebbles is discontinuous at the base and generally contains a dissemination of copper–iron sulfides. Overlying sediments are feldspathic arenites with sulfides and argillaceous material now represented by sericite and some biotite. Cross-bedding is common towards the margin of the orebody at Chibuluma and is from the east at the east margin and from the west or irregular on the west margin. The Chibuluma orebody occupies a channel, probably an abandoned river channel, eroded in the aqueous arkose. The channel became an elongated salt lake with anoxic conditions in the bottom waters. Water level fluctuated, with mud cracking and desiccation of emergent mud flats. Floods washed mud flakes and occasional pebbles into the lake. Cross-bedded microcline arkose on the margin passes into albitic sericitic arenite in the orebody.

At the base of and at several horizons in the orebody, are marker beds of sulfidite (rock with 33% or more sulfide), consisting of nearly massive cobaltiferous pyrite, carrollite, and minor chalcopyrite with detrital quartz and feldspar. They are from 7 to 120 cm thick, extend the whole strike of the orebody, and grade up into normal chalcopyrite dissemination in sericitic arenite. The base of the sulfidite is generally a scoured surface, in places with potholes up to 100 cm deep, filled by sulfidite with a few pebbles of quartz and Muva ortho-quartzite. The pebbles do not rest on the scoured surface, but "float" in the sulfidite. Below the scoured surface the pyritic or cupriferous sericitic arenite is blackened for 1–3 cm by development of microscopic crystals of tourmaline from the clayey coatings of both quartz and feldspar detrital grains, whilst the sulfidite has small stumpy tourmalines. Evidently, the flood waters carried borates which reacted with the clay minerals. The overlying sericitic arenites, commonly cross-bedded, are richer in copper than the sulfidite and the underlying arenite is commonly pyritic. Chibuluma thus exhibits a cyclic depositional sequence from top to bottom: pyritic, glassy albitic arenite–chalcopyrite-rich sericitic arenite, grading down into cobalt-rich sulfidite, with pebbles–scoured surface, with local potholes.

No mechanical process of detrital deposition can separate cobalt and iron sulfides from chalcopyrite. This cycle, unique to Chibuluma, has so far been explained only by a chemical precipitation model: flood waters dissolved iron, copper, and cobalt (and borate) efflorescences over extensive mud flats, sites of previous syngenetic sulfide deposition, and carried these metals as sulfate with detrital quartz, feldspar, and clay with some pebbles into the body of saline water. Iron and cobalt were immediately precipitated as hydroxides and with sand and pebbles filled pot-holes and covered the scoured surface. Less catastrophic sedimentation allowed small delta formation, with copper and iron with minor cobalt precipitated as sulfide in cross-bedded muddy sands. After elimination of most of the copper and mud, iron sulfide was precipitated from clearer water in albitic "quartzite", possibly formed in part by wind-blown sand settling through the water. The iron and cobalt hydroxides were gradually converted to sulfides by anaerobic bacteria in the pore water and microcline was wholly converted to albite by the saline water.

The above cycle is seen three times in the Chibuluma orebody and can just be recognized a fourth time in the hanging-wall pyritic "quartzites", but here the cobalt–iron–sulfide layer failed to make a "sulfidite" and the following copper precipitation was too weak to make ore.

Disregarding the sulfidite markers, there is a zonal pattern of the sulfide minerals, from barren arkose in the east, to bornite–chalcopyrite dissemination in cross-bedded arkose, mixed chalcopyrite–bornite–carrollite sericitic sub-arkose, chalcopyrite–carrollite–cobaltiferous pyrite in sericitic sub-arkose, to pyritic feldspathic arenite in the centre. The zones are then repeated in reverse order to the west margin. However, the initial cycle was so rich that cobalt and copper mineralization persist along the base of the orebody, giving the latter a saucer-shaped cross-section, the pyritic central zone filling the middle of the "saucer".

Immediately under the orebody and centrally placed in the channel, there is a thin section of pyritic arkose with disseminated scheelite and uranium. It has limited strike and dip dimensions, but is of interest in revealing the first elements to precipitate under reducing conditions, as is the case at No. 4 Shaft, Mindola, where the uranium was precipitated in shallower water, but in the same horizon as the copper.

Although the bulk of the Chibuluma ore is in sulfidite and sericitic arenite, there are numerous thin argillite layers and partings, commonly well mineralized, especially in the middle of the orebody, where there is over 1 m of interbedded siltstone and shale richly mineralized with chalcopyrite and carrollite.

Chibuluma West Orebody (Whyte and Green, 1971). The mineralization is very similar to that of Chibuluma, except for the absence of the iron–cobalt-rich sulfidite layers. The orebody occurs in a "moat" on the north, east, and south sides of a Basement hill projecting through the aeolian sands. The west side has yet to be explored. On the north and east, the ore rests on barren aqueous arenites, but on the south the moat was deep and narrow and the lower cupriferous arenites are flanked on the south by aeolianites and on the north by screes on the slope of the Basement hill. One drill hole here assayed more than 15% copper over 70 ft true thickness (25 m).

A stream or "wadi" flowed into the northern moat, making a deep embayment in the north margin of the orebody, and flowed south across the east extension of the Basement ridge, here buried under a thin cover of aeolian sand. The bottom of the channel was eroded in aeolian sands and infilled first with pebbly arkose, cross-bedded due to current from the north. The upper part of the channel was subsequently infilled with mostly barren sands, topped by sandy muds, contrasting with the cleaner mineralized sands of the orebody. A zonal sequence of sulfides was deposited around the debouchment of this stream into the moat, with chalcocite, bornite, and chalcopyrite deposited first, then cobalt as carrollite with dominant chalcopyrite, and finally, flanking the north side of the steep Basement hill, a zone of pyritic dissemination in clean feldspathic arenite.

Elsewhere and especially in the southern moat, sulfide zoning is not prominent. Evidently, hydrogen-sulfide generation by anearobes (see Chapter 6 by Trudinger, Vol. 2) was so intense that all metals were precipitated together to give an extremely rich ore of bornite, chalcopyrite, and carrollite, with many sulfidite layers of these minerals. Generally, there are three layers of richer cobalt mineralization that can be followed through the orebody here. Locally, even the Footwall aqueous formation in the moat developed anoxic conditions, with precipitation of cobaltiferous pyrite, carrollite, and minor chalcopyrite, thus doubling the usual thickness of ore. Well-mineralized shale layers occur in the central parts of the orebody, but more commonly they are pyritic. At the orebody fringes, distant from the Basement ridge, the mineralization grades rather abruptly into disseminated pyrite with lenses of chalcopyrite (or chalcocite) and within about 30 m fades out into barren feldspathic arenite, which on the southern fringe is a dark flinty-looking meta-arenite, or "quartzite".

In the southern moat, the matrix of scree-like conglomerate flanking the Basement ridge is well mineralized where overlain by orebody and, likewise, Basement schist is in places mineralized for as much as 7 m below the unconformity, but only when directly overlain by ore. Where barren footwall sediments intervene between the ore and the basement, the latter is barren except for the rare vein that intersects the orebody above.

Hanging-wall "Quartzite" (9 m). A hard pyritic feldspathic arenite overlies the Chibuluma orebody. At the top are some gritty arenites with thin argillaceous interbeds. At Chibuluma West only one to two metres of pyritic arenite overlies the ore and appears to be part of the ore formation as laterally it grades into chalcopyrite disseminations. This is overlain by 9–18 m of red-brown arkose interbedded with argillite and sandy argillite. The brown colour is due to weathering of disseminated pyrite and close to the underlying orebody there is in places considerable disseminated chalcocite, thus forming a second, but patchily developed, ore horizon.

Hanging-wall Conglomerate (up to 9 m). This consists of a thin bed with scattered pebbles in the west of Chibuluma, thickens to 9 m of compact conglomerate over the middle of the orebody, but is porous, weathered and water-bearing in the east, where most of the pyritic "quartzite" was eroded, so that the conglomerate in places directly overlies the east fringe of the orebody. It has been traced, nearly continuously, into the "Lower Conglomerate" of Nkana. It contains considerable disseminated pyrite, weathered to limonite in the east.

This formation has not been recognized at Chibuluma West, but it may possibly be equivalent to the talc–carbonate–albite breccia, with angular to rounded nodules and pebbles of dolomite, forming the base of the so-called Upper Roan.

Upper Roan

Chlorite and talc schists (up to 30 m). These schists, with calcareous arenite at the base, overlie the Hanging-wall Conglomerate and grade down dip into pyritic crystalline dolomite.

Carbonaceous Pyritic Shale (12–90 m). This is intersected by some of the deeper holes and eastwards has been traced into the Nkana ore-shale.

Dolomite and dolomitic siltstones (up to 240 m). These are interbedded with tremolite and scapolite dolomites and talc–chlorite schists.

POST-KATANGA INTRUSIVES

Sills of meta-gabbro inflate the Upper Roan dolomite succession at both Chibuluma and Chibuluma West. A lowermost sill of amphibolite schist up to 60 m thick occurs near the base of the Upper Roan, but multiple sills higher in the succession attain 300 m thickness and show internal chill contacts, basic pegmatites, chlorite–amphibolite shear zones, and narrow metamorphic aureoles of tremolite or actinolite at contacts with impure dolomite. Albite–quartz rocks are commonly developed at the contacts. The intrusive rock is ubiquitously converted to scapolite amphibolite. Pyrite and a little chalcopyrite occur in the shear zones. The multiple sill with internal chill contacts has gravity layering parallel to enclosing strata indicating that it was intruded before the Lufilian folding.

FOLDING AND METAMORPHISM

Folding

Lufilian folding has tilted the Chibuluma orebody so that it dips 38° north on the south flank of the Nkana synclinorium. A fold east of the mine plunges northwest and intersects the ore shoot at 450 m depth causing flattening and then steepening of the dip. Two other folds intersect the ore shoot at depths to 750 m, but the orebody continues over and beyond them. Axial planes are vertical to 70° northeast dip.

In the upper section to 450 m depth, there is little distortion of the orebody and adjacent beds, and the original channel shape, concave upwards, is preserved (see Fig. 30).

At Chibuluma West en-echelon folds plunge west-northwest and have telescoped the ore shoot and rotated it to a nearly east–west direction from its original northeasterly elongation parallel to the original Basement ridge. Highly competent aeolian arenites in synclines acted as the jaws of a vice and squeezed the intervening hill of Basement gneiss into a knife-edged screen of biotite schist, in places only 1 m thick although 100 m high. Anticlinal crests are arranged in right-hand echelon indicating an original direction north of east for the Basement hills.

Structure—ore relations at both mines demonstrate that the ore is pre-folding. At Chibuluma West, the Basement ridge controlled both the location of the ore-filled "moat" and the later anticlinal axes.

No appreciable foliation or cleavage has been developed, except in the isoclinal folds at Chibuluma West, where even the aeolianites have biotite clots parallel to cleavage and the Basement gneiss is converted to schist with near vertical cleavage. Hanging-wall argillite has a strain-slip cleavage.

Faulting

Faults, mostly reverse- or thrust-type, with throws of less than 1 m are fairly common and shatter zones oblique to strike occur in both mines. Where these intersect the ore-body, chalcocite replaces other sulfides, or all the sulfide may be leached. Calcite and cobalticalcite commonly fill fractures and cavities in the leached zone. At Chibuluma West, a shear zone assayed 6% MoS_2 over 15 cm.

Veins

Lateral secretion veins are common and contain quartz, dolomite, and anhydrite in the footwall beds and quartz, copper—iron and cobalt sulfides in the orebodies — usually the same minerals as occur in the adjacent wall rock. One large quartz vein at Chibuluma West contained appreciable galena as well as bornite, chalcopyrite, and carrolite.

Metamorphism

Quartz, biotite, chlorite, epidote schists formed in the isoclinal anticlines of Basement gneiss.

Katanga sediments are in the upper greenschist to epidote—amphibolite facies. Biotite and sericite are dominant in the argillaceous rocks, sericite especially in the sulfide-bearing arenites, whilst actinolite and tremolite are common in dolomitic arenites. Scapolite occurs in Upper Roan argillaceous sediments. Scapolite, hornblende, labradorite, and albite are the dominant minerals of the meta-gabbro.

Most of the microcline of Lower Roan aeolianites and aqueous arkose has been albitized.

SUPERGENE PROCESSES

The Chibuluma ore-shoot once extended to surface, but has been weathered to a gossany sandstone to 30 m below surface. The limonite derived from mainly pyrite and chalcopyrite contains no visible malachite or chrysocolla. Although this leached sub-outcrop is covered by 4—5 m of soil and laterite, it gave a distinct copper anomaly in the soil and the oxidizing sulfides gave a self-potential anomaly of minus 100 mV at surface. The orebody had been previously found by systematic drilling of the Lower Roan.

Chibuluma West was a blind orebody, also found by systematic drilling. It is practically unweathered, except in some shatter zones and has no distinct geochemical anomaly nor self-potential anomaly at surface. However, the overlying red arkoses are weathered to considerable depth, due to high porosity, and their patchy chalcocite mineralization gave a slight copper indication in surface soils.

MUFULIRA MINE, ZAMBIA (V.D.F.)

The Mufulira gross reserves in 1974 stand at 282 million tonnes at 3.47% total copper, about half of which have been mined out. Grouped with the Shaba deposits of Zaire, the vast Copperbelt region is unique in the sense that no similar orebodies of this magnitude and grade are known to occur anywhere in the world within a sedimentary pile correlatable with the Katanga System.

The Mufulira copper deposit (Figs. 24—33; Table III), occurs on the eastern side of the Kafue Anticline and, in contrast to the shale-type orebodies of Roan, Nkana, Chambishi, Nchanga, Karila Bomwe and Musoshi along an old shoreline to the west of this anticline, consists only of copper—iron sulfides in arenites. The shale deposits in contrast may contain cobalt sulfides in addition to copper. Other high-energy sediments with copper deposits along the Mufulira shoreline include Bwana Mkubwa, Mokambo and Kinsenda and several other areas of mineralization not fully explored.

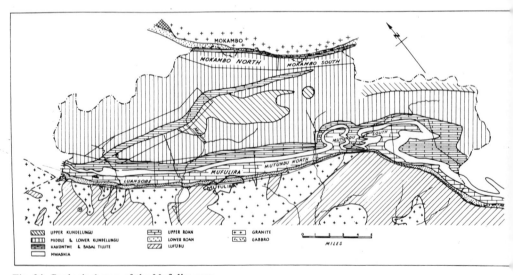

Fig. 24. Geological map of the Mufulira area.

TABLE III

Broad stratigraphic succession at Mulfulira

System, Supergroup or sequence	Series	Group	Formation	Thickness variation (in m)
	Kundelungu	Upper	Purple Sandstones	unknown
		Middle	Shales	unknown
		Lower	Shale	up to 900
			Kakontwe Limestone	
			Tillite	
Katanga		Mwashia	Shale, Upper	up to 600
			Xmas Sandstone	
			Shale, Lower	
	Mine	Upper Roan	Upper Dolomite	
			Interbedded Shales and quartzites	10*–800
			Intermediate Dolomite	
		Lower Roan	Hanging-wall	60–80
			Ore Formation	30–80
			----------- local disconformity -----------	
			Footwall Formation	nil to 150
----------------------------------- unconformity -----------------------------------				
Lufubu	not subdivided		part sedimentary, part igneous rock (intensely metamorphosed-chlorite stage)	
Old Granite	not subdivided		igneous rock (mainly granodiorite)	

* At the lower thickness much of this group is replaced by breccias (see text).

As mentioned above, all of the Zambian Copperbelt deposits are strikingly similar, being emplaced either in argillaceous or arenaceous rocks with varying amounts of dolomite and minor anhydrite in disseminated form or in distinct layers. The geology is very much the same also from property to property, especially the footwall succession. At Mufulira sand dunes rich in windblown gypsum have recently been discovered in this horizon. Gypsum or anhydrite constitutes up to 35% of the sand.

The deposits are interpreted as syngenetic or synsedimentary in origin, typical of continental-shelf types along a rugged coastline undergoing submergence.

BASEMENT COMPLEX

The Basement rocks comprise granodiorite intruded into Lufubu Schist. The latter formation is regarded as an igneous rock in the vicinity of the mine and not an altered

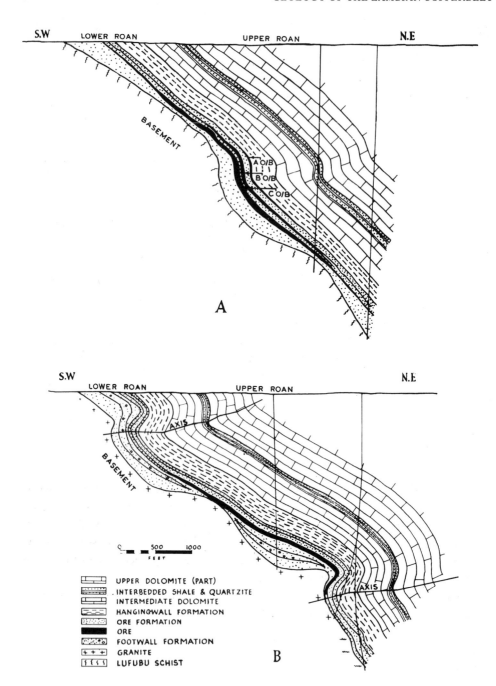

Fig. 25. A. Cross-section 28/29, Mufulira. B. Cross-section 55/56, Mufulira.

sedimentary pile as previously postulated, although sediments are known to occur in the district. No bedding has ever been observed underground and the rock here is akin to an acidic igneous rock in composition.

No consistent grain has been mapped in the Basement rocks, except in the large Lufilian folds where a pronounced schistosity is developed. The axial planes of these folds are nearly horizontal and an example where basement rocks are involved in such folding can be seen in Fig. 25.

The topography of the basement at the onset of Lower Roan sedimentation was quite irregular, the highest ridge known being about 150 m above the deepest valley floors.

KATANGA SUPERGROUP

Lower Roan Group (200–250 m); *Footwall Formation* (0–150 m); *Basal Conglomerate/ Breccia* (0–20 m)

A cobble-to-boulder conglomerate or scree breccia is poorly developed in the mine area but is at maximum thickness toward the northwest on the flank of a pronounced paleo-hill.

Variation in thickness of the footwall beds is controlled by the distribution of paleo-hills and ridges. As seen in Fig. 27, the mine is divided into three geological basins by these features which coincide with areas of maximum relief and, therefore, minimum development of footwall sediments. The composition of the footwall is similar to that described on the other Copperbelt properties with one exception. Anhydrite-rich dunes have recently been discovered in the central area of the mine. Current directions imply a source from evaporite lakes situated to the north, now down dip. One small stromatolite reef has recently been found right on the unconformity in a core from an exploration drill-hole where only a veneer of footwall is developed over the basement ridge separating the eastern and central geological basins. The most significant feature of the footwall is the evidence of a marine transgression at the base of the lowermost orebody. Cross-bedding indicates that clastics were deposited in essentially a high-energy aqueous environment over most areas of the mine. However, shale/arenite interbeds in this horizon near the flanks of paleo-hills suggest quieter interludes. Apart from sporadic footwall lenses of sulfide ore, the first signal of mineralization, albeit non-sulfidic, is found in the form of disseminated rhombohedral cavities after siderite. These lenses of iron carbonate minerali-zation are confined to particular beds and are also regarded as syngenetic in origin. In the deeper regions, siderite is unleached. The western basin is typical of fluviatile beds at the base of the C orebody, whereas in the eastern basin, aeolian quartzites abut against the footwall. The wind direction was persistently from the northwest and north.

These dunes, including the gypsiferous ones, display a maximum dihedral angle of 34° between the truncation planes and foresets. The foresets curve gently and almost imperceptibly into more closely banded bottomsets. The cross-bedding is large-scale and the aeolianites are usually well-bedded in thin light and dark grey seams. Significantly, no sulfides have ever been detected in these aeolian beds.

Ore formation (30–80 m thick)

General comments. The Basement topography appears to have exerted a significant control over mineralization, the main features being that pyritic interruptions in the ore horizon coincide with paleo-highs and the richest copper shoots occur on their flanks. Never has a paleo-hill been found to protrude through the lowermost C orebody, which suggests that the region was peneplaned immediately prior to or during ore deposition (see Fig. 26). This would imply that the area of provenance for the material of the ore

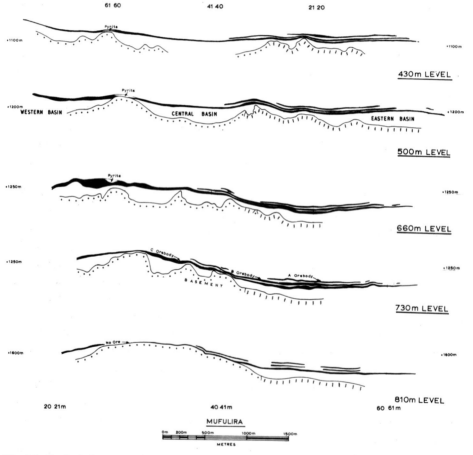

Fig. 26. Geological sections by main levels of Mufulira, showing the three orebodies in relation to the basement.

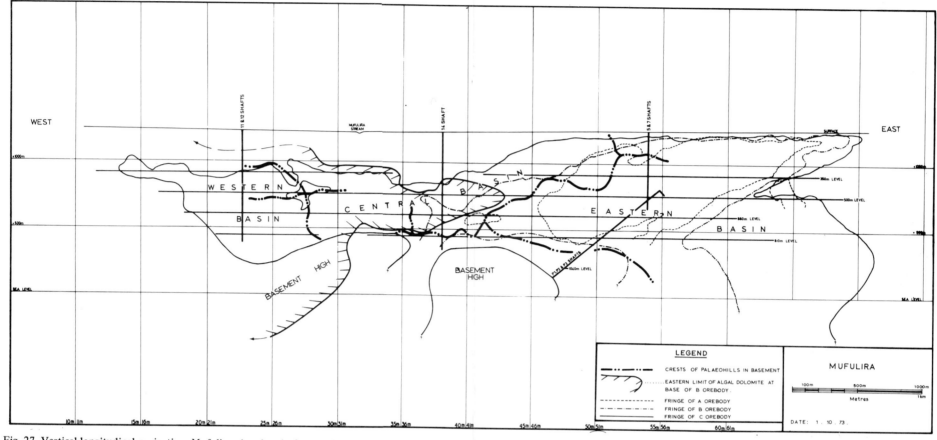

Fig. 27. Vertical longitudinal projection, Mufulira, showing the known limits of the three main orebodies, trend of paleo-hills and eastern limit of inter B/C algal dolomite.

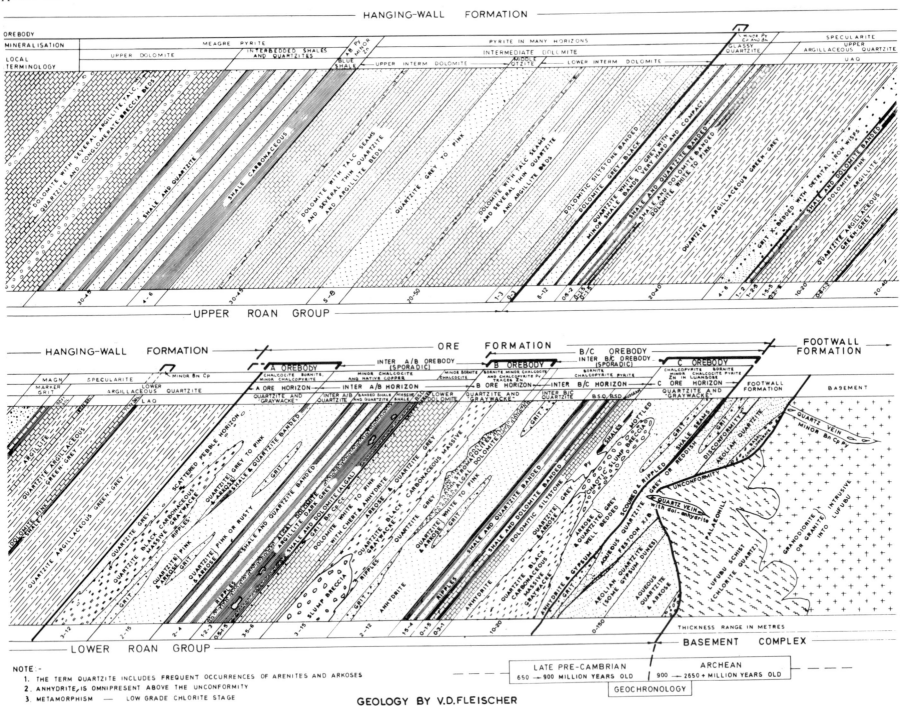

Fig. 28. Generalized stratigraphic column, Mufulira.

formation might have been well removed from the mine area. In each major ore unit there is a gradual decrease of feldspar from base to top and also an increase in roundness of the more resistant quartz grains which indicates a more prolonged sedimentary history for the upper sections.

Hence, all three main orebodies generally reveal a repetition of sedimentary environment, the lower parts of each represent high-energy conditions (rapid deposition) that gave way progressively to finer, well-worked clastics. The so-termed "graywackes", which occur as lenses in each orebody were deposited at a late stage in each case. These units are composed of carbon-rich arenites with much sericite. Mineralization ends abruptly at each hanging wall suggesting regression. This cut-out occurs at a change from arenite to dolomite and shallow-water muds in most cases.

The west fringes of the B and C orebodies are characterized by a facies change from fairly clean mineralized arenites to dirty argillaceous arenites that are barren. The finer grain of the latter testifies to deeper water where physicochemical conditions were presumably unsuited to sulfide formation.

The eastern fringe of the C orebody in the upper levels coincides with a change from sub-aqueous to aeolian arenite, whilst the east fringe of the B orebody and both fringes of the A orebody are marked by change from grey mineralized arkose to red arkose with minor red sandy shales of red bed environment.

Orebodies

Figure 28 is a generalized stratigraphic column showing the position of the orebodies in the geological column and their variations in thicknesses.

C orebody. This is the most extensive orebody extending 5800 m on strike and is continuous in dip at the east down to at least the 1500-m level. Chalcopyrite is the predominant sulfide mineral in this horizon although some areas contain significant bornite. The grade averages 2.5–3.5% copper.

The basal section is coarse with much festoon-type cross-bedding. Figure 29 represents

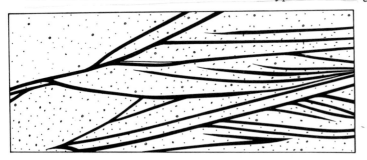

Fig. 29. Basal section of the C orebody at Mufulira West (half true scale). Cross-bedded feldspathic grits, the dark layers being rich in bornite which shows up the cross-bedding remarkably.

such features at the western section and illustrates an example of what is interpreted as detrital sulfide deposition in a high-energy environment. The darker layers are rich in copper sulfide and detrital zircons, most of the latter being well rounded. Refer also to Figs. 30 and 31 as the features detailed for Chibuluma are commonly exposed at the base.

The cross-bedding gives way upward to finer, and then to more massive, sediments. A characteristic horizon in the transition zone is a *chaotic slump breccia*. There is evidence that primary alternating dark- (mineralized) and light-coloured (non-mineralized) beds were intermixed by gravity sliding. Original interbeds, where preserved, indicate rhythmic deposition of copper sulfides in the darker layers which average about 1 cm thick. They are interpreted as seasonal, or some other cycle, in origin and strongly support syngenesis. Near the top of the C orebody, with the best development in the eastern basin, there occurs a black carbonaceous arenite locally termed a "graywacke". Carbon constitutes up to 2% of the rock and there appears to be an inverse relationship with copper within the graywacke itself. Higher copper values in overlying B and A graywacke lenses correspond to relatively lower carbon contents. For example, rich A graywacke with 20% copper sulfide contains only 0.5% carbon. Although massive with poorly developed graded bedding, it is noted that the lateral fringes exhibit strong bedding and in some localities cross-bedding. Near the base of the graywacke and in silty zones within the orebody, ripplemarks are sporadically dispersed. Concentrations of sulfides occur in the hollows of such features which cannot be readily explained by epigenetic processes. This is especially so where overlying sands are quite barren, although porous. Desiccation cracks occur commonly in argillite seams near the base and were presumably filled mechanically by copper sulfide and sand at the onset of subsequent sedimentation.

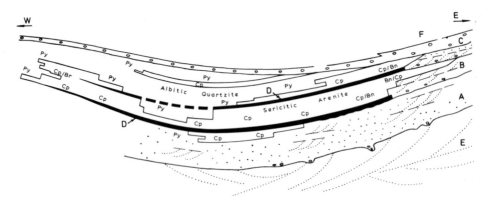

Fig. 30. Diagrammatic cross-section of the Chibuluma orebody, showing principal sulfidite marker horizons. Vertical scale is grossly exaggerated.
E = aeolian arenites; A = arkoses; B = pebbly arkoses grading westward into sericitic arenite; C = hanging-wall conglomerate; F = dolomitic sandstones, shales and dolomites of Upper Roan; D = sulfidite marker horizons traceable for long distances through orebody, locally with scours and pot-holes at base.

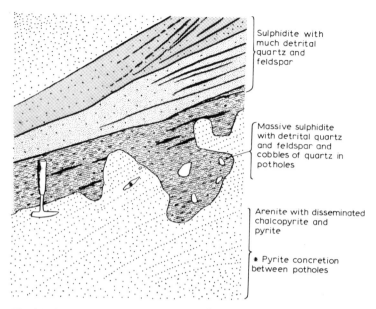

Sulphidite with
much detrital
quartz and
feldspar

Massive sulphidite
with detrital quartz
and feldspar and
cobbles of quartz in
potholes

Arenite with disseminated
chalcopyrite and
pyrite

* Pyrite concretion
between potholes

Fig. 31. Section 120, Chibuluma at 1340' level. Sulfidite layer is 80% cobaltiferous pyrite and carrollite with only a little chalcopyrite and, with cobbles of quartz, fills potholes eroded in cross-bedded arenite with disseminated chalcopyrite and pyrite assaying 2% copper. Cross-bedded sericitic arenite, assaying 6% copper and 0.2% cobalt, overlies sulfidite.

Sulfide-mineral zoning

Classical copper–iron sulfide zoning at Mufulira appears to have been controlled by the basement topography, to some extent, in that paleo-hills and ridges coincide with pyritic intervals in the ore formation whilst bornite-rich zones occur on the flanks and extend into the adjacent basins. The change from pyrite to bornite is complex and usually includes an interfingering transition zone through chalcopyrite. Resolution of the zonal patterns is further complicated by the imposition of chalcocite zones in the relatively more porous arenites, which are regarded as supergene. Deep into the basins of deposition, bornite gives way to chalcopyrite and eventually to non-pay zones followed by pyrite and ultimately barren, dirty looking argillaceous arenites. It is not known whether the pyritic lenses over paleo-hills within the orebodies are genetically related to the pyrite adjacent to the barren argillaceous arenite facies. The complexity of lateral interstratal zoning, as also noted across the dip in any one locality, does not necessarily oppose a syngenetic origin. Subtle facies changes imply differing environmental conditions during sedimentation. Distribution of the various sulfides would be a function of the availability of iron and copper fluxes at any particular time and place. As has been mentioned, there is evidence also that some of the sulfides appear to be detrital in origin, and in these regions the copper sulfides (if mono-mineralic) are well mixed. On the other hand, the bulk of the deposit displays subtle changes through thin laminae across the dip, for example fine alternating chalcopyrite and bornite. This rhythmic feature is ascribed to a response of mineralization to the prevailing physicochemical milieu and plain availability of iron and copper, as stated.

Apart from a geochemical distribution of many metals in the orebodies, only two others with the iron and copper are of any significance, albeit of academic interest. They are zinc as sphalerite which occurs in a 15—18 m thick C graywacke sequence about 5 km to the northwest of the mine intimately associated and disseminated with pyrite, and uranium on the western flank of the major paleo-ridge at 12 Shaft (western region). The uranium occurs as black disseminated uraninite in several horizons in the C arenite and is most conspicuous where concentrated and altered to the bright efflorescent secondary salts in leached areas on the flank of this paleo-ridge.

Inter B/C. The two most significant features of these beds are:

(1) The well-mineralized cupriferous part in the eastern and to a lesser extent the central and western basins. The mineralized part of the Inter B/C succession occurs in a spectacular sedimentary sequence comprising intercalated and finely laminated shale—silt and arenite. The buff silts are barren, whereas the interbeds of arenites are highly mineralized in bornite and/or chalcopyrite. Some of the thicker arenites may be barren and minor dolomite seams near the base are also devoid of sulfide. Detailed study of these beds provides sound evidence for syngenesis. One just cannot conceive of an epigenetic process which would "miss out" thin interbeds, inter alia porous sands and dolomites long argued to be ideal receptacles for trapping the valuable content of mineralizing solutions.

(2) A stromatolite reef that occurs at the top of the succession, the eastern limit of which coincides with the western fringe of the B orebody (Fig. 27). The stromatolite reef that was first recognized by Malan (1964) has subsequently been found to be extensive, occurring as a tabular sheet stretching sporadically from the western fringe of the B orebody to the Zaire border some 15 km to the northwest. The reef itself is barren, but sands that subsequently inhibited its growth near the B fringe contain copper sulfides and barren reef debris. The development of reef indicates shallow waters at the time and supports the theory that the bedded copper deposits were emplaced under shelf conditions during the interval between a marine transgression and eventual submergence of irregular shorelines into presumably deeper waters. The magnesites and dolomites in the far hanging-wall represent chemical precipitates in deeper water.

B orebody. The lithology is similar to the C horizon and represents a repetition of conditions that prevailed then. Differential compaction of the earlier sediments resulted in the spreading of a basin for the B cycle (Fig. 32). Bornite is the predominant mineral and the orebody grades between 3.5 and 4.5%.

Inter A/B. The base of this formation is characterized by the first major dolomite bed,

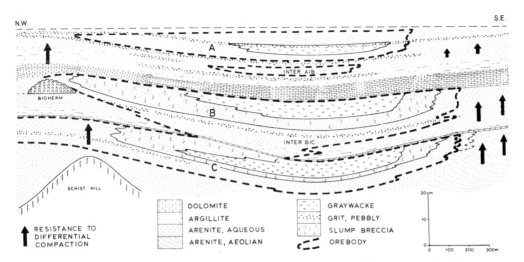

Fig. 32. Super-imposition of Mufulira "A", "B", and eastern part of "C" orebodies by compaction folding. Note development of algal bioherm on west flank of compaction fold over schist hill, thus enclosing lagoon in which "B" orebody developed.

termed the Lower Dolomite, which varies about 5 m thickness. It contains up to 30% anhydrite and for this reason is one of the main aquifers in the mine. The anhydrite is leached to below the 1000-m level in the eastern section of the deposit. This results in the dolomite being spongy, cavernous and water-charged.

Immediately overlying the Lower Dolomite is a thin grit–dolomite sequence which is well mineralized in copper sulfides. The grit also contains primary uraninite in places, usually in fringe areas. Copper grades up to 4%, but the seam is only 10–15 cm thick. Immediately over the grit, called the Gritty Marker, a thin algal (stromatolitic) reef is extensively developed but is thin as well, averaging only 10 cm. This dolomite is barren and appears to have inhibited the development of a potential orebody. Conditions favourable for sedimentation accompanied by a concomitant copper-sulfide mineralization gave way to stromatolite growth in a presumed lagoonal environment, momentarily barred from further sedimentation. A 5 m thick argillite overlying the reef suggests a deepening of the lagoon. Stromatolites are sporadically developed near the top of this argillite at the western extremity of the mine. Interbedded shale and sandstone was then deposited grading upwards into coarse gritty sandstones, which restored the eastern region to conditions similar to that for the deposition of the C and B orebodies, with the resultant thinner A then formed.

A orebody. As for the B orebody, the lithology is similar to the C horizon and represents a further repetition of sedimentary conditions. A most unusual feature in the A orebody is the development of conglomerate beds at two horizons, the one near the base and the other in the top third of the orebody. The upper conglomerate comprises well-rounded pebbles and boulders of essentially quartz, but also quartzite and granite. In one locality a granite boulder 0.8 m across was noted. The upper conglomerate can be traced through the graywacke facies. The grade of the A orebody is the highest, averaging between 4.5 and 5.5% copper. Chalcocite and bornite predominate.

A small pocket of uranium mineralization has been found at the very top of the A orebody in the central section, consisting of yellow uranium ochres.

Hanging-wall formation

Immediately overlying the A orebody is a sequence of anhydrite-rich argillaceous "quartzites" characterized by one 4—6 m thick grit and several thin dolomite beds. The lowermost is locally termed the 70 ft. Dolomite (being approximately 70 ft stratigraphically above the top of the A orebody) and recently it has been found to underlie a limited 1—1½% copper occurrence, 1—4 m thick at the western fringe area of the A orebody. Structures noted in the 70 ft. Dolomite suggest part algal origin.

A thin but persistent (over 100 km of strike) mineral zone occurs at the top of the Glassy Quartzite straddling the contact with the base of the Upper Roan which is represented by a 2—3 m thick dolomite, partly carbonaceous. No areas of economic significance have been found at this horizon in exploration drilling so far. Higher in the sequence, pyrite is ubiquitous in many dolomite and quartzite beds and several geochemically high zinc zones have been detected in the dolomites. One bed, locally called the Blue Shale due to its carbon content, contains abundant disseminated pyrite (3—10% by volume) and visible sphalerite.

A pronounced lithological feature of the Upper Roan sequence of sediments, i.e., between the Glassy Quartzite and the base of the Mwashia, is the irregular emplacement of breccias. Comprising mainly soft rock, these units are considered to represent products derived from rapid structural adjustments during the formation of the syncline. Thrusting on a regional scale may have produced some of these breccias which are discordant and reduce an 800 m thick succession of Upper Roan sediments and evaporites in the central area of the mine to only 10 m toward the southeast. The constituents of these breccias average about hen's-egg size and are made up of heterogenous rock types, mostly soft dolomite, talc, argillites, sandstones, magnesites and shale, rounded in shape with a rind of soft light-coloured argillaceous material, which suggests tectonic injection or rapid sedimentary deposition. Soft rock would not tolerate prolonged transport.

The Upper Roan Group is composed of dolomites, magnesites, shales, argillites, arenites, conglomerates and breccias in sandwich-type intercalations that are similar in a broad sense across the Copperbelt. Better core recoveries and new drilling has shown, contrary to popular belief, that dolomite makes up less than 30% of the column. The soluble dolomites are thinly sandwiched between insoluble strata, which mitigates against formation of karst-type country and very large reservoirs of water in the dolomite compartments, resulting in an amelioration of the drainage problem in mining.

PALEOGEOGRAPHY AND DEPOSITIONAL ENVIRONMENT

Geologists working at Mufulira have long been intrigued by the wide spectrum of sediment types found in the Mufulira syncline. Several have attempted to identify the conditions which prevailed at the time of sedimentation and a divergence of opinions has emerged. This is not unexpected as the decipherment of diverse fossil structures, perhaps over 650 m.y. old, is indeed a gigantic task. Modern processes of sedimentation are being applied in an effort to arrive at satisfactory explanations, but many structures are seen which cannot be resolved this way. We just do not have the keys at this stage, primarily because there are too many imponderables, the main ones being that very few geologists too for that matter, understand submarine/subaqueous processes well enough. This is understandable since no one can actually see these natural processes at work on a large scale. Recent literature stresses this drawback, but at the same time it is recognized that small strides are being made in the right direction. In other words, factual data from the maze of underground workings can be readily collected, yet universally accepted genetic interpretation is still too often in the "either–or" category. Examples are criteria for aeolian sand dunes, origin of ripples, water depths for stromatolite development, suggested sedimentary models to explain the varieties of cross-bedding, and a host of others.

The Late Precambrian Katanga rocks, as stated, contain a great diversity of sediments. Local interpretation of prevailing environments for their deposition range from deserts through marine transgressions and regressions and deep seas, to shallow deltas, large rivers, lagoons and possibly sabkha lakes. One horizon high in the hanging wall suggests that an ice age also occurred. On the northern limb of the Mufulira syncline, thin sheets of igneous extrusives have been identified. Since the first discovery of stromatolites at Mufulira, several other algal horizons have been recognized, three of which are within the ore formation itself. Ripple marks, desiccation cracks, festoon cross-bedding and stromatolites are all considered to point to a relatively shallow-water environment for the host rocks of the orebodies. The sand dunes of the footwall are the only non-aqueous beds in the succession.

Mineralization

Copper sulfides occur at no less than twenty-one distinct lithological horizons at Mufulira of which only three are not economically exploited. For simplicity in mining, the eighteen different ore beds are grouped into five broad mining units termed the C, Inter B/C, B, Inter A/B and the A from the oldest to youngest. A significant feature of the ore formation is the presence or absence of gypsum/anhydrite which testifies to sporadic saturation of $CaSO_4$ in the waters and, therefore, ready availability of sulphur for sulfide formation (see Chapter 6 by Trudinger, Vol. 2). Barren interbeds within the ore formation, the immediate hanging-wall argillaceous "quartzite" and barren "ore" horizons adjacent to the Mufulira deposit contain as much as 19—31% anhydrite as revealed by chemical analyses. On the other hand, the orebeds themselves contain interstitial dolomite with little or no anhydrite, revealing an interesting relationship between sulfide mineralization, dolomite and anhydrite. In passing, it may be of interest to readers that the drainage problems of all Copperbelt mines are considered to be related to the gypsum/anhydrite distribution in the surrounding sediments. A relative lack of anhydrite in the orebodies results in the unusual condition, highly acceptable to the miners, that the orebodies themselves are usually in unweathered competent ground in regions where some beds in the hanging wall and footwall may be weak and wet due to deep leaching of the water-soluble anhydrite.

It is expected that the geological genetic interrelationship between anhydrite and dolomite will provide valuable material for future research (see Renfro, 1974, for example).

Zonation of copper sulfides has been shown to be more complex than hitherto thought, due to more detailed information that is coming to hand, as the deposits are mined in depth. Low-grade pyritic areas over highs in the paleotopography remains a puzzle (see Fig. 33). Intricate interdigitation of bornite and chalcopyrite both on strike and across the dip may be explained by availability of copper and iron in select areas at that particular time of mineralization. Chalcocite unfortunately masks its true relationship and primary distribution, because of its presumed secondary overprint in enriched areas due to post-depositional processes attributed to weathering. Whatever arguments are put forward on genesis of the arenite ores, the following features, which strongly support a syngenetic origin, must be recognized: (1) The sulfides are characteristically dis-

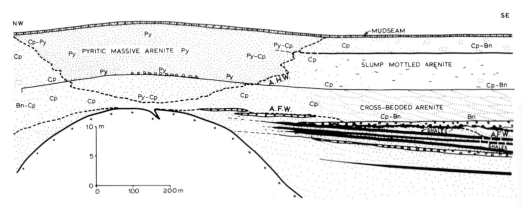

Fig. 33. Cross-section at Mufulira West. Hill of granite projects into base of "C" orebody and fissure in granite is filled with chalcopyrite-rich arkose. Aeolian beds flank eastside of hill and are overlain by pebbly grit, arkoses, and argillites with dessiccation cracks. A coarse pebbly arkose transgresses these beds and incorporates fragments of the barren argillite into the cupriferous sediments. In compaction fold over hill, chalcopyrite disseminations grades into pyritic dissemination. The pyritic cap is narrow close to the granite surface, but widens out upwards, reaching a maximum width at the mudseam.

seminated and occur in a particular bed or group of beds that stretch over several kilometres of strike. One thin persistent mineralized bed has been explored over 100 km
along strike from Mufulira. (2) Sulfides are concentrated on foresets and bottomsets of
cross-bedded units typical of the basal section of each major orebody. (3) Some finely
intercalated sediments within the ore formation are barren whilst others are well-mineralized, yet there is no marked lithological difference between them. (4) Mineralization is
closely related to other syngenetic sedimentary structures such as: (a) fossil desiccation
cracks; (b) sulfide infillings with detritus in potholes and scour features; (c) algal growths,
where fill is between the colonies indicates a detrital origin for the copper sulfide and is·
in contrast to the barren stromatolites themselves; and (d) absence of mineralization in
aeolian arenite.

New zone of mineralization at Mufulira (V.D.F.)

Subsequent to the write-up on Mufulira mine, drillhole DH 214 at the eastern fringe
area intersected and interesting zone of copper-cobalt mineralization in the Upper Roan
which is depicted in a cross-section in Fig. 34.

The host rocks are essentially crystalline dolomite (partly talcose at the base) and
dolomite–siltstone, finely interbedded, with the mineralization emplaced as lenses and
disseminations of essentially chalcopyrite comprising about 90% of the ore-type minerals.
Several silt seams in the mineralized zone are barren and a bioherm has been identified in
the immediate hanging-wall beds (stromatolitic) which indicates a shallow water deposi-

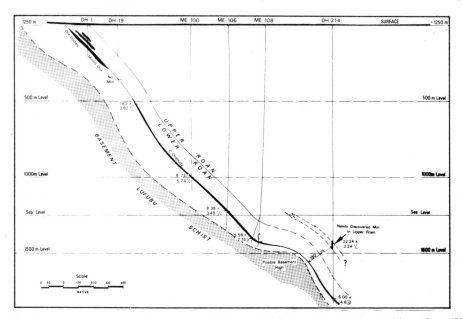

Fig. 34. Geological cross-section at the east fringe of Mufulira mine showing position of new Upper
Roan mineralization. Widths are true and in metres.

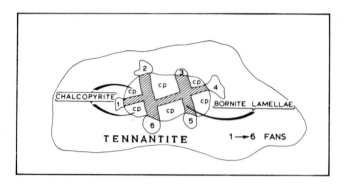

Fig. 35. Tracing of a micro-photograph (×300) of a tennantite crystal. The cobaltite/bornite "fans" peripheral to the chalcopyrite inclusion and at the outer extremeties of the bornite lamellae comprise colbaltite and bornite in myrmekitic intergrowth.

tional environment. The mineralization is not typical Copperbelt Lower Roan and structural deformation/remobilization is a notable feature. However, the sulfides are concentrated in bands along the bedding planes indicating strong stratigraphic control. Five main contiguous zones of mineralization have been identified but thicknesses are not for publication at this time. From base upwards they are:

(1) Overall 1.10% Cu zone. Chalcopyrite with minor bornite and rutile in talcose dolomite.

(2) Overall 1.92% Cu (+ 0.07% Co) zone (Cu—Co—As—S). Chalcopyrite with minor carrollite, bornite, *tennantite,* pyrite and rutile in dolomite.

This is a first discovery of the mineral tennantite on the Copperbelt, and was positively identified and described by Dr. B. Vink of the Research and Development Department in Kalulushi. A fascinating interrelationship between sulfides under the microscope is depicted in Fig. 35.

(3) Overall 3.50% Cu zone (+ 0.19% Ti). Chalcopyrite with minor bornite, considerable rutile in talcose dolomite-siltstone.

(4) Overall 2.23% Cu (+ 0.23% Co) zone (+ 0.12% As, + 0.23% Ti) (Cu—Co—As—S). Chalcopyrite with minor bornite, carrollite, tennantite, cobaltite, pyrite and considerable rutile in dolomite-siltstone.

(5) Overal 0.97% Cu zone. Chalcopyrite with minor pyrite and rutile in dolomite.

Description of minerals

Chalcopyrite often shows intergrowths with pyrite and carrollite.

Pyrite occurs as cataclastic crystals with chalcopyrite in the fractures or as allotriomorphic grains.

Bornite is usually associated with tennantite and colbaltite.

Cobaltite and Carrollite positively identified in zone 4.

Note: This is the first significant occurrence of cobalt found in basins to the northeast of the Kafue Anticline. However, it is positioned in the Upper Roan so previous comment on cobalt distribution in Lower Roan on a regional scale remains valid.

Tennantite positively identified in zones 2 and 4. Mainly occurs as large crystals up to 1 mm across.

Rutile occurs in zones 3 and 4 in considerable amounts, as small free grains 50 μm across.

Other minerals in adjacent rock. Albeit of academic interest, disseminations of *galena* and *sphalerite* in minor but visible quantities have been recorded in an anhydrite—dolomite sequence below the main zone of mineralization, separated from it by a breccia—conglomerate horizon common in the Upper Roan on the Copperbelt.

Exploration of this deep mineralization is planned over the next decade.

SUMMARY

Mufulira is one of the largest stratiform copper deposits in arenites known in the world and like the sister mines nearby commands a strong opinion amongst earth scientists who work (or worked) there, favouring a syngenetic or synsedimentary origin for the ore. There is new evidence coming to light that invokes processes of continued mineralization[1] covering the immediate post-depositional diagenetic period. Metamorphism during the Lufilian orogeny has left its imprint in the rearrangement of the sulfides — hence the formation of secondary secretion veins and coarse crystallization of the sulfides in select areas, but always within the ore formation. Thus, the uninitiated earth scientist falls into a trap and a small minority still attempt to explain ore genesis through epigenetic processes.

GENESIS OF THE ORE SHALE DEPOSITS (W.G.G.)

GRAIN OF THE BASEMENT

Folding of the Kibaran sediments and their intrusion by syntectonic granites, dated at 1300 m.y., resulted in the Kibarides striking northeasterly across the Katanga (or Shaba) Province. In the Bangweulu block of Zambia, the Plateau Series is mostly flat and is intruded by the laccolithic Lusenga syenite of 1390 m.y. age and locally folded in the Luongo belt on north-northeast axis. On the Copperbelt, Muva sediments are folded on easterly to northeasterly axes, possibly of Kibaran age. Post-tectonic tin-bearing granites and pegmatites in the Katanga are dated at between 850 and 950 m.y.

---

[1] For disscussion on multi-stage mineralization processes, see Wolf (1976), for example.

The Kibaride mountain chain was deeply eroded before onset of the Katanga System, so that post-tectonic granites were unroofed and on the Copperbelt pre-Muva granites and Lufubu schists (also with dominant northeasterly grain) were exposed over much of the landscape. Hills and ridges rose at least 100 m above the general level and the Nchanga granite towered at least 600 m above the country to the north. Erosion was rapid and potash feldspar was dominant in the sediments.

FOOTWALL TERRESTRIAL BEDS

On this rugged landscape, probably soon after 800 m.y., boulder conglomerates up to 30 m thickness were deposited in local hollows and valleys, but attained 300 m on the north side of the Nchanga granite and flanking some Muva quartzite ridges. The boulder beds were buried under fan deposits of pebbly arkose of up to 300 m and eventually much of the Nchanga granite was buried under red arkose.

Both to north and south of Nchanga, wind-blown sands covered the basal arkoses and gradually overwhelmed the lower hills and ridges. At Mufulira, hills of schist and of granite were steepened and undercut by the prevailing northerly wind. The dunes accumulated to a depth of only 50 m at Chambishi, 18 m at Roan Antelope, but attained 152 m at Nkana. At Mufulira, the dunes reached a thickness of 150 m at the east end of the mine, but westwards interdigitate with fluviatile arkose and conglomerate. The aeolian section at both Nkana and Roan is interrupted by dark pebbly layers representing sheet flood breccias. The steep foresets of the dunes are truncated by parallel originally horizontal bedding planes representing ablation down to the rising water table in the dune field. West of the Kafue, these interstitial waters were highly saline and potash feldspar was replaced by albite and anhydrite cement is common.

MARINE TRANSGRESSION

The dune field, with inselbergs of granite, schist, and quartzite, was inundated by a sea transgressing from the southwest. Arkosic deltaic deposits built out into the sea, to maintain shallow water. The advancing sea stabilized with a shore along a line through the eastern end of Roan, east of Nkana, Chambishi, round the west of the Nchanga granite salient, and back to the Chambishi—Karila Bomwe line. Immediately west of this line, a pebble-to-cobble beach accumulated to a thickness of 17 m and is the remarkably persistent Second Conglomerate of Roan, the Lower Conglomerate of Nkana—Mindola, and the Cobble Conglomerate of Chambishi. Both carbonate and anhydrite occur in the matrix between the cobbles. Leaching of the sulfate renders the formation porous.

During intermittent subsidence, southwest flowing streams built deltas of 7 m thickness of cross-bedded arkose, often red in colour, then 7 m of pro-delta muds, and finally another 6 m of reddish arkose in the Nkana–Chambishi strip. Similar sediments at Roan show considerable disturbance by preconsolidation slump. Carbonate and anhydrite occur in the cement of these clastics. The equivalent sandstone at Karila Bomwe is 30 m thick in the south, thinning to 3 m in the north against the old shore.

FOOTWALL CONGLOMERATE

The sea again transgressed from the southwest, scouring the underlying arkoses and blanketing them with the thin Footwall Conglomerate, usually under 1 m thick. The shoreline stabilized again for a long period a few kilometers east of the earlier shoreline, see Fig. 6. Basement ridges, mostly granite, formed salients projecting west into the sea, whilst in one of the intervening bays, between Mindola and Chambishi, pebbly beaches accumulated to a depth of 20 m. During this period the hills and headlands were worn down to smoother low profiles, in places lapped by or submerged under the saline waters of the encroaching sea.

BIOHERMS, PLANKTON, AND SULFIDES

In the turbulent waters against the headlands, in particular the Irwin Shaft gap at Roan Extension, the Kitwe gap between Nkana and Mindola, and the granite dome be- tween Mindola and Chambishi, algal deposits built up over gneiss or granite and spread over adjacent conglomerate to form bioherms of 10, rarely 20 m height. Adjacent to the bioherms extensive aprons of carbonate ooze were interlaminated with dark muds to a depth of 2–3 m. Muds and silts followed and, further out to sea, were carbonaceous. The surface waters were alive with plankton and their remains falling to the bottom resulted in anoxic conditions there. Bacterial reduction of sulfate with release of hydrogen sulfide caused precipitation of copper sulfides adjacent to the bioherms and in the shal- lower waters of the bays. In deeper waters farther from shore less copper, but cobalt and more and more iron, were precipitated as sulfides. During diagenesis and metamorphism, the mixtures of disseminated sulfides in the muds resulted in successive zones of chal- cocite, bornite, chalcopyrite (often with carrollite), and finally the deep-water precipitated pyrite, sweeping contour-like round the bays and wrapping round the headlands. Some of the latter projected through the shallower copper zones into the deep-water pyrite zone.

TRANSGRESSION AND REGRESSION

At Chambishi the basal banded dolomite and schist contains a rich and extensive bornite and chalcopyrite zone. Overlying muds contain zones which migrated shorewards with deepening of the water, so that in the upper beds the pyrite zone overlies the chalcopyrite zone and the latter overlies bornite in the lower beds (Fig. 36). This demonstrates transgression of the mineral zones towards the shore during deposition of the ore-shale. At Roan, the chalcocite and bornite zones are narrow in the basal schistose layer and even narrower in argillite immediately above, but become wider and even extensive in the higher beds. This results in pyritic disseminations in the lower beds, overlain by chalcopyrite, then bornite, and finally chalcocite. This regression of the zones is confirmed by further deposition of muds under oxygenated conditions and barren of sulfides. Small lenticles of arkose are further evidence of shallowing of the sea. Seawards the sharp assay cutoff on the hanging-wall steps upward across the lamination.

At Nkana, the zonal sequence is complicated by alternating regression and transgression. In the upper levels of the mine, Schistose ore is overlain by barren or Low-grade argillite, overlain by rich Banded ore and Cherty ore, succeeded by pyritic dolomite of the so-called Porous sandstone as well as by pyritic argillite. These pyritic beds contain narrow zones of chalcopyrite and bornite up-dip.

At both Roan and Nkana–Mindola, anhydrite nodules in the upper beds of the ore-shale testify to saturation of $CaSO_4$ in the saline waters.

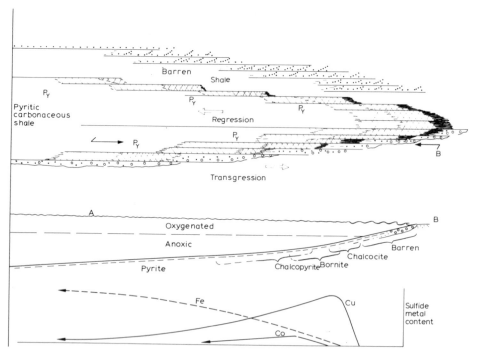

Fig. 36. Diagrams showing graphical representation of selective precipitation of copper, iron, and cobalt in near-shore sediments, section to show the corresponding mineral zones, and a top section to demonstrate stratigraphic arrangement of zones.

CARBONACEOUS MUDS

Westwards, usually 2–3 km from the shoreline, carbonaceous muds were deposited in deeper waters. They are ubiquitously pyritic west of Roan Extension and Baluba, between Nkana and Chibuluma, and west of Chambishi. However, at Nkana South, carbonaceous shales contain payable chalcopyrite and carrollite. Thus the mineral zones may transgress the carbonaceous facies of the ore-shale.

The massive Nchanga red granite formed a headland projecting 8 km west into the carbonaceous shale facies. Exfoliated slabs were buried in red arkose from the disintegration of the granite and harder varieties of the granite formed ridges overstepped by the shale, known as the Lower Banded shale. The blanket of Footwall conglomerate is absent and the copper mineralization extends 10 m down through the Transition sandstone and for as much as 20 m in the underlying arkose present in hollows of the granite. Ore occurs in shale at River Lode, at Nchanga West, and at Chingola, on the north and west flanks of the granite salient. On the south of the Nchanga dome, the carbonaceous shales are pyritic with only minor copper; the copper zones west of Chambishi have evidently been lost by post-orogenic erosion.

ORE IN BASEMENT

In Roan Extension, ore-shale directly overlies a truncated ridge of Basement schist and chalcopyrite plus pyrite occur as disseminations and veinlets for as much as 10 m below the unconformity. Where Footwall Conglomerate and arkose intervene between the ore-shale and the Basement, the latter is barren.

KALIRA BOMWE–MUSOSHI AREA

At Karila Bomwe, conditions of deposition were very similar to those of Nkana. The Porous conglomerate is 10 to 20 m thick and is overlain by 30 m of sandstone, grit, and shale, which thin to the north and pinch out east of the Karila Bomwe anticline. The Footwall Conglomerate forms a thin blanket under the ore-shale, but is absent in the far south. All three footwall units and the overlying ore-shale pinch out east of the anticline, a replica of the Lower Roan pinchout east of Chambishi. Anoxic conditions prevailed during deposition of the bottom 10 m of the Banded shale and sandstone formation. Five lithological units with varying content of siltstone and calcareous sandstone contain the highest copper concentration in the second and third units. Below the zone of surface oxidation and enrichment, there is a westward down-dip succession of bornite, chalcopyrite, and pyrite zones. This one to two km wide strip of copper mineralization plunges down the west flank of the Karila Bomwe anticline, crosses a saddle and comes to surface again on the southeast flank of the Konkola dome. Here, the host rock is a siliceous shale and the sequence of mineral zones is arranged along strike, with bornite in the east and pyrite in the west.

On the northeast flank of the Konkola dome, the continuation of the same mineral zones plunges down dip and constitutes the Musoshi mine in the Shaba (Katanga) province of Zaire.

HANGING-WALL SEDIMENTS

In the ore-shale at Roan, Nkana, Chambishi, Nchanga, Karila Bomwe, and Musoshi, there is a gradation up into shales with thin sandy lenses and layers and absence of sulfides, constituting a 5–10 m section of barren argillite. At all the mines, the ore-shale formation was then buried under deltaic arkoses brought in by currents from the northeast. Similar sands, silts, and thin dolomitic oozes accumulated to a total depth of 60 m at Karila Bomwe, 100 m at Nchanga, 70 m at Chambishi, 50 m at Nkana, but 150 m at Roan. The uppermost formation of the Lower Roan is a coarse arkose, pebbly at Roan, Chambishi and Nchanga, and was followed by a major marine transgression with deposition of shales, usually pyritic, and then by dolomitic oozes of the Upper Roan. At Nchanga these shales, the Upper Banded shales locally contain copper sulfide of ore grade.

CONTINUITY OF THE ORE SHALE (see Fig. 6)

Along the 120 km belt of copper-shale deposits, only 1–3 km beds can be matched with similar thicknesses from mine to mine. It is probably that mineralization was originally nearly continuous along this belt, small "gaps" being due to headlands, often with biohermal deposits, prcjecting west of the general shoreline. Folds are sub-parallel to the belt and to the shoreline at Nkana and Karila Bomwe, but at Roan, Baluba, Chambishi, and Nchanga the folds are across the belt, with foreshortening in plan.

Before mining started there was at least 30 million metric tonnes of copper metal in this belt. At least 50 km strike of the copper-shale has been eroded in the anticlinal areas between Baluba and Nkana and again between Chambishi and Nchanga, with a loss of probably over 20 million tonnes of copper.

In spite of local sinuousities, the belt has an overall straightness ascribable to a slight tilt of the early Roan landscape and a long continued halt of the transgressive sea, actually two halts, indicated by the approximate superposition of the Footwall Conglomerate over the Second or Lower cobble conglomerate 20 m lower in the sequence. Restriction of the rich copper zones to near shore must be a chemical control, probably a combination of limited supply of hydrogen sulfide and the very selective precipitation of highly insoluble copper sulfide. A bornite zone has not been recognized in the carbonaceous shale facies. Search for similar copper shale deposits should be based on tracing shorelines of carbonaceous shale formations.

GENESIS OF THE ARENACEOUS ORE DEPOSITS (W.G.G.)

TIME AND SPACE RELATIONS TO ORE-SHALE DEPOSITS

Copper, cobalt, and iron sulfides were deposited in sandy environments both before and after deposition of the ore-shale deposits just described. Most are distant from the site of the shale-type orebodies, but some immediately underlie ore-containing shale, in places with continuous mineralization from arenite into overlying shale.

For brevity throughout this discussion, the mineral zones will be referred to as chalcocite, bornite, chalcopyrite, and pyrite, respectively, although these mineral equilibria were probably established only during metamorphism from mixtures of fine-grained copper and iron sulfides precipitated during sedimentation.

Mimbula

Far to the southwest of the shale orebodies, at 12 km south of Nchanga, a chalcocite—bornite orebody formed as the lowermost lens in calcareous sands of the basal unit of the Lower Roan, containing conglomerate lenses and thin silty layers, lying on an irregular surface of Basement schist and paragneiss. Two other lenticular bodies of sulfide were formed in the succeeding 50 m of sands and silts before the B or uppermost orebody was laid down, more extensive and up to 10 m thick, in feldspathic sands and silt beds. This was overlain by nearly 30 m of similar barren arkosic sands and silts before deposition of the pyritic carbonaceous ore-shale.

Chibuluma West

This "Footwall Orebody" is 12 km west of the Nkana mine and may be contemporaneous with the Mimbula B orebody. Copper was precipitated during the first inundation of the windblown sands and, in fact, the surface of the dune field controlled the shape of the orebody. Wind eddies over a projecting ridge of Basement schist excavated a ring-like depression in the dunes, later transformed into a moat during marine inundation from the west and partly by a stream flowing into the northern segment. Erosion of the Basement schist provided boulders accumulating in gully floors leading into the moat, in the southern segment of which stagnant saline waters produced hydrogen sulfide sufficient to precipitate most of the iron, cobalt, and copper entering the depression. Thus was formed a narrow layered sandy deposit containing over 10% copper

and 2% cobalt with much pyrite, adjacent to aeolian sands on the south and conglomerate on the north. As the moat filled, the flooded area increased and cupriferous and
cobaltiferous sands overlapped first onto aeolian sands and then onto barren aqueous
beds laid down in oxygenated waters, especially in the northern segment of the moat.
With increased water area, precipitation of the metals became more selective, so that
bornite and chalcopyrite on the north margin pass into pyritic mineralization flanking
the north slope of the ridge. Similarly, there is a pyritic section formed in the southern
segment of the moat, surrounded by extremely rich chalcopyrite and bornite mineralization.

Throughout deposition of the ore, there is evidence of a stream entering the north
segment of the moat: first, as a scoured channel in the aeolian beds; then as boulders
of exotic granite, quartzite, and argillite in the lower part of the channel; strong cross-
bedding indicating current from north; marginal sun-cracked mud layers; embayment
of the north margin of the orebody by lean fluviatile sediments; and, finally, infilling
of the upper part of the silted-up channel by sandy muds contrasting with cleaner sands
of the levees (Whyte and Green, 1971).

The copper—cobalt orebody and adjacent barren aqueous arenites were then buried
under 10 m of sandy muds in a mud-flat environment. The lower section of this formation may contain ore-grade chalcocite adjacent to the underlying ore and representing minor depressions left from infilling of the moat. Generally, the muddy sands are
pyritic, with only minor copper.

Dolomitic breccia, dolomite, and dolomitic shales were deposited over the sandy
muds. Pyrite is sparsely distributed through all these hanging-wall formations.

Chibuluma

Three kilometres east of Chibuluma West, flooding of the dune field was probably
contemporaneous with that of the western deposit and strong currents caused scours
and potholes 45 cm deep filled with pebbles and coarse arkose. Barren arkose, with
gravel lenses and cross-bedding due to currents from the north, accumulated to a depth
of 15 m. A depression or channel about 700 m wide was formed by a south-flowing
stream, which later formed an elongated playa-type lake filled with brine with high
concentration of calcium and magnesium salts, which precipitated as the double carbonate and as calcium sulfate. Available soda converted microcline of the arkoses to
albite and boron reacted with the muds to eventually, during metamorphism, form
tourmaline. Flood waters from both sides, but more especially from the east, washed
arkosic sand, pebbles, and mud into the lake, causing local scouring of the sloping margins. A discontinuous layer of pebbles generally marks the bottom of the deposit.

Anaerobic bacteria in stagnant heavy saline bottom layers generated hydrogen sulfide, which precipitated copper, cobalt, and iron in the lower sediments, the latter of
which were arkosic in the east, but muddier in the centre. Upon metamorphism, they

formed sericitic meta-sandstone. Periodic desiccation of the shores allowed suncracking of mud layers with disseminated sulfides and farther from the lake formation of extensive mud flats, which on drying developed efflorescences of iron, cobalt, and copper sulfates. These were washed by storms into the main playa, the rush of muddy water causing scours and potholes in the previously deposited cupriferous arenite. Iron and cobalt were immediately precipitated as hydroxide gel admixed with sand and pebbles, later transformed into "sulfidite" layers with up to 20% cobalt. Copper was precipitated from the overlying water as sulfide to give with iron, chalcopyrite or bornite, in the overlying sands (see Fig. 31).

The sulfide deposit gradually widened by spreading of the waters over the flanks of the depression and the usual zonal sequence developed, with pyrite in the centre of the channel, flanked by chalcopyrite and meagre bornite near the margins of the channel. The sulfide-rich arenites attained a thickness of over 15 m in the centre of the channel and were then buried under 10 m of cleaner feldspathic sands with pyrite, but only minor chalcopyrite. The feldspar was albitized, probably by the saline waters.

A major marine transgression resulted in the Hanging-wall conglomerate, thin and poorly developed in the west, up to 10 m thick over the ore shoot and gradually thickening for 10 km to the east where it becomes the 15 m thick Lower cobble conglomerate of Nkana. On it were deposited calcareous sands, muds, and dolomitic oozes and then up to 50 m of pyritic carbonaceous shale, equivalent to the Nkana ore-shale. Up to 30 cm of grit, representing the Nkana Footwall conglomerate, occurs at the base of the carbonaceous shale. The arenaceous formations of the Nkana hanging wall are represented at Chibuluma mainly by argillites and dolomites.

The formations above the Hanging-wall conglomerate were intruded by gabbro sills and the whole complex folded on west-northwesterly axes. The gabbro was converted to scapolite amphibolite and biotite and tremolite developed in the sediments. At Chibuluma West the Basement ridges formed a series of en-echelon anticlines, so that the orebody lies on more competent Lower Roan aeolian quartzite on the flanks of and squeezed as synclines between the Basement ridges. At Chibuluma the aeolian quartzites are very thick and competent and Basement hills had apparently no control on the folds.

Footwall Orebody, Nkana South

Contemporaneously with the flooding of the dune field at Chibuluma, sands and muds with much anhydrite were deposited as playa lake beds above the aeolian beds at Nkana Central. To the south a Basement schist ridge projected above the dunes. In a depression eroded in the schist, sands with up to 8% carbonate and only minor mud were deposited with copper and iron sulfides. Cobalt and the sulfidite layers, as at Chibuluma, are not present and likewise there was little boron concentration. Some microcline survives the albitization. Most of the mineralized arenite shows no cross-bedding whatsoever, only parallel stratification with slightly muddy layers converted to phlogopite and chlorite during metamorphism. Mineralization is mainly bornite in the thick central area and western part, but grades into chalcopyrite in the eastern half and the marginal areas. Where the mineralized arenites rest on Basement schist the latter may be mineralized for 8 m below the unconformity, but generally only the thin paleosol is mineralized. In the eastern part, a pyritic arenite intervenes between the ore and barren Basement.

The cobble conglomerate of the marine transgression forming the extensive Lower conglomerate was then laid down, but is here very thin and westward grades into isolated cobbles. Above this were deposited about 20 m of arkose and mud of the Footwall sandstone. The next marine transgression laid down the thin Footwall conglomerate and then the Carbonaceous ore-shale with chalcopyrite, pyrite, and cobalt sulfides.

Folding is strongly reminiscent of that at Chibuluma West; a hill of Basement schist was squeezed up incompetently as a thin anticlinal wall on the west against the competent syncline of mineralized Footwall arenite, separating the latter from the isoclinally folded carbonaceous shales to the west.

Chambishi footwall mineralization

A little copper was precipitated as a chalcopyrite dissemination in the first muddy sands deposited on the eroded surface of the clean feldspathic sands of high-angle cross-bedding; these represent the dune field overlying the Old granite. A marine transgression formed the Lower boulder conglomerate, the barren Footwall sandstones and argillites, then the thin blanket of Footwall conglomerate heralding the transgression of the copper-shale of the Main Orebody. This footwall mineralization is not economic, but occurs at the same stratigraphic horizon as that of Nkana South and Chibuluma.

Exploration drilling below the east fringe of the Chambishi West orebody has found a small high-grade orebody at this horizon, about 15 m below the ore-shale. It rests directly on the Old granite.

Muliashi footwall mineralization

Throughout the Roan and Roan Extension the footwall beds of the ore-shale are consistently barren. The sulfides of the ore-shale rarely persist more than a few centimetres in to the Footwall conglomerate.

Entering the Muliashi area, 10 km from the end of the Roan syncline, the Footwall conglomerate is less well developed and in places is an arkose without pebbles and may be argillaceous. In this vicinity sulfides, generally chalcocite, persist for several metres through the Footwall arkose and underlying sandy argillite. This footwall mineralization is mined usually with the decreased development of copper in the basal ore-shale underlying up to 30 m of pyritic shale. Supergene processes and metamorphic migration do not adequately explain the presence of this footwall mineralization; it is probable that much of it represents syngenetic deposition of sulfides in marine beds at the western limit from which the sea started its gradual eastward transgression, forming the overlapping lenses of cupriferous shale, with pebbly conglomerate at its eastern migrating shore.

To date, no mineralization has been found below the Lower conglomerate in the Roan—Muliashi—Baluba basin.

Nchanga area

The Nchanga red granite formed a headland, which apparently held up the eastward advance of the ore-shale, allowing deposition of more coarse porous arkosic and pebbly deposits — the cause probably of considerable recent leaching and deep oxidation and enrichment.

At Mimbula, 12 km due south of Chingola, the B or Uppermost ore zone is up to 10 m thick in calcareous arenite overlain by 30 m of barren feldspathic sands and micaceous silts before deposition of the pyritic carbonaceous ore-shale.

The next copper deposit formed at Chingola "F" at the same horizon as Chibuluma and Chibuluma West, about 2 km west of the Nchanga granite on the western flank of a schist paleo-hill. The 25 m of mostly bornite and chalcopyrite dissemination in basal conglomerate and compact arkose was followed by deposition of 20 m of barren arkose and then pyritic carbonaceous shale of the ore-shale horizon. There is a little barren basal conglomerate under much of the orebody, but where the latter oversteps the Basement hill, mineralization extends for some distance into the Basement schist, probably mineralization of a paleo-sol as seen at Chibuluma West and under the ore-shale at Roan Extension.

From Mimbula's four mineralized horizons to Chingola F and A, and then to Nchanga West, the ore occurs generally close to the unconformity, but rising to higher levels in the arkose northeastward. In the last deposit, rich Arkose ore is contiguous through the Transition Quartzite with rich ore in the Lower Banded shale. However, this simple picture of ore lenses flanking the tilted margin of the transgressive geosynclinal sediments is complicated by the extremely irregular topography on the Basement Complex, paleo-ridges controlling the location of the basins of ore deposition, as shown in Fig. 13. Finally, at Nchanga West and in the Nchanga open-pit, sulfides were also precipitated in the Pink Quartzite, in The Feldspathic Quartzite (T.F.Q.), and the lower part of the Upper Banded shale. Geographical controls for these upper orebodies are little known, but compaction of underlying mineralized sediments, as at Mufulira, is suspected.

Mufulira

Precise correlation of sediments across the Kafue anticline is difficult and likely to remain controversial. Current-bedding at Mufulira is from northeast, as on the other side of the anticline, so the latter probably did not separate a Mufulira basin of deposition from a Chambishi-Nkana basin. More likely, the so-called Lower Roan arenites of Mufulira are the near-shore transgressive sediments contemporaneous with the basal part of the Upper Roan dolomites of Chambishi.

The picture is then of continued transgression of marine sediments northeastward up the gentle flank of the geosyncline and across the site of the future Kafue anticline. At Mufulira, streams deposited a basal conglomerate in the eastern basin eroded along an erosional surface along the Lufubu schist. Some gravels were also laid down in the two granite floored basins to the northwest. The central and east basins and much of the country to the east (in part now down dip), were blanketed by sand dunes built mainly by a north wind, which undercut the northern flanks of some of the granite ridges. The latter were gradually buried under the accumulation of sand, up to 150 m thick. The western basin was partly filled with festoon cross-bedded arkoses and grits by south-west flowing streams, from which the north wind lifted and winnowed sand for the dunes further south. In the central basin, aeolian interdigitated with fluviatile sediments, whilst in the west basin fluviatile arkose and grit accumulated probably in small deltas building out into deepening salt water, with anhydrite precipitation, and the first copper sulfides were deposited, mainly chalcocite, some being detrital.

The dune fields of the central and east basins were inundated and siderite was precipitated in aqueous sands, followed locally in some hollows by copper sulfide. As the marine transgression spread, wave action laid a blanket of pebbly grit over the floors of the three basins and sulfide precipitation became widespread. On the flanks of the now subdued and partially buried ridges, sands and muds were exposed to the sun and desiccation cracks formed. Barren mud flakes were washed down and incorporated into the cupriferous grit, becoming armoured by sulfide in the process (Garlick, 1967). The basins gradually filled with cupriferous (bornite and chalcopyrite) sands and minor muddy layers, which on the slopes slumped to form chaotic pre-consolidation breccias and turbidity currents. The latter carried sand and mud to the centre of the eastern basin, where with organic matter it accumulated to form the carbonaceous "C" graywacke, with chalcopyrite.

The bevelled tops of the Basement ridges were finally buried under shallower-water sands, cleaned of mud by winnowing and sorting and then — with carbonate and anhydrite — cemented, but no copper sulfide was precipitated. Probably after burial, iron sulfide was precipitated in the pores of the sand, thus making the pyritic "caps" over the ridges, narrow at the base, but flaring to greater widths at the top of the C horizon.

At the eastern margin of the C orebody, dunes accumulated above the beach, separated by a narrow band of barren aqueous beach sands from cupriferous sands grading west into carbonaceous graywacke. The four facies occur within a minimum distance of 300 m.

At the western margin of the C orebody, cupriferous arkose and breccia pass into pyritic arenites and graywacke, which persist to the Zaire border, 13 km distant. Zinc, as sphalerite, attains several percent grade locally in the graywacke and underlying arkose.

With continued transgression, the supply of sediment abruptly diminished and an extensive blanket of dolomitic siltstone covered all of the C orebodies and the aeolian sands to the east.

The waters shallowed again and mud layers with thin arkose beds were laid down generally under oxidizing conditions. However, in the deepening water over the compacting cupriferous C sediments of the east basin, intermittent copper precipitation continued, making spectacular "banded copperstones" similar to banded ironstones. An algal bioherm formed in the shallow water over the west margin of the east basin. Pyrite and a little chalcopyrite precipitated in the deeper waters west of the bioherm, whilst in the deepening lagoon to the east the rich B orebody developed in basal grit, silt, and cross-bedded arenite, neatly superimposed over the compacting eastern basin C orebody. Slumping took place along the flanks of the lagoon to form the usual breccias and the central carbonaceous graywacke, both rich in bornite. The east margin of the B orebody consists of red arkose and thin muds, deposited in shallow oxygenated waters under red-bed environment.

Abrupt subsidence again removed sediment sources to a great distance and the thick Lower Dolomite, with considerable anhydrite, was laid down. At the top of this bed a thin grit layer, up to 15 cm thick, contains copper sulfide and uraninite disseminations, but before an orebody could develop, an algal biostrome buried the grit. Deeper water was again evident over the compacting B and C orebodies of the eastern basin and muds accumulated to considerable depth, eventually compacting in turn to 5 m of argillite, barren except for a small width near the middle of the basin. The first arkosic sands laid on this mud layer sank and slid in the hydroplasic material to form a magnificent layer of load casts and pre-consolidation slump folds.

Coarse red arkosic sands and pebbly grits marked the third marine transgression over the compacting sediments in the eastern basin. Copper precipitation, now represented by chalcocite, started in the axis of the basin and gradually spread up the flanks, slumping from which caused turbidity currents to fill the centre of the basin with the chalcocite- and bornite-rich carbonaceous A graywacke. A small central area of the graywacke has chalcopyrite and pyrite mineralization. Near the top of the A orebody, a layer of glassy quartz pebbles with a few granite cobbles was washed into the cupriferous sediments and extends across the graywacke facies. This layer of pebbles appears incongruous in the graywacke, as the latter has no cross-bedding, only graded bedding and some lobate sole markings.

Copper deposition was terminated by blanketing of the A horizon arenites by a thick sheet of sandy muds with much anhydrite or gypsum, probably deposited under sabkha conditions. Dolomite oozes twice interrupted the deposition of the sandy muds, which accumulated to form the 60 m thick Hanging-wall Argillaceous "Quartzite". It contains a middle cross-bedded Marker Grit and another arkose termed the "Glassy Quartzite" at the top, with an extensive layer of pyritic mineralization in dolomitic arenite at the contact with the Upper Roan dolomite. This pyritic layer contains a little chalcopyrite and can be traced for 13 km to the southeast, where by change to chalcopyrite—bornite mineralization it attains payable grade.

At Bwana Mkubwa, 72 km southeast of Mufulira, there are five superimposed orebodies in feldspathic sands, graywacke, and dolomitic silts, all now metamorphosed at biotite grade. Similarities of the host rocks, mineralization, and stratigraphic succession are striking and almost identical. At Kinsenda, 35 km northwest of Mufulira and just over the Zaire border, three superimposed orebodies of spectacular grade occur in pebbly to gritty arkose. These three mines thus form a parallel belt to the ore-shale deposits on the other side of the Kafue anticline, but it appears that the continuity of the mineralization does not compare.

Special features of Mufulira for comparison with other arenaceous deposits and ore-shale deposits are: (1) Three orebodies are superimposed and associated with three nearly identical stratigraphical cycles of grit, arkose, slump breccia, graywacke, dolomite (silt), mud (Renfro, 1974). (2) Dolomite and mud are barren or poorly mineralized; aeolian sand is barren. (3) Orebodies occur in basins between Basement ridges, or in hollows formed by compaction of underlying orebody sediments. (4) Mineral zones from shore to deep water are: barren, chalcocite, bornite, chalcopyrite, pyrite with zinc. Over Basement ridges pyrite without zinc occurs with anhydrite in shallow-water sediment. (5) Cobalt is negligible, but anhydrite is abundant in hanging and footwall, whilst Chibuluma original reserve averaged 0.25% cobalt and sulfidite layers have 2—10% cobalt in high-soda and boron environment. (6) Washouts in the mineralized C orebody pre-consolidation slump breccia are filled by barren arkose. (7) In some cross-beds, sulfides show detrital concentration with tourmaline and zircon, with lighter bands enriched in quartz, feldspar, and anhydrite.

THE SYNGENETIC EXPLANATION (W.G.G.)

EVIDENCE OF SMALLER SEDIMENTARY FEATURES

Under the microscope it is evident that the sediments are extensively recrystallized by regional metamorphism during folding. The sulfide grains in the orebodies behave like the other minerals, but selective replacement by the sulfides could explain many of the relations. Crowding aside of sulfide grains by scapolite porphyroblasts, indicate a pre-metamorphic origin for the sulfides.

Restriction of sulfides to certain beds, especially the top of the Roan Upper Orebody, with a knife-edge contact against overlying barren argillite (identical in mineral composition except for absence of sulfides), is inexplicable by selective replacement.

Numerous small-scale sedimentary features favour a contemporaneous origin for the sulfides such as (Garlick, 1967): (a) concentration of sulfide with zircon and tourmaline on certain foresets and on bottomsets of cross-bedded arenites, e.g., Mufulira deposits (Garlick, 1972) (Fig. 29); (b) scour-and-fill structures marked by sulfide-rich laminae in argillites at Roan; (c) pre-consolidation slump structures in muddy arenites laterally between cupriferous cross-bedded arenites and carbonaceous turbidite graywacke at Mufulira; (d) concentration of pyrite and carrollite with pebbles in potholed, scoured surface of arenite with meagre chalcopyrite dissemination at Chibuluma (see Fig. 31); (e) concentration of sulfides in hollows of ripple marks at Mufulira; (f) sulfide-impregnated sandy fillings of sun-cracked muds at Chibuluma and Mufulira; (g) washed-in barren mud fragments, armoured by sulfide, in cupriferous grits at Mufulira; (h) barren washouts in ore which are small-scale, 20 cm across, in Roan ore-shale and large-scale, 3 m across, in Mufulira C orebody; (i) sulfidite (rock containing more than 33% metal-sulfide) layers, usually 1—3 cm thick, but at Chibuluma up to 1 m thick, form stratigraphic markers in the orebodies; for example at Chibuluma, in the Nchanga arkose; (j) sulfide concretions have barren haloes adjacent to the disseminated ore, such as at Mufulira and in the Nchanga Arkose.

The preservation of the above structures testifies to the very limited diffusion of sulfides and silicates during diagenesis and metamorphism. The following structures

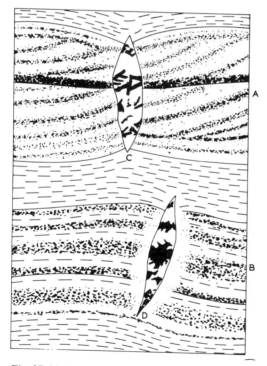

Fig. 37. Metamorphic veins.

indicate a slightly greater movement of sulfides during metamorphism (see Fig. 37): (a) metamorphic veins in barren argillite and in arkose consist of the following paragenesis of minerals: specularite, feldspar, anhydrite, dolomite, and quartz, all minerals in equilibrium with the host rock; (b) similar veins in ore-shale or arenite orebodies consist of feldspar, anhydrite, dolomite, quartz, pyrite, chalcopyrite, bornite, and/or chalcocite. The sulfide in the vein is almost invariably the same that occurs disseminated in the immediately adjacent ore. Commonly, there is a bleached halo or margin from 25 to 75 mm thick between the vein and the darker disseminated sulfides in the ore. The bleached margin is best developed in low-grade ore and is due to lateral secretion of the sulfides into the veins.

Major features of the orebodies, being little effected by metamorphism, assume critical importance in interpretation of genesis of the orebodies.

FOLDING

Folding on all scales has resulted in distortion of the orebodies. Analysis in 1940 of variation of copper grade between the expanded axial region and the squeezed limbs of a major drag-fold at Nkana provided proof of the prefolding origin of the mineralization. Figure 16 demonstrates that:

(a) The average grade of the high-grade section, consisting of the Cherty ore and the Banded ore, was 5.60% in the thickened and crumpled axial regions as against 5.62% in the squeezed limbs, that is practically the same in spite of the spectacular bornite- and chalcopyrite-bearing veins in the axial region.

(b) Although considerable copper has been expressed from the disseminated ore into the numerous veins, the average grade of the rock including veins has remained constant.

(c) The low-grade section of the orebody, consisting of Low-grade argillite plus Schistose ore is slightly higher at 2.06% copper in the axial region as against 1.87% in the squeezed limbs. This is probably a mechanical effect due to squeezing out of the richer Schistose ore from the limb below the fold and its accumulation in the greatly thickened and contorted schist overlying the rounded anticline of competent Footwall conglomerate and arkose. A quartz-sulfide saddle reef, up to 1 m thick, overlies the arch of Footwall conglomerate and introduces sampling difficulties.

At Roan Antelope the stratigraphic mineral zoning of the Upper Orebody leads to duplication of the bornite zone in the anticlinal crest with a grade of about 4.2% copper against a similar duplication of the chalcopyrite zone in the "V" trough with average grade of only 2.3%. However, each mineral zone maintains its normal grade whatever the position in the fold and the stratigraphic zonal sequence of pyrite–chalcopyrite–bornite–chalcocite–barren Hanging-wall argillite if maintained, even where the beds are extremely attenuated in overturned limbs; in fact, the sequence has been used since 1940 for structural mapping – without error.

MINERAL ZONES

The ore-shale deposits, excluding that in the Upper Banded Shale of Nchanga, all occur in a strip of less than 8 km width within one dolomitic argillite formation over-lying a Footwall conglomerate and extending for 130 km from Roan to Musoshi, except where eroded in anticlinal areas. Although the argillite is up to 55 m thick, usually only 7—10 m constitutes ore, the remainder being either barren or pyritic.

Upwards across the beds, there is commonly a mineral zoning from chalcocite, to bornite, chalcopyrite, and then pyrite, as at Chambishi and parts of Nkana, and the Lower orebody at Roan and Baluba. In the Upper orebody at Roan the sequence is reversed, with pyrite below and chalcocite on the top.

By contrast, the same mineral sequence may cover a lateral section along the bedding of several kilometres width. The sequence is usually from east to west: barren silty ar-gillite, then disseminated chalcocite, mixed chalcocite—bornite, then bornite, mixed bornite—chalcopyrite with or without carrollite, chalcopyrite, with less carrollite, mixed chalcopyrite—pyrite, and, finally, pyrite with meagre chalcopyrite and often a little sphalerite. The pyritic zone becomes carbonaceous westward and after 10—20 km may grade into less pyritic shaly dolomite. The above cupriferous zones may occupy a width of between 100 m and several km (see Figs. 6 and 34).

The above mineral sequence ranges from thin to thick ore-shale and is in the same di-rection as currents deduced from cross-bedded arenites above and below the ore-shale. There can be no doubt that the barren argillite is a near-shore feature and the pyritic zone a deep-water facies, with the cupriferous zones following approximately the contours of the sea floor between the two (Fig. 38).

The ore deposits in arenaceous hosts commonly show the same zonal sequence, with cross-bedding, slumping, and turbidite features confirming that the chalcocite and bornite zones are shallower-water and the chalcopyrite and pyritic zones deeper-water.

Several theoretical models, observable in modern seas, may explain syngenetic copper deposition: (1) Anaerobic production of H_2S *at or below* the water/sediment interface. This model holds for most marine sediments, even for inter-tidal muds and sands. It is probably the explanation for the pyrite in the shallow-water sands over granite hills at Mufulira. It apparently fails to explain the big-shale mineral zoning (see Rickard, 1973). (2) Anaerobic production of H_2S *above* the sediment, as in the Black Sea and at Walvis Bay. This model can explain the mineral zoning and also the absence of sulfide in the washouts recorded at Roan and Mufulira.

There is a near-shore zone of oxygenated water and no sulfide is formed, except where anaerobic conditions develop below the water/sediment interface, where the pore waters contain insufficient non-ferrous ions to form other than iron sulfide. Farther from the near-shore breakers, a wedge of anaerobic water, not necessarily stagnant, produces just sufficient H_2S to precipitate the very insoluble sulfides of copper and silver. There is sufficient mixing of the waters and slow fall of precipitate to allow any

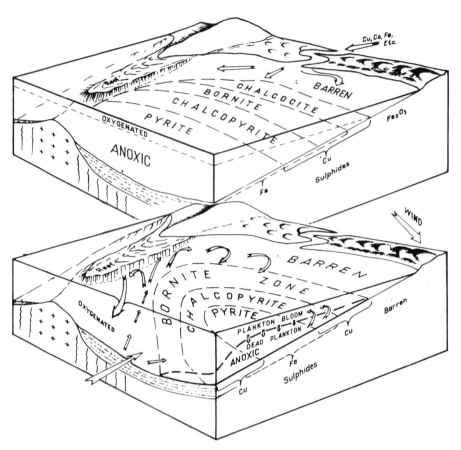

Fig. 38. Above. Block diagram showing diffusion of copper, cobalt–iron, and other metals in solution into a sea; copper only being precipitated in the near shore feather edge of lower layer, with increasing iron precipitation in depth, which after diagenesis or metamorphism gives the sedimentary zoning of the minerals as shown.

Below. Diagram to show arrangement of metal and mineral zones anticipated with an upwelling current bringing metals from ocean depths. The upwelling would be induced by a longshore or off-shore wind, forming sand dunes as observed along the eastern shore of the Mufulira "C" orebody. Plankton bloom and decay, with gradual accumulation of nutrients and metals, eventually leads to oxygen deficiency and development of anoxic zone.

iron sulfide to be replaced by copper ions before it reaches the bottom. The wedge of anaerobic water thickens seawards and more H_2S becomes available to precipitate iron as well as copper, and bornite develops, probably during diagenesis and metamorphism from the admixed copper and iron sulfides. In still deeper waters, due to depletion of copper by precipitation in the shallower waters, the proportion of iron sulfide to copper sulfide increases and chalcopyrite forms as well as bornite. Cobalt, if available, is now precipitated as well. In the next zone there is insufficient copper sulfide to form bornite, so chalcopyrite is the main sulfide and excess iron may form pyrite. Finally, in the extensive deep waters only iron sulfide with a little zinc and only traces of copper sulfide are precipitated. Carbon may be extensively preserved in these sediments, testifying to the anoxic nature of the overlying waters.

Drilling in the Chambishi—Nkana basin indicates that there may be yet another seaward zone in which a carbonate ooze is deposited with much less pyrite and no carbon. This may indicate a deep limit to the wedge or lens of anaerobic water lying above the sediment surface, in accordance with Brongersma-Sanders' (1967) theory of upwelling oceanic currents under the influence of steady trade winds, with plankton bloom and decay and development of anoxic zones (Fig. 38).

The concentration of cobalt in many of the Zambian and Shaba (Katanga) copper deposits is still a problem. As cobalt is an essential element in the vitamin B12 molecule and is a necessary trace element for development of dinoflagellates, it is one of the elements that with copper, zinc, nickel, and iron would be brought into the near-shore zone by upwelling currents and would be concentrated by multicycle plankton bloom and decay and eventual selective precipitation as sulfide.

Algal activity was an important contributor to the total organic matter essential for the anaerobic generation of hydrogen sulfide from the calcium sulfate, evidently supersaturated in the waters as shown by anhydrite nodules in Roan and Nkana ore-shale and in the Mufulira arenites. The growth of large algal bioherms against Basement headlands at Roan, Nkana, and south of Chambishi and the algal reef over an intra-formational breccia on the west margin of the Mufulira B orebody testify to the prolific growth of plant life in the shallow oxygenated waters. The abundant anhydrite probably indicates high salinity of the waters, which would be conducive to density layering and development of anoxic bottom waters.

TRANSGRESSION AND REGRESSION

With a transgressive shoreline, as exemplified by the overlapping lenses of beach gravel building the continuous blanket of Footwall Conglomerate, so the mineral zones in the overlying ore-shale exhibit a parallel overlap such that a chalcocite zone at the base is overlapped by bornite in the beds immediately above, and that in turn by chalcopyrite, and then by pyrite. This results in the stratigraphic zoning as at Chambishi, in parts of Nkana, and the central parts of the Mufulira orebodies.

Near the top of the ore-shale, there is evidence of shallowing of the waters and eventually arkosic deltas build out over the muds, but this happens at higher and higher levels seawards, resulting in a regressive offlap of arkose on mud. At the top of the sulfide dissemination in the shale or arenite orebodies, there is commonly a similar offlap of the mineral zones to give the stratigraphic zoning of the Upper orebody at Roan, with pyrite grading up through chalcopyrite, to bornite, to chalcocite, and then to barren sediment. With rapid regression this sequence becomes attenuated and fails to make ore (see Fig. 36).

DETRITAL REWORKING

Detrital copper sulfides in the bottoms of scour features at Roan and of the washouts of Mufulira C orebody and concentration of sulfides with tourmaline and zircon on certain foresets, whilst adjacent laminae are richer in anhydrite and feldspar, are evidence of rapid flocculation of sulfide to granules on the sea floor and their subsequent movement by bottom currents as grains heavier than quartz, feldspar, and anhydrite. The mineral zoning is explainable only by highly selective precipitation from dilute solution of the extremely insoluble copper sulfide as a mass chemical reaction in the presence of copper, iron, cobalt, and zinc ions in a large body of water. Slow diffusion of ions through the sediments would give the same reactions, but with varying permeability of muds and sands, the copper and iron sulfides would be patchily distributed in both coarse and fine detail, instead of showing the perfect development of evenly disseminated mono- or bi-mineralic zones parallel to the shore and the stratigraphic zoning.

SOURCE OF THE METALS

Upwelling ocean currents are conceivably capable of supplying the vast amounts of metals deposited with the sediments forming the disseminated orebodies (Brongersma-Sanders, 1967; see Fig. 38). The anoxic zone would be probably lenticular or confined by off-shore sand bar and some concentration of copper would be expected on the seaward side of the anoxic zone; such has not been observed on the Zambian Copperbelt to date.

Figure 38 also shows in the upper model the mineral zoning expected with metals, entering the sea by river, in solution or absorbed on clays in suspension. In addition, ground waters with metals leached from a hinterland of sand dunes and other terrestrial deposists would enter the sea as submarine springs in the oxygenated zone.

SYNGENESIS VERSUS EPIGENESIS (W.G.G.)

Since the first suggestion by Schneiderhöhn (1932) that the Northern Rhodesian copperbelt deposits be of syngenetic origin, much and overwhelming evidence has been collected to show that the mineralization is both pre-folding and pre-metamorphism and that the sulfides of copper, iron, and cobalt are original constituents of the sediments, modified in situ by the effects of moderate metamorphism and local migration into fractures.

This evidence, detailed previously for the individual deposits, mostly leaves ambiguous the exact process by which the metals were acquired by the enclosing sediment. The process may have been one or more of the following:

Syngenetic: deposition of the metals, mostly as sulfides, simultaneously with the enclosing sediment.

(1) Deposition as detrital sulfides carried by aqueous currents from *near* or far.

(2) Precipitation of the metals by hydrogen sulfide formed by anaerobic bacteria in the overlying body of water and settling of the precipitate into the accumulating sediment.

(3) Precipitation of the metals, especially iron and cobalt, as hydroxides on change of pH. The goethite layers in the deeps of the Red Sea are a modern example.[1] These could be later sulfidized before diagenetic compaction.

Diagenetic: precipitation of metals in the unconsolidated sediments.

(4) Precipitation from percolating superjacent cold waters of metals into the pores of the sediment by hydrogen sulfide.

(5) Precipitation by hydrogen sulfide in the pores of the sediment of metals from solutions — connate, convective meteoric, or hydrothermal — entering the unconsolidated sediment from below.

(6) Replacement of iron sulfide already in the sediment, by processes 2 or 4 above, by metal-bearing solutions from above or below.The work by Dechow and Jensen (1965) shows that the sulfur of the sulfides is of bacterial origin. The anaerobic bacteria have to be supplied with organic material as well as sulfate for their metabolism (see Chapters 6, Vol. 2 by Trudinger, and Chapter 5, Vol. 2 by Saxby). Plankton growth and decay, and especially algal activity in formation of stromatolitic bioherms and biostromes would provide ample food, and the abundant anhydrite nodules and beds indicate frequent sulfate-saturation of the water.

It is, therefore, probable that several of the six processes listed above operated in the formation of the Zambian copperbelt deposits. It is important to decide which was the most important. Much of the evidence is ambiguous, but the following features are of critical importance.

HOMOGENEITY OF THE MINERALIZATION

The orebodies are banded with richer and poorer layers that can be followed for hundreds of metres, both down dip and along strike. The persistence of either high copper values along the richer, or of low values along the poorer bands, is powerful evidence for a true syngenetic origin.

---

[1] In the Red Sea, however, mainly "hydrothermal" processes are responsible for the source of the metals; see Chapter 4 by Degens and Ross, Vol. 4.

METAL ZONING

The zonal arrangement of the metal sulfides, from barren near-shore sediments, to chalcocite in shallow water, to bornite, bornite with carrollite and chalcopyrite, then chalcopyrite, and finally pyrite, in places with sphalerite, in the deeper parts of marine lagoons and basins, can only be adequately explained by selective precipitation of the metals by limited amount of hydrogen sulfide in bodies of standing water. The supply of copper from streams or springs draining a hinterland of aeolian sands and terrestrial red beds, to form a chalcocite zone, must be sufficient to replace any iron sulfide before the latter can be buried in sediment.

In this model, gravity stratification of the water is envisaged, with the upper oxygenated layer carrying metals seawards, overlying a heavier more saline layer in which anoxic conditions prevail, either persistently or intermittently. All copper and later cobalt that enters the lower layer by mixing or diffusion from above, is selectively precipitated as sulfide, resulting in the upper layer being progressively depleted in these metals. Thus in deep waters, with more abundant hydrogen sulfide production, the more soluble sulfides of iron and zinc predominate in the precipitate.

Each layer of sediment contains this zonal sequence of metal sulfides, usually spread over a width of hundreds to a thousand metres. In superimposed layers, the zones may overlap or offlap in sympathy with marine transgression or regression.

It is unlikely that these wide mineral zones could be formed diagenetically by diffusion into sediment. Production of hydrogen sulfide in the sediment would be uneven and in areas of abundant H_2S, there would be dumping of all the metals, instead of the selective precipitation of copper. Lateral migration of solutions through the sediments could produce the zoning, by gradual replacement of iron sulfide by copper, but would not provide the layer by layer overlap or offlap. But lateral migration of solutions on the necessary scale through muds, is extremely unlikely. Upward diffusion of copper into the sediments, even by replacement of earlier iron sulfide, would not give the variable stratigraphic zoning, the overlap—offlap zonal relations, nor the regularity of the mineralization.

BARREN BIOHERMS

A stromatolite reef forms the seaward barrier enclosing the lagoon in which the rich Mufulira B orebody was formed. The carbonate-rich bioherm contains only a little copper sulfide in muddy fillings between the stromatolite columns, whilst adjacent arenites grade 3–6% copper as bornite and chalcopyrite. On the seaward side of the reef, argillaceous arenites with disseminated chalcopyrite and pyrite are rarely of ore grade and give way to pyritic sediments in deeper water. Evidently, copper was mostly precipitated in the anoxic waters of the lagoon and little in the shallow oxygenated water over the reef. After burial, reducing conditions should have developed in the algal-rich bioherm and would cause precipitation of copper from interstratal solutions, if there had been significant percolation. Evidently there was not.

The carbonate-rich bioherms at Irwin Shaft, Roan Antelope, of the Kitwe Gap, Nkana, and the huge algal reef over the granite dome between Mindola and Chambishi, are all barren of sulfides, but are flanked by ore-shale with arcuate zones of chalcocite and/or bornite, then chalcopyrite, and finally pyrite in deeper water. The shale adjacent to the bioherm contains layers rich in dolomite and sulfide, contrasting with the barren dolomite of the bioherm. The simplest explanation is that the bioherm projected above the anoxic layer to which sulfide precipitation was confined and that cupriferous solutions failed to penetrate the algal pile after its burial (see Wolf, 1976).

CROSS-BEDDING

Cross-bedding in arkose and sericitic arenite orebodies is emphasized by dark concentrations of copper sulfides along foresets and especially on truncation planes. In foresets at Chibuluma and Chibuluma West, conspicuous concentrations of bornite alternate with weaker disseminations of chalcopyrite, explainable by detrital copper sulfide being concentrated along the foreset laminae by stronger currents, whilst weaker currents dropped smaller amounts of copper sulfide which combined with the detrital iron oxides in the arkose to form chalcopyrite. In the Copperbelt orebodies, detrital iron oxides are practically absent, but are conspicuous in cross-bedded arenites above and below the ore.

In basal beds of the Mufulira C orebody thin-section and heavy-mineral studies show that chalcocite and bornite grains along dark foreset laminae are associated with increased tourmaline and zircon, whilst the white lean laminae are high in quartz, carbonate, and anhydrite. A chalcocite-rich layer at the base of a "washout", described later, is undoubtedly an alluvial concentration from the cupriferous mottled arenites eroded by the channel.

The copper sulfide concentrations on truncation planes, representing minor diastems and pauses in active sedimentation, may be the accumulation of falling precipitates from the overlying anoxic waters, but the concentrations on foreset laminae almost certainly represent current reworking of previously deposited cupriferous sediments.

Even in the ore-shale deposits, notably at the top of the upper orebody in the Roan basin, currents have scoured and eroded cupriferous shale and redeposited the sulfides as rich streaks along small-scale trough type cross-beds. A few metres higher in the section, lenses of cross-bedded arkose with magnetite occur in barren hanging-wall argillite.

SHARP ASSAY CUT-OFFS

Sharp assay cut-offs at hanging wall and footwall indicate a lack of migration of cupriferous solutions after the initial mineralization and generally favour a syngenetic rather than a diagenetic origin.

At Roan Antelope, the chalcocite-bearing layer at the top of the upper orebody assays between 3 and 6% copper and has an abrupt assay hanging wall with the overlying argillites assaying under 0.2% copper. This great change in grade occurs in a thickness of about 20 mm and there is usually no recognizable change in lithology across the contact, which steps up 0.3 m every 20 m along strike to the west. Evidently connate or other waters have not moved copper across this contact, either during diagenesis or metamorphism. This regression of the sulfide zones is a sedimentary feature parallel to the regressive shoreline, bringing in arkose above the ore-shale.

At Chambishi, Nkana, and Roan Antelope banded dolomite and schist at the base of the ore-shale commonly contain over 3% copper and at Chambishi locally contain 15%, resting on Footwall Conglomerate with usually less than 0.1% copper. The assay footwall is rarely more than 25 mm below the lithological contact. This makes a potent argument against permeability of the conglomerate, which is invoked as an aquifer for mineralizing solutions by many epigeneticists.

The Mufulira C orebody, in arenites of sub-aqueous origin, overlies in the eastern basin barren arenite with high-angle cross-bedding, no pebbles or grit grains, and no argillaceous layers. The maximum dihedral angle of 35° confirms an aeolian origin. The assay footwall is sharp and coincides with the change in angle of the cross-bedding, although there is little change in lithology of the arenites on either side. However, in places the dissemination of sulfides projects into the aeolian beds for as much as 30 cm as festoon-like pendants along the steep foresets. Weathering, promoted by leaching of gypsum or anhydrite in the aeolian beds, forms a conspicuous narrow band of red hematite against green malachite and black manganese oxides at this contact between barren aeolian and overlying mineralized rock. The aeolian beds evidently became an aquifer only during the present weathering cycle. During diagenesis and metamorphism it formed a relatively impermeable footwall to the orebody.

Westwards at Mufulira, the aeolian beds interdigitate with sub-aqueous sediments. Here the steep foresets, instead of curving tangentially into bottomsets of similar composition, have an abrupt inflexion into muddy sandy bottomsets, with anhydrite nodules. These dark arenites commonly show small siderite rhombs indicative of reducing conditions and some contain copper sulfide disseminations. These thin sub-aqueous cupriferous layers are overlain and underlain by barren aeolian sands.

A diagnostic occurrence on the west flank of the central basin of the Mufulira C orebody shows a sharp assay footwall of mineralized arenite resting on a sun-cracked pavement of barren argillite, with richly mineralized arenite filling the cracks between the argillite slabs. The latter are doubly concave due to compaction, the sand fillings of the cracks resisting the compression and protecting the ends of the slabs. A similar feature is seen at the east fringe of the Chibuluma orebody.

DIFFUSE CONTACTS

Some diffuse contacts indicate migration of cupriferous solutions during sedimentation or diagenesis, or during metamorphism.

Granite and gneiss pebbles and boulders up to 20 cm diameter in mineralized conglomerate at Chibuluma West contain copper sulfide disseminations or veinlets throughout, although the centres are usually leaner in grade.

Paleo-hills of the Basement, where blanketed by Footwall conglomerate, are barren, but where they project, if only a few centimetres into the ore-shale at Roan Antelope, Nkana, and Nchanga, they are usually mineralized for 5 to 15 m below the unconformity. Apparently, the anoxic marine waters penetrated the fractured and weathered Basement rocks and copper was deposited generally of lower grade than in the ore-shale and with diminishing grade downwards. At Nchanga, the perthitic lamellae of the microcline feldspar were destroyed by the weathering and not restored by the metamorphism. At Roan Antelope, the mineralization favoured the fissile schist whilst thin sills of granite gneiss between are usually barren. At Mufulira, a flaring fissure in the granite contains a muddy sand considerably richer in copper than the overlying arenite.

MINERALIZED AND BARREN CLASTS

A mineralized boulder of carbonaceous graywacke has been found in the graywacke hosting the Mufulira A orebody. The dissemination of chalcocite and bornite is evenly distributed, whereas, in the same horizon, mineralized granite pebbles show a diminution in sulfides towards the centre.

In the vicinity of the sun-cracked argillite pavement, previously described, a pebbly arkosic grit at the base of the C orebody has incorporated slabs of the barren argillite and carried them some hundreds of metres eastward towards the centre of the basin. Here the grit is mineralized, but these clasts remain barren, although encrusted by bornite and chalcopyrite. Farther east the pebbly grit overlies mineralized arenite and an argillite layer purplish with finely disseminated bornite. The currents tore up some fragments of the latter and these richly mineralized clasts contrast with the barren clasts from outside the orebody. The mineralized clasts have a 10–20 mm border impoverished in sulfide, proving marginal loss of copper sulfide during squeezing out of water during compaction. These clasts also have the external armour of sulfides as with the barren argillite clasts. Many have a typical "club-ended" shape due to compaction of the middle portion, whilst the ends of the slabs were protected by the less compressible sand (Garlick, 1967).

At both east and west ends of the Chibuluma orebody, the pebbly arkose contains sulfide armoured "club-ended" shale fragments.

These argillite or shale slabs (shaped as doubly concave with relatively thickened ends) attest to compaction and loss of both water and marginal copper sulfide after incorporation into pebbly arkoses. They were evidently mineralized before erosion and transport. Many such shale fragments were presumably completely abraded during transport and the copper sulfides, thus released, would contribute to the detrital sulfides concentrated along foresets and truncation planes of adjacent cross-bedded arenites.

SULFIDITE LAYERS

In all the Zambian copperbelt orebodies, there are layers from 2 cm to as much as 1 m thick consisting predominantly of sulfide. The terms mineralized shale, quartzite, or arenite are completely inappropriate and, hence, the term sulfidite has been coined, meaning sulfide rock. In the ore-shale, sulfidite layers are generally thin and many tend to open during folding to become the site of bedding metamorphic segregation veins, with coarse crystal growths of copper sulfide with quartz, dolomite, and anhydrite. In the arenaceous type orebodies, sulfidite layers are conspicuous and attain 3–5 cm thickness on truncation planes of cross-bedded arkose. They consist of bornite with chalcopyrite or chalcocite with a dissemination of detrital grains of quartz, feldspar, carbonate, and a little tourmaline and zircon, as at Mufulira, Nchanga, and Chibuluma West.

The most spectacular and unique sulfidite beds occur at Chibuluma, forming generally two marker horizons extending from the eastern to the western fringes of the orebody. They consist of cobaltiferous pyrite, carrollite, and less chalcopyrite with detrital quartz, albitized feldspar, sericite, small stumpy tourmalines, and, in places, small pebbles of quartz and quartzite. They grade up into cross-bedded sericitic arkose or arenite with much chalcopyrite and minor carrollite. By contrast, the basal contact is sharp, even when viewed under the microscope. It is a scoured surface, locally with potholes up to 1 m deep and filled with sulfidite with pebbles that, significantly, rarely rest on the bottom. The sericitic arkose below the contact is usually weakly mineralized with a pyrite or chalcopyrite dissemination, low grade compared to the beds above. For about 5 cm below the contact, the arenite is blackened by growth of microscopic blue acicular crystals around the detrital outlines of quartz and feldspar grains, subsequently enclosed in outgrowths of quartz and albite. This indicates that the flood waters forming the overlying bed were enriched in borates and sodium salts, which filtered down into the subjacent layer, although the cobalt and iron sulfides failed to make any penetration.

This unique occurrence provides the most emphatic evidence for a syngenetic origin, but raises further problems as to how iron and cobalt sulfides can be concentrated to say 15% iron and 5% cobalt in a basal pebbly layer grading up into cross-bedded sericitic arenite above with 4–6% copper as chalcopyrite and about 0.5% cobalt as carrollite. The boron concentration in the flood waters indicates erosion of a dried-out salt-lake or a sabkha, probably with efflorescence of sodium and borate salts and possibly iron, copper, and cobalt sulfates from oxidation of earlier syngenetic sulfides. This slightly acid water, on entering the marine lagoon or estuary in which the Chibuluma deposit was accumulating, would immediately precipitate the insoluble iron and cobalt hydroxides at the higher pH of the saline water. Most of the copper probably remained in solution to be precipitated as sulfide, as anoxic conditions returned to the bottom waters. The precipitate was incorporated into the muddy sands of foresets of small deltas building out into the stagnant water. The iron hydroxide, comparable to the goethite in the deeps of the Red Sea, could, with the cobalt, be sulfidized by anaerobic bacteria below the sediment/water interface. This invokes a diagenetic change of the negative radical, but much of the sulfur was already in the sediment as calcium sulfate. Erosion of a sabkha would probably introduce much algal material into the sediment to give the necessary nutrient for the anaerobic bacteria.

WASHOUTS

Between the mineralized cross-bedded arenites near the margins of the Mufulira orebodies and the graywackes occupying the centre of the eastern basin, there is always a mottled arenite due to slipping, stretching, breakage, folding, and flowage of wet sandy layers on muddy bottomsets, resulting in a chaotic slump breccia of blocks and wisps of white to grey carbonate-rich feldspathic arenite in a dark sandy argillite much richer in copper sulfides than the lighter arenite. Adjacent to undisturbed cross-bedded arenite, the breccia contains recognizable blocks of the latter, but down the initial slope the sands apparently became fluidized and drawn out as wisps in the sliding muds and sand, in which are commonly preserved, in aligned fragments, convolute folds and local inversions. At intervals through the breccia, there are layers of argillite up to nearly 1 m thick that are little disturbed, except for rare breaks filled with breccia and an upper uneven surface forming the base of the overlying breccia. This indicates that the slumping was not a catastrophic sliding of the whole mass, but was spasmodic and contemporaneous with the accumulation of superimposed small deltas. Moreover, the mineralization was concentrated in the more argillaceous bottomsets before slumping.

Sharply defined erosion channels, or washouts, occur near the base of and completely enveloped by the slump breccia. They are analogous to the "horse backs" of fluvial sandstone in a coal seam. They are up to 5 m wide and 1.5 m deep and are filled by parallel and cross-bedded white calcareous feldspathic arenite, practically barren of sulfides and contrasting with the surrounding darker cupriferous mottled slumped arenite. In the bottom of the concave base of one channel was a thin but rich layer of chalcocite, representing a detrital gravity concentrate from the erosion of the mineralized slump breccia.

No fragments of the slump breccia have been found in the washouts, indicating that the breccia had not been indurated before channeling; the strong current disintegrated the soft breccia, washed out the muddy matrix, and backfilled the channel with cleaned sand, with some of the sulfides settling to the bottom of the channel.

It is noteworthy that the channel sand was not mineralized during diagenesis, although totally enclosed in the orebody.

CONCRETIONS

The few sulfide concretions found on the Zambian Copperbelt, mostly at Mufulira, have a barren halo 20–30 mm wide, separating the concretion from the adjacent normal and uniform dissemination sulfides in the adjacent arenite. They are probably of diagenetic origin, whilst the even dissemination is probably syngenetic.

RIPPLE MARKS

Oscillatory wave ripple marks contain concentrations of copper sulfide in the hollows twice as rich as in the overlying sand. Copper sulfide was washed off the crests before deposition of more sand. The preservation of these small structures again shows lack of movement of sulfide during diagenesis and metamorphism.

REFERENCES

Bancroft, J.A. and Pelletier, R.A., 1929. Notes on the general geology of Northern Rhodesia. *Min. Mag.,* 41: 369–372; 42: 47–50; 117–120; 180–182.

Bateman, A.M., 1930. Ores of the Northern Rhodesian Copperbelt. *Econ. Geol.,* 25: 365–418.

Binda, P.L., 1972. Microfossils from the Lower Kundelungu (Late Precambrian) of Zambia. *Int. Geol. Congr., 24th, Sect. 1,* pp. 179–186.

Binda, P.L., 1975. Detrital bornite grains in the Late Precambrian B Graywacke of Mufulira, Zambia. *Mineral. Deposita.,* 10: 101–107.

Binda, P.L. and Mulgrew, J.R., 1974. Stratigraphy of copper occurrences in the Zambian Copperbelt. In: *Centenaire de la Société Géologique de Belgique, Gisements Stratiformes et Provinces Cuprifères,* Liège, pp. 215–233.

Brandt, R.T., 1962. Relationship of mineralization to sedimentation at Mufulira, Northern Rhodèsia. *Trans. Inst. Min. Metall., Lond.,* 71: 459–479.

Brongersma-Sanders, M., 1967. Permian wind and the occurrence of fish and metals in the Kupferschiefer and Marl Slate. *Proc. 15th Interuniv. Geol. Congr., Univ. Leicester,* pp. 61–71.

Brummer, J.J., 1955. The geology of the Roan Antelope orebody. *Trans. Inst. Min. Metall., Lond.,* 64: 257–318; 458–472; 581–590.

Cahen, L., Delhal, J., Deutsch, S., Grogler, N., Ledent, D. and Pasteels, P., 1970. Geochronology and petrogenesis of granitic rocks in the Copperbelt of Zambia and southeast Katanga Province (Report of the Congo). *Ann. Mus. P. Afr. Centr., Sci. Geol.,* 65.

Darnley, A.G., 1960. Petrology of some Rhodesian Copperbelt orebodies and associated rocks. *Trans. Inst. Min. Metall., Lond.,* 69: 137–173; 371–398; 540–569.

Davidson, D.M., 1931. The geology and ore deposits of Chambishi, Northern Rhodesia. *Econ. Geol.*, 26: 131–154.

Davis, G.R., 1954. The origin of the Roan Antelope copper deposit of Northern Rhodesia. *Econ. Geol.*, 49: 575–615.

Dechow, E. and Jensen. M.L., 1965. Sulfur isotopes of some central African sulfide deposits. *Econ. Geol.*, 60: 894–941.

De Swardt, A.M.J., 1962. Structural relationships in the Northern Rhodesian Copperbelt: an alternative explanation. *C.C.T.A., 4th Regional Comm. Geol.*, pp. 15–29.

Eskola, P.E., 1948. The problem of mantled gneiss domes. *Qt. J. Geol. Soc. Lond.*, 104: 461–476.

Fleischer, V.D., 1967. Relation between folding, mineralization and sub-Katanga topography at Mufulira Mine, Zambia. *Trans. Geol. Soc. S. Afr.*, 70: 1–44.

Garlick, W.G., 1940. Problems of the granites and location of the copper deposits of the Nkana Concession. (Unpubl. Company Report, Rhokana Corp. Ltd.)

Garlick, W.G., 1945. Zonal theory of sedimentary uranium–copper–cobalt–pyrite deposits. (Unpubl. Company Report, Rhokana Corp. Ltd.)

Garlick, W.G., 1953. Reflections on prospecting and ore genesis in Northern Rhodesia. *Trans. Inst. Min. Metall., Lond.*, 63: 9–20; 94–106.

Garlick, W.G., 1958. Structures of the Northern Rhodesian Copperbelt deposits. *C.C.T.A. Joint Meet., Leopoldville, Publ.*, 44: 159–179.

Garlick W.G., 1959. Geology of Chibuluma Mine with notes on lithology and sedimentation. (Contrib. Symp. Geol. Copper Afr. Assoc. Afr. Geol. Surv.)

Garlick, W.G., 1961. In: F. Mendelsohn (Editor), *The Geology of the Northern Rhodesian Copperbelt*. Macdonald, London, pp. 146–165.

Garlick, W.G., 1964a. Criteria for recognition of syngenetic sedimentary mineral deposits and veins formed by their remobilization, 6. *Proc. Aust. Inst. Min. Metall.*, 6.

Garlick, W.G., 1964b. Association of mineralization and algal structures on Northern Rhodesian Copperbelt, Katanga, and Australia. *Econ. Geol.*, 59: 416–427.

Garlick, W.G., 1967. Special features and sedimentary facies of stratiform sulfide deposits in arenites. *Proc. 15th Interuniv. Geol. Congr., Univ. Leicester*, pp. 107–169.

Garlick, W.G., 1973. The Nchanga granite. *Geol. Soc. S. Afr. Spec. Publ.*, 3: 455–474. (Symposium on Granites, Gneisses and Related Rocks, Salisbury, 1971.)

Garlick, W.G. and Brummer, J.J., 1951. The age of the granites of the Northern Rhodesian Copperbelt. *Econ. Geol.*, 46: 478–497.

Garlick, W.G. and Fleischer, V.D., 1972. Sedimentary environment of Zambian copper deposition. *Geol. Mijnbouw*, 51 (3): 277–298.

Gray, A., 1929. The outline of the geology and ore deposits of the Nkana Concession. *Int. Geol. Congr., 15th. Sess.*

Gray, A., 1930. The correlation of the ore-bearing sediments of the Katanga and Rhodesian Copperbelt. *Econ. Geol.*, 25: 783–801.

Gray, A., 1932. The Mufulira copper deposit. *Econ. Geol.*, 27: 315–343.

Gregory, J.W., 1930. The copper-shale (Kupferschiefer) of Mansfeld. *Trans. Inst. Min. Metall., Lond.*, 40: 3–30.

Jackson, G.C.A., 1932. The geology of the Nchanga district, Northern Rhodesia. *Qt. J. Geol. Soc. Lond.*, 88: 443–514.

Jensen, M.L. and Dechow, E., 1962. The bearing of sulphur isotopes on the origin of the Rhodesian copper deposits. *Geol. Soc. S. Afr., Trans.*, 61: 1–17.

Jolly, J.L.W., 1972. Recent contributions to Copperbelt geochemistry. *Geol. Minbouw*, 51 (3): 329–335.

Malan, S.P., 1964. Stromatolites and other algal structures at Mufulira, Northern Rhodesia. *Econ. Geol.*, 59: 397–415.

Maree, S.C., 1958. The geology and ore deposits of Mufulira, Northern Rhodesia. *C.C.T.A. Joint Meet., Leopoldville, Publ.*, 44: 147–158.

McGregor, J.A., 1964. *The Lumwana Copper Prospect in Zambia.* Thesis. Rhodes Univ., S. Afr., unpublished.

Mendelsohn, F., 1957. *The Structure and Metamorphism of the Roan Antelope deposit.* Thesis, Univ. of Witwatersrand, unpublished.

Mendelsohn, F., 1959. The structure of the Roan Antelope deposit. *Trans. Inst. Min. Metall., Lond.,* 68: 229–262; 415–423.

Mendelsohn, F., 1959. The lithology of the Roan Antelope deposit. (Contrib. Symp. Geol. Copper Afr., Assoc. Afr. Geol. Surv.)

Mendelsohn, F. (Editor), 1961. *The Geology of the Northern Rhodesian Copperbelt.* Macdonald, London, 523 pp.

Paltridge, I.M., 1968. An algal biostrome fringe and associated mineralization at Mufulira, Zambia. *Ecol. Geol.,* 63: 207–216.

Pettijohn, F.J., 1948. *Sedimentary Rocks.* Harper, New York, N.Y., 2nd ed., 718 pp.

Renfro, A.R., 1974. Genesis of evaporite-associated stratiform metalliferous deposits – a Sabkha process. *Econ. Geol.,* 69: 33–45.

Richard, G.W., 1964. Geology and Mineralization of the Copper–Cobalt Deposits of the South Orebody, Nkana, Northern Rhodesia. Thesis, Royal School of Mines, London, unpublished.

Rijken, J.H.A. and Clutten, J.M., 1972. The water problem in relation to mining at Konkola Division, Nchanga Consolidated Copper Mines Ltd. *Geol. Mijnbouw,* 51 (3): 399–408.

Sharpstone, D.C., 1929. An outline of the geology of the Roan Antelope deposit. *6th Int. Geol. Congr., 1930, Proc.*

Schmitz, J.G. and Askew, J.F.R., 1959. A facies change at the base of the Roan Antelope ore formation. (Paper presented at the 6th Inter-Terr. Geol. Conf., Lusaka.)

Schneiderhöhn, H., 1932. Mineralische Bodenschätze im südlichen Afrika (The geology of the Copperbelt, Northern Rhodesia; translation in abstract). *Mineral. Mag., Lond.,* 46: 241–245.

Schwellnus, J.E.G., 1961. A brief report on the geology and structure of Bancroft Mine.

Snelling, N.J., Hamilton, E.I., Drysdall, A.R. and Stillman, C.J., 1964. A review of age determinations from Northern Rhodesia. *Econ. Geol.,* 59: 961–981.

Stokes, W.L., 1968. Multiple parallel-truncation bedding planes – a feature of wind-deposited sandstone formations. *J. Sed. Petrol.,* 38: 510–515.

Trudinger, P.A., Lambert, I.B. and Skyring, G.W., 1972. Biogenic sulphide ores: a feasibility study. *Econ. Geol.,* 67: 1114–1127.

Van Doorninck, H.H., 1928. *Die Lufilische Plooiing in den Boven Katanga.* G. Naeff, The Hague.

Van Eden, J.G., 1974. Depositional and diagenetic environment related to sulfide mineralization, Mufulira, Zambia. *Econ. Geol.,* 69: 59–79.

Van Eden, J.G. and Binda, P.L., 1972. Scope of stratigraphic and sedimentologic analysis of the Katanga Sequence, Zambia. *Geol. Mijnbouw,* 51 (3): 321–328.

Vink, B.W., 1972. Sulphide mineral zoning in the Baluba orebody, Zambia. *Geol. Mijnbouw,* 51 (3): 309–313.

Voet, H.W. and Freeman, P.V., 1972. Copper orebodies in the basal Lower Roan meta-sediments of the Chingola open-pit area, Zambian Copperbelt. *Geol. Mijnbouw,* 51 (3): 299–308.

Whyte, R.J. and Green, M.E., 1971. Geology and paleogeography of Chibuluma West orebody, Zambian Copperbelt. *Econ. Geol.,* 66: 400–424.

Woakes, M.E., 1959. Mineral zoning in the B orebody at Mufulira mine, Northern Rhodesia. (Unpubl. Company Report, M.C.M. Ltd., presented at the Inter-Terr. Geol. Conf., Lusaka.)

Wolf, K.H., 1976. Ore genesis influenced by compaction. In: G.V. Chilingarian and K.H. Wolf (Editors), *Compaction of Coarse-grained Sediments.* Elsevier, Amsterdam, II, in press.

Chapter 7

KUPFERSCHIEFER[1] IN THE GERMAN DEMOCRATIC REPUBLIC (GDR) WITH SPECIAL REFERENCE TO THE KUPFERSCHIEFER DEPOSIT IN THE SOUTHEASTERN HARZ FORELAND

WOLFGANG JUNG and GERHARD KNITZSCHKE

INTRODUCTION

The Kupferschiefer of the Central European marine Upper Permian is known as a horizontally persistent sediment extending over an area exceeding 600,000 km^2 (see Fig. 1). Certain metal concentrations in this layer and also in the directly underlying bed and partly also in the directly overlying unit are always present, but only in few places has the metal accumulation reached such a concentration that one can speak of economically utilizable and exploitable deposits. These are the following areas:
(1) the Richelsdorfer Gebirge (synclines of Solz-Sontra and Ronshausen-Höhnebach); (2) the southeastern Harz foreland (region of Mansfeld-Eisleben-Sangerhausen); (3) the Mulkwitz structure at Spremberg; (4) the Inner Sudetic syncline (region of Boleslawiec); and (5) the subsudetic monocline (region of Lubin).

The mines in the synclines of Solz-Sontra and Ronshausen-Höhnebach were closed down after 1950. Mining in the syncline of Mansfeld (Mansfeld-Eisleben region) was discontinued in 1970. During a period of approximately 770 years, a total of about 1.5 million tons of copper was mined in this district. This corresponds to 1.1 million tons of metallurgical copper. Some 45,000 tons of this amount are calculated to have been produced during the period between 1200 and 1687, whilst from 1688 to the present time nearly 1.1 million tons of copper were produced. The annual production for the period from 1779 to 1970, including the newly opened district of Sangerhausen can be seen in Fig. 2. The considerable declines in 1918 and 1945 are the result of the two world wars. Mining is at present only carried out in the region of Sangerhausen. The Mulkwitz structure can be considered as geologically explored. Mining in the Inner Sudetic syncline (region of Zajaczek and Grodzice) has been carried out since the nineteen-thirties. A

---

[1] The German term "Kupferschiefer", meaning "copper slate", is used in English when reference is made to this world-known deposit. The so-called "Schiefer" has attributes of shales so that the latter term is appropriate.

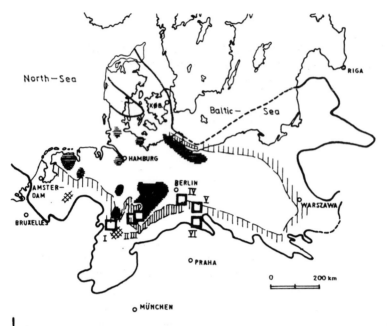

North—Sea

RIGA

Baltic — Sea

HAMBURG

AMSTER-
DAM

BERLIN
IV

V

WARSZAWA

BRUXELLES

I   II III   VI

o PRAHA

0   200 km

o MÜNCHEN

⊟ evident, supposed distribution of sea

moundfacies of $Z1^x$

fine classification of $A1^{xx}$, proved directly

fine classification of $A1^{xx}$, proved indirectly

pottassic marginal deeps in $Z1^x$

Most important copper ore deposits:
  I   Richelsdorf mountains

  II Ore district of Sangerhausen

  III Mansfeld Syncline

  IV Structure Mulkwitz

  V   Presudetic Monocline

  VI Centralsudetic Syncline

^{x}Z1 Zechstein 1
^{xx}A1 Werra anhydrite

Fig. 1. Distribution of the Zechstein Formation in Central Europe.

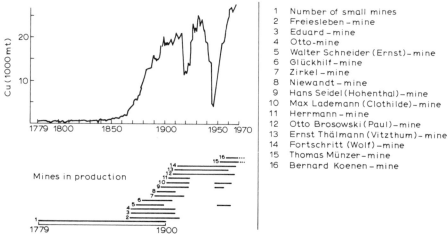

Fig. 2. Copper production (metallurgical) from ores of the Mansfeld and Sangerhausen districts.

metal concentration along the base of the Zechstein, believed to be the largest copper deposit in Europe, has been discovered in the nineteen-sixties in the foreland of the Sudetic mountains (region of Lubin-Polkowice-Sieroszowice and of Wroclaw-Glogow). A series of shafts has been sunk and at present a yearly production of approximately 180,000 tons of copper is reached.

Apart from certain lithologic and thickness changes, the conditions in the above-mentioned regions, including the entire distribution area, are the same for the whole ore horizon. For this reason a pars-pro-toto representation is selected, i.e. we describe chiefly the well explored situation in the 1200 km^2 large depositional area of Mansfeld-Sangerhausen, about 200 km^2 of which have been opened for mining and as a result of which this area is covered by a network of approximately 1000 boreholes. In the light of investigations carried out on core material in the past years, we can make statements concerning the territory of the German Democratic Republic which we consider as being generally applicable to the entire depositional area of the Kupferschiefer.

On the basis of these investigations, it was possible to increase considerably, and to define exactly, the knowledge gained so far with regard to the palaeogeographic development of the basal part of the Zechstein as well as the non-ferrous metal distribution in the Kupferschiefer, including the associated host rocks which are genetically related to the ore. Progress has been made especially in solving the lithological, paragenetical and geochemical problems of Kupferschiefer mineralization.

GEOLOGICAL-TECTONIC SURVEY OF THE KUPFERSCHIEFER DEPOSITS OF THE SOUTHEASTERN HARZ FORELAND

The following generalized stratigraphic column has been established:

		thickness (m)
Quaternary	coarse gravel	up to 100
Tertiary	sands, gravels, clays, lignite	up to 100
Lower Trias (so-called variegated sandstone)	sandstones, slaty shales	up to 350
Upper Permian (Zechstein)	shales, limestones, anhydrites, halitites, potassites	200–800
Lower Permian (Oberrotliegendes)	conglomerates, slaty shales	up to 25
	unconformity	
Stephanian and Lower Permian	conglomerates, sandstones, slaty shales	up to 1000
	unconformity	
Old Palaeozoic	slates	

The Old Palaeozoic has been intensely and isoclinally folded in the Variscan mountain formation (main phase between the Lower and Upper Carboniferous).

The sedimentation of the molasse above the folded basement starts only in the Stephanian. The molasse itself fluctuates rapidly with regard to both facies and thickness. The Lower Rotliegendes is in part not developed, so that the Upper Rotliegendes transgresses over the Stephanian. The strongly varying thickness of the sediments of the Upper Permian (Zechstein) is mainly caused by the varying thickness of the halites. Primary facies differentiations are partly responsible for this and to some extent is also the result of leaching processes which began with tectonic fault movements at the end of the Mesozoic. Faulting and periods of erosion or salt leaching explain the gaps in sedimentation and also the deposition of Tertiary and Quaternary sediments into differently sized basins. It is appropriate to present in greater detail that part of the examined profile which includes the ore horizon and in which most of the mining activities have been carried out:

		thickness (m)
Zechstein 1	Upper Werra anhydrite	20–30
	Werra rock salt/anhydrite	0–15/0.2–1.2
	Lower Werra anhydrite	25–35
	Zechsteinkalk (limestone)	2–6
	Kupferschiefer	0.30–0.40
Lower Permian (Oberrotliegendes so-called Eislebener layers)	conglomerate	0–12
	or sandstones	or 0–10
	Sandsteinschiefer (slaty sandstone)	0– 4
	Porphyrkonglomerat (porphyry conglomerate)	0–20
Stephanian	conglomerates, sandstones, slaty shales	>500

Fig. 3. Generalized map of the groups of the sub-saline section between the Halle Market Place fault and Finne dislocation or displacement.

1   mines, discovered   2   faults of different intensity with information of the down thrown blocks

3   --500-- isohypses of $T1^x$- basis   ×T1   Kupferschiefer

The structural conditions of the ore horizon can be seen in Fig. 3. In general, the layers dip from the outcrop at the Harz border towards the south or southeast, with an average dip of 5 to 8°. The region is intensely broken by a number of faults which run mainly NW–SE and partly NE–SW. The throws of the faults vary between a few centimeters and more than 1000 meters. Altogether the unfaulted areas between the faults differ in size and have approximately the same distance, reflecting a more or less equal spacing of the faults. The faulted blocks have been tilted. Details will not be given here and we refer to the publication by Jung (1965). It should be pointed out, however, that this graduated, equidistant faulting probably applies to greater areas. Larger fault zones have been reactivated several times. Remarkable is the fact that the Kupferschiefer deposits are within the range or above the points of intersection from NW–SE or NE–SW of such large geofractures.

LITHOLOGY AND PALAEOGEOGRAPHY

Base of the Kuperschiefer

The lowest unit of the Kupferschiefer in the district of ore deposition of the southeastern Harz foreland is formed by basal conglomerates, sandstones and siltstones. The conglomerate facies is mainly fine- to medium-grained, but seldom coarse-grained. Lithic grains (= "rubble stones") of lydite, quartzite, vein quartz, slate, graywacke and mainly of porphyry are present. The grains have rounded corners or edges, the matrix is sandy, to a lesser degree clayey or limy. The colour is grey to pale red. The sandy facies is present in the form of fine- to medium-grained quartz sand with limy cement, and the roundness and sorting of the quartz granules are characteristic. Besides quartz, there are also granules of quartzite, lydite, feldspar and slate. The silty-clayey facies is mainly in the form of an alternating stratification of light-grey sandstone beds with somewhat darker coloured layers of slate. Lithic grains (= rock fragments) of quartzite, lydite, porphyry and granite have been found in addition to well-rounded quartz grains. Especially zircon, tourmaline and apatite are present in the form of heavy-mineral accessories.

The basic chemical composition of the three facies is compared in Table I. Very clear is the decrease of the SiO_2-quartz/sand component and the increase of the clay substance

TABLE I

Chemical composition (in wt.%) of the layer immediately underlying the Kupferschiefer

Facies	SiO_2	Al_2O_3	CaO	MgO	FeO	CO_2	C	SO_3	S
Conglomeratic	60.8	10.3	9.2	1.3	2.6	7.9	0.4	1.4	0.9
Sandy	59.1	11.5	9.2	1.2	2.4	7.6	0.4	1.4	1.2
Silty-clayey	50.8	21.1	7.9	0.6	2.2	7.6	0.4	1.8	1.3

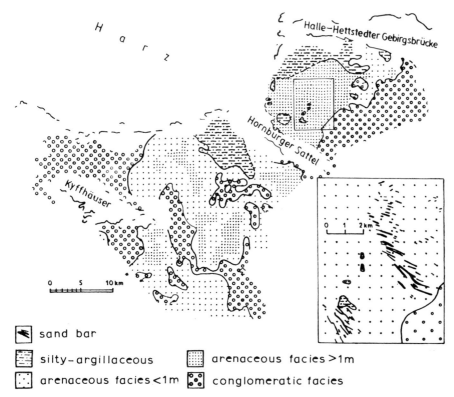

Fig. 4. Distribution and formation of the underlying layer of the Kupferschiefer in the depositional area of the southeastern Harz foreland.

(Al_2O_3-value) at the transgression from the coarser to the finer clastic facies. A proof of the origin of the silty-clayey facies in deeper basinal areas, and at greater distance from basement highs, is also the slight decrease of the carbonate components in the silty-clayey facies composition.

The age of the conglomeratic sandstones and siltstones cannot be stated with certainty. The basal clastic sediments of the Zechstein are found there with few exceptions. A clear demarcation from the underlying sedimentary sandstones of the Lower Permian (Rotliegendes) cannot be made because of a gradational contact. The sandy facies has a peculiarity in that the thickness which in general varies between 20 cm and 2 m, is known to increase to as much as 15 m. Such sand bars are mainly in the central part of the Mansfeld syncline. Figure 4 gives a detailed picture of these conditions. These sand bars have formerly been considered as "dunes"; according to the latest sedimentological investigations, however, they originated as shelf sands.

The petrographic differentiation of the underlying sediments of the Kupferschiefer in the depositional region corresponds completely with the entire area of occurrence of the Kupferschiefer in the German Democratic Republic as shown in Fig. 5.

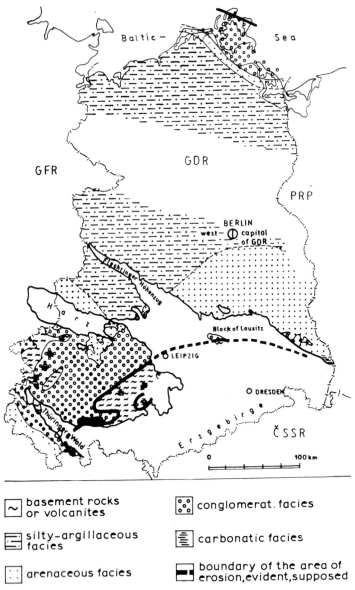

Fig. 5. Distribution and characteristics of the underlying layer of the Kupferschiefer (according to unpublished data of H. Rentzsch).

The base of the Kupferschiefer or Zechsteinkalk (limestone) over the basin borders is characterized by the occurrence of crystalline rocks, slightly metamorphic sediments and volcanites. Sedimentary breccia of the underlying rocks and sedimentary carbonate rocks, mostly with a moderate thickness (4 m at the most), are locally present between these rocks and the Kupferschiefer. This carbonate rock, the so-called mother seam, transgresses over to the clastic sediments along the border localities of the basin. Beds of conglomerate rocks extend along the northern and southern borders of the basin and individual basement-rock elevations. They grade into sandstones in the direction of the basin, as demonstrated for example, in the southwest of the Thuringian Forest, in the southeastern Harz foreland, at the northern border of the Harz, and between the Lausitzer block and the region of Berlin. The area of distribution of pure sandy sediments is only very small parallel to the northern border of the Upper Permian (Zechstein) basin. Sand bars with a considerably greater extent than those found in the southeastern Harz foreland occur to the north and northeast of the Lausitzer block. A sandy-clayey formation must be assumed for the entire central part of the basin on the basis of the drilling results obtained up to now.

For the period before the Kupferschiefer sedimentation, it can be deduced from the rock distribution that the sedimentation area had an almost completely even relief. Only in the southwestern part there are some not completely eroded basement rock highs in the Lower Permian (Rotliegendes) running NNE–WSW and NE–SW (sill of Ruhla-Langensalza, Upper Harz sill, east-Thuringian sill and others). The interbedding of sediments which occasionally occurs in narrow stratigraphic belts reflects the topographic differences along the original sea floor.

Characteristics of the Kupferschiefer

The Kupferschiefer is a fine-grained, finely laminated clayey marl or marlstone containing carbonaceous constituents. Its bedding characteristics are determined by the distribution of the carbonates, clay minerals, clastic constituents and the carbonaceous substance. In general, it can be said that the lamination and the bitumen content of the Kupferschiefer are decreasing with increasing clastic particles, reflecting increasing rate of sediment accumulation. The grain sizes of the rock-forming minerals ($100-200$ μm in diameter) and the carbonate proportion increase towards the hanging wall. Calcite, dolomite, quartz, feldspars, biotite, sericite, kaolinite, illite, chlorite, gypsum, anhydrite, soluble bitumen, solid bitumen and glauconite are constituents of the Kupferschiefer.

The Kupferschiefer seam in the Sangerhausen-Mansfeld depositional area is subdivided into the following strata according to colour, hardness and microscopic textures:

	strata	average thickness (cm)
Schwarze Berge	(T1ϵ)	12–17
Schieferkopf	(T1δ)	10–12
Kammschale	(T1γ)	2.5–4
Grobe Lette	(T1β)	6– 9
Feine Lette	(T1α)	2– 4

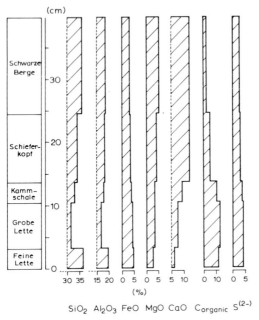

Fig. 6. Geochemical evolution of the Kupferschiefer in vertical profile.

Chemical investigations and petrographic standardization indicate for the Kupferschiefer that in the above-mentioned profile, from the underlying layer to the hanging wall, the petrographic composition changes from strongly carbonaceous-bituminous, carbonate-containing claystones and clayey marlstones to slightly to moderately carbonaceous-bituminous clayey marlstones and marlstones. The petrographic change in the individual strata is reflected in their chemical composition (Fig. 6).

Two geochemical cycles can be derived for the vertical profile from the chemistry of the units; these take place in the contrary sense and in a chronological succession. The progressive cycle (1st cycle) in the Feine Lette and Grobe Lette — characterized by a decrease of SiO_2 and Al_2O_3 and a simultaneous increase of the C_{org} — and S^{2-} content — is followed in the Kammschale by a regressive cycle (2nd cycle). The latter is characterized by a continuous increase of SiO_2 and Al_2O_3 and a continuous decrease of chemical reducing conditions.

Evidently, very subtle processes were involved which did not influence the general trend of the gradually increasing accumulation of carbonate sediments, characterized by the continuously increasing values of CaO, MgO and FeO. The progressive first geochemical cycle is identical with the first petrographic cycle of Luge (1965), whilst the second, regressive geochemical cycle covers the petrographic cycles 2 and 3 of Luge.

The lateral geochemical-petrographic facies differentiation of the rock types of the Kupferschiefer is of importance in addition to the lithologic differentiation in the vertical

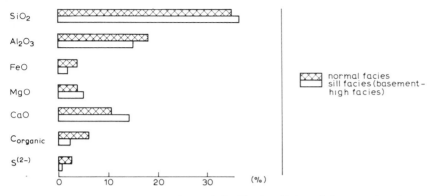

Fig. 7. Comparison of the geochemistry of the Kupferschiefer of the normal environment and the facies developed on the basement high.

stratigraphic sections both of which are considerably influenced by the basin morphology. In general, a differentiation should be made between normal basin facies and the facies on the topographic high (often called sill-facies = "Rote Fäule" facies; Fig. 7). Typical characteristic values of the different developments of both these facies are the contents of CaO, MgO, C_{org} and S^{2-}.

The reduzate components C_{org} and S^{2-} of the sill facies attain only 25 or 30% of the content determined in the normal facies, whilst, on the other hand, the proportion of the precipitate components CaO and MgO are 25–30% higher than in the normal facies. Also remarkable is the decrease of FeO and Al_2O_3 and the increase of SiO_2 in the sill facies,

Fig. 8. Areal and thickness distribution of the Kupferschiefer in the depositional area of the southeastern Harz foreland.

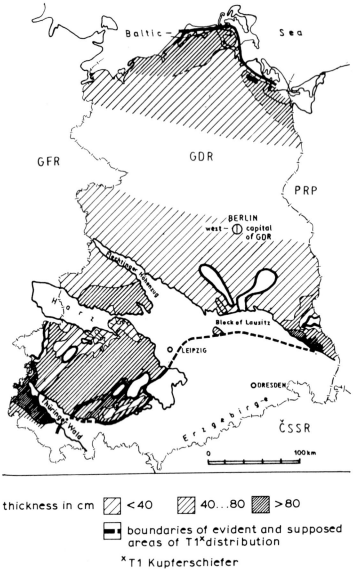

thickness in cm ▨ <40   ▨ 40...80 ▨ >80

▬ boundaries of evident and supposed
areas of T1^{x}distribution

^{x}T1 Kupferschiefer

Fig. 9. Areal and thickness distribution of the Kupferschiefer in the GDR (according to J. Rentzsch).

which indicates a decrease of the clayey substance and an increase of the fine sand components.

Figure 8 gives an areal view of the thickness distribution. It should especially be pointed out that the thickness of the Kupferschiefer can change from 0 to 58 cm within the area of the sand bars. The change takes place over a very limited areal extent (sometimes measuring only a few meters).

The average composition of the German Kupferschiefer fluctuates in its distribution area between carbonaceous-bituminous clayey marlstones, marlstones, and clayey carbonate rocks all of which are more or less finely laminated. Marlstone changes towards the basin to clayey marlstones because the total carbonate content decreases from the border and from the basement highs. Clayey carbonate rocks occur only locally within the border facies. Fine to coarse sandy inclusions in the Kupferschiefer occur at the northern border of the Lausitzer block and southeast of the Thuringian Forest.

The subdivision of the strata of Mansfeld cannot be carried out over a wide area. It has been proven, however, that from the lithological point of view the fine-stratigraphic structure of the Kupferschiefer is uniform in the entire southern part of the Upper Permian basin. Three small rhythms are always detectable in spite of strong thickness fluctuations in the vertical profile, each of which has a clayey fine-sandy basis and a carbonate-rich overlying part. The first cycle corresponds in the depositional area with the Feine and Grobe Lette and the lowest section of the Kammschale. The overlying part of the Kammschale and the Schieferkopf represents the second, and the Schwarze Berge and Dachklotz the third cycle.

In large parts of the Upper Permian basin, the thickness of the Kupferschiefer fluctuates between 25 and 45 cm. At the southern border of the basin, thicknesses in general exceed 80 cm with a maximum of approximately 2 m (Fig. 9). The Kupferschiefer at the northern border of the Lausitzer block is either absent in the immediate vicinity of sand bars or it has thicknesses of >80 cm.

The above-mentioned indicates that the relief of the basin floor was even or only slightly undulating during the origin of the Kupferschiefer. The basement highs and sand bars were at that time also active elements. The Kupferschiefer either thins out over the basement sills depending on the water depth, or it can exhibit reduced thicknesses above the sill top and increased thickness on the flanks, or the thickness is greater above the sill top. On the sand bars, the Kupferschiefer has either a reduced thickness or it thins out completely. The influence of the sand bars upon sedimentation indicates that the Kupferschiefer sea must have been very shallow. Continuous investigations at the northern border of the Upper Permian basin indicate a higher-energy sedimentation.

The Upper Permian limestone (Zechsteinkalk)

The Upper Permian limestone (Zechsteinkalk) is the hanging wall (= overlying bed) of the Kupferschiefer. It can be divided into the following five units:

thickness (m)

Cal δ contaminated to strongly clayey limestone with anhydrite beads and clouds 0.60–0.80

Cal γ less contaminated to clayey limestone, brecciated, pseudo-oolitic and oolitic, sometimes containing numerous anhydrite micro-nodules 0.15–0.20

Cal β moderately contaminated to clayey limestone, clearly bedded   0.20—4.50

Cal α_2 marlstone to sandy marlstone with feather or sickle-shaped anhydrite
              inclusions   1.20—1.40

Cal α_1 sandy-clayey limestone to sandy marlstone with pyrite and anhydrite
              micro-nodules   0.15—0.20

It should be pointed out that the fluctuations in thickness are in general only in the β zone. Only above the sand bars, the zones Cal α_1 and Cal α_2 are not present. Thickness increases otherwise from south to north from <3 m to >6 m, apart from the above-mentioned areas (see Fig. 10). The known Upper Permian limestone (Zechsteinkalk) subdivision, and its refinement on the basis of the latest investigations, can be proved for the southern part of the basin, provided thicknesses are below 10 m. Correlations of the thicker carbonate units with the inclusion of the partly stronger clastic intercalations, are partly possible and are present over a wide area. Sandy beds occur also in the Zechsteinkalk along basin borders and on basement highs as has been noted already with regard to the Kupferschiefer. They are found in the northern border of the Lausitzer block, the Upper Harz sill and at the northern border of the basin. In the latter region also conglomeratic horizons are present.

The Zechsteinkalk (Fig. 11) is thin (1.5—4.0 m) in the central part of the basin and near its borders, i.e., in the larger part of the sedimentation area. The same as for the Kupferschiefer applies also to the thickness variations within the range of the basement

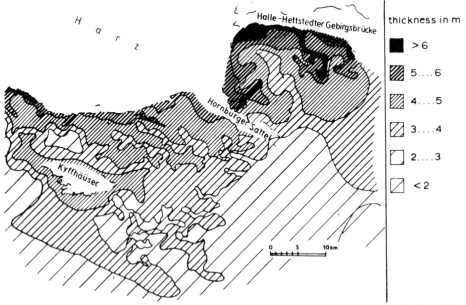

Fig. 10. Areal and thickness distribution of the Werra carbonate (Zechsteinkalk) in the depositional area of the southeastern Harz foreland.

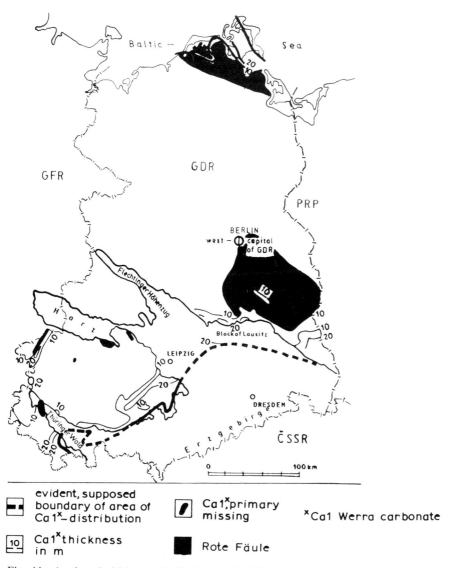

Fig. 11. Areal and thickness distribution of the Werra carbonate in the GDR (according to J. Rentzsch).

rock elevations. Contrary to earlier opinions, the thick Permian limestone (>6 m) is a shallow-water formation possibly formed on a high. The frequent occurrence of thick Zechsteinkalk directly above the basement rocks speaks against its sedimentation in border basins. It is noteworthy that individual basement elevations in the western part of the Thuringian Forest, which are in themselves important features as far as the formation and thickness of the Kupferschiefer is concerned, have no longer been active tectonic elements during the sedimentation of the Upper Permian limestone (Zechsteinkalk).

The interpretation of the greater thickness of the Zechsteinkalk as being an expression of lesser water depth leads, in the case of the depositional area, to the relief inversion within the range of the Lower Harz and, thus, to the assumption of early Upper Permian epeirogenic movements.

The "Rote Fäule"

The succession of beds from the directly underlying unit of the Kupferschiefer to the hanging wall of the Zechsteinkalk is normally grey to partly blackish-grey in colour. Red colouring appears here and there in the form of dots, spots, streaks, clouds and lenses. At the time these features were first observed in the beds of the "Fäule" ($Ca1\alpha_2$), miners named them "Rote Fäule". This term is today applied to all red colourations from the directly underlying bed of the Kupferschiefer to the Lower Werra anhydrite. It has been proved that at least one to four cycles of red colouring are present. Most important is the cycle formed in the underlying bed of the Kupferschiefer, followed by the cycle beginning at the base of the Fäule ($Ca1\alpha_2$). The Rote Fäule appears at the northern border of the basin and in the foreland of the Lausitzer block whilst more or less large patches thereof are found in the area of deposition and southwest of the Thuringian Forest.

The average thickness of Kupferschiefer is lower in areas with Rote Fäule than in sapropel facies. An increased fossil sequence has also been found in the Rote Fäule and on its margins. The depositional structures also speak in favour of a shallow-water sedimentation. The red-coloured rocks are notable for an extreme deficiency in non-ferrous metal and for varying quantities of haematite which gives them the colouring. This indicates conditions of formation within the range of pH 6–7 and Eh 0.1–0.3. It can be deduced that these red colourings appear mainly in sediments that accumulated in shoals and over highs, and especially on their flanks.

METAL TYPES OF THE KUPFERSCHIEFER AND THEIR DISTRIBUTION IN RELATION TO THE PALAEOGEOGRAPHICAL SITUATION

The general designations "Kupferschiefer" and "lead–zinc shale" were no longer sufficient for exact designation of the metal types in the non-ferrous metal-bearing deposits at the base of the Zechstein (directly underlying the Kupferschiefer base of Kupferschiefer of the Upper Permian limestone (Zechsteinkalk). For this reason the basic Cu–Pb–Zn diagram (Fig. 12) has been designed, in which the non-ferrous metal content can be classified independently of the metal type.

The distribution of the Cu-type decreases strongly from the underlying to the overlying beds in favour of the Pb-type and especially of the Zn-type. Figure 13 shows the metal and metal-type distribution in the depositional area. It can be clearly seen that the

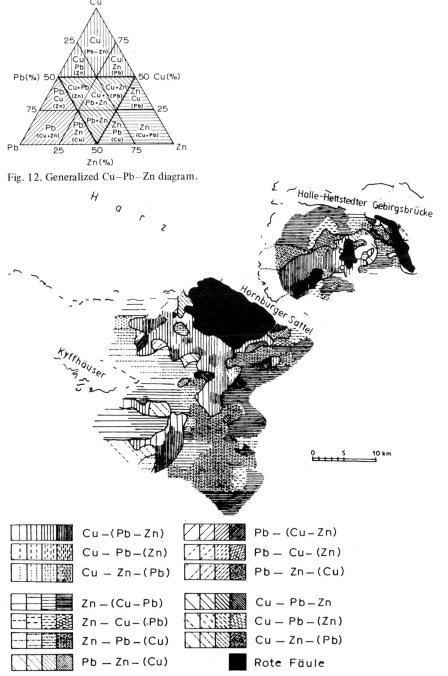

Fig. 12. Generalized Cu–Pb–Zn diagram.

Fig. 13. Distribution of the metal types of the Kupferschiefer in the depositional area of the southeastern Harz foreland. An increasing metal content (ΣCu + Pb + Zn) illustrated in the squares 1 to 4 is represented by an increase in the density of the hatching.

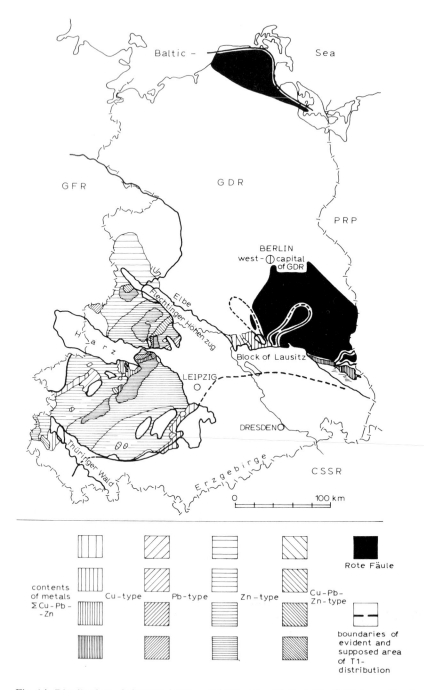

Fig. 14. Distribution of the metal types of the Kupferschiefer in the GDR (according to Rentzsch and Knitzschke, 1968). The metal content is portrayed as in Fig. 13.

main area of copper precipitation is in the central part of the deposition area and that its distribution is in general NE–SW (Erzgebirgian "core line" in the sense of Gillitzer, 1936). The boundary lines are here and there strongly indented. They show besides the Erzgebirgian extension also a NW–SE (Hercynian) and a N–S (Rhaenian) trend. The zinc-shale type is distributed over large areas and mainly flanks the Kupferschiefer areas. Only in the area of the Mansfeld basin is the Cu-type bordered in the north by Cu–Pb–Zn containing shale [(sub-type Cu–Zn (Pb)]. Pb-containing shale and Cu–Pb–Zn shale occur otherwise in the form of "patches" and often mark the transition between the two predominantly occurring metal-types.

Figure 14 shows that in the entire distribution area non-ferrous metal concentrations worth mentioning are only present in a bed at the southern border of the Upper Permian basin which has a width of up to 150 km. Only various intensely red-coloured rocks, with or without Cu-mineralization in traces, have been detected up to now at the northern border of the Upper Permian basin. The Cu-shale in the central parts of the basin, in the upper portions of deposits overlying basement highs and along the basin borders contains very little non-ferrous metals.

Apart from the deposition area, the Cu-type with a higher metal content occurs only to a greater extent in the foreland of the Lausitzer Block. Considering the respective deposits in Poland and in West Germany, one can say in general that all Cu-concentrations of significance and which are known up to now, are confined to shallow-water sediments containing extensive areas of red-coloured basal rocks of the Zechstein in the foreland located on basement highs. The combined occurrence of an economically important copper content and of Rote Fäule-facies has been known for a long time. In addition to the lateral dependence, there is also a vertical dependence between red colouring and copper-concentration. This has been established only recently in as far as the deposit in the southeastern Harz foreland is concerned.

The Zn-type predominates by far in the entire Upper Permian basin, as shown in Fig. 14.

The horizontal zonality of the mineralization, with the metal sequence Cu–Pb–Zn from the borders of the basin or the individual shallows towards the basin, and also the vertical zonation which has the same metal sequence from the underlying to the overlying beds as well as the diminution of the Cu-type in the same directions in favour of the Pb and Zn-types, corresponds to a normal transgressive sequence. Deviations from this vertical metal sequence (for example Cu–Pb–Cu, Cu–Zn–Cu) indicate regressive tendencies which can be particularly well determined in the transitional area from the sapropel to the Rote Fäule facies, i.e., where the two interfinger.

The general trend of the individual metal-types west of the Elbe is directed SW–NE in correspondence with the palaeogeographic situation prior to and during the Cu-shale sedimentation and also with the formation of the lower underground, and it gradually curves east of the Elbe into a NW–SE direction.

If one tries to characterize the total geologic situation within the non-ferrous metal-bearing bed of the areas with a higher Cu-content and with the red rocks extremely poor in non-ferrous metals, one will find that they are lying over the clastic Lower Permian sediments which, in their turn, are situated over the so-called Central German crystalline high (which has been tectonically active during the Lower Palaeozoic), or are situated along their border areas.

It must be pointed out here that contrary to previous opinions, the proof now furnished with regard to the confinement of the Rote Fäule and the Cu-types to the flanks of former shallow-marine areas is based on results obtained from the latest palaeogeographic and physicochemical investigations.

ORE-MINERAL PARAGENESES OF THE KUPFERSCHIEFER AND THEIR REGIONAL DISTRIBUTION

The ore minerals proved in the Kupferschiefer and in its directly underlying and overlying beds can be combined into ten ore-mineral parageneses (Fig. 15). In the following sections we shall briefly describe these primary ore-mineral parageneses of the Kupferschiefer without going into further details regarding all occurring forms of intergrowth and replacements.

Hematite type (paragenesis 1)

Hematite occurs as the main ore-mineral[1] in the red-coloured parts of the ore hori-

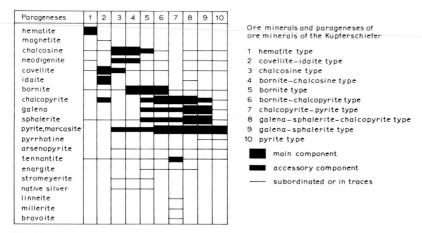

Fig. 15. Ore minerals and ore-mineral parageneses of the Kupferschiefer (according to Rentzsch and Knitzschke, 1968).

---

[1] Although hematite is not economically mined, it has been included among the "ore" minerals.

zon. It is mainly found in the carbonates and their grain boundaries in the form of finely dispersed crystallites. The grain sizes are less than 5 μm. Pseudomorphoses of hematite occur more rarely after pyrite flasers and pyrite globules (30–100 μm). In this paragenesis we were able to prove graphite as the final product of carbon-rich substances.

Covellite-idaite type (paragenesis 2)

This type of paragenesis occurs locally in the form of a lamellar decomposition product of bornite. The idaite contains numerous very small chalcopyrite spindles or is surrounded by chalcopyrite. The chalcopyrite aggregates are displaced by the covellite lamellae with differing intensities. All sulphides replace in particular the carbonates of the ore in sandstones and of the Kupferschiefer and also of the carbonated silicate detritus.

Chalcosine type (paragenesis 3)

The chalcosine type occurs in general only in the vicinity of the red-coloured sediments in the tectonically undisturbed localities. Chalcosine replaces the carbonates and the pyrite present only in small quantities in this paragenesis (ball pyrite, blackberry pyrite, pyrite dust). Irregular intercrescences of chalcosine with neodigenite are present. Both chalcosine and neodigenite have partly covellite lamellae. Native silver and more rarely stromeyerite are, on the margin, grown together with the chalcosine. Bornite occurs relatively seldom with the chalcosine, alternatively replacing each other.

Bornite-chalcosine type (paragenesis 4)

The bornite-chalcosine type is more widespread than the pure chalcosine type. Chalcosine and bornite are frequently myrmekitically intergrown. Mostly covellite, neodigenite and chalcosine replace the bornite. The latter contains chalcopyrite replacement lamellae in varying quantities. Native silver and stromeyerite are intergrown both with chalcosine and bornite, as is the case with the bornite type described next.

Bornite type (paragenesis 5)

The quantity of chalcosine, neodigenite and covellite decreases greatly in the bornite type. Chalcopyrite, sphalerite, galena and pyrite or marcasite gain to some extent in importance. The chalcopyrite replacement lamellae in the bornite increase. Drop-like inclusions of bornite, chalcosine and covellite occur in the sphalerite.

Bornite-chalcopyrite type (paragenesis 6)

The quantity of chalcopyrite increases considerable, together with that of the bornite. Pyrite and marcasite also increase. Replacements of bornite by chalcopyrite gain importance, with the exception of the chalcopyrite flame-like replacements in bornite. Chalco-sine, neodigenite and covellite are only present to a minor extent or in traces having the form of the intergrowth described above.

Chalcopyrite-pyrite type (paragenesis 7)

The chalcopyrite-pyrite type is relatively widespread. The quantity of pyrite can great-ly exceed that of the chalcopyrite. Replacements of pyrite by chalcopyrite occur fre-quently in addition to the generally occurring replacements of the carbonates and sul-phates by chalcopyrite. Tennantite has been observed in *all* sulphide parageneses (see Fig. 15) in varying but mostly small quantities. Its proportion can increase in this paragenesis to such an extent that it locally forms the main copper mineral. Mutual replacements of chalcopyrite and tennantite were observed.

Galena-sphalerite-chalcopyrite type (paragenesis 8)

The Pb–Zn content increases in this paragenesis to such an extent that Pb + Zn is >50% of ΣCu, Pb, Zn, whilst in the parageneses 2 to 7, Cu always amounts to >50% of ΣCu, Pb, Zn. This paragenesis can be subdivided again because a galena-chalcopyrite paragenesis (paragenesis 8a) results when galena is predominant, and a sphalerite-chalco-pyrite paragenesis (paragenesis 8b) when sphalerite is predominant. Chalcosine, neo-digenite, covellite, bornite and idaite in addition to galena, sphalerite and chalcopyrite can be present to a minor extent when the parageneses 2 to 6 are present in the vicinity of paragenesis 8. Pyrite and marcasite are some of the most widely occurring minerals of this ore-type. The most varied alternating mutual replacements of the sulphides were also demonstrated in this paragenesis. Replacements of the iron sulphides by the non-ferrous metal sulphides is ubiquitous apart from younger pyrite formations in capillary fractures.

Galena-sphalerite type (paragenesis 9)

Only small amounts of chalcopyrite or tennantite are present with an increasingly strong decrease of the Cu-portion. The local occurrence of Cu-rich sulphides in traces is very rare and can be traced back to a secondary influence. The paragenesis 9 can, like type 8, also be subdivided in a galena type (paragenesis 9a) and a sphalerite type (para-genesis 9b). The sphalerite type is most widely occurring non-ferrous metal paragenesis of the Kupferschiefer. Some pyrrhotite was noticed locally in addition to great quantities of pyrite and marcasite. The above-mentioned applies to the observed replacement phe-nomena. In this paragenesis, Pb is >50% of ΣCu, Pb, Zn; or Zn is >50% of ΣCu, Pb, Zn.

Pyrite type (paragenesis 10)

The pyrite type is the most widely occurring in the overlying parts of the Kupferschiefer and along the base of the Upper Permian limestone (Zechsteinkalk). The pyrite and the abundant marcasite often mineralize the organo-detritus and mainly form concretions, the so-called "Hieken". The remaining sulphide minerals, present to a minor extent, in general replace pyrite and marcasite.

The distribution of the individual ore-mineral parageneses and the forms of mineralization depend on the vertical profile and also on the horizontal profile of lithofacies and physicochemical conditions of precipitation (H_2S supply).

The non-ferrous metal mineralizations in the area of the sandy or sandy-conglomeratic underlying facies of the Kupferschiefer are bound to the grain boundaries of the host rock and fill the fractures and crevices, whereby a large part of the limy dolomitic cement has been secondarily replaced by the ore minerals. The grain sizes of the ore minerals are, on the average, 20 to 50 μm. Ore minerals penetrate only now and then from the outside into the lithic pebbles. The ore minerals often occur in an intergrown condition. Crystallization sequences can frequently be clearly determined in the ore-mineral parageneses whereby, in principle, the difficult-soluble copper sulphides occur as the oldest mineral formations, followed in the order of solubility by galena and finally by sphalerite, the latter being the most readily soluble sulphide. The copper mineralization reveals the age sequence chalcosine–bornite–chalcopyrite.

The basal parts of the Kupferschiefer are characterized by a fine-rhythmical alternation of clayey-bituminous and carbonaceous layers. The ore minerals are aligned well with this fine bedding. The most commonly occurring ore forms are orientated granules, flasers and lentiles parallel to the bedding. The average grain sizes are, in the case of granule mineralization, 10–30 μm and in the case of lentil and flaser mineralization, 20–70 μm. Occasionally, coarse ore mineralizations occur with grain sizes of 50–250 μm.

The ore minerals very often replace the calcite of the so-called calcareous algae or globular calcites. There are also frequent intergrowths of the ore minerals and, consequently, no age relations can be detected, so that a simultaneous separation must be assumed.

The upper parts of the Kupferschiefer are characterized by a massive texture. The strongly decreasing ore mineralization is mostly in the form of irregularly distributed granules having a size of 5–20 μm; lentils and flasers occur only seldom. Intergrowths and mutual replacements of the ore minerals and of the calcite can also be observed. Altogether, the formation of the ore minerals can be considered as typical syndiagenetic.

Non-ferrous metal sulphides in the directly overlying bed of the Kupferschiefer occur mostly in granules of 1–10 μm. These sulphides are mainly enriched in the so-called "Erzhieken" (= ore concretions) in combination with pyrite which is abundantly represented in all samples. The Erzhieken attain diameters ranging from 500 to 20,000 μm and often indicate signs of tectonic deformation on the microscopic scale.

A horizontal zonation of the ore-mineral distribution is present in the mineralization zone of the Kupferschiefer: hematite–chalcosine + neodigenite + covellite–bornite–

chalcopyrite + pyrite—pyrite + pyrrhotite. Galena and sphalerite appear already within the occurrence of the bornite. The maximum of galena is generally between those of the Cu-containing sulphides and of the sphalerite, which confirms the horizontal metal sequence Cu—Pb—Zn.

The parageneses 2—7 of the Cu-type (Cu >50% of ΣCu, Pb, Zn) are the most widespread in the directly underlying bed of the Kupferschiefer. This applies also to the distribution areas of paragenesis 1.

The distribution areas of the parageneses types 1—7 are to a great extent diminished from the sand ore above the Kupferschiefer to the directly overlying bed. Three types of parageneses distributions are present. The first type represents the distribution areas of paragenesis 1 in which the cupriferous sulphide parageneses 2—5 are present. The parageneses 4 and 5 appear mainly in the second distribution type, whilst parageneses 6—10 with a poor copper-ore mineralization are combined in the third type. The ore-rich areas indicate in general an Erzgebirgian to Rhaenish trend and are influenced by Hercynian tectonic cross-elements. Their distribution pattern in the area of the GDR can be seen from Figs. 16 and 17.

Comparative investigations of the Kupferschiefer of the GDR indicate that the zonation of the paragenesis types apply here in general. However, the details as to the degree of completeness of the zonal sequences show regional differences which are dependent on the regional metal distribution patterns and the original basin morphology and its influence on the mechanisms of precipitation. The vertical sequence of the types of paragenesis corresponds to the horizontal distribution. The paragenesis of the Cu-containing sulphides dominate in the directly overlying unit and in the basal parts of the Kupferschiefer. Above it, there follow the copper-deficient sulphide parageneses with galena or sphalerite, so that here also the principle remains the same, i.e., the precipitation sequence Cu—Pb—Zn, with the corresponding metal supply. Interrupted sequences are due either to the absence of supply or as yet remain to be explained, due to the lack of precipitation.

A horizontal and vertical "zoning" of the metal distribution takes place at the border of the Rote Fäule areas, as has been mentioned above. A Cu-rich bed following the border of the Rote Fäule areas is situated diagonally across the bedding planes, which at some distance from the border of the Rote Fäule is overlapped by a Pb-rich bed, the latter changing rapidly to a Zn-rich unit with a greater areal extent. The dependence of the metal zoning on physicochemical factors has been mentioned already. The more complex "zoning" of the ore-mineral parageneses corresponds with the "zoning" of the metals:

(1) The hematite type (paragenesis 1) is on and along the boundary of basement highs which have been active in the Kupferschiefer basin. Beds with parageneses 2—6 of the copper-mineralization type are situated along the boundary of paragenesis 1, i.e., on the boundary of the highs.

(2) Parageneses 5, 6, 7 and 10 of which paragenesis 7 is the most frequent, are present along the boundary of the basement highs along the basin limits.

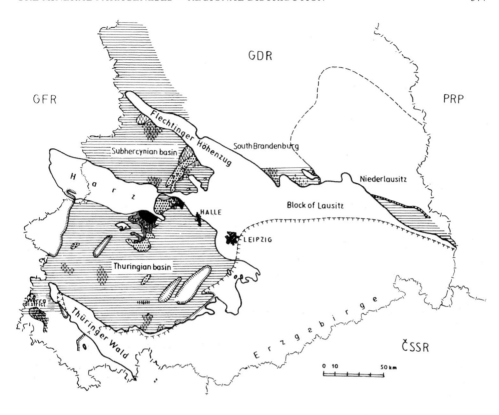

1   —ı—ı—   state border

2   ———   margin of pre – Tertiar and younger areas of erosion

3   ⲙⲙⲙⲙ   evident margin of area of distribution of Kupferschiefer and its equivalents

4   ⊤⊤⊤   supposed margin of ....

5   ⁞⁞⁞⁞⁞   chalcosine, neodigenite, covellite

6   ··········   bornite, idaite

7   ≡≡   chalcopyrite

8   ⌇⌇⌇⌇   tennantite

9   ▦▦   chalcopyrite $\pm$ bornite

10   ▦▦   chalcopyrite $\pm$ tennantite

11   ▮   hematite

Fig. 16. Copper mineral distribution in the ore-impregnated sandstone (according to Rentzsch and Knitzschke, 1968).

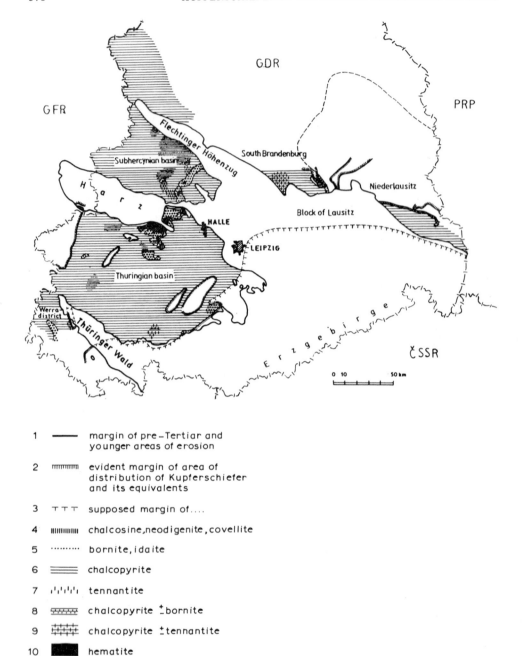

1	——	margin of pre–Tertiar and younger areas of erosion
2	ⲙⲙⲙⲙ	evident margin of area of distribution of Kupferschiefer and its equivalents
3	T T T	supposed margin of....
4	ⲓⲓⲓⲓⲓⲓⲓ	chalcosine,neodigenite,covellite
5	·········	bornite,idaite
6	═══	chalcopyrite
7	⳿ⵏⵏⵏⵏ⳿	tennantite
8	▦▦▦	chalcopyrite $\pm$ bornite
9	▦▦▦	chalcopyrite $\pm$ tennantite
10	▓▓	hematite

Fig. 17. Copper mineral distribution in the Kupferschiefer and its equivalents (according to Rentzsch and Knitzschke, 1968).

(3) Parageneses 8 and 9 are typical for the troughs relatively close near the borders of the basin whilst paragenesis 10 is predominant in the deepest parts of the basin.

The transgression (diminishing with progressing sedimentation) corresponds with the vertical sequence chalcosine–bornite–chalcopyrite–pyrite. The same zoning is present from the border to the basin.

Stratigraphic regressions are connected with a reversal of the metal sequence. The presence of regressive tendencies in the Kupferschiefer and the Zechsteinkalk sedimentation is, aside from the sandy detritus, also proved by parageneses 3–10 with the occurrence of paragenesis 1 above the Kupferschiefer.

The primary cores of the highs covered with Rote Fäule are surely synsedimentary. This is, for example, also demonstrated by the decrease of the thickness or by the absence of the Kupferschiefer near the top of the basement highs. The Rote Fäule facies, however, spreads out on the flanks of these cores, conditioned by syndiagenetic changes in the Eh-Ph conditions, whereby the sapropel facies of Kupferschiefer was syndiagenetically transformed into Rote Fäule facies, resulting in pseudomorphoses of hematite after primary ore minerals. This gave rise to a migration of the non-ferrous metal components which, in the subsequent reduction zone, are reprecipitated in the form of economically rich ores of parageneses 2 to 6.

REGULARITIES IN THE OCCURRENCE OF NON-FERROUS METALS, TRACE METALS, AND ORE MINERALS (explained on the basis of the chemical composition of the Kupferschiefer in the southeastern Harz foreland)

Investigation of major and trace metals

Of the very numerous qualitatively identified elements in the Kupferschiefer, a great number have also been determined quantitatively in the past decade, because of their positive or negative influences during metallurgical processing of the ore, and by considering the chemical and physicochemical aspects. The quantitative data obtained are compiled in Table II which shows that the larger part of the metals has a maximum content in the lower portion of the Kupferschiefer within the area of the transgressive cycle, whilst concentration decreases clearly take place in the regressive cycle and the underlying and hanging walls. Additional quantitative information on other trace metals in the Kupferschiefer are of earlier date and should only be used with reservations (Cr = 150 g/t; In = 0.03 g/t; Ga = 0.09 g/t).

Figure 18 shows the concentration of the most important metals in the Kupferschiefer and in limestone, slate and sandstone in comparison with the Clarke concentration (earth's crust). The metals are compiled in the element associations lithophile, lithophile-chalcophile and chalcophile. Figure 18 shows clearly that in the Kupferschiefer, and in contrast to the earth's crust and sedimentary rocks (such as limestone, shale and sand-

TABLE II

Quantitative data for the most important major and trace metals in the Kupferschiefer

	Major metals (kg/t)			Trace metals (g/t)					(Rare metals)						(Economically undesirable)			
	Cu	Pb	Zn	Ag	Co	Ni	V	Mo	Se	Cd	Tl	Ge	Re	Te	As	Sb	Hg	Bi
Dachklotz	1.4	1.4	1.8	9	16	37	74	43	5	5	3	8	20	<3	22	10	<2	<2
Schwarze Berge	2.3	4.0	5.0	14	28	61	141	73	8	8	6	8	21	<3	74	10	<3	<2
Schieferkopf	6.9	5.7	12.5	36	46	78	315	119	20	20	7	8	21	3	87	11	3	2
Kammschale	17.9	7.6	16.7	107	86	111	751	253	34	29	15	8	22	3	146	12	3	2
Grobe Lette	29.0	8.6	18.5	191	144	140	914	308	48	34	13	9	21	3	327	20	3	3
Feine Lette	25.7	6.1	9.6	183	159	147	877	251	48	18	20	9	21	3	855	44	3	2
Sand ore	29.5	8.4	10.2	147	102	90	115	79	13	45	14	8	22	<3	162	45	<2	<2

stone) the metals with pronounced lithophile characteristics are only slightly concentrated, whereas those with pronounced chalcophile characteristics are very strongly enriched.

Physicochemical regularities

Metal concentration as possible indicators of Eh. The metal content of a marine sediment which accumulated under relatively constant pH-values and near-surface pressure and temperature conditions is controlled by: (a) mechanical transposition and deposition mechanisms; (b) chemical and biological precipitation; and (c) adsorption on clay minerals, organic matter and/or sulphide gels.

The mechanisms of mechanical concentration are of no importance to the Kupferschiefer environment, in contrast to chemical processes. The direct precipitation of metals under marine conditions at constant pH-value, constant pressure and constant temperature and in the presence of small quantities of H_2S as precipitation agent, is a complex function of the concentration of the metal ions to be precipitated, of the solubility of the

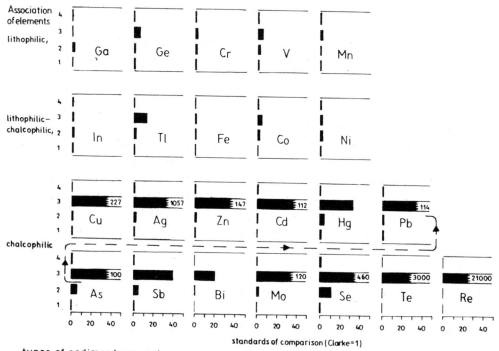

types of sedimentary rocks:

1 sandstone   3 Kupferschiefer
2 shale   4 limestone

Fig. 18. Concentration of major and trace elements of the Kupferschiefer and the most important sedimentary rock types in comparison with the Clarke-value.

possible compounds, of the solvents and the redox potentials (= Eh) present. The concentration of the metal ions and solvents in the sea water is a geological unknown factor, but results from physicochemical investigation using model experiments are available with regard to the solubility of the possible compounds and the redox potentials (see Chapter 10 by Vaughan, Vol. 2).

The syngenetic lateral precipitation sequence $Fe^{3+} \rightarrow Cu^+ \, Pb^{2+} \rightarrow Zn^{2+}$, or in other words the horizontal zoning of the major metals of the ore minerals of the Kupferschiefer, is a double function of the solubility products of the possible compounds and the prevailing Eh-potentials in the case of H_2S deficiency.

Table III shows very clearly the interrelation between metal facies, Eh-value and solubility product. The deposition of metals took place with a H_2S-deficiency, which in turn determined the sequence of precipitation because it also controlled the solubility of the sulphide components. A specific precipitation sequence was the result. The more readily soluble sulphide compounds evidently require higher negative Eh-values and especially a higher H_2S-supply in order to be precipitated directly. The minimum required concentrations of the major metals were surely above the limiting values stated here, as the proven syngenetic precipitation indicates, but the low H_2S quantity available was only sufficient for precipitating the more insoluble compounds and thus led to the formation of facies zonation. Apart from the main metals Cu, Pb, Zn, and Fe, the concentrations of the other metals were evidently below those required for a direct precipitation.

Other processes – adsorptions, such as the enrichment with organic substances, sulphidic minerals, and clay minerals – consequently played a considerable role in their concentration (see Chapter 5, Vol. 2). Adsorptive concentrations of trace metals are mainly effected by co-precipitation, ion exchange or isomorphic substitutions, whereby in each case more or less strong electrostatic forces or attraction become effective. The degree of the adsorption depends mainly on the concentration of the solution, adsorbents (formation of isomorphic and not readily soluble compounds), the valency, ionic radius, and degree of polarization of the ions to be adsorbed (Noll's laws).

The complexity of the above-mentioned factors shows that a classification of the

TABLE III

Correlations between metal facies, Eh-value and solubility products

Metal facies	Redox potential	Possible precipitate	Solubility product
Rote Fäule	positive	$Fe(OH)_3$	$4 \cdot 10^{-38}$
Copper belt	slightly negative	Cu_2S	$2.5 \cdot 10^{-50}$
Lead–zinc belt	stronger negative	PbS	$1 \cdot 10^{-29}$
		ZnS	$8 \cdot 10^{-26}$
		FeS	$4 \cdot 10^{-19}$

factors that lead to the trace-metal concentration in the Kupferschiefer is still quite problematic and only partly possible. By evaluating the mathematic-statistical investigations carried out with regard to certain metal combinations, in the following section an attempt is made to group the metals until now quantitatively determined, according to equal geochemical behaviour.

It is postulated in this case that the formation of the copper belt can be traced back to slightly negative, and the formation of the lead—zinc belt to more strongly negative Eh-potentials. The following groups are accordingly distinguished: one group of chemical elements which separates out in an environment with a slightly negative Eh-range and of which Cu is the dominant element; a second group, dominated by Pb and Zn, which is preferably concentrated at more strongly negative Eh-potentials; and a third Eh-independent group (Table IV).

In addition, it should be noted that the indication of a "preferred separation environment" does not exclude the occurrence of lower concentrations of the same element outside the preferred redox-range, which again confirms the complexity of the concentra-

TABLE IV

Metal concentration at various redox potentials

Element	Ion radius (Å)	Ion potential	Solubility product of the most important sulphides
Group 1: Metal concentrations of slightly negative redox potentials (copper belt)			
Cu^+	0.96	1.04	Cu_2S $2.5 \cdot 10^{-50}$
Ag^+	1.26	0.79	Ag_2S $1 \cdot 10^{-51}$
Co^{2+}	0.72	2.78	CoS $2.0 \cdot 10^{-27}$
Ni^{2+}	0.69	2.90	NiS $1.4 \cdot 10^{-24}$
Fe^{2+}	0.74	2.70	FeS $4 \cdot 10^{-19}$
Se^{2-}	1.93	−1.04	
S^{2-}	1.82	−1.09	
Group 2. Metal concentrations of stronger negative redox potentials (lead—zinc belt)			
Pb^{2+}	1.20	1.67	PbS $1 \cdot 10^{-29}$
Zn^{2+}	0.74	2.70	ZnS $8 \cdot 10^{-26}$
Fe^{2+}	0.74	2.70	FeS $4 \cdot 10^{-19}$
Cd^{2+}	0.97	2.06	CdS $1 \cdot 10^{-29}$
Tl^{1+}	1.47	0.68	
As^{3+}	0.58	5.17	
Sb^{3+}	0.76	3.94	
Te^{2-}	2.12	−0.94	
S^{2-}	1.82	−1.09	
Group 3: Eh-indifferent metal concentrations			
$V^{3+;\,5+}$	0.74; 0.59	4.05; 8.47	
$Mo^{4+;\,6+}$	0.70; 0.62	5.71; 9.68	
$Ge^{2+;\,4+}$	0.73; 0.53	2.74; 7.55	
$Fe^{4+;\,7+}$	0.72; 0.56	5.56; 12.50	

tion processes that became active in the Kupferschiefer environment. Elements with different geochemical behaviour are compiled within the element groups of Table IV, each having similar preferred redox environments. Different environmental factors conditioned their concentration.

The most important adsorbents of the Kupferschiefer environment are, without doubt, the sulphide minerals, the organic substances, and the clay minerals.

Effectiveness of some adsorbents in the metal concentration in the case of slightly negative Eh-potentials. The fact that mineral phases data of the Kupferschiefer are at present only obtainable at considerable expense and with insufficient quality, due to difficulties involved in mineral dressing, induced us to find indirectly by means of mathematic-statistical methods (correlation calculation) a measure of the dependence between the trace-metal associations and the most important adsorbents. For the sequence $Cu \to Ag \to Ni \to Co \to Se$, which is typical of the slightly negative Eh-range of environment, a decreasing dependence of Ag, Ni, Co and Se on the copper sulphide as an adsorbent was established. In our opinion, the predominant concentration mechanism is the isomorphic substitution which, however, is influential to different degrees from element to element. Thus, the elements Cu and Ag indicate, for example, a similar solubility behaviour in their sulphide compounds, i.e., they are precipitated at the same time so that the conditions of the co-precipitation are fulfilled when there is sufficient silver concentration in the initial solution. The reverse can also be the case, namely, instead of direct precipitation, the isomorphic substitution takes place in the case of too low a silver concentration, i.e., the incorporation of Ag^+-ions in the copper-sulphide minerals instead of Cu^+. The selective or preferential incorporation of Ag takes place in the cupreous sulphides and is reciprocal to the incorporation of Fe-ions into the copper sulphides. By constantly keeping the parallelism from Cu to Ag, the Ag/Cu ratio changes to the disfavour of the Ag with progressing sedimentation, as is shown in Fig. 19. The more insoluble Ag_2S is relatively quickly extracted from the solution, or the Ag removed from the solution has been replenished at a lower rate than the Cu.

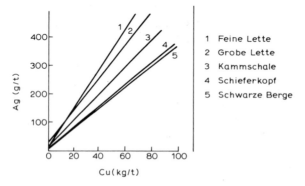

Fig. 19. Correlation diagram Cu/Ag.

The direct precipitation in the form of sulphides due to the considerably higher solubility of the cobalt and nickel sulphide would, in the case of Co and Ni, take place only at a relatively high concentration of these elements and only after precipitation of the more insoluble copper sulphides so that the process of co-precipitation is ruled out here. The isomorphic substitution of Ni^{2+} and Co^{2+} for Fe^{2+} in the copper–iron sulphides, on the other hand, is possible, due to the similarity of the ionic radii. A confirmation of this assumption is the maximum concentration of Ni and Co in the bornite and chalcopyrite mineralization types in comparison with the chalcosine type, as we have demonstrated.

Contrary to the Ag, the incorporation of Ni and Co into the copper sulphides is consequently parallel to the iron concentration in these sulphides. The iron concentration again occurs selectively in the more negative range of the redox intervals which is characterized by the copper precipitation, so that an environmental differentiation takes place here in which more readily soluble components are obtained in higher redox ranges.

The correlative relationships between Cu, on the one hand, and Co and Ni, on the other (Fig. 20), are especially pronounced in the lower beds containing the most ore, whereby similarly to Ag, the ratio of Cu/(Co + Ni) changes in favour of Cu in direction of the overlying parts of the seam. This indicates that the nickel–cobalt content in the sulphide precipitation has been consumed in the lower beds and that a considerable supply of these elements no longer took place, thus resulting in a deficiency of Co and Ni in the overlying parts of the beds. This strongly supports the opinion of Wedepohl (1964) that the required high cobalt and nickel concentrations were removed from the Lower Permian sediments (Rotliegendes) during the transgression of the Upper Permian sea. We shall deal later with the considerable importance of C_{org}.

The opinion of Wojciechowska and Serkies (1968), which assumes for the Kupferschiefer of the lower Sudetic deposits (on the basis of only a very few samples, however) a dependency between the copper and selenium content, cannot be confirmed for the Kupferschiefer deposits of the southeastern Harz foreland. Although a maximum selenium concentration is present in the copper belt as a whole, and here again in the copper-rich ore facies, it is not possible to prove a correlative relationship between Cu and Se. Thus, the incorporation of Se into the copper sulphides is only of secondary impor-

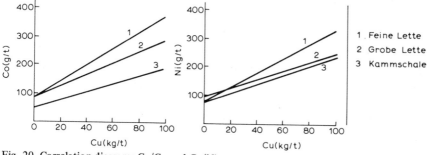

Fig. 20. Correlation diagrams Cu/Co and Cu/Ni.

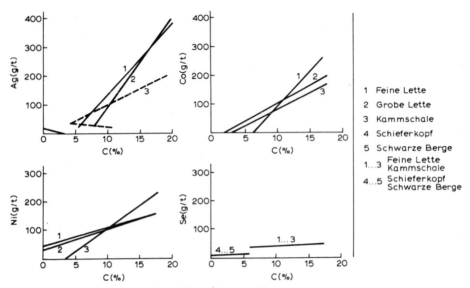

Fig. 21. Correlation diagrams C/Ag, C/Co, C/Ni and C/Se.

tance. The isomorphic substitution of Se^{2-} for S^{2-} in the copper-sulphide minerals is, in this case, likely due to the similarity of the ionic radii of Se^{2-} and S^{2-} which, on the basis of our investigations, takes place preferentially or selectively in a slightly negative to slightly positive Eh-range within the vicinity of the basement high (in the chalcosine and bornite facies). Copper selenides as minerals on their own were evidently not formed due to the very low selenide ion-concentration. Lead–zinc sulphides were not formed in the slightly negative Eh-range and are, therefore, not considered to be trace-element carriers.

Organic substances as adsorbents.[1] The established correlations of certain trace elements and C_{org} can be used as a measure of the effectiveness of organic adsorbents in the concentration of metals. The element association Cu → Ag → Co → Ni → Se indicates correlative relations with the C_{org} of the Kupferschiefer (Fig. 21). The dependence of the above-mentioned metals on C_{org} is, as a whole, less pronounced than that on Cu; it increases in the represented sequence and thus acts exactly in the opposite manner to the correlation sequence Cu → trace metals. Therefore, it can be inferred that:

(1) the fixation of Ag by metal-organic reactions, or its direct adsorption to the organic matter, is of less importance than the silver concentration in the copper sulphides;

(2) the direct fixation by sulphide precipitation simultaneous with the copper separation is of lesser importance for Co and Ni than the direct (physiological) or indirect (adsorptive) concentration to or with the organic substance; and

(3) the predominant form of concentration of the Se in the Kupferschiefer is the adsorption on organic-carbonaceous substances, whereby the adsorption probably took

---

[1] For a detail, more general, discussion, see Chapter 5 by Saxby, Vol. 2.

place most readily within that Eh-potential which was the most favourable for the copper precipitation.

This interpretation agrees with the finding of Tischendorf (1966) that the highest Se content occurs in shales which are relatively rich in carbonaceous substances, and with the parallelism between the selenium and bitumen contents proven by Trelease and Beath (1949) in the Morrison Formation (Utah). In general, it can be assumed that the concentration of trace metals by organic matter had a complementary effect on the isomorphic substitutions in sulphidic minerals, whereby it is not clear whether the mechanisms of concentrations are strictly adsorptive or organo-metallic.

Clayey substances as adsorbents. Because of the chemico-petrographic subdivision of the Kupferschiefer, it was possible to consider the metal content of the clayey and of the carbonaceous-marly parts of the Kupferschiefer as evidence of the influence of clayey substances as adsorbents. An increased concentration of metals was thereby found in the sequence of Co → Cu → Ag → Ni within the clay- and carbon-rich sections, whereas the remaining metals varied independently.

Effectiveness of some adsorbents in the metal concentration at stronger negative Eh-potentials

Sulphide minerals as adsorbents. The precipitation sequence PbS → ZnS → FeS can be expected to occur as a result of their increasing solubility products under conditions of H_2S-deficiency within the more strongly negative Eh-range, because only the metals Pb, Zn and Fe occur in concentrations necessary for direct precipitation, if one disregards the residual Cu^+ ions. It has now to be examined which sulphide minerals are adsorbents and in which manner they are effective.

We were able to prove a clear linear correlation between Cd and Zn and thus a preferred sulphidic precipitation of Cd (Fig. 22), in accordance with the diadochy relations between the Cu and Zn on the basis of similar ionic radii and similar solubility products of these two metal sulphides. The statement already made with regard to the magmatic cycle and also to the sedimentation in the black-shale environment is therefore confirmed. The downward trend of the Cd/Zn ratio in the upper beds should be considered to be similar to that of Ag, Co and Ni as a function of decreasing metal supply.

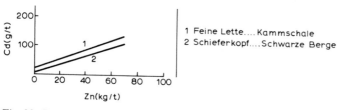

Fig. 22. Correlation diagram Zn/Cd.

An evident relationship which, due to only a few samples being available, cannot be proved by mathematical-statistical methods, also exists, however, between the Pb–Zn content, on the one hand, and the As and Sb content, on the other. It can be assumed that As and Sb are adsorptively bound to the lead–zinc–sulphide gels.

A co-precipitation of As and Cu in chalcopyrite under conditions of stronger Eh-potentials in the lead–zinc shale is also possible. We refer in this connection to the tennantite-arsenopyrite-pyrite-chalcopyrite paragenesis in the lead–zinc shale facies described by Rentzsch and Knitzschke (1968).

Although a high lead content (>25 kg/t) is always accompanied by an increased thallium content (which confirms the diadochy Pb^{2+}/Tl^+ very often postulated in the literature) it seems that other concentration principles, such as the adsorptive bonding to C_{org}, involve similarly high thallium contents. Hence, a linear correlation Pb/Tl could not be deduced.

The direct relationship between a higher tellurium content and increased sulphide sulphur values is considered to be an indication of the isomorphic substitution of Te^{2+} for S^{2-} in the non-ferrous metal sulphides.

Organic substances as adsorbents. The correlation with or relationship to C_{org} decreases for the element sequence Tl → As → Sb → Te → Cd.

The dependency on organic substances is predominant (Fig. 23) in the case of Tl, with a clear correlation to C_{org}, whereas for the elements As, Sb, and Te the influence of C_{org} is less than was found for the mechanisms that resulted in their concentration in sulphides. A regular relationship between Cd and C_{org}, however, cannot be determined.

Clayey substances as adsorbents. No information with regard to the effectiveness of clay minerals as adsorbents can be given here, due to the lack of reference values.

Effectiveness of some adsorbents in the metal concentration independent of Eh-potentials

No clear dependence on a preferred Eh-range is established for the elements V, Mo, Ge, Re.

There are no observations which indicate a preferred incorporation of the above-mentioned metals into the structure of sulphide minerals or clay substances. Only for Mo and V was it possible to determine a correlative relationship with organic carbon. The sugges-

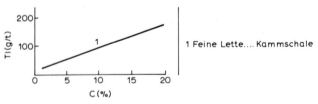

Fig. 23. Correlation diagram C/Tl.

tion of Wedepohl (1964) that V and Mo are two of the elements which are to a great extent indirectly concentrated in the Kupferschiefer environment as a result of reduction or adsorption, or are physiologically bound to the organic substance, can thus be confirmed. The vanadium and molybdenum contents locally reach such values that a partial direct precipitation of these metals in the form of sulphide, in the sense of Rankama and Sahama (1950), is quite possible.

No exact information is available on the mechanisms of concentration of Ge and Re in the Kupferschiefer.

HISTORY AND PRESENT POSITION CONCERNING THE GENESIS OF THE KUPFERSCHIEFER

The here accepted genetic concepts of the Kupferschiefer were based on the research done up to the present and include all essential points of view from many researchers.

A great variety of geological concepts have developed on the origin of the Kupferschiefer deposits as a consequence of the mining activity during a period of 775 years in the southeastern Harz foreland, which includes 400 years of more precise scientific research. The great variety of hypotheses are a reflection of the complexity of problems encountered.

The history of the geological hypotheses can basically be divided into four phases of development. The description of isolated phenomena was in the foreground in the period between 1546 to 1800. Freiesleben gave the first detailed description of a Kupferschiefer deposit. The end of the 18th century and the first half of the 19th century were under the influence of the teachings of Werner who interpreted the Kupferschiefer as a sedimentary deposit within stratified rocks and proposed that the metals came from a basement rock as a result of surface erosion.

Geological works especially of a descriptive character dominated in the period between 1800 and 1900, which was a phase of collecting data comprehensively and extensively. The emphasis of scientific activity was directed to the solving of technical problems. The geological data-collecting was concluded around 1900 in accordance with the then available research possibilities.

The synthetizing and compilation that resulted in many individual publications took place between 1900 and 1945. Progress in systematic hypothesing on the origin of the Kupferschiefer is to the merit of many researchers although, and inspite of, the fact that their individual contributions were originally considered beyond direct economic applicability, but offered an advance in purely theoretical terms.

The work of Walther (1910, 1919, 1921) should especially be mentioned in this connection. He placed the Kupferschiefer within the historical framework of the development of its overlying and underlying rocks. Based on the Variscan rock formations, he visualized the copper concentration as taking place immediately at the earth's surface in arid sedimentary basins of the Lower Permian. This was followed by transgression of the

Kupferschiefer sea and the dissolution of part of the earlier precipitated copper and the reprecipitation (syngenetically again) of the same with the sapropel under chemically reducing conditions. Walther's visualisation or concept also takes into account the immediate influences of tectonic-orogenic, climatic-biological, chemical, mechanical and epeirogenic effects during the formation of the Kupferschiefer deposit.

In the years from 1900 to 1925 the theory of Walther was further developed and interpreted in various directions. By means of an actualistic and a comparative approach, Pompecki (1914, 1921) tried to deduce evidence for the formation of Kupferschiefer-type sediments in the bottom sediments of the Black Sea and to establish from their prevailing sedimentological–biological–chemical conditions the formation of the sapropel facies of the Kupferschiefer sea. The Black Sea is characterized as a recent example of the formation of sulphidic sapropels.

The following problems have to be considered by syngenetisists, once the mechanism(s) of metal accumulation and the reasons for the wide areal distribution of the marine clastic Kupferschiefer sediments have been determined: (1) the discrepancy between the large regional occurrence of the Kupferschiefer beds per se, on one hand, and the areal limitations of the economic non-ferrous metal concentrations within the Kupferschiefer, on the other; and (2) the derivation of the metals.

According to Lier et al. (1922), the weathering of eruptive rocks — especially of the Lower Permian eruptiva — and of the ore deposits (Rammelsberg) was an essential starting point of the non-ferrous metal supply. The mechanism of the concentration of Cu and Ag under arid conditions, even from Cu-deficient rocks, has especially been studied by Harrassowitz (1922).

Geochemical investigation by Von Wolff (1922) supports the above view. He proved that the acid Lower Permian eruptiva, with a copper content of <0.05%, cannot have been the primary metal carriers, but instead only the basic volcanic rocks and older eroded deposits of the Harz Mountains. Von Freyberg (1923) presented the first palaeogeographical map of the Kupferschiefer and this emphasizes the importance of the Variscan saddle axes and their capacity as suppliers of primary metal, whereby the copper content of the Mansfeld deposit is brought into relation with the Harz area.

A different view with regard to the origin of the metals was taken by Lang (1921) who deduced that the metal contents of the Kupferschiefer came from the Erzgebirge. From the beginning this concept was contrary to the knowledge of that time about the palaeogeography of the Lower Permian basin, because the transport of the metals to the Mansfeld deposit is prevented by the intermediary Schwarzburg saddle and the granulite rocks. Another problem was that the characteristics of the copper-deficient Variscan metal province of the Erzgebirge do not correspond to those of the metal contents of the Kupferschiefer deposits.

Speaking from the purely chemical point of view, Erdmann (1919) was of the opinion that the copper content of sea water was also sufficient without an additional supply of non-ferrous metals from the Lower Permian continent in order to reach non-ferrous metal concentrations in the sapropel facies. This could be via adsorption processes in the clay sludges by means of the concentration of organic substance due to a purely chemical precipitation mechanism. In 1922, Erdmann proposed a physicochemical precipitation mechanism, which had as a prerequisite for the purely chemical precipitation of non-ferrous metals directly from sea water, the sulphate reduction of organic substances. The latter reduction was in connection with the formation of ammonia during the decomposition of organic substances.

Correns (1924) proved in model tests the adsorption power of clay sludges towards copper.

Based on palaeontological results, Freygang (1923/24) expressed an opinion as to the development of ecological conditions during the sedimentation of the Kupferschiefer.

Fulda (1928) was a further exponent of the theory of preliminary concentration of non-ferrous metals under arid climatic conditions. In the evaluation of new exploration results in the eastern part of the Mansfeld deposit, he established a correlation between the rich ore occurrences and the sandstone schist distribution. He considered the presence of basins in arid areas with fine clastic sedimentation, especially suitable for the concentration of non-ferrous metals, to be a prerequisite for the syndiagenetic accumulation of the bog copper precipitates due to adsorption or adhesion metasomatosis.

Stelzner and Bergeat (1904), as well as Köhler (1905) and Geipel (1919), advocated a syngenetic accumulation for the primary metal sulphide of the Kupferschiefer deposit. The problem of the secondary remobilisation of the metals in the immediate vicinity of tectonic faults takes a special place within the scope of this theory. The influence of faulting on the composition of the beds in the vicinity of the faults especially within the downthrown block, has already been discussed by earlier authors, such as Freiesleben, Von Veltheim, Plümicke, Bäumler, and Schrader. The economically unproductivity of beds in the vicinity of faults and the fact that the ore content is neither influenced nor changed by the faulting process itself was also pointed out. The observed concentrations and depletion of copper in the Kupferschiefer in the immediate vicinity of faults, on the other hand, were the subjects of study by Posepny (1895), Beyschlag (1900), and Kruson (1904). They proposed in general, a secondary, i.e., epigenetic and ascending supply of non-ferrous metals into the Kupferschiefer bed by solutions. The greatest advocate of this opinion was, without doubt, Beyschlag (1900, 1920). He subdivided the genesis of the deposits into the following phases: (1) syngenetic accumulation of pyrite- and copper-containing shale; (2) epigenetic alteration of pyrite into chalcopyrite by ascending acid solutions which moved into the bed via the faults along the contact of the Lower Permian (Zechstein); (3) alteration of chalcopyrite by descending waters from the faults into the sulphide-rich beds.

Beyschlag calls the non-ferrous metal mineralization a cementative one and he considers the Permian eruptive centers to be the metal suppliers, whereby the faults are regarded as circulation channels for the ascending solutions. One of the main reasons for this concept was the often observed concentration gradient of the non-ferrous metals in a horizontal direction away from the fault into the unfaulted block or rock and the occurrence of non-ferrous metal-bearing faults also in areas where the Kupferschiefer bed was not exploitable.

Geological and also chemical arguments advanced by the syngeneticists (Walther, Erdmann, Lang, and others) are opposed to Beyschlag's hypothesis: the neighbouring zone, reaching horizontally but a few meters from the faults, and the occurrence of rich ores even in the tectonically faulted areas are especially mentioned in their reasoning against Beyschlag's proposals.

According to the data gained from geochemical investigations it is furthermore incomprehensible how the acidic solutions required for the transport of Cu were able to cause precipitation of non-ferrous metal sulphides without causing a stronger reaction with the highly carbonaceous Kupferschiefer rocks. Also, why were rocks of a similar composition (i.e., bituminous shale) not mineralized in the same fashion in spite of having been faulted?

The secondary metal remobilization is differently interpreted by the advocates of syngenesis. All the hypotheses have the same basic concept of the displacement of the primarily existing metal content without supply from below.

Different and contrasting opinions exist only with regard to the origin of the solutions and the extent of the mobilization processes. Lang (1921) traced the metal remobilization in areas beneath rock-salt covering back to lateral-secretional processes caused by the solutions liberated from the evaporite rock during thermometamorphosis. Köhler (1905) and Geipel (1919), on the other hand, speak of descending and very slowly circulating waters which, on the edges of the groundwater level, penetrate into the faults and from there into the ore horizon. In addition to this "hypohaline depth zone", Lang also defined an "epihaline zone" in which larger displacements develop due to the influence of the groundwaters.

Geipel already pointed out in 1919 that the metal content of the faults decreases in an easterly direction, i.e., with increasing distance from the outcrops. He traced this back to the fact that the waters of the faults in the dip of the deposit migrate into the inside of the basin, thereby losing their metal content which had originated from the leaching of the ore horizon within the range of the groundwater level or from the surface waters of arid areas.

It must be underlined here, that a solution of this problem was not possible, due to the inadequate level of geological knowledge and working methods at that time. It was only with the introduction of ore microscopy that a decisive solution was reached which was in favour of syngenesis. Pioneer work in this field was done by Schneiderhöhn (1921, 1926) who carried out the first microscopic chalcographical investigation of the Mansfeld Kupferschiefer. He was able to prove a syngenetic precipitation of bornite and chalcocite as well as the host rock for the tectonically unfaulted area.

There appears to have been no metal remobilization in host rocks containing cementation-type of ore impregnations or concentrations. The textures and paragenesis indicate a somewhat simultaneous sedimentary formation of the carbonates and of the bornite, whereas chalcopyrite was clearly formed secondarily. Primary and secondary depth differences are absent in the unfaulted area.

Schneiderhöhn was also the first to detect mineralized micro-organisms which he interpreted as mineralized sulphur bacteria. The suggestion of Pompecki (1914) concerning the formation of H_2S in the sapropel environment was thus confirmed.

The investigations of Hoffmann (1923/24) proved for the unfaulted rock units a syngenetic ore-mineral content, i.e., criteria suggesting remobilization were not found inasmuch as the textures of the ore and carbonate minerals indicated penecontemporaneous precipitation.

Epigenetic processes should have been evident from relicts of older sulphides in the cementation ores. Cementation by acidic solutions (basic solutions are not suitable for the transport of Cu) is physicochemically impossible in basic rock.

Schlossmacher (1923) was entirely in favour of the epigenetic formation of copper mineralization; this was on the basis of a unilateral ore-microscopic investigation of ore samples from a tectonically faulted area, in parts of which he was able to prove clear cementation paragenesis such as chalcopyrite → bornite → chalcocite, and characteristic displacements of the carbonates by the ore minerals.

Copper cementation in fault zones by descending solutions was also confirmed by Schneiderhöhn (1921, 1926) and Hoffmann (1923/24). The latter states, on the basis of comparison of chemical model tests and ore-microscopic results, that descending acidic solutions containing $CuSO_4$ are the reason for the cementation.

Wienert (1923) dealt also with the physicochemical fundamentals of copper cementation by magnesium chloride solutions, especially with the effects of tectonic pressure-releases, i.e., the recrystallization of chalcopyrite as a result of a change of the chemical equilibrium.

Schouten (1937) and Siegl (1940) tried to argue in favour of the epigenetic-metasomatic formation of the entire non-ferrous metal mineralization of the Kupferschiefer, whereby a syngenetic origin of pyrite and a subsequent metasomatic displacement by copper sulphides was proved microscopically.

After performing microscopic investigations, Hoffmann (1923/24) tried for the first time to explain genetically the metal-impoverishment within the area of the southern flank of the Mansfeld basin, and he writes as follows: "For explaining the presence of the Rote Fäule, it must be assumed that acid ferrisulphate-rich solutions effected (in the bed under red colouring) post-humous remobilization of the ore in such a way that these solutions dissolved the primary copper sulphides and precipitated the same again in other places after the main quantity of the iron content was precipitated in the zones of the Rote Fäule. The Rote Fäule occurs independent of the faults, so that the character of the remobilization remains unexplained".

Additional progress was due to the application of physicochemical analytical methods, especially spectrology, for the determination of the trace-element content of the ore horizon.

Geochemical and genetic comparative investigations were based on detailed micro-chemical analyses, although geochemistry was still only sporadically used as exemplified by the work done by Wagenmann (1926), Cissarz (1929, 1930), Noddack and Noddack (1931), and Kautzsch (1942).

It can be seen from the foregoing that already at the beginning of the nineteen thirties the techniques and methods developed in the geological sciences permitted a complex geological–minerological–geochemical solution for certain problems of the Kupferschie-fer genesis. However, the available data was insufficiently integrated so that the genetic interpretations remained very speculative.

In connection with the great need for non-ferrous metals important for German strate-gic purposes in the nineteen-thirties, the geological reconnaissance of new, exploitable Kupferschiefer areas suddenly gained great importance because of the limited reserves. The problems of the natural occurrence of the metals copper, lead and zinc became the focal point of geological research and considerations. Under direct political and economic influences, the Mansfeld AG took over the reconnaissance of new Kupferschiefer mining areas. Connected with this was the beginning of industrial participation in solving basic geological questions. The basic concept of the geological reconnaissance was developed by Gillitzer (1936) who can thus be considered the first exploration geologist employed in this area.

On the basis of his first comprehensive documentation of the copper content in the Mansfeld basin for the subsurface levels No. 1 to 9, Gillitzer deduced the following basic theses with regard to the metal content of the Kupferschiefer:

(1) All copper concentrations are located on the flanks of Palaeozoic rocks.

(2) All ore-bearing parts of the Mansfeld basin are in areas whose strikes or trends are of Variscan age, accompanied in the southeast by similarly striking uneconomic seams of Rote Fäule.

(3) The main Erzgebirgian seam can be followed in a SW (Sangerhäuser Revier) and also in a NE direction (Golbitzer Revier). Economic deposits can be expected within these areas.

(4) There is a genetic relationship between the distribution of Rote Fäule and that of the parts rich in ore. In its origin, the Rote Fäule was affected by the supply of oxygen from the depositional environment and by the amount of copper and iron from the continental source area(s). During sediment accumulation, the copper accumulation was replaced by ferrisulphate precipitation, but iron compounds reached beyond the Rote Fäule environment into the areas where the copper became economically concentrated.

(5) On the other hand, the copper precipitation of the Kupferschiefer basin took place in an arid-climate bog milieu, which is in contrast to the minor uneconomic accumulation described under 4.

Further progress with regard to the estimation of the metallization of the Kupfer-schiefer came from the studies by Eisentraut (1939), Richter (1941) and Kautzsch (1942). They carried out a large-scale palaeogeographic evaluation of the exploratory borehole information for Kupferschiefer obtained during 1936 to 1940.

Richter recorded the relationships between the metal-containing facies and the original palaeogeographic setting of the beds over- and underlying the ore horizons as well as the regional (i.e., large-scale) thickness variations of the individual units of both the Lower Permian sandstone and Zechstein. However, satisfactory results could not be expected inasmuch as the data was obtained from an insufficient number of localities. Nevertheless, he was able to prove a direct relationship between the occurrence of greater copper concentrations, on one hand, and the rapidly changing thicknesses of the Zechstein, i.e., in channels and depressions within the major sedimentary basin, on the other.

The correlation already established by Fulda (1928) from the ore-rich areas to the adjacent sediment, especially to the Upper Permian sandstone as based on data obtained from cores, was thus confirmed by Richter. The areally widespread vertical zoning of Cu, Pb, Zn was documented for the first time and was supposed to have been the result of shallowing of the sedimentary basin during the sedimentation of the Kupferschiefer. Richter was of the opinion that Cu was precipitated in the deeper parts of the basin and Pb and Zn in the more shallow near-shore environment. The change of the metal content is considered to be related to palaeofacies and not to the variation of the solution supplied, i.e., according to Richter, the solutions, did not exhibit any areal compositional variations at any particular instant in geologic time.

The basic relationship between the palaeomorphology of the surface of the Lower Permian sandstone, the facies and the thickness of the layers which accompany the Kupferschiefer directly in the underlying and overlying beds, and the metal content was also recognized and described by Kautzsch (1942). Like Gillitzer, Kautzsch proved the genetic connection between the ore-rich parts and the distribution of the Rote Fäule. The Rote Fäule impoverishment zones are considered to be sediments of basins of continental, copper-bearing fresh waters on the flanks of which a preferred fixation of rich ores was possible, due to abundant non-ferrous metal supply.

The beginning of a progressive development in the knowledge of the laws of metal fixation in Kupferschiefer and its directly adjacent foot and hanging walls were unfortunately interrupted in World War II due to the limited possibilities for research.

By means of a general geochemical investigation, Krauskopf (1955, 1956), in particular, created the basis for comparing the geochemistry of the black-shale facies, whilst Garrels (1960) and his students developed basic physicochemical diagrams for the precipitation of non-ferrous metal sulphides from marine waters depending on the Eh and the pH conditions. The precisely delineated environmental conditions of the Kupferschiefer with regard to the Eh and the pH given by Tischendorf and Ungethüm (1964, 1965) must also be mentioned here.

Fleischer (1955) compiled the first results of geochemical phase analysis and determined among other things also the typical trace-element concentrations for low and high temperatures.

A comprehensive reconnaissance survey on a large scale was started during the same period around 1955 in the area between Finne and the Halle-Hettstedt mountain bridge which led to the detection of new, exploitable Kupferschiefer areas in the regions of Sangerhausen, Allstedt, and Heldrungen and thus produced a replacement for the nearly exhausted Mansfeld basin. Thus, a new situation arose in the science of mineral deposits, and this was coupled with the fact that a number of existing geological–mineralogical studies of the Kupferschiefer were based only on investigations of small scope, and then only on the basis of a few samples, or on geological and mineralogical data from which only general conclusions could be made. This made it necessary for the geologists of the Mansfeld Kombinat to organize a more comprehensive ore-microscopical geochemical study to carry out first a general review, with the aim of formulating some conclusive geologic synthesis. The investigations of Jung and Knitzsche in the nineteen-sixties and -seventies (Jung, 1960, 1965, 1966, 1968; Jung et al., 1971, 1973a,b, 1974) especially should be mentioned in this connection. They led to a detailed knowledge of the Zechstein Formation and resulted in important concepts related to the geochemical behaviour of the main and trace elements, the development of the metalliferous facies, and the mineralization mechanisms of the ore horizon in the southeastern Harz foreland.

Similar investigations were also carried out on the newly discovered Kupferschiefer deposit of Spremberg. Franz (1965) worked on problems concerning the lithologic and the metal facies, whilst Schüller (1959) and Rentzsch (Rentzsch and Langer, 1963) carried out ore-microscopical investigations. The reconnaissance and documentation of the Kupferschiefer deposit of Lubin-Sieroszowice was undertaken at the same time in the northern foreland of the Sudeten; in this connection the works of Ekiert (1960) and Haranczyk (1961) are of importance. Ekiert developed new thoughts on the Kupferschiefer, whilst Haranczyk dealt especially with the mathematical statistics for deducing geochemical laws that controlled the origin of the Kupferschiefer facies. Ekiert was of the opinion that basic magmatic differentiates with increased Cu-content were formed during the Lower Permian; these were at a considerable depth (approx. 15 km) below the surface of the Lower Permian. In the course of the cooling period up to the beginning of the Zechstein transgression, these magmas formed highly concentrated aqueous hydrothermal solutions. The Variscan zones, already weakened by epeirogenic movements which at the same time caused a marine transgression, were again tectonically stressed and forced to the surface during the Lower Permian. The hydrothermal solutions moved upward along new tectonically formed channels. The main part of the hydrothermal solutions was able to pass directly into the surface water which spread out and at least partly penetrated into the porous Lower Permian sediments, because the epeirogenetically lowered rocks were to a great extent loosened and faulted by larger or smaller dislocations. The metals in the solution were precipitated by the bacterially formed hydrogen sulphide within stagnant waters.[1] Subsequently, an accretive crystallization in the μm-range also took place during diagenesis to form micro-aggregations or concentrations of the metals.

The localization of all exploitable non-ferrous metal deposits of the Kupferschiefer

---

[1] See Chapter 6 by Trudinger, Vol. 2, for a comprehensive review of bacterial processes.

within the vicinity of basement highs, as well as their occurrence within large-scale tectonics of the Hercynian and Erzbergian (i.e., locations offering tectonically produced conduites for ascending hydrothermal fluids), was given as evidence to support the above concept.

This hypothesis is also supported by the fact that hydrothermal activity during the post-Carboniferous period led to the formation of several copper deposits in numerous places in the Eurasian metallogenetic province as, for example, the copper-sandstone deposits of Dseskasgan and the "disseminated copper ores"—porphyry deposit Kuonrad in Central Asia.

Similar mechanisms of hydrothermal supply and synsedimentary precipitation of the metal are also known from the other ages, for example, the Devonian iron, zinc and barite deposit of Meggen and the zinc—lead deposits of the Silesian Trias may be included in this category.

A solution of all the problems regarding the origin of the metals is not as yet available, so that all proposals presented here still have the character of hypotheses.

Another manner of the supply of abnormally high metal contents in the Zechstein ocean or in the bottom sediments of the Zechstein sea was seen by Hoyningen-Huene (1963) in the paleohydrological migration processes. Non-ferrous metal-containing groundwaters above flat-lying Lower Permian sandstones as well as from coastal areas seeped down into the subsurface sections of the porous Lower Permian rocks. The hydrostatic pressure of these intraformational fluids caused them to move horizontally and upwards into the permeable Lower Permian sandstones and conglomerates. Precipitation of non-ferrous metal sulphides took place along the boundary between the sapropelitic, reducing H_2S-rich groundwaters of the Zechstein sea and the acid-metal and oxygen-rich groundwater of the Lower Permian sandstones, depending on the solubility and the H_2S and metal supply. The horizons along which precipitation took place depended on the paleohydrological pressure condition between the sea waters and the groundwaters in the Lower Permian sandstone pressing downwards. On the slope of the basement highs, with a relatively low hydrostatic pressure of the "loading" sea waters, the groundwaters of the Lower Permian sandstones entered already at a moderate hydrostatic pressure from the Lower Permian sediments into the sea water and produced there, due to their acidity and oxygen content, positive Eh-values which led to the precipitation of iron hydroxide. Thus, the Rote Fäule originated, whilst the non-ferrous metals remained in solution and were only precipitated and concentrated along the borders of this zone where accordingly low Eh-values of the sapropel facies are present.

A diffusion-like penetration[1] of the sediments (i.e., of the Kupferschiefer) took place at a very low artesian pressure of the Lower Permian groundwaters. As a result, the syndiagenetic precipitation level (vertical zoning) of the metals with regard to their affinity to the sulphur is within the scope of the alkaline reaction of the pore solutions

---

[1] See Chapter 2 by Duursma and Hoede, Vol. 2, for discussion of diffusion processes.

(NH_3-production from the albumen decomposition in the sapropel).[1]

Some precipitation occurred along the sediment/water interface, whereas the main metal deposition took place at some depth within the sapropel sediment, when finally the hydrostatic pressure of the H_2S-containing sea water became higher than that of the oxygen-containing groundwaters of the Lower Permian layers.

A considerable contribution towards solving basic genetic problems was given by Freese and Jung (1965) with their special documentation concerning the problem of the Rote Fäule. Besides a complete examination of all exposures, the interrelations between Rote Fäule and metal distribution, lithofacies and rock thicknesses were investigated. It was clearly shown that the Rote Fäule originated syngenetically to diagenetically and formed on the flanks of highs and in shallow depressions under specific conditions of Eh-potential, whilst the non-ferrous metals were precipitated in deeper-water environments in the form of sulphides. Rich ores are not always present along the flanks of the Rote Fäule, i.e., the origin of the metals is not directly in genetic connection with the Rote Fäule.

The following evidence, among other, suggests that the Rote Fäule was formed as facies on basement highs: the decrease of thickness and the carbon deficiency of the bed and the confinement of the Rote Fäule to areas with relatively stronger fluctuations in the thickness of the Lower Permian limestone as well as the occurrence of euryhaline coral and brachiopod fauna which are completely absent in the sapropelitic environment. This geological interpretation of the Rote Fäule and the known horizontal zoning Rote Fäule → copper seam → lead–zinc unit, is very well confirmed physicochemically as being of sedimentary origin as already mentioned by Garrels (1960). According to him, these phenomena originated depending on the Eh-gradient and pH-value and the existing convection currents directed downwards along the flanks; furthermore, positive Eh-values were present on the highs and consequently precipitation of Fe-III took place instead of that of non-ferrous metals. Downwards along the flanks, first only the copper sulphides (copper seam) were precipitated at slightly negative Eh-values and then, subsequently, the lead–zinc sulphides (lead–zinc seam) were formed at strongly negative Eh-values.

In spite of the apparent convergence of field observations and physicochemical model tests, it still remains a problem as to how the origin of the metals can be explained. On the basis of the latest geochemical knowledge, Wedepohl (1964) developed a hypothesis for the syngenetic precipitation of non-ferrous metal sulphides in the non-ferrous metal concentrations which are normally present in the sea water. Detrital minerals, residual organic matter, precipitated carbonates and sulphides were brought to the area of sedimentation by the sea water, according to Wedepohl, and accumulated at the start of the bacterial sulphate reduction in the sea water. The quantitative relationship of these sediment components to one another depended on the depth of the basin, water exchange, distance to the coast and supplies from the continent. The quantity of the directly

---

[1] For a review of the role of organic matter in the precipitation of metallic ions, see Chapter 5 by Saxby in Vol. 2.

precipitable sulphides of Cu, Ag, Fe, Pb and Zn in the stagnant sea basin, as well as in the normal sea water, depended especially on the sulphur sulphide production and the amount of water influx.

Large quantities of non-ferrous metal sulphides are only precipitated upon frequent mixing of oxygen- and metal-containing water with the stagnant oxygen- and metal-free, H_2S-containing stagnant waters.

The continuous flow and frequent supply of oxygen-containing waters acted against the formation of anaerobic conditions of the bacterial sulphide production, so that the existing water column was not very large (established from the geomorphology of the Zechstein). The basin depths were of a maximum of a few hundreds of meters, and assuming the rate of exchange to be very slow (several years for a water change), the total amount of water can then be estimated. The concentration of the precipitated sulphides in the sediment depends at the same time greatly upon the dilution by simultaneously accumulating detrital particles, especially carbonate grains. Such a dilution is particularly extensive close to the coast and where a strong continental relief supplies clastic material.

Wedepohl considered that the composition of normal sea water is insufficient to account for the high non-ferrous metal concentration of the Kupferschiefer if a sedimentation period of 1000–10,000 years, a water exchange or removal of 1–10 times per year, and a water depth of 100–300 m are assumed.

Brongersma-Sanders (1965) pointed out the possibility that upwelling along continental margins can form rhythmic or cyclic onshore currents rich in organic nutrients and non-ferrous metals. Upon precipitation, so she reasoned, the latter can form sediments rich in metal and organic matter. Furthermore, the Lower Permian bog copper accumulations may have served as a copper source during the transgression of the Zechstein sea that was accompanied by reworking of the earlier formed deposits.

The divergent genetic view-points expressed by the individual geologists, as well as the necessity to establish a common basis for agreement to further meaningful future research, resulted in a symposium on the Rote Fäule in 1964 to which many leading Kupferschiefer specialists made contributions. The large amount of data, almost without exceptions previously unpublished material from the German and Polish parts of the Kupferschiefer, truly offered for the first time a genuine basis for large-scale regional comparisons related to the complex palaeogeographical, geochemical and mineralogical conditions. Almost unanimous agreement was reached that the Rote Fäule reflects the primary conditions that prevailed in the environment on the basement highs and that the metal-rich facies were the product of an interaction between the Eh-gradient and the supply of metals directly supplied along the basement slopes to the deeper parts of the basin.

Only Borchert (1965) accepted the proposals of Hoyningen-Huene (1963) that epigenetic metal remobilization was involved and emphasized the great importance of the changes that took place as a consequence of Saxonian tectonism and large-scale leaching and metasomatism. The latter was due to alkaline saline groundwater movements within the lower stockworks as well as above the fault systems.

Following a comprehensive accumulation of geological data of the Kupferschiefer in the GDR, the researchers Rentzsch, Jung, Knitzschke and co-workers prepared regional geochemical, geological (delineating lithologies and the palaeography, for example), and metallogenic maps. It is suggested that these offer significant contributions to the formulation of a genetic conceptual model useful in geological prognosis in the exploration for and study of stratified copper–lead–zinc deposits.

Especially Kautzsch (1953) and Messer (1955) considered the problem of the "Rücken" ("faults") in the Mansfeld and Richelsdorf deposits. Both opined that the fault-zone mineralization is of hydrothermal origin, that the different mineral parageneses occurred during several stages, and that the Cu-mineral paragenesis was formed by descending solutions.

Secondary modifications of the primary syngenetic ore by solutions passing along the faults occurred only at certain localities. By comparing sulphur and oxygen-isotope data and employing absolute-age determinations, it was established that the hydrothermal mineralization, by fluids moving along the fault conduits in the Harz and Thuringian forests, are of Saxonian age (Baumann, 1967; Baumann and Rösler, 1967; Baumann and Werner, 1968; Harzer and Pilot, 1969).

Rentzsch et al. (1973) defined concepts concerning the origin of the metals in the Kupferschiefer, and they proposed that the metals were supplied by non-ferrous metal-bearing intraformational waters from the Variscan molasse sedimentary pile and ascended into the surface waters during the Zechstein transgression.

The non-ferrous metal contents of the plutonic and sedimentary rocks of the molasse stockwork and the folded Variscan geosynclinal rocks were determined. It was found that the Cu–Pb–Zn concentrations (expressed as Clarke values) of the plutonic rocks and the sediments, the latter of which originated in an intramontane molasse basin environment, exhibit great similarity within areas having the size of southern Brandenburg or the Lausitz. The lead anomalies at the base of the Zechstein coincide with the areas of lead anomalies of the Permian, so that in this particular instance a genetic relationship was proven to have existed between the non-ferrous metal contents of the magmatites, the molasse sediments and the base of the Zechstein. To explain this, Hoyningen-Huene (1963) and Borchert (1965) proposed that the mobilization, and transportation of the non-ferrous metals from the molasse sedimentary pile into the Zechstein sea, was accomplished by intraformational connate waters.

The restriction of higher non-ferrous metal concentrations to tectonically disturbed localities is explained by the increased mobility of the connate fluids as a consequence of the formation of fault zones and fracture systems.

According to Rentzsch et al. (1973), diagenetic remobilization, dissolution and migration of the solutions from the Rote Fäule (formed as an oxide facies over basement highs) toward the euxinic sapropelic Kupferschiefer, formed upon reprecipitation the important economic concentrations of non-ferrous metals. This secondary mechanism was also suggested by the diagenetic transformation of iron in the same direction, i.e., towards the basin. The iron deficiency, characterizing the chalcocite-rich belt paralleling the Rote Fäule, is accompanied by an increased Fe-content of the bornite zone in the basin.

CONCLUSIONS

As to the origin of the Kupferschiefer in the southern part of the GDR, the following generalized conclusions have been drawn:

(1) The Kupferschiefer consists of three similar petrographic and lithologic cyclic deposits which occur over the entire southern part of the GDR. Each cycle consists of a clayey, very fine-grained sandy base and a carbonate-rich overlying unit. The Kupferschiefer depositional environment was a shallow sea with an euxinic sapropelic facies. Its characteristics, e.g., petrography and thickness variations, were controlled by the surface morphology of the underlying units. The total carbonate content decreases towards the margin of the basin whilst the amount of clay increases.

(2) The Rote Fäule (= "red rot", when translated literally, is due to ion oxide impregnations of syngenetic and diagenetic origin) occurs predominantly in areas of shallow environments and on basement highs as well as along their flanks. This interpretation is supported by the non-ferrous metal impoverishment, hematite mineralization, relative carbon-deficiency of the Kupferschiefer and the decrease in thickness of both the Kupferschiefer and the overlying Zechstein limestone.

(3) All copper concentrations of some significance are within deposits that are time-equivalents of the red-coloured basal Zechstein sediments in the foreland beyond the rise of the basement. There is an interrelationship between the red colouration, on one hand, and the copper precipitation, on the other.

The non-ferrous metal mineralization of the Kupferschiefer originated as syngenetic to diagenetic precipitates, especially as non-ferrous metal sulphides. As a consequence of the regional Eh-gradient present in the marine sedimentary milieux and the accompanying variation in the H_2S, precipitation of the non-ferrous metal sulphides was in accordance to their different solubilities, thus forming mineral and metal zonation. The differences in the palaeo-relief of the depositional environment was a prerequisite for the establishment of the Eh-gradient along the ocean floor. High Eh-potentials correspond to the top of basement highs which are free from non-ferrous metals, whereas deposits that originated along the slope from the highs accumulated under slightly negative to neutral Eh-values leading to predominantly Cu-rich sulphide precipitates. Strongly negative Eh in the surface milieux lead to Pb–Zn mineralization.

Reversal of the topographic relief during geologic evolution of the depositional area and while sulphide accumulation occurred, gave rise to an increase of the Eh-potential and a consequent syndiagenetic migration of non-ferrous metal-containing fluids. The direction of movement was controlled by the established gradient, and reprecipitation from the migrated solutions was determined by the newly encountered lower Eh-values. The horizontal and vertical sequence of Cu → Pb → Zn was formed by a normal marine transgression, and any deviation from such an ideal sequence, especially in localities of interbedding of the sapropel sediments and the Rote Fäule, are the result of regressive tendencies.

The chemical agents that were responsible for the precipitation of the metals were H_2S-released during decomposition of organic matter, but in particular from the reduction of sulphates during the metabolism of thio-bacteria under stagnant euxinic conditions (see Chapters 6 and 5 by Trudinger and Saxby, respectively, Vol. 2). The required low Eh-value was favoured by the de-carbonization processes. The present amount of organic matter can be considered to be an indication of the effectiveness of oxydation during syndiagenesis.

(4) Ten types of paragenetic ore mineralization sequences were established for the Kupferschiefer deposits which reflect the original Eh–pH conditions and metal concentrations. Of particular importance was the Fe/Cu ratio during the precipitation of the various Cu-containing minerals. Lateral changes of the paragenetic sequences within the facies from the basement high to the deepest part of the basin are as follows: hematite type → chalcocine type → bornite/chalcocine type → bornite type → bornite/chalcopyrite type → galena/chalcopyrite type → sphalerite type → pyrite type. This sequence corresponds to the solubility products of the sulphides listed above and a decreasing Eh-value within the sedimentary milieux. Thus, the metal-facies belt of the Kupferschiefer and its zonation is based on a geochemical differentiation. The syngenetic–syndiagenetic concepts developed are, therefore, supported by the synthesis of geological and physico-chemical model studies.

The large-scale comparative work related to tectonism and the localization of the economic Kupferschiefer deposits has given new insight into the origin of the metals as well as into their source on the continent. As a result of the restriction of exploitable Kupferschiefer deposits to tectonically reworked zones of the Variscan and Hercynian tectonics, additional or supplemental metal sources are suggested: (a) submarine emanations or exhalations of hydrothermal solutions into the early Upper Permian (Zechstein); (b) non-ferrous metal-carrying intraformational connate waters of the Late Variscan molasse stockwork which emerged on the sea bottom during the transgression of the Zechstein sea (Rentzsch et al., 1973).

REFERENCES

Baumann, L., 1967. Zur Frage der varistischen und post-varistischen Mineralisation im sächsischen Erzgebirge. *Freiberg. Forschungsh., C,* 209: 15–38.
Baumann, L. and Rösler, H.J., 1967. Zur genetischen Einstufung varistischer und post-varistischer Mineralisation in Mitteleuropa. *Bergakademie,* 19: 660–664.

Baumann, L. and Werner, C.D., 1968. Die Gangmineralisation des Harzes und ihre Analogien zum Erzgebirge und zu Thüringen. *Ber. Dtsch. Ges. Geol. Wiss., Reihe B*, 13:528–548.

Beyschlag, F., 1900. Beitrag zur Genesis des Kupferschiefers. *Z. Prakt. Geol.*, 8: 115–117.

Beyschlag, F., 1920. Zur Frage der Entstehung des Kupferschiefers. *Z. Dtsch. Geol. Ges.*, 72: 318–328.

Borchert, H., 1965. Fazieswechsel in den Zechstein-Salzlagerstätten in ihrer Beziehung zur Entwicklung der "Roten Fäule" des Kupferschiefers und deren Entstehung. *Freiberg. Forschungsh., C,* 193: 197–226.

Brongersma-Sanders, M., 1965. Metals of Kupferschiefer supplied by normal sea water. *Geol. Rundsch.*, 55: 365–375.

Cissarz, A., 1929. Die Metallverteilung in einem Profil des Mansfelder Kupferschiefers. *Centralbl. Mineral., Abt. A*, pp. 425–427.

Cissarz, A., 1930a. Quantitativ-spektralanalytische Untersuchungen eines Mansfelder Kupferschiefer-profils. *Chem. Erde*, 5: 48–75.

Cissarz, A., 1930b. Die durchschnittliche Zusammensetzung des Mansfelder Kupferschiefers. *Metall Erz*, 27: 316–319.

Correns, C.W., 1924. Adsorptionsversuche mit verdünnten Kupfer- und Bleilösungen und ihre Bedeutung für die Erzlagerstättenkunde. *Kolloid-Z.*, 34: 341–349.

Dunham, K.C., 1961. Black shale, oil and sulfid ore. *Adv. Sci.*, 18: 284–299.

Dunham, K.C. and Hirst, D.M., 1963. Chemistry and petrography of the Marl slate of southeastern Durham, England. *Econ. Geol.*, 58: 912–940.

Eisenhuth, K.H. and Kautzsch, E., 1954. *Handbuch für den Kupferschieferbergbau.* Fachbuch, Leipzig.

Eisentraut, O., 1939. Der niederschlesische Zechstein und seine Kupferlagerstätte. *Arch. Lagerstätten-forsch.*, 71.

Ekiert, F., 1960. Neue Anschauungen über die Herkunft des in den Sedimenten des Unteren Zechsteins auftretenden Kupfers. *Freiberg. Forschungsh., C*, 79: 190–201.

Erdmann, E., 1919. Diskussion zu Geipel: über die Rücken in Mansfeld. *Jahrb. Hallesches Verb.*, 1: 27–29.

Erdmann, K., Von Wolff, F., Lier, K. and Walther, J., 1919. Diskussionen über die Entstehung des Mansfeldschen Kupferschiefers. *Jahrb. Hallesches Verb.*, 1: 29–40.

Erzberger, R., Franz, R., Jung, W., Knitzschke, G., Langer, M., Luge, J., Rentzsch, H. and Rentzsch, J., 1968. Lithologie, Paläogeographie und Metallführung des Kupferschiefers in der Deutschen Demokratischen Republik. *Geologie*, 17: 776–791.

Fleischer, M., 1955. Minor elements in some sulfide minerals. *Econ. Geol.*, 50: 970–1024.

Franz, R., 1965. Metallfazies und Rote Fäule im Unteren Zechstein bei Spremberg-Weisswasser. *Freiberg. Forschungsh., C*, 193: 41–54.

Freese, C. and Jung, W., 1965. Über die Rotfärbung des Basaltschichten des Zechsteins (Rote Fäule) und ihre Beziehungen zum Nebengestein im südöstlichen Harzvorland. *Freiberg. Forschungsh., C,* 193: 9–23.

Freygang, J., 1923/24. Gliederung und Fossilgehalt des Kupferschiefers. *Jahrb. Hallesches Verb.*, 4: 183–192.

Fulda, E., 1928. Zum Problem des Kupferschiefers. *Jahrb. Preuss. Geol. Landesanst.*, 49: 995–1002.

Garrels, R.M., 1960. *Mineral Equilibria.* New York, N.Y.

Geipel, M., 1919. Über die Rücken im Mansfeldschen Revier und ihren Einfluss auf die Erzführung des Kupferschiefers. *Jahrb. Hallesches Verb.*, 2: 21–40.

Gillitzer, G., 1936. Die Geologie der Erzanreicherungen im mitteldeutschen Kupferschiefer. *Jahrb. Hallesches Verb.*, 15: 1–19.

Haranczyk, C., 1961. Correlation between organic carbon, copper and silver content in Zechstein copper-bearing shales from the Lubin-Siereszowice Region (Lower Silesia). *Bull. Acad. Pol. Ser. Sci. Geol. Geogr.*, 9.

Harrassowitz, H., 1922. *Aride Erzanreicherungen und die Entstehung des Kupferschiefers.* Bonn.

Harzer, D. and Pilot, J., 1969. Isotopengeochemische Untersuchungen an Ganglagerstätten des Harzes. *Ber. Dtsch. Ges. Geol. Wiss., Reihe B*, 14: 129–138.

Hoffmann, W., 1923/24. Erzführung und Erzverteilung des Mansfelder Kupferschiefers und die sich hieraus ergebenden mineralbildenden und umbildenden Vorgänge im Kupferschiefer. *Jahrb. Hallesches Verb.*, 4: 278–324.

Jung, W., 1960. Die Sedimentationsverhältnisse während des Oberrotliegenden und Zechsteins im SE-Harzvorland. Einige Bemerkungen zu Arbeiten von E. Kautzsch und H. Steinbrecher. *Z. Angew. Geol.*, 6: 598–604.

Jung, W., 1965. Zum subsalinaren Schollenbau im südöstlichen Harzvorland (Mit einigen Gedanken zur Äquidistanz von Schwächezonen). *Geologie*, 14: 254–271.

Jung, W., 1966. Zechstein. In: *Grundriss der Geologie der D.D.R.* Akademie, Berlin, pp. 219–237.

Jung, W., 1968. Über Gesteinstypen, Faziesdifferenzierungen und zyklisch-rhythmische Sedimentation im deutsch-polnischen Zechstein. *Proc. 23rd Int. Geol. Congr.*, 8: 211–225.

Jung, W., Knitzschke, G. and Gerlach, R., 1971. Entwicklungsgeschichte der geologischen Anschauungen über den Mansfelder Kupferschiefer. *Geologie*, 20: 462–484.

Jung, W., Knitzschke, G. and Gerlach, R., 1973a. Zur Selenführung des Kupferschiefers in SE-Harzvorland. *Z. Angew. Geol.*, 19: 57–67.

Jung, W., Knitzschke, G. and Gerlach, R., 1973b. Zur Kadmium- und Thalliumführung des Kupferschiefers im SE-Harzvorland. *Z. Angew. Geol.*, 19: 1047–1053.

Jung, W., Knitzschke, G. and Gerlach, R., 1974. Zur geochemischen Stoffbilanz des Kupferschiefers im SE-Harzvorland. *Z. Angew. Geol.*, 20: 248–256.

Kautzsch, E., 1942. Untersuchungsergebnisse über die Metallverteilung im Kupferschiefer. *Arch. Lagerstättenforsch.*, 74.

Kautzsch, E., 1953. Tektonik und Paragenese der Rücken im Mansfelder und Sangerhäuser Kupferschiefer. *Geologie*, 2: 4–24.

Knitzschke, G., 1965. Die wichtigsten Erzminerale des Kupferschiefers sowie seines unmittelbaren Liegenden und Hangenden im südöstlichen Harzvorland. *Z. Angew. Geol.*, 11: 626–637.

Knitzschke, G., 1966. Zur Erzmineralisation, Petrographie, Hauptmetall- und Spurenelementführung des Kupferschiefers im SE-Harzvorland. *Freiberg. Forschungsh., C*, 207: 1–147.

Köhler, G., 1905. *Die Rücken in Mansfeld und in Thüringen*. Engelmann, Leipzig, 29 pp.

Konstantynowicz, E., 1965. Fleckenmergel und Erzführung der Zechsteinsedimente in Dolny Slask (V.R. Polen). *Freiberg. Forschungsh., C*, 193: 95–99.

Krauskopf, K.B., 1955. Sedimentary deposits of rare metals. *Econ. Geol.*, 50: 411–463.

Krauskopf, K.B., 1956. Factors controlling the concentrations of thirteen rare metals in sea water. *Geochim. Cosmochim. Acta*, 9: 1–32.

Krusch, P., 1919. Die Verteilung der Metallgehalte (Kupfer, Silber, Molybdän und Vanadin) im Richelsdorfer Kupferschiefer — ein Beitrag zur Genesis des Flözes. *Z. Prakt. Geol.*, 27: 76–84.

Lang, R., 1921. Der mitteldeutsche Kupferschiefer als Sediment und Lagerstätte. *Jahrb. Hallesches Verb.*, 3: 1–135.

Langer, M., 1963. Über die erzmikroskopischen Untersuchungen des Kupferschiefers und seines unmittelbaren Liegenden im Raum Spremberg-Weisswasser. *Z. Angew. Geol.*, 9: 449–452.

Ludwig, H. and Rentzsch, J., 1967. Das Sanderzproblem—ein Beitrag zur Genese des Kupferschiefers. *Ber. Dtsch. Ges. Geol. Wiss., Reihe B*: 3–12.

Messer, E., 1955. Kupferschiefer, Sanderz und Kobaltrücken im Richelsdorfer Gebirge. *Hess. Lagerstättenarch.*, 3.

Noddack, J. and Noddack, W., 1931. Die Geochemie des Rheniums. *Z. Phys. Chem. Abt. A*, 154: 207–244.

Oelsner, O., 1959. Bemerkungen zur Herkunft der Metalle im Kupferschiefer. *Freiberg. Forschungsh., C*, 58: 106–113.

Pompecki, J.F., 1914. Das Meer des Kupferschiefers. *Branca-Festschr.*, pp. 444–494.

Rankama, K. and Sahama, T.G., 1950. *Geochemistry*. Chicago University Press, Chicago, Ill., 912 pp.

Rentzsch, J., 1964. Der Kenntnisstand über die Metall- und Erzmineralverteilung im Kupferschiefer. *Z. Angew. Geol.*, 10: 281–288.

Rentzsch, J., 1965. Fazielle Gesetzmässigkeiten beim Auftreten der Roten Fäule. *Freiberg. Forschungsh., C,* 193: 99–106.

Rentzsch, J. and Knitzschke, G., 1968. Die Erzmineralparagenesen des Kupferschiefers und ihre regionale Verbreitung. *Freiberg. Forschungsh., C,* 231: 189–211.

Rentzsch, J. and Langer, M., 1963. Fazielle Probleme des Kupferschiefers von Spremberg-Weisswasser. *Z. Angew. Geol.,* 9: 507–513.

Rentzsch, J., Schirmer, B. and Röllig, G., 1973. Zur Metallherkunft und Genese der Buntmetallvererzungen an der Zechsteinbasis, Typ Kupferschiefer. (Vortrag, gehalten auf der Tagung der GGW der DDR in Erfurt.)

Richter, G., 1941a. Geologische Gesetzmässigkeiten in der Metallführung des Kupferschiefers. *Arch. Lagerstättenforsch., Neue Folge,* 73.

Richter, G., 1941b. Paläogeographisch bedingte Eigentümlichkeiten im Metallgehalt des Kupferschiefers zwischen Kyffhäuser und Harz. *Z. Prakt. Geol.,* 49: 113–124.

Richter, G., 1947. Paläogeographische Grundlagen für die Erschliessung des deutschen Kupferschiefers. *Dtsch. Tech.,* 1947: 366–368.

Rösler, H.J. and Lange, H., 1965. *Geochemische Tabellen.* Grundstoffindustrie, Leipzig, 328 pp.

Rydzewski, A., 1965. Die Petrographie und die Vererzung der Zechsteinsedimente in der vorsudetischen Monokline und der Periklinale von Zary (V.R. Polen). *Freiberg. Forschungsh., C,* 193: 157–168.

Schlossmacher, K., 1923. Die sekundäre Erzmineralparagenesis des Kupferschiefers. *Centralbl. Mineral.,* 9: 257–264.

Schneiderhöhn, A., 1921. Chalkographische Untersuchung des Mansfelder Kupferschiefers. *Neues Jahrb. Mineral.,* 41: 1–38.

Schneiderhöhn, A., 1926. Erzführung und Gefüge des Mansfelder Kupferschiefers. *Metall Erz,* 23: 143–146.

Schouten, C., 1937. *Metasomatische Probleme.* Amsterdam.

Schüller, A., 1959. Metallisation und Genese des Kupferschiefers von Mansfeld. *Abh. Dtsch. Akad. Wiss. Berlin, Kl. Chem., Geol., Biol.,* 6.

Siegl, W., 1940. Zur Genesis des Kupferschiefers. *Mineral. Petrogr. Mitt.,* 52: 347–362.

Tischendorf, G., 1966. Zur Verteilung des Selens in Sulfiden. *Freiberg. Forschungsh., C,* 208.

Tischendorf, G. and Ungethüm, H., 1964. Über die Bedeutung des Reduktions-Oxydationspotentials (Eh) und der Wasserstoffionenkonzentration (pH) für Geochemie und Lagerstättenkunde. *Geologie,* 13: 125–158.

Tischendorf, G. and Ungethüm, H., 1965. Zur Anwendung von Eh-pH-Beziehungen in der geologischen Praxis. *Z. Angew. Geol.,* 11: 57–67.

Trelease, S.F. and Beath, O.A., 1949. Selenium, its geological occurrence and its biological effects in relation to botany, chemistry, agriculture, nutrition and medicine.

Von Freyberg, B., 1923/24. Paläogeographische Karte des Kupferschieferbeckens. *Jahrb. Hallesches Verb.,* 4: 266–278.

Von Hoyningen-Huene, E., 1963. Zur Paläohydrologie des Oberrotliegenden und des Zechsteins im Harzvorland. *Ber. Geol. Ges. DDR, S-H.,* 1: 201–220.

Von Wolff, F., 1922. Der Kupfergehalt der rotliegenden Eruptivgesteine Mitteldeutschlands. *Jahrb. Hallesches Verb.,* 3: 135–139.

Wagenmann, K., 1926. Einige Grundlagen und wesentliche Gesichtspunkte zur Frage einer günstigeren Verarbeitung Mansfelder Minern unter besonderer Berücksichtigung eines Aufbereitungsprozesses. *Metall Erz,* 24: 149–154.

Walther, J., 1910. *Geschichte der Erde und des Lebens.* Geestmann und Portig, Leipzig.

Walther, J., 1919. Diskussionsbeitrag zu Geipel "Die Rücken im Mansfeldschen . . . Begründung der Syngenese". *Jahrb. Hallesches Verb.,* 2: 38–40.

Walther, J., 1921. *Geologie von Deutschland.* Geestmann und Portig, Leipzig.

Wedepohl, K.H., 1964. Untersuchungen am Kupferschiefer in Nordwestdeutschland, ein Beitrag zur Deutung der Genese bituminöser Sedimente. *Geochim. Cosmochim. Acta,* 28: 305–364.

Wedepohl, K.H., 1969. *Handbook of Geochemistry*. Springer, Berlin—Heidelberg—New York.

Wienert, F., 1923. Die Bildungsbedingungen der sulfidischen Kupfer- und Eisenerze, mit besonderer Berücksichtigung des mitteldeutschen Kupferschiefers. *Jahrb. Hallesches Verb.*, 4: 192—224.

Wojciechowska, J. and Serkies, J., 1968. Selenium in the foresudetic copper deposit (region of Lubin). *Bull. Acad. Pol. Sci.*, 17: 91—96.

Chapter 8

SABKHA AND TIDAL-FLAT FACIES CONTROL OF STRATIFORM COPPER DEPOSITS IN NORTH TEXAS

GARY E. SMITH

INTRODUCTION

Stratiform copper mineralization occurs in three general stratigraphic zones in the Permian of North Texas, Oklahoma, and southern Kansas (Fig. 1). At the Creta Copper Mine in southwestern Oklahoma, a 6-inch (15.0 cm) layer of chalcocite-bearing gray shale is currently being mined by open-pit mining methods from the Mid-Permian Flowerpot Shale. In North Texas, copper occurs in the lower parts of the Pease River and Clear Fork Groups (Fig. 2), and in the lower part of the Wichita—Albany Group (Barnes, 1972).

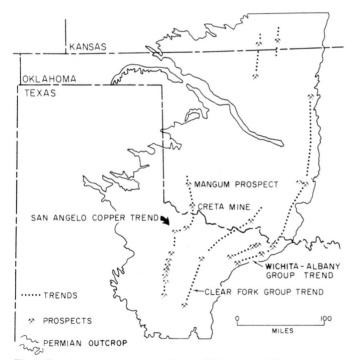

Fig. 1. Permian copper prospects, Texas, Oklahoma and Kansas.

SERIES	GROUP	NORTH TEXAS		SOUTHWEST OKLAHOMA	
		FORMATIONS AND MEMBERS		GROUP	FORMATIONS
GUADALUPIAN	PEASE RIVER	DOG CREEK SHALE		EL RENO	DOG CREEK SHALE
		BLAINE FORMATION			BLAINE FORMATION
		SAN ANGELO FORMATION	FLOWERPOT MEMBER		FLOWERPOT SHALE
			DUNCAN MEMBER		DUNCAN SANDSTONE
LEONARDIAN	CLEAR FORK	CHOZA FORMATION		HENNESSEY SHALE	
		VALE FORMATION			
		ARROYO FORMATION			

Fig. 2. Mid-Permian stratigraphic section, North Texas and southwestern Oklahoma.

Mineralization in the Pease River Group is restricted to the Duncan and Flowerpot Members of the San Angelo Formation and the basal part of the Blaine Formation (Fig. 3). Copper occurs principally in localized channel-fill sandstone lenses associated with masses of woody material, and in laterally persistent shale beds. It rarely occurs in thin dolomite beds, at the base of gypsum layers, and as disseminations or coatings associated with thin ripple-bedded sandstone units.

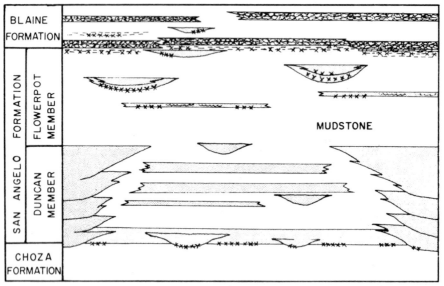

Fig. 3. Distribution of copper minerals, San Angelo Formation, North Texas. X = copper mineral zones.

The contents of this chapter dwell on the mode of occurrence and origin of copper mineralization in, and associated with, the San Angelo Formation. The study included an integration of surface observations, measured sections, subsurface information from oil-well electric logs, and petrologic data from thin and polished sections (Smith, 1974). The San Angelo outcrop area investigated extends over a linear distance of 135 km in North Texas (Fig. 4) and was chosen to include all known San Angelo copper occurrences in Texas. Subsurface study extended 32 km southward and 42 km westward beyond the limits of the surface investigation. An understanding of the origin and host rock rela-

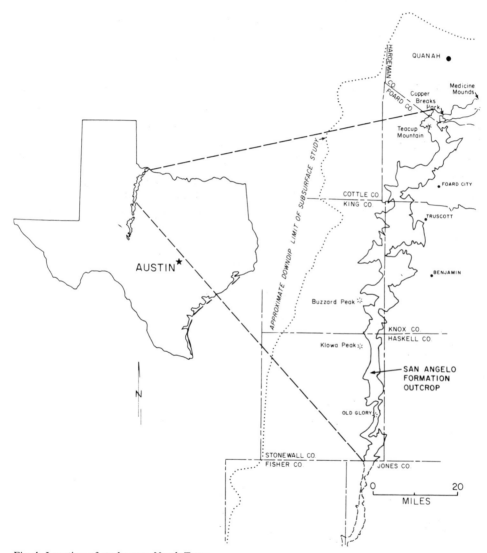

Fig. 4. Location of study area, North Texas.

tionships of the stratiform copper mineralization requires a thorough understanding of the facies relationships and depositional history of the San Angelo and Blaine Formations. For that reason, an important segment of this paper will be concerned with a review of the depositional systems and lithologic facies that comprise the San Angelo and lower part of the Blaine.

Regional setting

The San Angelo Formation is a Mid-Permian wedge of terrigenous clastic sediments of diverse composition that can be traced westward into the Midland Basin where it intertongues with red mudstone and evaporite beds. Along strike, the San Angelo is characterized by numerous facies changes that reflect complex relationships between diverse depositional environments.

The San Angelo Formation was deposited as a progradational unit that built westward across a restricted shelf which formed the eastern margin of the Midland Basin (Fig. 5). By Early Guadalupian time the Eastern Shelf of the Midland Basin was shallow and lacked a distinct basinward depositional shelf-edge (Oriel et al., 1967). Restricted shelf refers to a specific depositional environment characterized by hypersaline conditions with a limited biota.

East of the San Angelo shoreline was a wide coastal plain bordered on the east by the Ouachita fold belt (Flawn et al., 1961). This complex of Paleozoic rocks had been

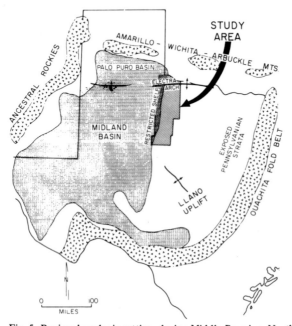

Fig. 5. Regional geologic setting, during Middle Permian, North Texas and southwestern Oklahoma.

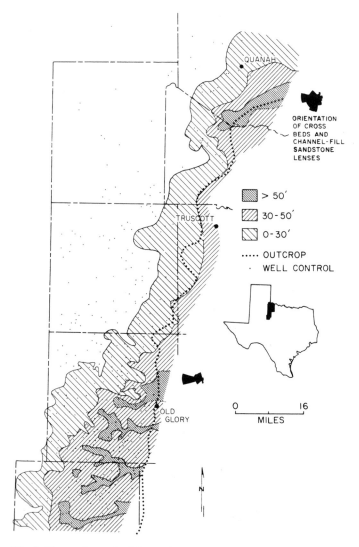

Fig. 6. Net sandstone isolith map, San Angelo Formation. Rose diagrams, based on measurements of cross-bedding and channel-fill directions, indicate paleocurrent directions.

extensively eroded, but was probably still shedding some sediments westward. North of the study area in the Texas panhandle and southwestern Oklahoma, parts of the Amarillo—Wichita—Arbuckle Mountain system were still exposed. The Wichita Mountains may have been buried or greatly subdued by this time (Ham and Johnson, 1964); the Arbuckle Mountains were probably supplying some sediments to the northern San Angelo coastal plain. The presence of the Arbuckle-derived Chickasha Formation in Oklahoma (Green,

1937), and an average S45°W paleocurrent direction in the northern San Angelo outcrop (Fig. 6) supports an Arbuckle source during San Angelo deposition.

The Electra Arch, an elongate, E—W positive structural axis that traverses the southern half of Foard County (Fig. 4) exerted some effect upon San Angelo deposition by causing depositional thinning in the vicinity of the arch. The San Angelo dips westward at about 15 ft/mile (7.2 m/km) throughout most of the study area.

STRATIGRAPHY

In this study the San Angelo has been divided into two informal members: (1) the basal Duncan Sandstone Member, which is from 14 to 21 m thick; and (2) the upper Flowerpot Mudstone Member, which is from 8.4 to 19 m thick (Fig. 7). The Duncan Member of the San Angelo Formation lies conformably on the underlying Choza Formation, although in places channeling during deposition of the Duncan Member resulted in local unconformities. The contact with the overlying Flowerpot Member is transitional and is characterized by an upward decrease in sandstone content and a concurrent increase in mudstone. The contact between the Flowerpot Member and the overlying Blaine Formation is gradational both upwards and downdip. For mapping purposes the base of the first regionally widespread, approximately 30.5 cm (1 ft) thick gypsum bed was defined as the base of the Blaine Formation. Descriptions on lithologies present in these stratigraphic units are included in Table I.

DEPOSITIONAL SYSTEMS

The San Angelo Formation (Fig. 8) is a sequence comprised of the following depositional systems: (1) Old Glory fluvial—deltaic system; (2) Copper Breaks deltaic system; (3) Buzzard Peak sand-rich tidal-flat system; and (4) Cedar Mountain mud-rich tidal-flat system. The upper 9.2 m of the underlying Choza Formation examined in this study is interpreted to represent a depositional system composed of subtidal shelf, mud-rich tidal-flat, and sabkha facies that existed in the area prior to initiation of San Angelo deposition. The lower 30.8 m of the overlying Blaine Formation is likewise interpreted to represent a system of sabkha and mud-rich tidal-flat facies in the area covered by this study. These depositional systems are comprised of complex facies associations deposited within sedimentary environments that changed both spatially and temporally. Each facies displays distinctive rock types and sedimentary characteristics (Table I), which are a function of depositional processes that were active in Late Leonardian and Early Guadalupian depositional environments of North Texas.

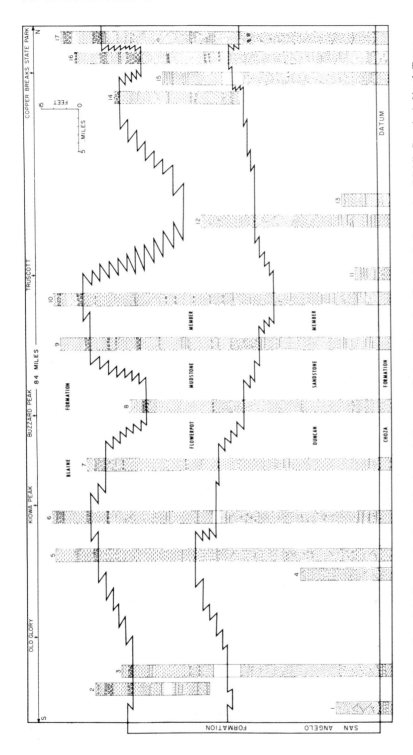

Fig. 7. Strike section of the Duncan Sandstone Member and Flowerpot Mudstone Member, San Angelo Formation (Middle Permian), North Texas.

TABLE I

Summary of stratigraphic and facies characteristics

FORMATIONS AND MEMBERS	DEPOSITIONAL SYSTEM	FACIES	LITHOLOGIC DESCRIPTION	SEDIMENTARY STRUCTURES AND BEDDING TYPES	BIOGENIC FEATURES
Choza Formation (upper 9.2 m)	Subtidal shelf tidal-flat and sabkha	Subtidal, tidal-flat and sabkha (undiff. in study)	Red and gray mudstone (d); siltstone, very fine grained sandstone, gypsum beds and nodules, and dolomitic mudstone (m).	Massive to horizontal bedded mudstone and siltstone (d); lenticular, flaser bedded sandstone bodies (m); mud-cracks present.	✕
Duncan Sandstone Member	Copper Breaks Deltaic System	Prodelta	Reddish brown mudstone and sandy mudstone.	Thin to medium, parallel laminated beds.	No burrowing
		Delta front	White, fine grained, well to moderately sorted quartz-arenite and sub-litharenite; CO_3 cement (d).	Horizontal bedding with thin clay laminae separating thicker sand laminae.	No burrowing
		Distributary Mouth bar	White, fine grained, well sorted quartzarenite. Carbonate cementation is dominant with minor silica.	Horizontal bedding (d); medium-scale trough cross bedding and climbing ripple cross-stratification (m).	No burrowing
		Distributary channels	Reddish brown to tan, fine grained, well to moderately well sorted sub-litharenite and quartz-arenite. Cementation: carbonate and silica.	Medium to large scale trough cross bedding (d); small scale ripple cross-stratification, horizontal bedding and climbing ripples (m).	Burrowing is common
		Delta plain and associated facies	Reddish brown siltstone, mudstone, and thin sandstone units.	Horizontal and trough cross bedding are dominant and are facies and process dependant.	Vertebrate bones present
Flowerpot Mudstone Member	Old Glory Fluvial-Deltaic system		Characteristic facies are similar to those in the Copper Breaks system except for the introduction of pebbles of metamorphic quartz and rock fragments and a scarcity of quartz-arenites in the distributary channel facies. There is also a general increase in the size of trough cross-stratification.		
	Buzzard Peak Sand-rich Tidal-flat System	Tidal sand-flat	Very fine to fine grained, well sorted, white to tan quartz-arenite with weak silica cement.	Ripple cross-stratification, flaser and horizontal bedding (d); current lineations and clay laminae (m).	Minor burrowing
		Tidal channels	Similar to tidal channels described in Cedar Mountain mud-rich tidal-flat system (see below) except for a predominance of trough cross-stratification and the presence of calcite and silica cementation.		Minor burrowing
	Cedar Mountain Mud-rich Tidal-flat System	Tidal mud-flat	Reddish brown mudstone (d); variable silt and sand content (m). Gray sandy mudstone and discoidal gypsum and barite nodules (m).	Thin, parallel laminated beds to massive beds. Beds break into small (1/2 to 1-inch) blocks.	Vertebrate bones present
		Tidal channels: Lower segment	Very fine grained, gray, muddy sandstone to sandy mudstone.	Indistinct, small-scale ripple cross-stratification.	Carbonaceous material
		Tidal channels: Middle segment	Very fine grained, gray quartz-arenite; well cemented by gypsum.	Medium to large-scale trough cross-stratification and horizontal bedding (d); flaser bedding (m)	Minor carbon. material
		Tidal channels: Upper segment	Very fine grained, gray quartz-arenite; well cemented by gypsum.	Sequence (bottom to top - idealized): bimodal bedding (r), low angle trough cross-stratification (m), horizontal bedding (m) and flaser bedding (d).	Vertebrate bones (?)
		Swash-zone	Light gray, well sorted, very fine quartz-arenite with gypsum cement.	Small scale ripple cross-stratification.	none
		Algal mat	Dolomite (dolomicrite, oölitic dolomite, and intra-clastic dolomite), dolomitic shale, and well laminated gray shale.	Dolomite is thin bedded and exhibits mud cracks, syneresis cracks (r), mud clasts, and crenulations (m); flaser bedding (d). Graded laminae are present in the dolomitic shale.	Cephalopod shells (rare-storm derived)
Blaine Formation	Blaine Sabkha and Tidal-flat System	Sabkha	Nodular gypsum in discrete beds.	Nodular texture with thin red or gray clay coatings separating tne nodules (d); enterolithic bedding (m).	None
		Tidal-flat (some undiff. sabkha sediments present)	Pale reddish brown and gray mudstone, claystone and gray dolomite (d); fine grained, gray, well sorted quartz-arenite present as lenticular channel-fill (m)	Salt hopper casts (m).	✕

SAN ANGELO FORMATION

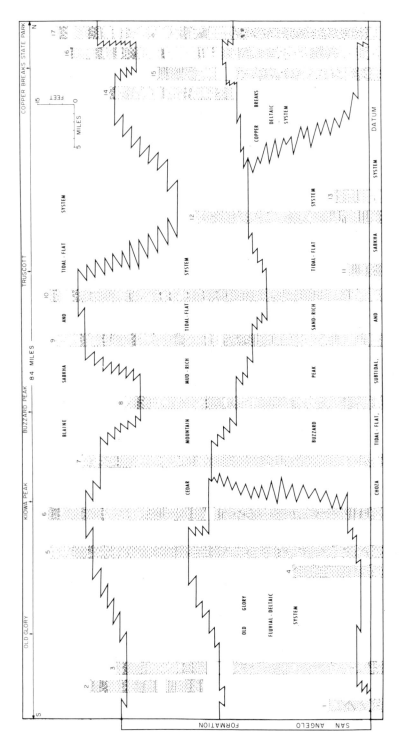

Fig. 8. Strike section of the uppermost Choza, San Angelo and basal Blaine Formations (Middle Permian) with inferred depositional environments.

Copper Breaks and Old Glory fluvial–deltaic systems

Fluvial–deltaic facies develop when terrigenous clastic sediments are deposited in an open body of water by a fluvial channel. The delta plain is a complex of interrelated distributary channel, natural levee, interdistributary marsh and bay, and crevasse splay sediments. Distributary channel-fill sand deposits form the framework of this facies complex and mud and silt facies of the delta plain serve to fill in interdistributary areas. The sands and silts of delta-front and distributary-mouth-bar facies form basinward of the delta plain complex and represent areas of extensive deposition and, in some cases, reworking of sediments being transported through the fluvial–deltaic network. Prodelta mud deposits are gradational upward and landward with the delta-front facies; these muds and silts form distally on the marine shelf as a result of deposition of suspended load introduced by the fluvial system.

Continuous fluvial–deltaic sedimentation results in progradation or regression of fluvial–deltaic facies. Progradation leads to development of a vertical sequence composed (upward) of lowermost prodelta facies overlain by delta-front and distributary-mouth-bar facies, and capped by facies of the delta plain complex (Scruton, 1960; Coleman and Gagliano, 1964).

Fluvial–deltaic facies of the Duncan Sandstone Member are interpreted to have been deposited by high constructive lobate deltas (Fisher et al., 1969), based on analysis of subsurface sand trends (Fig. 6) and outcrop examination of the deltaic facies.

Copper Breaks deltaic system

The Copper Breaks deltaic system crops out from Medicine Mounds southwestward into the Teacup Mountain area where the deltaic sediments disappear into the subsurface (Fig. 4). The Copper Breaks system displays in outcrop a spectrum of facies from upslope upper-delta plain to downslope deltaic. Deltaic facies, in addition, fit well into a tract from delta plain through delta front to distal prodelta facies, which, because of the progradational nature of the deltaic system, has resulted in a vertical sequence of superposed deltaic facies.

Prodelta facies. Prodelta facies occupy a thin zone about 1.85 m thick beneath delta-front and distributary-mouth-bar facies; prodelta mudstone facies rest conformably on the tidal-flat facies of the Buzzard Peak system. The well bedded appearance and lack of gypsum nodules and selenite seams in the prodelta facies make it possible to differentiate it from underlying mudstone facies of the Choza Formation and the Buzzard Peak system.

Delta front and distributary mouth bar facies. Sandstone of the delta-front and distributary-mouth-bar facies conformably overlies mudstone of the prodelta facies. This facies is about 3.1 m thick and is composed of closely spaced distributary-mouth-bar deposits and laterally reworked delta front sheet sands which form a sheetlike body. A characteristic feature of the delta-front sandstone is the presence of thin clay laminae separating thicker, horizontal sand laminae. On a shallow restricted shelf, wind-generated, as well as astronomic tides, may have periodically reworked the delta-front sands; clay laminae may represent deposition during periods of very low physical energy.

Distributary channel and associated delta plain facies. Distributary channel-fill sandstone facies form the bulk of the sediments in the Copper Breaks system. Individual lenticular sandstone units of the distributary channel-fill facies up to 9.2 m thick and 61.5 m wide were observed on outcrop. Channel deposits have erosional bases, and they normally display evidence of reoccupation and partial erosion by later streams. Contorted, flow-roll sandstone structures, are common throughout this facies, indicating sediment overloading and subsequent subsidence and soft-sediment deformation. Distributary channel-fill sandstone facies grade laterally and upward into poorly developed siltstone, mudstone, and thin sandstone units that constitute other delta plain facies, such as interdistributary, crevasse-splay and levee deposits.

Old Glory fluvial–deltaic system

The Old Glory fluvial–deltaic system crops out in the southern third of the study area. Subsurface study indicates four areas of E–W trending high net sand values (Fig. 6) that are interpreted to represent individual fluvial–deltaic lobes. South of the town of Old Glory (Fig. 4), the fluvial–deltaic sandstone becomes slightly conglomeratic indicating a closer source area, different source rocks, and/or a steeper paleostream gradient. The appearance of pebbles in the Old Glory system corresponds with a local increase in dip of the strata to 30 ft/mile (14.5 m/km), possibly supporting a hypothesis that a steeper paleogradient aided in bringing coarser material into the coastal area.

Buzzard Peak and Cedar Mountain tidal-flat systems

Tidal-flat facies result from deposition in the intertidal zone between mean low-tide and mean high-tide levels. Tidal-flat deposition occurs in response to complex factors directly related to coastal energy and sediment supply (Beall, 1968). Factors include: (1) wave energy; (2) width of the inner shelf; and (3) tidal range. Sediment supply factors include: (1) ratio of sand to mud supplied to the tidal-flat system; (2) quantity of each sediment type; and (3) proximity of sediment source.

Facies developed in a tidal-flat environment exhibit a sheet-like geometry elongate parallel with the coastline. In a dip oriented cross-section, tidal-flat facies are wedge-shaped (Reineck, 1972) and their width perpendicular to the coastline is determined by tidal range, rate of sediment supply, and the topographic slope of the coastal zone. Tidal channels exhibiting a lenticular shape in tranverse cross-section and a meandering pattern in plan view drain the tidal flats which may be composed of mud, silt, sand, or biogenic (algal mats, foraminiferal tests, shell hash) sediments.

Tidal-flat systems in the San Angelo Formation were formed by depositional regression or seaward accretion and aggradation of terrigenous sediments attendant with compaction and subsidence of previously deposited sediments; the early tidal flats (Buzzard Peak system) were sand-rich; later tidal flats (Cedar Mountain system) became mud-rich.

Buzzard Peak sand-rich tidal-flat system

The Buzzard Peak sand-rich tidal-flat system (Fig. 8) is present as a thin veneer beneath the San Angelo fluvial–deltaic systems and is coincident with the Duncan Member through the middle third of the study area. The Buzzard Peak system developed as a result of intermittent longshore transport of sand, silt, and mud from adjacent fluvial–deltaic systems in a manner analogous to the modern formation of mud flats and cheniers along the southwestern Louisiana coast (Gould and McFarlan, 1959). Tidal flats that develop under conditions of relatively low sediment influx and weak to moderate tidal energy tend to be dominated by tidal processes that result in a winnowing of tidal-flat sediments and separation of sand-sized grains from clay-sized particles. In the Buzzard Peak system this resulted in the formation of chenier-like lenses of sand oriented parallel with the coastline. During periods of high sediment influx mud-flats formed in front of, and over, the sand lenses that underwent subsidence into subjacent muds. The Buzzard Peak system is composed of layers of sandstone interbedded with sandy mudstone and averages about 16 m thick. Sandstone extends 6.5 km downdip into the subsurface where it pinches out into mudstone (Fig. 6).

Tidal sand-flat facies. Sand-flat facies form the framework of the Buzzard Peak system and are separated by beds of sandy mudstone. Individual tidal-flat sandstone beds are 1–3.1 m thick, thicken and thin along outcrop, and are locally continuous; sandstone beds regionally pinch out laterally and are replaced by beds at slightly different stratigraphic levels. Clay drapes may have originated on the tidal-flat during the periods of slack water when the tides were changing (Reineck, 1967), or as a result of wind tides (Fisk, 1959) that pushed thin sheets of turbid water over the tidal flats. The clay would have been deposited from suspension as parallel laminae when the wind-induced current slackened. Reddish brown sandy mudstone interbedded with the sandstone of the sand-flat facies is 0.6–2.5 m thick; some gray sandy mudstone is also present.

Tidal channel facies. Lenticular tidal channel-fill deposits are present throughout the Buzzard Peak system; individual channel-fill sandstone lenses grade laterally into sandstone beds of the sand-flat facies or into surrounding mudstones.

Cedar Mountain mud-rich tidal-flat system

The Cedar Mountain mud-rich tidal-flat system is gradational with the subjacent Buzzard Peak system (Fig. 8) and is formed as a regressive unit by accretion and vertical aggradation of mud supplied by active deltaic systems centered possibly 130 km to the south.

Depositional processes active along southwestern Louisiana coastal mudflats, on Texas Laguna Madre wind tidal flats, and on intertidal algal flats along the Trucial Coast of the Persian Gulf provide a composite modern analog of the Cedar Mountain tidal-flat system. Mud was being supplied to the Cedar Mountain system in essentially the same way that the modern Atchafalaya River is supplying the coastal Louisiana mudflats. Laguna Madre, Texas, is characterized by low-energy, subtidal, intertidal (tidal-flat), and supratidal environments (Fisk, 1959). Deposition on the flats is by wind and wind—tidal processes; precipitation of evaporites, and biochemical action of bacteria and algal mats are other significant processes. Transport of sediment across the Laguna Madre flats by wind-tides is analogous to the processes that are inferred to have been active on the Cedar Mountain tidal flats. Tidal channels with mud—pebble channel-lags, and primarily low flow regime bedding types typify many modern tidal flats (Van Straaten, 1961); similar channels existed on the Cedar Mountain-system flats.

Tidal mud-flat facies. Mud-flat facies constitute the principal deposits within the Cedar Mountain system (Fig. 9). Gypsum nodules in the Cedar Mountain system resemble nodules in modern tidal-flat sediments that develop in response to interstitial diagenetic precipitation (Kendall and Skipwith, 1969; Fisk, 1959). Formation of the gypsum is related to high evaporation and low rainfall. On the Trucial Coast the gypsum nodules are forming in tidal-flat facies under supratidal sabkha facies which have prograded seaward over the intertidal flats (Fig. 10).

Tidal channel facies. Tidal channel-fill deposits are present throughout the tidal-flat systems, but are more common and better exposed in the Cedar Mountain system. Cedar Mountain tidal channel-fill deposits exhibit a facies tract that is divided in this study into lower, middle and upper tidal channel-fill segments in reference to relative position on the tidal-flat (Table I).

Lower tidal channel-fill deposits are 1.8—2.5 m thick and occupy channels eroded into red mudstone. Organic matter often replaced by copper minerals is common in the lower tidal channel-fill facies. Middle tidal channel-fill deposits are lenticular shaped sandstone bodies that average 1.2 m thick and 30.8 m wide. The middle tidal channel-fill deposits form a cap over lower tidal channel-fill deposits suggesting that progradation or accretion of tidal flats resulted in a vertical stacking of tidal channel-fill segments. Upper tidal channel-fill deposits are lenticular shaped quartzarenite bodies 0.3—1.0 m thick and 9.2 m wide. Modern erosion of tidal flat mudstone surrounding upper tidal channel-fill sandstone has locally exhumed the channel-fill bodies which exhibit a meandering pattern.

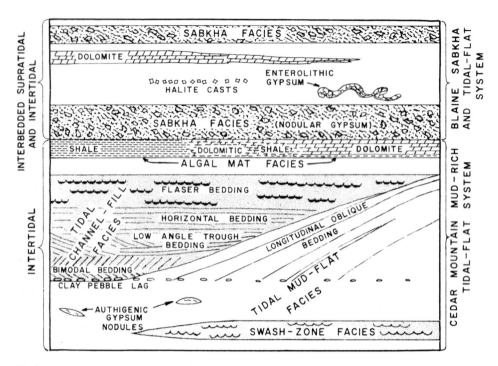

Fig. 9. Idealized sequence of Cedar Mountain mud-rich tidal-flat and Blaine sabkha and tidal-flat systems.

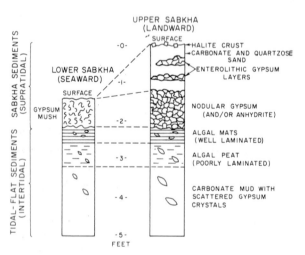

Fig. 10. Vertical sequence through modern tidal-flat and sabkha deposits. Trucial Coast. Modified from Kendall and Skipwith (1969) and Butler (1969).

Swash-zone facies. Sandstone beds of the swash-zone facies are commonly 5.0–15.2 cm thick tabular bodies that have a relatively sharp basal contact with underlying mudstone and locally pinch out into red or gray mudstone. During periods of low sediment supply wave and tidal energy may have winnowed out mud and concentrated sand in localized areas on tidal flats of the Cedar Mountain system.

Algal mat facies. Dolomite, dolomitic shale and well laminated gray shale in the Cedar Mountain system (Fig. 9) grade laterally and vertically into each other and are inferred to have formed in association with algal mat zones on the mud-rich tidal flats. Discontinuous brown laminae in thin sections of the dolomite are indicative of an organic, algal mat, origin. The dolomite and dolomitic shale may have been caused by sulfate-reducing or ammonifying bacteria producing aragonite during decomposition of the algal mats (Purdy, 1963). Aragonite may also have formed by either direct evaporation or respiration of the living algae (Dalrymple, 1965); later diagenesis altered the aragonite to dolomite. Cedar Mountain system algal mats were characterized by a smooth surface with only minor stromatolitic structures. Smooth algal mats are described by Davies (1970), who attributes them to frequent submergence of algal mats by thin sheets of water. Small cavities in some dolomite samples may have been formed by methane or hydrogen sulfide gas caused by decomposition of algae prior to lithification; others may be molds of small gypsum crystals.

Beds of the algal mat facies are 7.6–35 cm thick and usually underlie the first nodular gypsum of the Blaine sabkha and tidal-flat system, although algal-mat facies shale is sometimes present up to 6 m below the base of the Blaine system.

Blaine sabkha and tidal-flat system

Along many coastlines with very low relief, wind- or tide-generated currents may periodically flood vast coastal areas to elevations several feet above the reach of normal daily tides. Such broad tidal flats are called supratidal flats, and modern examples provide insight into the depositional nature of the Blaine Formation in the area of this study.

Salt-encrusted supratidal surfaces that are only occasionally inundated are known in Arabic countries as sabkhas (Kinsman, 1969). Local climate dictates whether a supratidal surface will be a marsh environment, as along the Western Louisiana coast (Morgan et al., 1953), or a sabkha environment as along the Persian Gulf Trucial Coast (Kinsman, 1969). The climate on the Trucial Coast is hot and arid, and the net rate of evaporation is high (Butler, 1969). Trucial sabkhas, which form a linear coastal zone with a gradual slope of 1/1,000 and average about 0.6 m thick and 9 km wide, have formed in approximately the last 3,000 years (Evans et al., 1969). The barren sabkhas have prograded over the algally-bound sediment of the intertidal zone by interstitial precipitation of evaporites and vertical accretion of wind blown and tidally introduced sand and calcareous mud. The seaward edge of the sabkha facies is a gypsum mush zone up to 30.5 cm thick; landward, the gypsum mush develops a nodular appearance (Fig. 10) and may be replaced by anhydrite (Butler, 1969). Evaporite minerals forming in and on the sabkha are gypsum, anhydrite, halite, celestite, and dolomite (Kinsman, 1969). The gypsum and anhydrite are the result of precipitation and crystal growth within the sediment, and are, thus, the result of diagenesis rather than primary sedimentation. Displacement of the surrounding sediment is the most common mechanism involved in the diagenetic process on sabkhas (Lucia, 1972).

The Blaine Formation in the area of this study is interpreted to be a complex repetitive sequence of regressive, off-lapping, sabkha facies and transgressive tidal-flat facies at least 30.8 m thick. Each regressive and transgressive process formed a thin layer of sediments; repetition of these processes caused aggradation of a thick sabkha and tidal-flat sequence. Transgression of tidal-flat facies landward over the sabkha was the result of compaction and subsidence of the subjacent facies. This may have occurred when a critical weight of off-lapping sabkha sediments had accumulated. Persian Gulf sabkhas are analogous to the Blaine system except for the presence of terrigenous (silicic) mudstone in the Blaine system. In Persian Gulf sabkhas carbonate mud and biogenic sediment predominates in the underlying intertidal sediments and calcareous mud and quartzose sand is present on the sabkha surface. Relatively large amounts of terrigenous mud were strike-fed from clastic sources south of the area of this study into the intertidal system during offlap of the Blaine sediments; in addition, hypersaline conditions on the restricted Blaine shelf precluded extensive marine faunas.

Sabkha facies. Laterally continuous nodular gypsum beds of the sabkha facies are from 20 cm to 1.8 m thick, with a median thickness of 0.46 m; they form the framework elements of the Blaine system (Fig. 9). Individual gypsum beds thicken and thin along outcrop and pinch out regionally to be replaced at slightly different stratigraphic levels. Nodular gypsum beds have a muddy appearance and, in general, appear to lack bedding. On cut surfaces, the beds exhibit an intergrowth of 0.6–3.8 cm diameter gypsum nodules that have coalesced to differing degrees set in a red or gray mudstone matrix. Gypsum beds may exhibit sequences of enterolithic layers as much as 1.2 m thick, similar to small crenulations and tight folds of ptygmatic quartz veins. Analogous beds of gypsum up to 2.6 m thick containing enterolithic bedding are forming on the landward margin of the Trucial Coast sabkhas (Butler, 1969).

Tidal-flat facies. Brown and gray mudstone and claystone form the bulk of the tidal-flat facies. Laminated and intraclastic dolomicrite beds in this facies are thin to medium bedded and up to 0.3 m thick. Some of the dolomite beds and fine grained clastics were probably deposited in a sabkha environment, but differentiation between a tidal-flat or sabkha environment of deposition for individual thin beds was not attempted. Salt hopper casts and molds are present on dolomite beds in the Blaine system; similar layers of halite in the form of hopper-shaped crystals form a patchy crust over much of the Trucial Coast sabkha (Kendall and Skipwith, 1969).

DEPOSITIONAL HISTORY

Mid-Permian depositional conditions in North Texas consisted of a broad, low, almost featureless coastal zone characterized by a hot and arid climate. West of this coastal zone was the broad, restricted, shallow shelf of the Midland Basin, which served to dampen incoming waves; to the east was a broad alluvial plain of low relief. Bordering the alluvial plain on the east were the subdued mountains of the Ouachita Fold Belt; to the northeast and southeast, respectively, were the Arbuckle Mountains and the Llano Highlands (Fig. 5). These mountains and exposed Pennsylvanian and older Paleozoic rocks provided sediment to Mid-Permian fluvial systems.

During Late Choza time, the area of this study was a low coastal zone in which subtidal, tidal-flat, and thin sabkha facies were being deposited.

Initiation of nearby fluvial–deltaic activity increased the input of sand into the coastal zone causing the development of a thin veneer of off-lapping, sand-rich tidal flats of the Buzzard Peak system that marked the beginning of San Angelo deposition. The increase of terrigenous clastics was soon followed by progradation of thin, but aerially widespread, high constructive, lobate deltas of the Copper Breaks deltaic system and the Old Glory fluvial–deltaic system (Fig. 11) over deposits of the lowermost part of the Buzzard Peak system which had undergone compaction and subsidence.

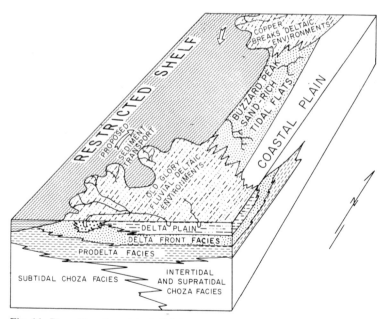

Fig. 11. Depositional model of Old Glory, Copper Breaks, and Buzzard Peak systems. These systems compose the Duncan Sandstone Member.

Sand and mud were transported laterally by longshore drift from the fluvial–deltaic systems into the intermediate Buzzard Peak tidal system (Fig. 11). Tidal processes generated by astronomic and wind tidal action, resulted in regression of the Buzzard Peak sand-rich tidal-flat system. During relatively short periods of high sediment influx, tidal-flat regression was rapid and sandy mudstones were deposited; during intervening periods of low sediment influx, tidal reworking resulted in the winnowing and removal of clay-sized particles and the concentration of sand in layers. Compaction and subsidence contemporaneous with progradation created a relatively thick sequence of facies.

The Copper Breaks deltaic system, Old Glory fluvial–deltaic system, and Buzzard Peak sand-rich tidal-flat system together comprise the Duncan Sandstone Member of the San Angelo Formation (Fig. 11).

Upstream avulsion of rivers supplying sediment to the fluvial–deltaic systems ended deposition by the Old Glory and Copper Breaks systems; the fluvial–deltaic deposits slowly compacted and subsided into underlying muds. Sand supply for the San Angelo depositional systems was drastically reduced when fluvial–deltaic progradation ended, but was replaced eventually by mud-rich sediment transported by longshore drift from a fluvial–deltaic complex active south of the area. Influx of mud initiated accretion of mud-rich tidal flats of the Cedar Mountain system (Fig. 12) that eventually covered the

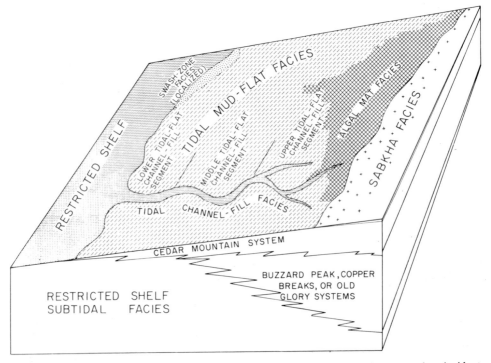

Fig. 12. Depositional model of Cedar Mountain mud-rich tidal-flat system. This system is coincident with the Flowerpot Mudstone Member.

previously deposited depositional systems. The Cedar Mountain mud-rich tidal-flat system prograded over the restricted shelf deposits; contemporaneous compaction and subsidence created a relatively thick sequence of mud-rich tidal-flat facies. The Cedar Mountain tidal flats were drained by tidal channels that became progressively smaller up the tidal-flat; portions of the middle and upper tidal flats were covered by algal mat facies.

As coastal tidal-flat progradation continued, the inner, landward portion of the Cedar Mountain tidal-flat system was gradually raised above mean high tide and a thin supratidal sabkha facies began to offlap the intertidal deposits (Fig. 12). The end of extensive tidal-flat deposition concluded deposition of the Cedar Mountain system and led to development of the Blaine sabkha and tidal-flat system. Termination of Cedar Mountain deposition marked the close of San Angelo deposition in the area.

Continued offlap of the initial Blaine sabkha facies covered the Cedar Mountain system (Fig. 8). Each regressive episode of this sabkha facies was followed by compaction, subsidence of the sabkha deposits, and onlap of thin tidal-flat facies. Continued repetition of these processes created a relatively thick sequence of Blaine facies.

COPPER MINERALIZATION[1]

Stratiform copper mineralization has been recognized in the Permian rocks of North Texas for over 120 years. Many mining ventures were attempted but because of poor mining practices, primitive technology, erratic fluctations in the price of copper, limited ore grade reserves in individual mines, and a lack of understanding of the geologic relations between the mineralization and the host rocks, attempts at exploitation have been unsuccessful. Total production from all the prospects in North Texas has exceeded 5,000 tons of crude ore (Stroud et al., 1970).

Mineral assemblage

Malachite, azurite, covellite and chalcocite were the only copper minerals found during this study. Malachite is the most common mineral at outcrop, whereas in unweathered samples, chalcocite predominates. Pyrite is present in minor quantities as an accessory mineral.

Distribution of copper mineralization

The lowermost sandstone bed of the Buzzard Peak, Old Glory, and Copper Breaks systems (Fig. 8) is commonly bounded at its base by a zone of light gray mudstone or sandy mudstone averaging 5.0–30.5 cm thick. Copper mineralization in the Duncan

---

[1] Editor's note: For contributions giving related information, see chapters 5, 6 and 10 in Vol. 2, chapters 2 and 7 in Vol. 6, and chapter 3 in Vol. 7 on uranium and copper mineralization, and on organic matter and bacterial processes.

Sandstone Member of the San Angelo Formation is restricted to this gray layer and the bottom 15 cm of the overlying sandstone. The highest copper concentration is normally found at the contact between the overlying sandstone and the gray zone. The copper occurs as widely disseminated blebs of sulfide and malachite, and as localized interstitial malachite cement. Some possible charcoal and tar-like organic material occurs in the copper-bearing layer.

The bulk of the copper minerals occur in the Cedar Mountain mud-rich tidal-flat system of the San Angelo Formation (Flowerpot Member), and in the basal 3.1 m of the overlying Blaine sabkha and tidal-flat system of the Blaine Formation (Figs. 3 and 9).

Copper occurs in the following four facies of the Cedar Mountain and Blaine systems: (1) swash-zone facies; (2) algal mat facies; (3) sabkha facies; and (4) tidal channel-fill facies (Figs. 9 and 12).

Ripple cross-stratified sandstone beds of the swash-zone facies are thin destructive units that are distributed randomly through the Cedar Mountain system. Copper, in the form of malachite, can be found weathering out of the base of the sandstone beds, and as thin coatings on weathered fragments of the sandstone. Copper sulfides and organic matter were not observed in the swash-zone facies on outcrop. During formation of the sandstone, organic matter may have accumulated in the swash-zone as flotsam; later post-depositional decay may have removed traces of the organics.

Mineralization in the gray, well laminated shale, dolomitic shale, and dolomite of the algal mat facies at the top of the Cedar Mountain system and in the basal 3.1 m of the Blaine system exhibits the best lateral continuity. The copper-bearing gray shale beds range from 7.6 to 35.6 cm thick, and there is commonly a 5.0–7.6 cm interval of a gray well laminated shale near the middle or upper part of the zone which contains the highest concentration of copper. Copper values in the mineralized shale beds run as high as 4.4% through 5 cm intervals and 2.2% from 17.8 cm intervals. Copper is present in these thin extensive algal mat shale beds as malachite and chalcocite; these minerals are present as platelets of malachite on shale laminae, and as blebs of chalcocite. In outcrop only malachite is observable unless the shale is artifically exposed. Black specks of what may be bituminous matter are present in the shale. No recognizable organic matter was actual-ly noted in the algal mat facies, but their origin as algally-bound sediment, and their dark color suggests that a significant amount of organic matter was present in these sediments at the time of burial. Dolomitic gray shale beds of the algal mat facies that are commonly found with and laterally continuous to the gray well laminated shale beds also contain malachite stains.

Dolomite layers in the algal mat facies at the top of the Cedar Mountain system and in the basal 3.1 m of the Blaine system contain chalcocite and rarely malachite. These minerals occur as scattered blebs, partial fillings in small vugs, and small fracture fillings. The dolomite is thin bedded and ranges from less than 2.5 cm up to 30.5 cm thick. Thin sections show that copper was preferentially deposited in pore space, and as a replace-ment of organic matter.

Copper may also occur in trace amounts at the base of the first nodular gypsum bed of the sabkha facies in the Blaine system, immediately above the algal mat facies of the Cedar Mountain system (Fig. 9); the copper occurs as both malachite and chalcocite. 2 mm cubic crystals of chalcocite, possibly pseudomorphic after pyrite, were observed imbedded at the base of a gypsum bed.

The thickest ore, and the ore with the highest copper concentration, occurs in tidal channel-fill facies in the Cedar Mountain mud-rich tidal-flat system and in the basal portion of the Blaine system. Copper in these channel-fill deposits has been mined intermittently since 1877. The bulk of the copper comes from lower tidal channel-fill sandy mudstone and muddy sandstone. Some copper has also been mined from middle tidal channel-fill sandstone. Based on observations made by Phillips (1917a), it is apparent that in many of the copper mines, the hard sandstone of the middle tidal channel-fill facies was not mined even though it contained some copper. This was done in order that the sandstone could provide roof support for the mining of the underlying copper-bearing sandy mudstone and muddy sandstone of the lower tidal channel-fill facies. Phillips (1917b) noted that mineralized zones in the lower tidal channel-fill facies range from 0.6 to 3.4 m thick, 5.0–14.0 m wide and about 46 m in length. Copper minerals present in the tidal channel-fill deposits include malachite, azurite, covellite, and chalcocite. At the Farris prospect near Old Glory, there are pyrite nodules exposed at the entrance to the abandoned adit. The ore in the tidal channell-fill facies occurs as copper nodules, replacements of wood fragments, disseminated blebs of copper, and as interstitial cement. Copper nodules are 2.5–5.1 cm long, 1.3–2.5 cm in diameter and are found in a variety of shapes. Some have exterior textures that suggest that they grew around a wood nucleus, whereas other nodules have an amorphous appearance.

Paragenesis

The paragenetic sequence has apparently been a four-step process, outlined as follows: (1) pyrite replacing organic material; (2) chalcocite replacing pyrite; (3) covellite replacing chalcocite; and (4) malachite and azurite. The precise time relationship between the first two steps depends on the origin of the mineralization, which has not been definitely established. The position that covellite occupies in this sequence is equivocal since the textures may have formed under a variety of conditions.

The essential parts of this sequence are shown in polished thin sections of copper nodules (Fig. 13). Pyrite has replaced woody material and preserved the cellular structure of the woody host material (Fig. 13A,B). Contortion of cell geometry, inferred to have been caused by lithostatic pressure, is more extreme in some examples (Fig. 13A) than others (Fig. 13B). Rims of clear, noncellular pyrite have formed around some of the pyrite blebs that contain an inner portion with the cellular structure preserved (Fig. 13A). Chalcocite replaces pyrite (Fig. 13C), and in some instances (Fig. 13C, upper right), ghosts of cells of woody material are preserved by chalcocite. Replacement textures of

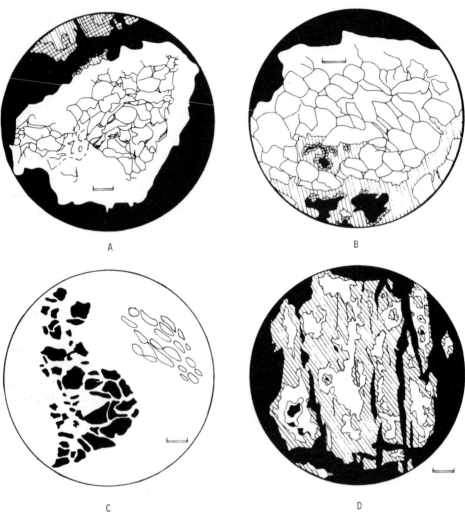

Fig. 13. Camera lucida sketches of polished thin section. Bar scales are 0.1 mm long. A. Pyrite with relict wood structure and "clear" rim. B. Pyrite with relict wood structure. (Fig. 13 A, B: white = pyrite; black = woody matter; cross-hatch = covellite; parallel lines = chalcocite.) C. Chalcocite (white) replacing pyrite (black) with relict wood structure preferentially along cell walls; ghosts of wood structure in upper right. D. Covellite replacing chalcocite; white = chalcocite; parallel lines = covellite; black = woody matter.

pyrite by chalcocite (Fig. 13C, lower left) suggest preferential replacement along cell boundaries. Covellite also replaces chalcocite (Fig. 13D). This replacement has advanced from the outer margins of the chalcocite and from numerous cracks that are present in the chalcocite bleb (Fig. 13D). Nodules of copper minerals also form by enveloping quartz grains (Fig. 14A). Some of the quartz grains exhibit evidence of solution and mechanical disintegration during precipitation of the sulfides.

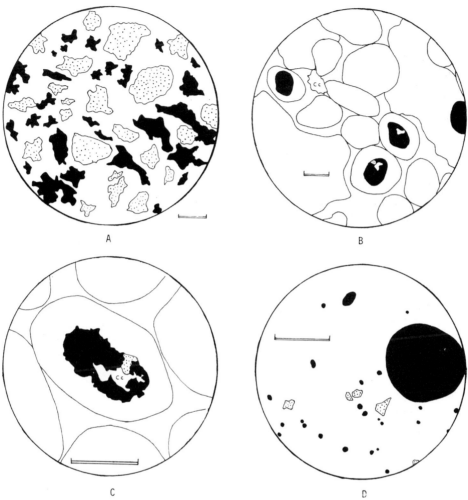

Fig. 14. Camera lucida sketch of (A) polished thin section, copper nodule (B, C) tradings of photo-micrographs, polished thin section of copper-bearing dolomite, and (D) tracing of photomicrograph, thin section of Creta copper-bearing claystone. Bar scales are 0.1 mm long. A. Solution and mechanical disintegration of quartz (dot pattern); white = chalcocite; black = pyrite. B. Precipitation of chalcocite in pore space between oolites in dolomite. Cc = chalcocite; white = dolomite; black = organic matter. C. Chalcocite replacing organic matter in the nucleus of an oolite; Cc = chalcocite; dots = quartz; white = dolomite; black = organic matter. D. Mineralized spore (0.15 mm sphere) framboidal spherules (4–10 μ spheres) in copper-bearing claystone. Creta mine; black = chalcocite; white = claystone.

Polished thin sections of mineralized dolomite exhibit preferential replacement of organic matter by chalcocite and crystallization of chalcocite in pore spaces (Fig. 14B,C). In an oolitic dolomite (Fig. 14B), a bleb of chalcocite is present in the pore space among four oolites. The black centers in some of the oolites are organic matter. In some cases this organic matter is being replaced by chalcocite (Fig. 14C).

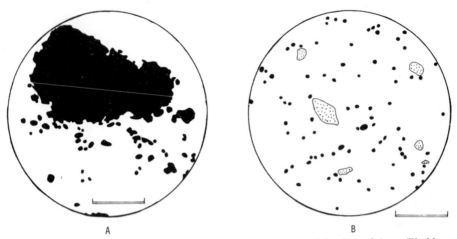

Fig. 15. Tracing of photomicrograph of (A) thin section from Medicine Mound Area, (B) thin-section of Creta copper-bearing claystone. Bar scales are 0.1 mm long. A. Mineralized polyframboidal structure (black) in dolomite (white). B. Mineralized fromboidal spherules (black) in copper-bearing claystone (white), Creta mine; black = chalcocite; dots = quartz.

Thin sections of mineralized gray shale and mudstone from the Creta copper mine in Southwest Oklahoma and from near Medicine Mounds (Figs. 1, 4) contain scattered circular blebs of chalcocite, 0.10—0.15 mm in diameter (Fig. 14D) and oblong blebs of chalcocite 0.05—0.27 mm in length (Fig. 15A); numerous smaller circular blebs of chalcocite from 4—10 μ in diameter are also present (Fig. 14D; 15A,B). The size and shape of the 4—10 μ blebs suggest that they are framboids (Rust, 1935) of chalcocite pseudomorphic after pyrite; the oblong 0.05—0.27 mm chalcocite blebs may be polyframboidal structures (Love, 1971). The 0.10—0.15 mm chalcocite spheres (Fig. 14D) from the copper-bearing shale at the Creta Mine are interpreted as mineralized spores. The gray, relatively unmineralized mudstone under the copper-bearing shale at the Creta Mine contains orange spheres 0.08—0.15 mm in diameter that have been identified as being spores (Clair R. Ossian, personal communication, 1973). As the strongly mineralized shale is approached, the spores are mineralized from the inside to the outside in a progressive manner, until, in the ore zone, only spheres of chalcocite can be observed. In almost all of the exposed copper-bearing shale beds, malachite is present as a recent oxidation product on shale laminae because of solution of copper sulfide in ground water, slight movement of the copper in solution, and reprecipitation of the copper as malachite.

Textural relationships among the ore minerals, as observed in polished thin sections by the writer and as reported by other workers who have studied diagenesis and oxidation processes in modern sediments, allow some comments to be made concerning the time of ore deposition. Replacement of organics by pyrite may be an early diagenetic phenomenon that occurs soon after burial; pyrite is currently forming under reducing conditions at a depth of 40–70 cm in tidal-marsh sediments on the west coast of Florida because of anaerobic bacterial processes related to decomposing organic matter (Swanson et al., 1972). In Florida tidal-marsh sediments, Swanson et al. (1972) found that total sulfur, primarily as iron sulfide, increases with depth and that mobile sulfur (mostly the reduced HS^--ion or gaseous H_2S) decreases with depth; strongest H_2S odor came from samples having the highest organic matter content. Swanson et al. (1972) suggest that a distinctive, strong odor of H_2S in samples from any depth in the tidal-marsh sediments may indicate that excess reduced sulfur is available in the sediment to bind any metals that form insoluble sulfides, provided the metals are present in the sediment. Part of the reduced sulfur in Florida tidal-marsh sediments goes through a transition from largely dispersed HS^- and H_2S in the upper 10 cm, to a very unstable hydrous iron sulfide phase in the 10–25 cm interval, a relatively stable amorphous iron sulfide in the 25–40 cm interval, and finally to a stable crystalline pyrite below 40 cm. The rim of clear pyrite around pyrite with relict cell structure (Fig. 13A) may indicate that bacterial decomposition of organic matter in the Permian sediments was forming an H_2S or HS^- halo around the organic matter, causing precipitation of iron sulfide to extend beyond the confines of the organic particle.

The age of the chalcocite, which is now the dominant sulfide present, is less definite. Depending on the origin of the mineralization, it may have formed diagenetically before lithification of the sediment (but later than the pyrite), or after lithification of the enclosing sediments. This question is considered further in the section on the origin of the copper. Covellite may have formed a little later as an early replacement mineral or during the Holocene as an oxidation product of chalcocite. If covellite formed as an early replacement of chalcocite, its formation may have been caused by a relative decrease in reducing conditions present in the concentrating environment as H_2S and HS^- were consumed in precipitation of copper sulfide. Covellite would then represent slightly higher Eh conditions, but still probably negative.

There is also a possibility that some of the covellite formed at the same time as the chalcocite. A polished section of a copper sulfide nodule contains small, apparently isolated, masses of covellite within chalcocite, which may indicate segregation of cupric and cuprous ions during formation of the copper sulfides. Some of the covellite exhibits the permanent blue (or blue-remaining) characteristic described by Ramdohr (1969). Ramdohr (1969) listed similar occurrences of primary chalcocite and covellite of the "arid alluvial basin" and "sulfur cycle" type from Mansfeld in Germany and from the Colorado Plateau region of the United States. He admits that the distinction between primary and secondary covellite is difficult to make and that both are probably present. Replacement of organic matter in oolites (Fig. 14C) by chalcocite indicates that some diffusion over at least short distances has occurred. Malachite and azurite probably formed during the Holocene as the result of oxidation of sulfides by surface and ground water, followed by reprecipitation in the presence of carbonate ions.

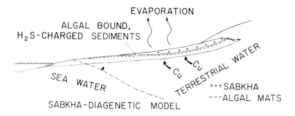

Fig. 16. Sabkha-diagenetic model for copper mineralization. Modified from Renfro (1974).

Origin of copper mineralization

The depositional history of the host rocks, the obvious facies control of the mineralization, the importance of organic matter, a lack of mineral zonation, and absence of
known igneous activity point to one of three origins for these deposits. A sabkha–
diagenetic origin for the copper deposits is proposed that involves evaporative discharge
through the sabkha surface (Fig. 16). Alternate epigenetic origins involve vertical movement of an aqueous solution containing copper chloride complexes along basin-rimming
growth faults (Fig. 17A), or metasomatic intergranular diffusion of copper over relatively
long distances (Fig. 17B), respectively. Copper mineralization appears to be unrelated to
any structural influence, and there is no apparent faulting in or near the study area.

Sabkha–diagenetic model

Diagenetic processes associated with the sabkha environment may concentrate copper
into economic deposits provided that the following conditions are met:
(1) Adequate copper in the ground water.
(2) High evaporation rate.

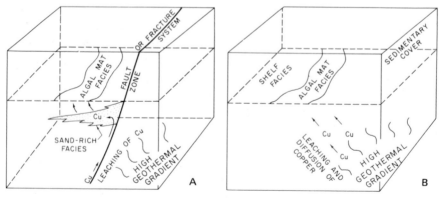

Fig. 17. A. Epigenetic model; fault or fracture-controlled mineralization. B. Epigenetic model; diffu-
sion-controlled mineralization.

(3) Sufficient amount of reductant present.

(4) Persistence of sabkha and intertidal sedimentation.

In the sabkha—diagenetic model (Fig. 16), evaporative discharge at the sabkha surface induced by evaporation may create an upward decrease in hydrodynamic potential with the result that ground water of primarily terrestrial origin would be induced to flow upwards through the sabkha. Hydrogen sulfide formed as a metabolic by-product of sulfate-reducing bacteria in decomposing algal mats, and in decomposing organic matter in tidal channels and ripple-bedded sands, should be capable of stripping out the copper by causing it to precipitate as copper sulfide. Hydrogen sulfide associated with decomposing organic matter at the base of the Duncan Member would likewise precipitate some copper. It has been demonstrated (Baas Becking and Moore, 1961) that digenite (Cu_9S_5), which is very similar to chalcocite (Cu_2S), can be produced by bacteria in a medium of lactate and cuprous oxide. Lactate is one of the two main organic compounds utilized by sulfate-reducing bacteria in natural environments (Trudinger et al., 1972; see also chapter 6, vol. 2). In such a model, the lack of copper in the bulk of the Duncan Member could be explained by oxidation of organic matter during and shortly after deposition. Later, small-scale diffusion and diagenesis could further concentrate available copper. Pyrite, which forms by early diagenesis prior to any significant concentration of copper, is partially replaced by copper sulfide during concentration of the copper. The copper content of the ground water, which is elaborated on later, need only be of approximately average value. A condition of depositional equilibrium allowing sabkha processes to exist in the same area for a relatively long period of time would allow the diagenetic copper concentrating process to continue long enough to allow a potentially economic concentration of copper to accumulate.

Evaporation rates in the Persian Gulf area are as much as 128 cm/yr and rainfall averages only 3.8 cm/yr (Butler, 1969). Hsü and Schneider (1973) showed that this high rate of evaporation causes a vertical hydraulic gradient to form in the sabkha. In response to the upward decrease in hydrodynamic potential caused by the interstitial water loss, evaporative pumping, or discharge, takes place. Saline lagoon-water and terrestrial water are "pumped" laterally, and then vertically, to be discharged at the sabkha surface. The relative amount of terrestrial ground-water and saline lagoon-water that is evaporated depends on the dynamic equilibrium that exists between the two. The writer proposes that most of the water evaporated in the proposed model was terrestrial in origin (Fig. 16). This would occur if the slope of the water-table in the coastal zone created a hydraulic head sufficient to push the salt-water wedge of the Midland Basin away from the area immediately under the sabkha. Along the arid Trucial Coast today, terrestrial ground-water has been found in the "high supratidal zone" of the sabkha (Butler, 1969). This suggests that a hydraulic head sufficient to force a salt-water wedge away from the area under the sabkha facies may be able to form in an arid region, provided there is a sufficiently elevated aquifer recharge area to create the hydraulic head. During concentration of the San Angelo copper deposits, subdued highlands in the

Ouachita fold belt may have provided the elevation necessary to produce a sufficient hydraulic head. Hsü and Siegenthaler (1969) stressed that this process is governed by evaporation and not permeability and would not take place in a humid climate where ground-water movement is induced by a large gravity head. Hydrogen sulfide gas, sulfate ions, gypsum, and anhydrite are common features found in sabkha and intertidal sediments of the Trucial Coast (Butler, 1970). Anaerobic, sulfate-reducing bacteria utilize organic material present in algal mats to provide energy with which to reduce sulfate, which is in the form of sulfate ions. The reduction is a metabolic process in which oxygen is provided by sulfate breakdown with H_2S or HS^- formed as a by-product. Presence of these gases, and a lack of oxygen, results in a low pH and a negative redox potential in the sediments.

With hydrogen sulfide acting as a reductant, and evaporative discharge serving as a driving force, the stage may be set for the accumulation of copper. Copper is contained in the ground water as dissolved copper in ionic form, copper adsorbed on clay-sized particles, and as colloidal and soluble organo—copper complexes which are pumped through the sabkha. As water goes through the zone of H_2S and HS^-, copper is extracted and precipitated as chalcocite or perhaps covellite. Breakdown of the various organic complexes and desorption of the copper may be facilitated by the strongly ionic (Na^+ and Cl^-) conditions under the sabkha facies, and the low Eh and slightly acidic pH in the organic-rich facies (Butler, 1969). Desorption or release of copper may partly occur by substitution of Na^+ for Cu^+ (Rickard, 1970; Kharkar et al., 1968). Studies by Temple and LeRoux (1964) show that metal toxicity would limit the continued action of sulfate-reducing bacteria only if the rate of introduction of metals to the environment exceeds the rate of H_2S production. In the sabkha—diagenetic model, it is proposed by the writer that copper was passing through organic-rich zones at a relatively average concentration and that production of H_2S was keeping pace with the introduction of copper. The gypsum layer above the algal mat zone may have acted as a partial seal to upward dispersal of H_2S. The presence of some copper mineralization in the gypsum indicates that some H_2S did seep into the base of the gypsum layer. The gypsum layer of the sabkha would also serve as a seal to oxygen, thus preventing oxidation of the algal mats. The pH values in the intertidal-sabkha environment are zoned (Butler, 1969). Interstitial algal-flat waters have a pH of 7.5; landward across the sabkha and vertically downward through the sabkha sediments, the pH decreases to the 6.0—6.8 range. This decrease in pH is due to bacterial decomposition of organics (i.e., algal mats) that are being prograded over by sabkha sediments.

Information on the Eh values is essentially lacking; values in the −270mV or lower range (Swanson et al., 1972; Butler, 1970) should be expected. Around pieces of decaying organics, the Eh may be lower than −270 mV. A consideration of the stability fields of chalcocite and covellite (Garrels and Christ, 1965; see also chapters 10 and 7, vols. 2 and 6), along with the Eh and pH conditions present in sabkha sediments, indicate that copper sulfides could form in the algal mat zone under the sabkha facies gypsum. A theoretical study by Trudinger et al. (1972) supports the hypothesis that biogenic processes may take part in forming stratiform copper deposits. Lockwood (1972) reported that chalcocite from Creta exhibits a spread in the relative ratio of sulfur isotopes ($^{34}S/^{32}S$) that suggests a biogenic origin for the sulfide anion.

The amount of copper present in the ground water and the length of time in which the concentration process is active are critical factors in the sabkha—diagenetic theory. Hem (1970) reported that sea water contains 0.003 ppm copper. The mean copper content of fresh water streams is 0.010 ppm (Livingstone, 1963). The fresh water figure is undoubtedly too low since most analysis do not consider copper that has been complexed by naturally occurring chelators such as humic acids, copper in metallo—organic compounds, or copper in suspension adsorbed on inorganic and organic detrital particles (see chapter 5, vol. 2). Volfson and Arkhangel'skaya (1970) reported that streams in the Azov—Black Sea and Caspian Sea basins carry 0.06 ppm and 0.038 ppm of copper, respectively; much of the copper is carried in suspension. From data by Kharkar et al. (1968), it can be seen that adsorption of cobalt by particles in streams causes a four-fold increase (average) in total cobalt content. Collins (1973) suggested that similar figures may be obtained from a study of copper because of their similarities in behavior. Rashid and Leonard (1973) have demonstrated the importance of organic compounds (e.g., humic and amino acids) in dissolving insoluble metallic salts and in keeping them in solution by preventing their precipitation under conditions that would otherwise lead to immobilization. Humic acids have a base-exchange capacity and tend to sequester or complex metal ions (Martin et al., 1971).

Baker (1973) in a laboratory study using humic acid extracted from a podzolic soil in Tasmania, has shown that "metal humates" are readily dissolved and mobilized in the presence of humic acids. Unfortunately, very few quantitative data are available on copper adsorbed on particles or as organo—copper complexes in ground water. White et al. (1963) reported as much as 0.12 ppm copper in ground water from a shale and siltstone terrane. Most values of copper in ground water from all terranes composed of unmineralized rocks are apparently under 0.010 ppm (based on study of charts in White et al., 1963). No mention is made by White and co-workers of copper present as part of soluble or colloidal organic compounds, or adsorbed on particles that may travel in ground water.

The length of time during which a sabkha environment can exist depends on the stability of depositional conditions. On the Trucial Coast an average of 0.70 m of upper intertidal (algal mat) and sabkha sediments have accumulated in about 3,500 years (Evans et al., 1969). Depositional processes in the Trucial Coast sabkha are such that the sabkha is slowly prograding across intertidal algal flats and into a lagoon connected to the Persian Gulf. Under alternate conditions suggested for the Blaine sabkha and tidal-flat system, a sabkha could prograde over an intertidal flat, subsidence could then occur forming a wide intertidal algal flat over the sabkha, and finally a new sabkha could again prograde over the intertidal zone. This process could be repeated many times. This mechanism is proposed elsewhere in this paper to explain the deposition of the Blaine Formation in North

Texas. Bosellini and Hardie (1973) and D.G. Bebout (personal communication, 1973) proposed a similar process for thick, ancient sabkha sediments in Italy and Canada. Near Old Glory 30.8 m of the Blaine appears to be sabkha—intertidal in origin. At a deposition rate of 0.70 m/3,500 yr, it would take 154,000 years of sabkha deposition to accumulate 30.8 m of the Blaine Formation.

Calculations (Table II) were made to determine the feasibility of concentrating economic copper deposits by the diagenetic process just described. From figures supplied by Hsü and Siegenthaler (1969) and using the 154,000 years estimated to deposit the Blaine sabkha and tidal-flat deposits, it can be seen that the sabkha—diagenetic theory may be valid provided there are approximately 0.06 ppm (or slightly less) copper in the ground water, the rate of evaporation from the sabkha is between 50 and 100 cm/yr, and ground-water is almost totally terrestrial in origin.

An important aspect of the sabkha—diagenetic theory of copper concentration is the necessity for having most of the groundwater pulled through the lowermost sabkha even when sabkha—intertidal sediments are 30.8 m thick. This requirement may be acceptable for the following reasons. Sabkha sediments form as a wedge of low permeability sediments with a slight seaward tilt that have prograded over intertidal sediments (Evans et al., 1969; figs. 4, 5). In general, the sediments seaward of the sabkha have a low permeability, whereas sediments under and landward of the sabkha have a relatively high permeability; this generalization holds true for the San Angelo Formation even though there is a higher mud content in the San Angelo than in Trucial Coast sediments. The presence of some copper in dolomite and gypsum layers above the main copper zone at the top of the Cedar Mountain system indicates that some copper did either pass through the bottom gypsum layer or come in above the bottom layer.

TABLE II

Number of years required to concentrate a 25 cm column of 1% copper over 1 cm^2 of surface[1]

Sabkha evaporation rates (cm per year)	ppm copper in ground water[4]	Number of years required
100[2]	0.06	100,000
100	0.01	600,000
100	0.003	2,000,000
50[3]	0.06	200,000
50	0.01	1,200,000
50	0.003	4,000,000

[1] 1% copper and 25 cm thickness are personal estimates of the average thickness and tenor of copper bearing shale in the study area.

[2] From Hsü and Siegenthaler (1969), evaporation from standing surface of water.

[3] From Hsü and Siegenthaler (1969), evaporation loss of interstitial water for sediments of 40% porosity.

[4] Previously explained in text.

Epigenetic models

The discovery of mineralized brines in the Red Sea (Degens and Ross, 1969; see also chapter 4, vol. 4) and in the Salton Sea sediments (White, 1968) has recently focused attention on the importance of saline solutions as being a transporting medium for a variety of elements (Tarling, 1973). The importance of chloride complexes in the transportation and deposition of ore-forming metals has been pointed out by Helgeson (1964), and the formation of complex ions has been shown to increase the solubility of some metals by several orders of magnitude (Barton, 1959; see also chapter 10, vol. 2).

Transport of copper in chloride complexes (e.g., $Cu(H_2O)_{n-3}Cl_3$, or simply $CuCl_3$) vertically along faults or fractures, and then laterally along permeable horizons with precipitation taking place in favorable chemical zones is proposed by the writer as an alternate, epigenetic origin for the copper deposits in the San Angelo Formation (Fig. 17A). Faulting or fracturing may have occurred in or near the area of this report in response to sedimentary loading in the Midland Basin, or as a result of separation of the North American and South American plates during the Late Paleozoic and Early Mesozoic (see Walper and Rowett, 1972). Apparent alignment of copper deposits along the Midland Basin and other basins in Oklahoma and southern Kansas (Fig. 1) may possibly result from growth-faulting along the margin of these basins. Copper in the form of chloride complexes may have traveled vertically along faults or fractures and spread out laterally along porous sandstone beds. Deposition could have occurred when the mineralizing solutions came in contact with hydrogen sulfide in organic-rich zones such as tidal channel-fill facies and algal mat facies. Reduction of solubility as a result of a lowering of temperature may have had some influence on concentrating the copper, although the effect is probably minimal (Barton, 1959). The copper could have been derived from the leaching of deeply buried copper-rich detrital minerals (i.e., biotite, and possibly hornblende and augite), connate water in older Paleozoic rocks, and possibly some copper ions from deep-seated, unidentified magmatic sources. A high geothermal gradient along the eastern side of the Midland Basin, perhaps related to the separation of lithospheric plates may have provided additional heat for leaching copper and increasing its solubility. Tarling (1973) discussed the application of plate tectonic concepts to ore genesis and stressed the importance of high heat flow connected with plate movements and organic-rich zones as places for deposition of metals.

A second epigenetic theory involving thermal diffusion and metasomatic processes (Fig. 17B) was investigated in order to obviate the necessity of a fluid moving upward along a fault or fracture. Garrels et al. (1949) stressed the possible importance, theoretically, of diffusion in transferring significant quantities of materials over relatively long distances. Ramberg (1952) proposed that a high geothermal gradient may cause intergranular diffusion of atomic, ionic, or molecular-sized particles over relatively long distances, giving rise to a low-temperature metasomatic type of ore deposit. Micro-porosity may be developed in lithified and perhaps semi-lithified, deeply buried rocks as a result of intragranular micro-fissuration (Dandurand et al., 1972). This porosity would be intermittently formed and destroyed as a result of uneven lithostatic stresses. The multitude of healed fractures noted in quartz grains (Sipple, 1968) may be partly caused by these processes. During the micro-fissuration process, fluids are released and the solubility of many minerals is increased (Dandurand et al., 1972).

Copper may be able to move out of its position in the lattice of minerals bearing trace amounts of copper, and move into intergranular, or if micro-fractures in the grains develop, into an intragranular position. In this intergranular or intragranular position, the copper ion may go into solution as perhaps a chloride complex and be physically transported a short distance by fluid movement, or the ion itself may diffuse a short distance through the solution and, perhaps, along grain boundaries. In a sufficiently long period of time, an integration of this process over a large area and a general upward movement of copper in the direction of the inferred concentration gradient may lead to the concentration of an economic copper deposit when an H_2S or HS^- bearing strata is reached, and the micro-fissuration process ceases. A delicate balance among these three factors may be necessary, thus providing indirect evidence for an apparent lack of copper mineralization in rocks older than the Permian. An abnormally high geothermal gradient might increase the rate and efficiency of this diffusion process. As in the previous epigenetic theory involving mass movement of aqueous solutions along fault or fracture zones, the source for the copper could be deeply buried copper-rich detrital minerals, connate water, or an as yet unidentified magmatic source. The emplacement of ore bodies by either epigenetic theory could take place after deposition of the host rocks and after an unspecified thickness of sediments had been deposited over the host rocks.

Evaluation of mineralization models

Both the sabkha–diagenetic model and the epigenetic models for the origin of copper deposits in the San Angelo Formation stress the importance of H_2S and HS^- that are produced in organic-rich zones by sulfate-reducing bacteria for the deposition of copper. These models differ, though, in the emphasis that is placed on facies control and time of copper deposition. The sabkha–diagenetic model is closely linked spatially and temporally to specific facies. Without the relatively rapid progradation of sabkha facies over the Cedar Mountain algal mat facies and continued deposition of sabkha and tidal-flat cycles (thus forming the Blaine system), the mechanism of evaporative discharge, which regulated the concentration of copper in the sabkha–diagenetic theory, would not have been able to operate. In the epigenetic models only an H_2S or HS^- rich facies is required. For movement of copper chloride complexes laterally from a fault or fracture, the H_2S should be in or adjacent to a permeable facies such as sandstone. Diffusion of copper through a few meters could transport the copper from a permeable sandstone facies through relatively impermeable mudstone strata to a chemically favorable facies (Garrels et al., 1949), such as the algal mat facies.

The evaporative discharge process involves diagenetic concentration of copper in the host facies during deposition of the overlying Blaine sabkha and tidal-flat system, and may occur in a short interval of geologic time; perhaps 100,000–200,000 yr. The epigenetic process involving movement of fluids along faults or fractures could occur in a very short period of time shortly after deposition, or sometime during or perhaps after lithification of the host facies. A time limit on when this could occur may be imposed by the length of time during which the sulfate-reducing bacteria can remain active, (which is partially dependant on the amount of organic matter present), or the length of time during which H_2S can remain in or near the host facies. The epigenetic process involving principally diffusion would involve very slow movement of copper, and presence of H_2S for a much longer time.

The writer favors the sabkha–diagenetic model rather than the epigenetic models for the origin of the San Angelo copper deposits. Facies control of the copper deposits observed on outcrop fits well with the vertical succession of facies and the inferred paleogeographic history in the area of this study. Lack of copper mineralization in underlying coal deposits and other organic-rich facies of Pennsylvanian Age, and an absence of igneous or metamorphic activity in the region of this study, suggests that perhaps a diagenetic process, such as outlined in the sabkha-diagenetic model, was active in concentrating the copper.

An epigenetic model involving movement of copper chloride complexes in solution along faults or fractures cannot be ruled out, but until evidence of faulting or fracturing is detected in or near the area of this study, facies control as described in the sabkha–diagenetic model seems to be the more plausible of these theoretical models for concentrating San Angelo copper deposits. Movement of solutions along fault or fracture zones would appear to be a suitable explanation for the fact that there are roughly three copper zones in the Permian of North Texas (Fig. 1) but dense drilling for oil, especially near the zones east of the San Angelo Formation, has failed to detect faults or anomalous fracture patterns. It is remotely possible that the movement that occurred along any faults or fractures was solely of an opening and closing nature, making it extremely hard, if not impossible, to detect the existence of these former zones.

Calculations were made (see appendix) using the diffusion equation (Weast, 1971; cf., chapter 2, vol. 2) to see how long it would take to concentrate a 25.4 cm layer of 1% copper. By using 1,840 m, which is the approximate depth to the basement (Flawn, 1956), as the length distance through which diffusion occurred, and an initial concentration of 100 ppm copper, it was determined that it would take 3.25 m.y. to concentrate the copper by diffusion. The diffusion coefficient was for the diffusion of $CuCl_2$ in an aqueous solution (Bruins, 1929); the length of time derived by using this diffusion coefficient value may be substantially too low for the following reasons. According to Manheim (1970), diffusion in unconsolidated sediments generally occurs at rates ranging from one-half to one-twentieth of those occurring in free solution (see chapter 2, vol. 2). Part of this slower rate of net ion flux is caused by reactions between diffusing ions and the enclosing sediments (ibid.). The 100 ppm initial concentration may also be too high since in the Salton Sea geothermal area the concentration of copper in pore waters is only 8 ppm (White, 1968). It would appear from these calculations that diffusion may not be a satisfactory theory for explaining the copper mineralization in the San Angelo Formation, although diffusion may be useful in explaining concentration phenomenon that may require movement of copper over distances of only a few centimeters or meters (e.g., nodules of copper minerals).

The copper deposits at Creta and near Mangum (Fig. 1) in southwestern Oklahoma, although not part of this study, were generally examined since they are at approximately the same stratigraphic level as the copper-bearing shales in Texas. The cap gypsum bed on top of the copper shale at Creta exhibits on cut surfaces a nodular texture, suggesting a sabkha origin. Nodules in the slabs are more tightly packed, more highly distorted, and have less matrix present than gypsum beds from Texas. The high boron content (X2) of the copper-bearing shale, as compared to other shale in the Creta area (Ham and Johnson, 1964), may point to a lagoonal origin for the shale as they propose, or the high concentration may result from bathing of upper tidal-flat algal mats by waters of the restricted shelf during supratidal flooding.

SUMMARY AND CONCLUSIONS

The San Angelo Formation is a Mid-Permian complex of progradational facies composed of terrigenous clastics derived primarily from the Ouachita folded belt and older Paleozoic rocks. The San Angelo has been divided into two members: a lower sand-rich Duncan Member, and an upper mud-rich Flowerpot Member (Fig. 7).

The Duncan Member is composed of deltaic facies of the Copper Breaks deltaic system and the Old Glory fluvial–deltaic system (Fig. 11) that prograded across Choza subtidal facies and submerged tidal-flat and sabkha facies on the eastern side of the Midland Basin. Intermediate between these two depositional systems, and as a thin veneer under them, is the strike-fed, Buzzard Peak sand-rich tidal-flat system (Figs. 8, 11).

Transitionally above the Duncan Member is the Cedar Mountain mud-rich tidal-flat system which is essentially stratigraphically equivalent to the Flowerpot Member (Figs. 7, 8). With gradual cessation of deltaic outbuilding and subsidence of the deltaic sediments, the shoreline was prograded by the Cedar Mountain system (Fig. 12) supplied by mud transported along strike from active deltas to the south.

Coastal tidal-flat progradation led to a gradual raising of the landward portion of the Cedar Mountain system above mean high tide and subsequent offlap of sabkha facies of the Blaine sabkha and tidal-flat system (Lower Blaine Formation). The Blaine system is composed of diagenetically formed nodular gypsum beds of the sabkha facies (Figs. 9, 10) interbedded with thin dolomite and mudstone beds of both tidal-flat and sabkha origin.

The principal stratiform copper deposits are present in relatively widespread mineralized zones in laminated, algal-bound shale beds of the algal mat facies at the top of the Cedar Mountain system; copper also occurs as discontinuous deposits in sandstone lenses filling tidal channels in the Cedar Mountain system (Fig. 3). Minor amounts of copper occur in the basal part of the Duncan Member, at the base of thin ripple bedded sandstone beds of the swash-zone facies in the Cedar Mountain system, and in thin dolomite and nodular gypsum beds at the top of the Cedar Mountain system and in the lower part of the Blaine system. The most important sulfide mineral is chalcocite, which shows evidence of having replaced pyrite (Fig. 13C). Minor amounts of covellite, primarily present as a replacement of chalcocite, are also present. Where recent oxygen-rich waters containing CO_2 have permeated the mineralized facies, malachite and azurite have formed. Organic matter (wood and algal mats) served as loci for the copper mineralization.

Three theories for the origin of San Angelo copper mineralization are proposed. The sabkha—diagenetic theory (Fig. 16) involves evaporative pumping of copper-bearing ground water through the sabkha surface. As ground water passed through algal mats buried beneath the sabkha sediments, copper was stripped out by H_2S and HS^- generated by sulfate-reducing anaerobic bacteria. Some of the copper also replaced pyrite that was formed earlier. Alternate, epigenetic theories involve:

(1) The ascension of copper chloride solutions along faults and fractures with lateral movement to stratigraphic zones favorable for concentration (Fig. 17A)

(2) Ascension of copper by intergranular and intragranular metasomatic diffusion (Fig. 17B). The source of the copper in the epigenetic theories would have been primarily from leaching of the thick underlying sedimentary pile in the Midland Basin with deposition occurring upon contact with H_2S or HS^- in organic-rich facies.

Progradation of the Blaine sabkhas and tidal flats over the Cedar Mountain algal mats, the process of evaporative discharge through the sabkha facies, and the length of time inferred for deposition of the Blaine system provide the essential elements for a viable diagenetic theory.

ACKNOWLEDGEMENTS

L.F. Brown Jr., of the Texas Bureau of Economic Geology suggested this study and provided assistance and guidance throughout the course of my studies. Appreciation is extended to L.B. Gustafson of the Anaconda Company who encouraged publication of this paper.

APPENDIX

Methods

Calculation for Table II: approximate time required to concentrate the required amount of copper from ground water containg 0.06 ppm copper:

$$0.06 \text{ ppm of Cu} = \frac{0.06 \text{ mg Cu}}{1000 \text{ g of } H_2O}$$

$$\frac{6 \cdot 10^{-2} \text{ mg Cu}}{1 \cdot 10^3 \text{ ml of } H_2O} \qquad 1 \text{ ml} = 1 \text{ cc}$$

$$\frac{6 \cdot 10^{-2} \text{ mg Cu}}{1 \cdot 10^3 \text{ cc}} \qquad \frac{6 \cdot 10^{-5} \text{ mg Cu}}{\text{cm}^3 \ H_2O}$$

In one year the evaporation is 100 cm^3/cm^2.

$$(1 \cdot 10^2 \text{ cm}^3/\text{cm}^2) \, (6 \cdot 10^{-5} \text{ mg Cu/cm}^3)$$

In one year = $6 \cdot 10^{-3}$ mg Cu, or in one year $6 \cdot 10^{-6}$ g Cu.
Therefore, in one year $6 \cdot 10^{-3}$ mg of Cu will be brought up.
$8.93 \cdot 10^3$ g/cm^3 density of Cu.
$2.5 \cdot 10^3$ g/cm^3 for shale (approximate)
$2.5 \cdot 10^3$ g shale/cm^3
We have a 25 cc (cm^3) volume.

$$\frac{2.5 \cdot 10^3 \text{ g shale}}{1000 \text{ cm}^3} \times 25 \text{ cm}^3 = \frac{2.5 \cdot 10^3 \text{ g shale}}{4 \cdot 10^1} = 6.25 \cdot 10^1 \text{ g shale in prism 25 cm deep}$$

We assume that the $6.25 \cdot 10^1$ g includes all trace elements, therefore, addition of copper is taken into account and does not contribute appreciably to weight of shale. Therefore, 1% copper in shale would equal a weight of $6.25 \cdot 10^1$ g $\times 1 \cdot 10^{-2} = 6.25 \cdot 10^{-1}$ g of copper. Therefore, we need to know how long it would take to deposit $6.25 \cdot 10^{-1}$ g of copper.

$$\frac{6.25 \cdot 10^{-1} \text{ g of Cu for 25 cm}^3 \text{ prism}}{6 \cdot 10^{-6} \text{ g Cu/yr}} = 1.04 \cdot 10^5 \text{ yr (or 100,000) to deposit enough copper to make}$$

25 cm^3 of shale with 1% Cu. This is assuming that the evaporation rate on the surface equals the amount that can be drawn from the shale. 25 cm of shale was chosen since this is the approximate average thickness of the copper-bearing shale in the area of this report.

Formula, data, and method used for calculating the time for diffusion to concentrate 1% copper in a 25.4 cm thick shale.

$$t = \frac{mh}{\Delta \cdot A(d_2 - d_1)} \qquad \text{(Weast, 1971)}$$

t = time for diffusion to concentrate the copper.
m = mass of substance which diffuses through a specific cross section in a specific time (t).
A = cross sectional area which is diffused through.
Δ = diffusion coefficient.
h = distance through which diffusion occurs.
$d_2 - d_1$ = initial concentration of copper.

Data used: $\Delta = 0.50 \cdot 10^{-5}$ cm^2/sec (Bruins, 1929)

$h = 1{,}840$ m

$A = 1$ cm^2

$m = 0.265$ g (1% of 26.5 g of shale)

$d_2 - d_1 = 100$ ppm (0.1 mg/cc or $1 \cdot 10^{-4}$ g/cm^3)

Additional data: sec/yr = $3 \cdot 10^7$; 25.4 cm of shale @ 2.65 specific gravity = 26.5 g of shale.

$$ t = \frac{0.265 \text{ g} \cdot 1.84 \cdot 10^5 \text{ cm}}{0.5 \cdot 10^{-5} \text{ cm}^2/\text{sec} \times 1 \text{ cm}^2 \times 1 \cdot 10^{-4} \text{ g/cm}^3} = \frac{0.4876 \cdot 10^5 \text{ g cm}}{0.5 \cdot 10^{-9} \text{ g cm/sec}} = $$

$$ \frac{9.75 \cdot 10^{13}}{3.0 \cdot 10^7} = 3.25 \cdot 10^6 \text{ yr} = 3{,}250{,}000 \text{ yr} $$

Editor's note: The reader should consult Chapter 2 by Duursma and Hoede, Vol. 2, for a detailed discussion of diffusion.

REFERENCES

Baas Becking, L.G.M. and Moore, D., 1961. Biogenic sulfides. *Econ. Geol.*, 56 (5): 259–272.

Baker, W.E., 1973. The role of humic acids from Tazmanian podzolic soils in mineral degradation and metal mobilization. *Geochim. Cosmochim. Acta*, 37: 269–281.

Barnes, V.E., 1972. Geologic atlas of Texas, Abilene sheet. *Texas Univ. Bur. Econ. Geol.*, map with explanation.

Barton Jr., P.B., 1959. The chemical environment of ore deposition and the problem of low temperature ore transport. In: P.H. Abelson (Editor), *Researches in Geochemistry*. Wiley, New York, N.Y., pp. 279–300.

Beall Jr., A.O., 1968. Sedimentary processes operative along the Western Louisiana shoreline. *J. Sediment. Petrol.*, 38 (3): 869–877.

Bosellini, A. and Hardie, L.A., 1973. Depositional theme of a marginal marine evaporite. *Sedimentology*, 20 (1): 5–27.

Bruins, H.R., 1929. Coefficient of diffusion in liquids. In: E.W. Washburn (Editor), *International Critical Tables of Numerical Data*. McGraw-Hill, New York, N.Y., 5: 63–76.

Butler, G.P., 1969. Modern evaporite deposition and geochemistry of coexisting brines, the sabkha, Trucial Coast, Arabian Gulf. *J. Sediment. Petrol.*, 39 (1): 70–89.

Butler, G.P., 1970. Holocene gypsum and anhydrite of the Abu Dhabi sabkha, Trucial Coast: an alternative explanation of origin. J.L. Rau and L.F. Dellwig (Editors), *Third Symposium on Salt*, 1. Northern Ohio Geol. Soc., Cleveland, Ohio, pp. 120–152.

Coleman, J. and Gagliano, S.M., 1964. Cyclic sedimentation in the Mississippi River deltaic plain. *Gulf Coast Assoc. Geol. Soc. Trans.*, 14: 67–80.

Collins, B.I., 1973. The concentration control of soluble copper in a mine tailing stream. *Geochim. Cosmochim. Acta*, 37 (1): 69–75.

Dalrymple, D.W., 1965. Calcium carbonate deposition associated with blue-green algal mats, Baffin Bay, Texas. *Inst. Mar. Sci. Publ., Univ. Texas*, 10: 187–200.

Dandurand, J.L., Fortune, J.P., Rerami, R., Schott, J. and Tollon, F., 1972. On the importance of mechanical action and thermal gradient in the formation of metal-bearing deposits. *Miner. Deposita*, 7 (4): 339–350.

Davies, G.R., 1970. Algal-laminated sediments, Gladstone embayment, Shark Bay, Western Australia. In: G.W. Logan et al. (Editors), *Carbonate Sedimentation and Environments, Shark Bay, Western Australia. Am. Assoc. Pet. Geol. Mem.*, 13: 169–205.

Degens, E.T. and Ross, D.A., 1969. *Hot Brines and Recent Heavy Mineral Deposits in the Red Sea*, Springer, New York, N.Y., 600 pp.

Evans, G., Schmidt, V., Bush, P. and Nelson, H., 1969. Stratigraphy and geologic history of the sabkha, Abu Dhabi, Persian Gulf. *Sedimentology*, 12 (1/2): 145–159.

Fisher, W.L., Brown Jr., L.F., Scott, Alan J. and McGowen, J.H., 1969. Delta systems in the exploration for oil and gas — a research colloquim. *Texas Univ. Bur. Econ. Geol. Rep. Invest.*, 77 pp.

Fisk, H.N., 1959. Padre Island and the Laguna Madre flats, coastal South Texas. In: R.J. Russell (Editor), *Second Coastal Geography Conference*. U.S. Off. Nav. Res., Washington, D.C., pp. 103–151.

Flawn, P.T., 1956. Basement rocks of Texas and Southeast New Mexico. *Texas Univ. Bur. Econ. Geol. Publ.*, 5605: 261 pp.

Flawn, P.T., Goldstein Jr., A., King, P.G. and Weaver, C.E., 1961. The Ouachita system. *Texas Univ. Bur. Econ. Geol. Rept. Invest. Publ.*, 6120: 401 pp.

Garrels, R.M. and Christ, C.L., 1965. *Solutions, Minerals and Equilibria*. Harper and Row, New York, N.Y., 450 pp.

Garrels, R.M., Dreyer, R.M. and Howland, A.L., 1949. Diffusion of ions through intergranular speces in water-saturated rocks. *Geol. Soc. Am. Bull.*, 60 (12): 1809–1828.

Gould, H.R. and McFarlan Jr., 1959. Geologic history of the chenier plain, Southwestern Louisiana. *Gulf Coast Assoc. Geol. Soc. Trans.*, 9: 261–270.

Green, D.A., 1937. Major divisions of Permian in Oklahoma and Southern Kansas. *Am. Assoc. Pet. Geol. Bull.*, 21 (12): 1515–1533.

Ham, W.E. and Johnson, K.S., 1964. Copper in the Flowerpot shale (Permian) of the Creta area, Jackson County, Oklahoma. *Okla. Geol. Surv. Circ.*, 64: 32 pp.

Helgeson, H.C., 1964. *Complexing and Hydrothermal Ore Deposition*. Macmillan., New York, N.Y., 28 pp.

Hem, J.D., 1970. Study and interpretation of the chemical characteristics of natural water. *U.S. Geol. Surv. Water-Supply Pap.*, 1473: 363 pp.

Hsü, K.J. and Schneider, J., 1973. Progress report on dolomitization — Hydrology of Abu Dhabi sabkhaz, Arabian Gulf. In: B.H. Purser (Editor), *The Persian Gulf*. Springer, New York, N.Y., pp. 409–422.

Hsü, K.J. and Siegenthaler, C., 1969. Preliminary experiments on hydrodynamic movement induced by evaporation and their bearing on the dolomite problem. *Sedimentology*, 12 (1/2): 12–35.

Kendall, C.G.St.C. and Skipwith, P.A.s'E., 1969. Holocene shallow-water carbonate and evaporite sediments of Khor al Bazam, Abu Dhabi, Southwest Persian Gulf. *Am. Assoc. Pet. Geol. Bull.*, 53 (4): 841–869.

Kharkar, D.P., Turekian, K.K. and Bertine, K.K., 1968. Stream supply of dissolved silver, molybdenum, antimony, selenium, chromiam, cobalt, rubidium and cesium to the oceans. *Geochim. Cosmochim. Acta*, 32 (3): 285–298.

Kinsman, D.J.J., 1969. Model of formation, sedimentary associations and diagnostic features of shallow-water and supratidal evaporites. *Am. Assoc. Pet. Geol. Bull.*, 53 (4): 830–840.

Livingstone, D.A., 1963. Data of geochemistry, Chapter G. Chemical composition of rivers and lakes. *U.S. Geol. Surv. Prof. Pap.*, 440-G, 64 pp.

Lockwood, R.P., 1972. *Geochemistry and Petrology of Some Oklahoma Redbed Copper Occurrences*, Ph.D. Diss., Univ. Okla., 58 pp.

Love, L.G., 1971. Early diagenetic polyframboidal pyrite, primary and redeposited, from the Wenlockian Denbigh Grit Group, Conway, North Wales, U.K. *J. Sediment. Petrol.*, 41 (4): 1038–1044.

Lucia, F.C., 1972. Recognition of evaporite–carbonate shoreline sedimentation. In: J.K. Rigby and K.W. Hamlin (Editors), *Recognition of Ancient Sedimentary Environments*. Soc. Econ. Paleontol. Mineral., Spec. Publ., 16: 160–191.

Manheim, F.T., 1970. The diffusion of ions in unconsolidated sediments. *Earth Planet. Sci. Lett.*, 9: 307–309.

Martin, D.F., Doig, M.T., III and Pierce Jr., R.H., 1971. Distribution of naturally occurring chelators (humic acids) and selected trace metals in some west coast Florida streams, 1968–1969. *Fla Dept. Nat. Resour., Univ. S. Fla, Prof. Pap. Ser.*, 12: 52 pp.

Morgan, J.P., Van Lopik, J.R. and Nichols, L.G., 1953. Occurrence and development of mudflats along the Western Louisiana Coast. *La State Univ. Coastal Stud. Inst. Tech. Rept.*, 2: 34 pp.

Olson, E.C., 1962. Late Permian terrestrial vertebrates, U.S.A. and U.S.S.R. *Am. Philos. Soc. Trans., New Ser.*, 52 (2): 224 pp.

Oriel, S.S., Myers, D.A. and Crosby, E.J., 1967. West Texas Permian basin region. In: E.D. McKee, S.S. Oriel et al. (Editors), *Paleotectonic Investigations of the Permian System in the United States. U.S. Geol. Surv. Prof. Pap.*, 515: 21–60.

Phillips, L.C., 1917a. The copper beds of Knox County. *Texas Univ. Bur. Econ. Geol. Open-file Rept.*, 14 pp.

Phillips, L.C., 1917b. Report on the property of the Foard County Copper Company. *Texas Univ. Bur. Econ. Geol. Open-file Rept.*, 6 pp.

Purdy, E.G., 1963. Recent calcium carbonate facies of the Great Bahama Bank. 2. Sedimentary Facies. *J. Geol.*, 71 (4): 472–497.

Ramberg, H., 1952. *The Origin of Metamorphic and Metasomatic Rocks.* Univ. Chicago Press, Chicago, 317 pp.

Ramdohr, P., 1969. *The Ore Minerals and Their Intergrowths.* Pergamon, New York, N.Y., 1174 pp.

Rashid, M.A. and Leonard, J.D., 1973. Modifications in the solubility and precipitation behavior of various metals as a result of their interaction with sedimentary humic acids. *Chem. Geol.*, 11 (2): 89–97.

Reineck, H.E., 1967. Layered sediments of tidal flats, beaches, and shelf bottoms of the North Sea. In: G.H. Lauff (Editor), *Estuaries.* Am. Assoc. Adv. Sci. Publ. No. 83, Wash., pp. 191–205.

Reineck, H.E., 1972. Tidal flats. In: J.K. Rigby and W.K. Hamblin (Editors), *Recognition of Ancient Sedimentary Environments.* Soc. Econ. Paleontol. Mineral. Spec. Publ., 16: 146–159.

Renfro, A.R., 1974. Genesis of evaporite-associated metalliferous deposits: a sabkha process. *Econ. Geol.*, 69 (1): 33–45.

Rickard, D.T., 1970. The chemistry of copper in natural aqueous solutions. *Stockh. Contrib. Geol.*, 23 (1): 64 pp.

Rust, G.W., 1935. Colloidal primary copper ores at Cornwall mines Southeastern Missouri. *J. Geol.*, 43: 398–426.

Scruton, P.C., 1960. Delta building and the deltaic sequence. In: F.P. Shepard et al. (Editors), *Recent Sediments, Northwest Gulf of Mexico Bull. Am. Assoc. Pet. Geol.*, 44: 82–102.

Sipple, R.F., 1968. Sandstone petrology, evidence from luminescence petrography. *J. Sediment. Petrol.*, 38 (3): 530–554.

Smith, G.E., 1974. *Depositional Systems and Facies Control of Copper Mineralization – San Angelo Formation (Permian), North Texas.* Thesis, Univ. Texas, Austin, Texas, 178 pp.

Stroud, R.B., McMahan, A.B., Stroup, R.K. and Hibpshman, M.H., 1970. Production potential of copper deposits associated with Permian red-bed formations in Texas, Oklahoma, and Kansas, *U.S. Bur. Mines. Rept. Ivest.*, 7422: 103 pp.

Swanson, V.E., Love, A.H. and Frost, I.C., 1972. Geochemistry and diagenesis of tidal-marsh sediment, Northwestern Gulf of Mexico. *U.S. Geol. Surv. Bull.*, 1360: 83 pp.

Tarling, D.H., 1973. Metallic ore deposits and continental drift. *Nature*, 243 (5404): 193–196.

Temple, K.L. and LeRoux, 1964. Syngenesis of sulfide ores: sulfate-reducing bacteria and copper toxicity. *Econ. Geol.*, 59 (2): 271–278.

Trudinger, P.A., Lambert, I.B. and Skyring, G.W., 1972. Biogenic sulfide ores: a feasibility study. *Econ. Geol.*, 67 (8): 1114–1127.

Van Straaten, L.M.J.U., 1961. Sedimentation in tidal flat areas. *Alta. Soc. Pet. Geol. J.*, 9 (7): 203–226.

Volfson, F.I. and Arkhangel'skaya, V.V., 1971. The conditions of formation of cupriferous sandstone deposits. *Lithol. Miner. Resour.*, 7 (3): 268–279.

Walper, J.L. and Rowett, C.L., 1972. Plate tectonics and the origin of the Caribbean Sea and the Gulf of Mexico. *Gulf Coast Assoc. Geol. Soc. Trans.*, 22: 105–116.

Weast, R.C., 1971. *Handbook of Chemistry and Physics.* Chem. Rubber Co., Cleveland, Ohio, 1144 pp.

White, D.E., 1968. Environments of generation of some base-metal ore deposits. *Econ. Geol.*, 63 (4): 301–335.

White, D.E., Hem, J.D. and Waring, G.A., 1963. Data of geochemistry, Chapter F. Chemical composition of subsurface waters. *U.S. Geol. Surv., Prof. Pap.*, 440-F: 67 pp.

Chapter 9

CARBONATE-HOSTED LEAD–ZINC DEPOSITS

D.F. SANGSTER

INTRODUCTION

Carbonate-hosted lead–zinc deposits constitute one of the world's great sources of these two metals. This type of deposit, the principal source of lead and zinc in both the United States and Europe, is represented in the famous Appalachian, Tri-State, southeast Missouri, and upper Mississippi Valley districts in the U.S., and in the Alpine, Silesian, Central Irish Plain, and Pennines districts of Europe and Great Britain. In the papers that follow, descriptions of some of these better-known mining districts are presented as well as a review of the internal features of this intriguing class of deposits.

As a group, carbonate-hosted lead–zinc deposits exclude skarns and other replacement deposits, such as the Pb–Zn bodies at Bingham, Utah, which could normally be related to nearby intrusions. By far the majority of these deposits occur in dolomite; magnesian limestone and limestone are less-common hosts. Although many districts can be shown to be spatially related to large-scale sedimentary features such as reefs, facies changes, basin margins, and basement topography, the major factor necessary for development of deposits of this type appears to be the presence of a thick carbonate sequence. Thin carbonate layers in shales, however, seldom contain significant deposits of this type. Formation of carbonate sediments is, of course, enhanced by proper climatic conditions, particularly warm water, and therefore it is not surprising that a plot of major carbonate-hosted lead–zinc deposits on available paleolatitude maps shows a preference of these deposits for low paleolatitudes. The warmer climates of low latitudes also encourage the development of reefs and so an association between reefs and base-metal deposits should not be surprising. This general reef–base metal association has been emphasized recently by Monseur and Pel (1973) and is particularly evident in such well-known districts as Pine Point, Canada (Skall, 1975) and southeast Missouri, U.S.A. (Snyder and Gerdemann, 1968; Gerdemann and Meyers, 1972). The open framework of a carbonate reef, however, is not the *only* site for these deposits. For example, the importance of the back-reef environment has been stressed by Schneider (1964).

As pointed out by Stanton (1972, p. 543), important carbonate-hosted lead–zinc deposits occur in rocks of every age from Precambrian to Cretaceous with the curious apparent exception of the Silurian (the Silurian anomaly with respect to lead–zinc is also shown in Strakhov's (1970) diagram (fig. 26) of the stratigraphic distribution of sedimentary lead and zinc ores). In the Proterozoic, however, relative to the abundance of carbonate rocks, lead–zinc deposits appear to be rather uncommon. Only six come to mind, the Broken Hill Mine in Zambia (Whyte, 1968); the Nanisivik Mine, Canada (Geldsetzer, 1973); the Black Angel Mines, Greenland (Fish, 1974); the Balmat–Edwards district, U.S.A. (Lea and Dill Jr., 1968); the Zeerust district, South Africa (Willemse et al., 1944); and the unique zinc-oxide deposits of Franklin and Stirling Hill, U.S.A. (Frondel, 1972). Numerous smaller deposits in Proterozoic carbonates are known, to be sure, but relative to the number of significant ore bodies in Phanerozoic rocks, the much longer time range represented by the Proterozoic, and the abundance of thick carbonates in the latter, the Proterozoic appears to be under-represented in terms of carbonate-hosted lead–zinc deposits.

For a variety of reasons, data on grade and tonnage of individual deposits are extremely difficult to compile. For the older districts, production records may have been lost or perhaps never kept, mine operators frequently do not publish even production figures let alone reserve data, and the geology of these deposits is often such that the ore does not occur in individual, discrete bodies but, instead, may occur as irregular and ill-defined zones connecting higher-grade pockets. However, in the author's experience, average ore grades in the larger districts range between 3 and 10% combined lead–zinc with individual ore bodies or zones running sometimes up to 50% Pb–Zn. Furthermore, where it is possible to distinguish individual deposits, tonnages range between a few tens of thousands to perhaps 10–20 million tons. Some of the average grade and tonnage data for a few of the better-known districts are presented in Table I.

In previous publications, the author (Sangster, 1970, 1975) has attempted to distinguish between what appear to be two major classes of carbonate-hosted lead–zinc deposits: (1) Mississippi Valley type, which are strata-bound, were emplaced after lithification of the host rocks, and were largely controlled by pre-ore structures, and (2) Alpine type, which appear to be stratiform and synsedimentary relative to their host rocks.

TABLE I

Approximate grade and tonnage of several carbonate-hosted lead–zinc districts

District	% Pb	% Zn	Tons (short)
Tri-State, U.S.A.	0.6	2.3	500,000,000
Eastern Tenn., U.S.A.	–	4	50,000,000
Old Lead Belt, U.S.A.	3		370,000,000
Upper Miss. Valley, U.S.A.		4	50,000,000
Pine Point, Canada *	3	7	65,000,000
Central Irish Plain, Ireland	3	10	115,000,000

* The Pine Point data include 1.4 million tons of direct shipping ore at 19.3% Pb and 26.7% Zn.

MISSISSIPPI VALLEY TYPE[1]

As the name implies, classical deposits of this type occur in the drainage basin of the Mississippi Valley of the central United States. Similar deposits, however, also occur in the Appalachian Valley district, U.S.A. and the Mackenzie Valley (Pine Point) district of Canada.

The presence or absence of structural controls to this class of carbonate-hosted lead—zind deposits and the age of these structural features relative to that of the mineralization has led, and will undoubtedly continue to lead, to a bewildering array of genetic models for these deposits. It has been the author's experience, however, that a majority of carbonate-hosted lead—zinc deposits of the type considered here are the result of open-space filling, i.e., that mineralization was emplaced into pre-existing rocks. This emplacement took place regardless of the original depositional environment of the carbonate. While reefs are mentioned as possible important factors in southeast Missouri (Snyder and Gerdemann, 1968), the Central Irish Plain (Morrissey et al., 1971), and Pine Point (Skall, 1975) districts, widespread, non-reef, platform carbonates of uniform thickness and composition, lacking rapid lateral facies changes, are the host rocks in upper Mississippi Valley (Heyl, 1968), Tri-State (Brockie et al., 1968), eastern Tennessee (Crawford and Hoagland, 1968), and Silesia (Galkiewicz, 1967; Gruszczyk, 1967). The characteristic widespread host-rock brecciation, albeit of diverse origins, followed by infilling by zinc and lead sulphides, accounts for the epigenetic character of many districts. In some areas this structural control appears to be largely due to what may be referred to as mild tectonic stresses (e.g. the English Pennines?). In other areas the brecciation is apparently only indirectly related to structure in that broad upwards have resulted in erosion and karsting of the carbonate rocks by meteoric waters. As with modern-day caves, carbonate solution is often preferentially directed along strong joints or small faults in the host rocks. Shallow-water carbonates are naturally subject to widespread development of disconformities and/or minor unconformities caused by small-scale perturbations in sea level during deposition. Unconformity-related deposits enjoy an enthusiastic school of proponents, chiefly because of the variety of secondary-porosity situations (paleoaquifers), e.g., collapse breccias, solution breccias, karst caves, etc. generated by the circulation of oxygenated meteoric water through the carbonate during periods of emergence (see, for example, Callahan, 1964; Wedow Jr., 1971; Bernard, 1973).

In addition to that provided by structural means, porosity in carbonate rocks can also be brought about by normal sedimentary or diagenetic means such as the primary porosity of a reef, or by secondary means such as dolomitization. In many instances, of course, the distinction between structurally controlled and sedimentary-controlled brecciation is not clear because processes such as secondary dolomitization or carbonate

---

[1] Editor's note: For a discussion on the sub-types of lead—zinc ore deposits, see also Amstutz (1972).

solution are instigated or directionally controlled by subtle structural features.

Thus, with all these variations and combinations of primary and secondary porosity available, it is perhaps not surprising that some of them would serve as traps to base-metal mineralizing fluids, just as they are traps to other subsurface fluids such as petroleum and associated brines. In fact, comparison of the compositions of oil-field brines and fluid inclusions in Mississippi Valley ores has resulted in widespread advocation of some form of connate brine as the transporting medium for these deposits (see, for example, Jackson and Beales, 1967; Ford, 1969; Bush, 1970; Dunham, 1970; Dozy, 1970). Furthermore, the close spatial association of Mississippi Valley deposits with oil fields is well known and the relation brine–oil–metal was recently presented by Dozy (1970, p. 163): "There is now general acceptance of the view that the metallic elements in the ores have been supplied, at least to a large extent, by brines– connate or formation waters from adjacent basins. Ore deposition took place from these brines in the ore districts ... These formation waters have also been the medium with which oil has been intimately associated from its genesis through migration to accumulation." The oil (or hydrocarbon) itself has also been postulated to have played an active role in ore formation by acting as the necessary reductant to convert dissolved sulphate to sulphide (hydrogen sulphide). Reduction may be brought about either through the agency of bacteria at low temperatures or non-biologically at higher temperature (Dunsmore, 1973). (The oil–metal association was also emphasized recently at a "Forum on oil and ore in sediments" held at Imperial College, London, March, 1975). The presence or absence of hydrogen sulphide (sour gas) may be the controlling factor in the formation of this type of ore deposit; in fact, the predominance of sour-gas fields in carbonate rocks as opposed to sandstones has been pointed out by Beales and Jackson (1968). In the clastic rocks, the H_2S may have been stripped out by reaction with iron in the sediment and precipitated as iron sulphides.[1]

The result of these analogies between hydrocarbons, brines, basins, etc., has been the establishment of a genetic model for Mississippi Valley type deposits which involves more-or-less independent migration of metal-bearing chloride brines (Nriagu and Anderson, 1971) and H_2S-bearing fluids to a common site where precipitation of the metals as sulphides takes place. This depositional site may be controlled by porosity brought about by secondary processes such as solution collapse and dolomitization or it may be primary porosity such as found within a reef, reef talus or other sedimentary breccias, all of which are also common hydrocarbon traps. Whether it is hydrocarbons or metals, the model presented requires emplacement of the sulphides into *pre-existing rocks and traps,* presumably some time after lithification of the host and at some depth to produce the fluid-inclusion temperatures recorded (assuming 25°C surface temperature and a gradient of 3°C/100 m, they can be attained at a depth of 2500 m).

---

[1] Editor's note: See related synopsis with detailed discussion by Wolf (1976) and Wolf and Chilingarian (1976).

High-quality laboratory studies on fluid-inclusion compositions and temperatures, experimental studies on the solubility of metal-chloride complexes, isotopic studies of several types, combined with diligent field observations, have all led to a better understanding of this type of deposit in almost every aspect save one – the age of ore emplacement. Many Mississippi Valley type ore districts have not undergone significant tectonic disturbance since the deposition of the host rocks and, for these, age of mineralization is difficult, if not impossible, to determine with any degree of certainty. In the East Tennessee district, for example, zinc sulphides and white dolomite fill interstices between fragments of collapse breccias developed in Lower Ordovician dolomite. The collapse breccias, formed as a result of a paleoaquifer which was established during a period of uplift and erosion during Middle Ordovician time, were once thought to be of tectonic origin, related to the Late Paleozoic Appalachian orogeny. However, Kendall (1960, 1961) pointed out the presence of detrital sphalerite and dolomite, exhibiting graded bedding, in certain of the ore structures. The sphalerite–dolomite sand is clearly post-ore in age and represents "in-cave" re-sedimentation of these two components as a result of continued, post-ore partial dissolution of gangue dolomite with the resultant freeing of sphalerite grains which were then re-deposited as sand grains. As Crawford and Hoagland state (1968, p. 249): "The most significant observation with respect to this sand is that the laminations are parallel to the strike and dip of the country rock. This parallelism of the bedding of the formation and the sand laminations indicates that the formations were essentially horizontal when the sand was deposited. Hence the ore must have been deposited before any tilting of the strata occurred. It follows, therefore, that the ore deposition must have taken place before the period of the Appalachian orogeny." Hence, while the spectacular collapse breccias, so characteristic of this district, appear to be related to an early Middle Ordovician unconformity (Hill and Wedow Jr., 1971), the time of emplacement of zinc sulphides *into* these permeable zones must be in the interval between the early Middle Ordovician and Late Paleozoic, over 200 million years! For the upper Mississippi Valley district, Heyl (1968, p. 442) states: "The age of the deposits cannot be definitely determined from any of the known relations in the Mississippi Valley". Similarly for the Tri-State district, hosted in Mississippian (Lower Carboniferous) carbonates, Brockie et al. (1968, p. 413) concluded that the ores "were emplaced in the Mississippian sediments as late as (or even much later than) post-Cherokee (Pennsylvanian) time". Age of mineralization for other well-known Mississippi Valley type ore districts such as, for example, Pine Point (Skall, 1975) and southeast Missouri (Snyder and Gerdemann, 1968) are all similarly unknown.

An attempt to date mineralization by paleomagnetism is a refreshingly new approach and for two deposits, that of Newfoundland Zinc hosted in Lower Ordovician dolomite and the St. Joe Minerals #8 Mine hosted in Upper Cambrian dolomite, Beales et al. (1974, p. 222) concluded that "hosts and ores have statistically indistinguishable remnant magnetization ... therefore the emplacement of the ores seems to be close, in a geological time sense, to the deposition of the host. A time span of less than 25 m.y. is likely to encompass both the formation of the host and the ore." Two other deposits, the Pine Point orefield in Middle Devonian host rocks and the Magmont Mine (in Upper Cambrian), had magnetization intensities too low to be measured. Nevertheless, in view of the inability of lead isotopes in Mississippi Valley type deposits to accurately define the age of mineralization (Sangster, Vol. 2, Chapter 8), the results presented by Beales et al. (1974) appear sufficiently encouraging to warrant further study on other districts of similar type. A better definition of the age of mineralization in Mississippi Valley type deposits would do much to dispel the plethora of multiple genetic hypotheses surrounding these controversial deposits.

ALPINE TYPE

The Alpine type lead–zinc ores are named after deposits of the Alpine Mesozoic geosyncline of central Europe, particularly those in the eastern Alps (Maucher and Schneider, 1967). As with the Mississippi Valley type, the Alpine-type deposits also show epigenetic features such as vein fillings, breccia cementation, and massive replacement bodies. The main difference however, as pointed out by Schneider (1964, p. 73), is that while the ore bodies, *in senso stricto,* have some epigenetic features, the non-economic equivalents, i.e. mineralization *in senso lato,* are clearly syngenetic or syn-sedimentary. In reference to the Alpine lead–zinc deposits of the Triassic, he states: "As described in the literature, the ore occurs predominantly as more-or-less massive accumulations in short unconformable veins and also in larger 'replacement bodies'. These types of ore bodies are of major economic interest. In addition, however, the ore occurs in far lesser concentration on conformable layers and lenticular bodies which, in general, contain a sprinkling of ore mixed with various types of country rock". Schneider then proceeds to describe and illustrate with photographs several examples of lead–zinc mineralization exhibiting many features of primary sedimentation (1964, pp. 37–38; Maucher and Schneider, 1967, pp. 80–81; also Schulz, 1964, pp. 49–51). Implicit in the Schneider–Maucher school of thought is that the epigenetic ore zones merely represent secondary remobilization and concentration of elements originally present in the essentially syngenetic (–diagenetic) protore. This, they maintain, is the reason for the restriction of the "epigenetic" (in their present position) ore bodies to narrow stratigraphic zones in the otherwise thick Triassic carbonates. In addition to those in the Alps, several Irish deposits, notably Tynagh (Derry et al., 1965), Mogul (Silvermines) (Graham, 1970) and Navan are, for the most part, conformable with host-rock stratigraphy except where they are in fault contact with the pre-Carboniferous basement. Because limestone, rather than dolomite, is host to some of the Irish deposits, in contrast to most Mississippi Valley type deposits, a secondary mineralization process somehow related to dolomitization cannot be invoked to account for their presence. The essentially stratiform nature of such deposits as Tynagh and Navan and their close association with Carboniferous volcanic centres (Morrissey et al., 1971, fig. 1) are, in these respects, similar to the Alpine deposits which are also associated with contemporaneous volcanism (Maucher and Schneider, 1967). The main difference between the two districts would appear to be that the Alpine deposits, in their original form, are too low-grade to be economic and require post-depositional remobilization and concentration to upgrade the original protores whereas the Irish deposits may simply represent a higher-grade "protore."

The difference between Mississippi Valley and Alpine type deposits requires *slightly* different genetic models for the two types because of the difference in the time of sulphide deposition relative to host rock in the two types. A model involving emplacement of sulphides into pre-existing traps, whether primary or secondary, may well explain most of the observed features of the Mississippi Valley type of carbonate-hosted lead—zinc deposits. It cannot, however, explain the many syn-sedimentary ore textures described by Schneider (1964) and Schulz (1964), for example, in Alpine type lead—zinc deposits of the "Calcareous Alps". Intercalated laminations of sulphides and clay marls, together with "ore and limemud breccia" (Schneider, 1964, fig. 7; Maucher and Schneider, 1967, figs. 4 and 5) are presented as textural evidence that mineralization was essentially contemporaneous with the host sediments in contrast to the typical Mississippi Valley deposits. Similarly for certain carbonate-hosted lead—zinc deposits in Canada, Sangster (1970, 1975) distinguished between a Mississippi Valley type (occurring in post-host permeable zones) and a Remac type (stratiform deposits consisting of alternating layers of carbonate and sulphides). In their form relative to host rocks, the Alpine (Remac; Irish) type of deposit is more akin to such shale-hosted deposits as Mt. Isa, Sullivan, etc. Contemporaneous volcanism is a feature of the Triassic Alpine deposits, the Carboniferous Irish deposits and possibly the Cambrian Remac deposits and, as has been suggested for the first two of these (Schneider, 1964; Schulz, 1964; Morrissey et al., 1971), may constitute a contemporaneous source of metal-bearing fluids for these deposits as opposed to the relatively later metal-bearing fluids of the typical Mississippi Valley type deposits.

DISCUSSION

The above-mentioned distinction between typical Mississippi Valley type deposits (characteristically discordant, open-space filling) and Alpine (—Irish) type deposits is important because most studies of such features as fluid inclusions, temperature of formation, isotopes, etc., have been performed on the former and hence have profoundly influenced genetic hypotheses with regard to carbonate-hosted lead—zinc deposits in general. If the Alpine type deposits are of submarine-exhalative origin (Schulz, 1964), then migration of ore fluids through sea-water with consequent sulphide precipitation in carbonate muds is a genetic model requiring further research. Petroleum geologists have developed the concept of a source rock (containing a low content of syngenetic hydrocarbons) and a reservoir rock (into which a relatively higher content of epigenetic hydrocarbons is emplaced). In the present context, the possibly syngenetic, usually low-grade Alpine type deposits would be equivalent to the source rock (protore). In some cases, such as Ireland, this source rock (protore) is of sufficiently high grade that it can be profitably exploited, just as syngenetic oil shales can occasionally be exploited. In other cases, the low-grade protore must undergo local epigenetic concentration to become commercially attractive and, in this sense, becomes equivalent to the epigenetic petroleum deposits. Identification of the source rock to epigenetic deposits is as much a problem to the petroleum geologist as it is to the mineral-deposits geologist. Both groups, in addition, may be dealing with similar processes involving initial depositions of low-grade material, followed by migration (either local or long distance) and redeposition into higher-grade deposits. (Cf. Wolf, 1976.)

Whether they be of the Mississippi Valley or Alpine type, however, carbonate-hosted lead–zinc deposits (among others) appear to owe their origin and distribution to normal sedimentary (and diagenic) processes rather than igneous. The association with reefs, facies changes, hydrocarbons, unconformities, subsurface brine, etc., are all normal features and products of sedimentary basins and rocks. Even where minor volcanism may be a factor, the distribution of the Alpine ores, for example, is restricted to the "back-reef" carbonate facies, i.e., a sedimentary control. To exmphasize the strong sedimentary aspect of these deposits, the author proposes the term "sedimentogenic" to stress the genetic connection between these deposits and sedimentation and/or sedimentary processes (including continued movement of basinal fluids). Many other types of sediment-hosted deposits could conceivably be covered by this term but the author has found it useful to collectively refer to this class of deposits just as the term "volcanogenic" is used in a similar sense (Sangster, 1972) for another class (see also Sangster and Scott, this volume, Chapter 5).

REFERENCES

Amstutz, G.C., 1972. Observation criteria for the classification of Mississippi Valley–Bleiberg–Silesia type of deposits. 2nd. Int. Symp. Min. Deposits Alps, Ljubljana, pp. 207–215.

Beales, F.W. and Jackson, S.A., 1966. Precipitation of lead–zinc ores in carbonate reservoirs as illustrated by Pine Point ore field. *Trans Inst. Min. Metall., Sec. B*, 75: 278–285.

Beales, F.W. and Onazick, E.P., 1970. Stratigraphic habitat of Mississippi Valley type orebodies. *Trans. Inst. Min. Metall., Sec. B*, 79: 145–154.

Beales, F.W., Carracedo, J.C. and Strangway, D.W., 1974. Paleomagnetism and the origin of Mississippi Valley-type ore deposits. *Can. J. Earth Sci.*, 11: 211–223.

Bernard, A.J., 1973. Metallogenic processes of intra-karstic sedimentation. In: G.C. Amstutz and A.J. Bernard (Editors), *Ores in Sediments*. Springer, Berlin, pp. 43–57.

Brockie, D.C., Hare Jr., E.H. and Dingess, P.R., 1968. The geology and ore deposits of the Tri-State district of Missouri, Kansas, and Oklahoma. In: J.D. Ridge (Editor), *Ore Deposits in the United States, Vol. 1*. American Institute of Mining, Metallurgical and Petroleum Engineers, New York, N.Y., pp. 400–430.

Bush, P.R., 1970. Chloride-rich brines from sabkha sediments and their possible role in ore formation. *Trans. Inst. Min. Metall., Sec. B*, 79: 137–144.

Callahan, W.H., 1964. Paleophysiographic premises for prospecting for strata bound base metal deposits in carbonate rocks. CENTO Symposium on mining geology and base metals, Ankara, Turkey, pp. 191–248.

Crawford, J. and Hoagland, A.D., 1968. The Mascot–Jefferson City zinc district, Tennessee. In: J.D. Ridge (Editor), *Ore deposits in the United States, Vol. 1*. American Institute of Mining, Metallurgical and Petroleum Engineers, New York, N.Y., pp. 242–256.

Derry, D.R., Clark, G.C. and Gillat, N., 1965. The Northgate base metal deposit at Tynagh, County Galway, Ireland. *Econ Geol.*, 60: 1218–1237.

Dozy, J.J., 1970. A geological model for the genesis of the lead–zinc ores of the Mississippi Valley, U.S.A. *Trans. Inst. Min. Metall., Sec. B*, 79: 163–170.

Dunham, K.C., 1970. Mineralization by deep formation waters: a review. *Trans. Inst. Min. Metall., Sec. B*, 79: 127–136.

Dunsmore, H.E., 1973. Diagenetic processes of lead–zinc emplacement in carbonates. *Trans. Inst. Min. Metall., Sec. B*, 82: 168–173.

Fish, R., 1974. Mining in Arctic lands: The Black Angel experience. *Can. Min. J.*, No. 8: 24–36.

Ford, T.D., 1969. The stratiform ore deposits of Derbyshire. In: C.H. James (Editor), *Sedimentary Ores, Ancient and Modern*. Proc. 15th Inter-University Geol. Congr. Univ. of Leicester, Spec. Publ. No. 1: 73–96.

Frondel, C., 1972. *The Minerals of Franklin and Sterling Hill: A Check List*. Wiley-Interscience, New York, N.Y., 94 pp.

Galkiewicz, T., 1967. Genesis of Silesian–Cracovian zinc–lead deposits. In: J.S. Brown (Editor), *Genesis of Stratiform Lead–Zinc–Barite–Fluorite Deposits. Econ. Geol. Monogr.*, 3: 156–168.

Geldsetzer, H., 1973. Syngenetic dolomitization and sulfide mineralization. In: G.C. Amstutz and A.J. Bernard (Editors), *Ores in Sediments*. Springer, Berlin, pp. 115–127.

Gerdemann, P.E. and Myers, H.E., 1972. Relationship of carbonate facies patterns to ore distribution and to ore genesis in the southeast Missouri lead district. *Econ. Geol.*, 67: 426–433.

Graham, F., 1970. *The Mogul-Base Metal Deposits, Co. Tipperary, Ireland*. Thesis, Univ. of Western Ontario, London, Ont., 227 pp., unpublished.

Gruszczyk, H., 1967. The genesis of the Silesian–Cracow deposits of lead–zinc ores. In: J.S. Brown (Editor), *Genesis of Stratiform Lead–Zinc–Barite–Fluorite Deposits. Econ. Geol. Monogr.*, 3: 169–177.

Heyl, A.V., 1968. The upper Mississippi Valley base-metal district. In: J.D. Ridge (Editor), *Ore Deposits in the United States, Vol. 1*. American Institute of Mining, Metallurgical and Petroleum Engineers, New York, N.Y., pp. 431–459.

Hill, W.T. and Wedow Jr., H., 1971. An early Middle Ordovician age for collapse breccias in the East Tennessee zinc districts as indicated by compaction and porosity features. *Econ. Geol.*, 66: 725–734.

Jackson, S.A. and Beales, F.W., 1967. An aspect of sedimentary basin evolution: the concentration of Mississippi Valley-type ores during late stages of diagenesis. *Bull. Can. Pet. Geol.*, 15: 383–433.

Kendall, D.L., 1960. Ore deposits and sedimentary features – Jefferson City Mine, Tennessee. *Econ. Geol.*, 55: 985–1003.

Kendall, D.L., 1961. Ore deposits and sedimentary features in Tennessee. *Econ. Geol.*, 56: 1137–1138.

Lea, E.R. and Dill Jr., D.B., 1968. Zinc deposits of the Balmat–Edwards district, New York. In: J.D. Ridge (Editor), *Ore Deposits in the United States, Vol. 1*. American Institute of Mining, Metallurgical and Petroleum Engineers, New York, N.Y., pp. 20–48.

Maucher, A. and Schneider, H.J., 1967. The Alpine lead–zinc ores. In: J.S. Brown (Editor), *Genesis of Stratiform Lead–Zinc–Barite–Fluorite Deposits. Econ. Geol. Monogr.,* 3: 71–81.

Monseur, G. and Pel, J., 1973. Reef environment and stratiform ore deposits (Essay of a synthesis of the relationship between them). In: G.C. Amstutz and A.J. Bernard (Editors), *Ores in Sediments.* Springer, Berlin, pp. 195–207.

Morrissey, C.J., Davis, G.R. and Steed, R.M., 1971. Mineralization in the Lower Carboniferous of central Ireland. *Trans. Inst. Min. Metall., Sec. B,* 80: 174–185.

Nriagu, J.D. and Anderson, G.M., 1971. Stability of the lead chloride complexes at elevated temperatures. *Chem. Geol.,* 7: 171–183.

Sangster, D.F., 1970. Metallogenesis of some Canadian lead–zinc deposits in carbonate rocks. *Geol. Assoc. Can. Proc.,* 22: 27–36.

Sangster, D.F., 1972. Precambrian volcanogenic massive sulphide deposits in Canada: A review. *Geol. Surv. Can. Pap.,* 72–22: 44 pp.

Sangster, D.F., 1975. Canadian carbonate-hosted lead–zinc deposits: A summary. *Geol. Soc. Am. – Geol. Assoc. Can. – Min. Assoc. Can., Abstr. with Programs,* 7 (6): p. 848.

Schneider, H.J., 1964. Facies differentiation and controlling factors for the depositional lead–zinc concentration in the Ladinian geosyncline of the eastern Alps. In: G.C. Amstutz (Editor), *Developments in Sedimentology, 2. Sedimentology and Ore Genesis.* Elsevier, Amsterdam, pp. 29–45.

Schulz, O., 1964. Lead–zinc deposits in the calcareous Alps as an example of submarine-hydrothermal formation of mineral deposits. In: G.C. Amstutz (Editor), *Sedimentology and Ore Genesis. Developments in Sedimentology, 2.* Elsevier, Amsterdam, pp. 47–52.

Skall, H., 1975. The paleoenvironment of the Pine Point lead–zinc district, *Econ. Geol.,* 70: 22–47.

Snyder, F.R. and Gerdemann, P.E., 1968. Geology of the southeast Missouri lead district. In: J.E. Ridge (Editor), *Ore Deposits in the United States, Vol. 1.* American Institute of Mining, Metallurgical and Petroleum Engineers, New York, N.Y., pp. 326–358.

Stanton, R.L., 1972. *Ore Petrology.* McGraw-Hill, New York, N.Y., 713 pp.

Strakhov, N.M., 1970. Accumulations of Cu–Pb–Zn; their origin and distribution in arid regions. In: *Principles of Lithogenesis, Vol. 3.* Plenum, New York, N.Y., 577 pp. (also published by Oliver & Boyd, Edinburgh).

Taupitz, K.C., 1967. Textures in some stratiform lead–zinc deposits. In: J.S. Brown (Editor), *Genesis of Stratiform Lead–Zinc–Barite–Fluorite Deposits. Econ. Geol. Monogr.,* 3: 71–81.

Wedow Jr., W.H. (Chairman), 1971. A paleoaquifer and its relation to economic mineral deposits: The Lower Ordovician Kingsport Formation and Mascot dolomite. *Econ. Geol.,* 66: 695–810.

Whyte, W.J., 1968. Geology of the Broken Hill Mine, Zambia. In: P. Nicolini (Editor), *Lead–Zinc Deposits in Africa, Ann. Mines Geol. (Tunis),* No. 23: 393–426.

Willemse, J., Schwellnus, C.M., Brandt, J.W., Russell, H.D. and Van Rooyen, D.P., 1944. Lead deposits in the Union of South Africa and South West Africa with some notes on associated ores. *Mem. Geol. Surv. S. Afr.,* 39: 177 pp.

Wolf, K.H., 1976. Ore genesis influenced by compaction. In: G.V. Chilingarian and K.H. Wolf (Editors), *Compaction of Coarse-Grained Sediments,* II. Elsevier, Amsterdam, in press.

Wolf, K.H. and Chilingarian, G.V., 1976. Diagenesis of sandstones and compaction. In: G.V. Chilingarian and K.H. Wolf (Editors), *Compaction of Coarse-Grained Sediments,* II. Elsevier, Amsterdam, in press.

Chapter 10

TRI-STATE ORE DEPOSITS: THE CHARACTER OF THEIR HOST ROCKS AND THEIR GENESIS

RICHARD D. HAGNI

INTRODUCTION

The Tri-State zinc–lead district in southwest Missouri, northeast Oklahoma, and southeast Kansas consists of an area of about 5000 km^2. Its longest axis from Springfield (Mo.) to Miami (Okla.), is about 160 km.

Mining in the Tri-State district began near Joplin (Mo.), in 1848. During much of its history, the district was the world's greatest zinc producer and ranked third or fourth in world lead production. The district's combined production of zinc and lead concentrates, which has been valued at over two billion dollars (Brockie et al., 1968), places it in a class with few peers.

The district contains a number of mining fields (Fig. 1). The Picher field, located at the west edge, was the last to be discovered but has been the principal producer. It has accounted for nearly two-thirds of the district's total production. Fields located at Webb City (Mo.), Joplin (Mo.), Granby (Mo.) and Galena (Kans.) have provided the bulk of the remaining production.

STRATIGRAPHY

The exposed sedimentary formations in the Tri-State district are of Mississippian and Pennsylvanian age. Older formations are not exposed but have been encountered in deep drill holes; these are of Precambrian, Cambrian, Ordovician, and Devonian age. The Precambrian basement complex exhibits a pattern of rock-type distribution, which is related to a southwest Missouri high that trends southwesterly through the Tri-State district (Kisvarsanyi, 1974). A granite prophyry (Spavinaw-type), which occurs approximately in the center of the high, is flanked on the northwest and southeast by rhyolite. Both rock types are surrounded by granite. Isotopic age dating of the granite porphyry and the granite by Denison et al. (1969) indicates that these rocks are Late Precambrian and are roughly equivalent in age to the Precambrian rocks that are exposed in the St. Francois Mountains of southeastern Missouri. Andesite porphyry has been encountered in some

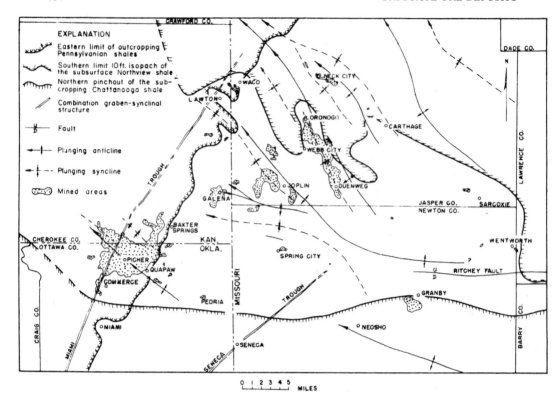

Fig. 1. Map of the Tri-State district showing the major structural and geological features. (After Brockie et al., 1968. Courtesy of Society of Mining Engineers of AIME.)

deep drill holes in the Picher field and in a single hole in the Missouri portion of the district. This porphyry is petrologically similar to the Precambrian hypabyssal intrusions in the St. Francois Mountains (Kisvarsanyi, personal communication, 1974). The surface of the Precambrian is very irregular, and although it usually lies about 520–550 m below the surface, the tops of isolated knobs lie as close as 88 m from the present surface in the area west of Carterville (Okla.), which is about 8 km southwest of the main Picher field. A Precambrian knob outcrops at Spavinaw (Okla.), which is about 65 km south of the district. The combined stratigraphic thickness of the overlying Cambrian and Ordovician formations can be as much as 2120 m in some places (McKnight and Fischer, 1970). The Cambrian, Bonneterre Formation, the host rock for the ore deposits in the southeast Missouri Lead Belt, is thin or absent in the Tri-State district (Brockie et al., 1968). The succeeding Chattanooga Shale, which is Late Devonian in age, is present to the south of the district. This unit is followed by the Carboniferous succession.

The stratigraphic column for the Carboniferous formations in the Tri-State district is given in Table I. McKnight and Fischer (1970) have subdivided the Boone Formation into members; the lettering of the beds is that of Fowler and Lyden (1932). The thickness, general lithology, and relative importance of each unit as an ore horizon are indicated.

Table I

Stratigraphic column of Carboniferous formations in the Tri-State District

System	Series	Fm./Mbr.	Bed	Thickness (m.)	Lithology	Importance as an Ore Horizon
Pennsylvanian	Desmoinesian Morrowan & Atokan	Several Fms. & Mbrs.		0–90	Dark sh. & ss.	Minor importance
Mississippian	Chesterian	Several Fms. & Mbrs.		0–30	Ls. cong., sh. & ss.	Minor importance
	Meramecian	Quapaw Ls. Fm.		0–9	Ls.	Unimportant
		Moccasin Bend Mbr.	B	0–6	Ls.	Minor
			C	0–10	Ls. & chert nod.	Minor importance
			D	5.5–7	"Cotton rock" & chert	Unimportant
			E	1.5–2.5	Ls. and chert nod.	Important
			F	3.5–4.5	"Cotton rock" & chert	Unimportant
			G H	9–12	Thin bedded chert & ls.	Important
		Baxter Springs Mbr.	J	0–12	Glauconitic, shaly ls. & chert	Minor importance
			K	0–12	Ls. & chert nod.	Very important
			L	0–11	Chert	Unimportant
		Short Creek Oolite Mbr.		0–3	Oolitic ls.	Unimportant
	Osagean	Joplin Mbr.	M	0–21	Ls. & chert nod.	Most important
		Grand Falls Chert Mbr.	N	6–9	Chert	Unimportant
			O	2.5–3	Thin bedded chert & ls.	Important
			P	0–3	Chert	Unimportant
			Q	0–3	Thin bedded chert & ls.	Unimportant
		Reed Spring Mbr.	R	15–30	Ls. & dark chert nod.	Important
		St. Joe Ls. Mbr		4.5–20	Ls. & chert nod.	Unimportant
	Kinderhookian	Northview Fm.		1.5–3	Sh.	Unimportant
		Compton Fm.		1.5–3	Ls.	Unimportant

Note: Fm = Formation; Mbr = Member; sh = shale; ss = sandstone; ls = limestone.

STRUCTURE

Structurally, the Tri-State district is located on the northwest flank of the Ozark uplift, a broad elliptical dome. The regional dip in the district is toward the northwest at about 3–5 m/km. Gentle folds and normal faults locally produce steeper dips. Mild tectonic stress has been an important factor in producing brecciation of the brittle chert beds in the Boone Formation (McKnight and Fischer, 1970).

The trends of the major folds and faults indicate that the Tri-State district is a region characterized by intersecting structural elements (Fig. 1). A series of gentle folds which trend northwesterly and plunge gently down the flank of the Ozark uplift are prominent within and to the north of the district. Intersecting these structures are northeast trending structures, particularly faults, which are characteristic of the structural pattern to the south of the district in Oklahoma and Arkansas.

The Joplin anticline is one of the best known northwest trending structures. The axis of this asymmetrical fold lies just to the northeast of the mineralized Joplin field. Structure contours on top of the Short Creek Oolite Member show that the crest of the anticline rises as much as 60 m (Smith and Siebenthal, 1907, p. 9). The Picher anticline, named by Brockie et al. (1968, p. 412), trends northwesterly through the southwest portion of the mineralized Picher field. The Bendelari monocline, which trends northwesterly, coincides with the northwest extension of the Picher field. The northeast side of this structure is displaced downward as much as 45 m (McKnight and Fischer, 1970). Other northwest trending folds are shown in Fig. 1. Drilling indicates that the Joplin anticline and the Bendelari monocline are near-surface expressions of faults at depth (Brockie et al., 1968).

The two principal northeasterly trending structures are the Miami trough and the Seneca trough. Both structures are narrow linear breaks, each of which consists of a combination syncline and graben structure (Siebenthal, 1908; Weidman, 1932; Ireland, 1930; Pierce, 1935; Fowler, 1938; Lyden, 1950 and Bieber, 1955). Although their widths and displacements vary from place to place, they are about 300 m or less wide and exhibit stratigraphic displacements of 90 m or less. Deep drilling has shown that the Miami trough extends into the Precambrian basement (Brockie et al., 1968).

The Ritchey fault is an easterly trending normal fault. Its south block in down thrown as much as 45 m (Bieber, 1955). McCracken (1971) has noted that its trace can be followed on aeromagnetic maps. The fault lies in the southern portion of the district away from most of the ore deposits.

The major structures in the Tri-State district appear to reflect a pattern of basement faulting. The faults, which may be Precambrian in age, apparently have experienced subsequent movement and periodic rejuvination during repeated uplifts of the Ozark uplift. They are important because they have played a significant roll in determining the location and extent of many of the district's ore fields. The Picher field, for example, is located at the intersection of the Miami trough and the Bendelari monocline. Mined areas extend several kilometers to the southwest along the Miami structure, beyond the main Picher field, and to the northwest along the Bendelari structure. The folds, faults, brecciation, and openings in the rocks that are associated with these structures have served to provide channelways for solutions to gain access to soluble limestone. The host carbonate rocks near these structures have suffered extensive underground solution (Siebenthal, 1925 and Pierce, 1935). Unique solutional features, such as pipe slumps, which are nearly circular blocks 30–90 m in diameter, that have been lowered by solution, tend to be concentrated in the vicinity of major graben structures, such as the Miami trough (McKnight and Fischer, 1970). The additional structures in the Missouri portion of the Tri-State district that are shown in Fig. 1 are catalogued and described by McCracken (1971).

A major east-trending structure, the 38th parallel lineament, occurs about 40 km north of the Tri-State district (Snyder, 1970 and Heyl, 1972). Although this structure may be an important factor in some ore districts, it bears no evident genetic relationship to the Tri-State deposits.

Smaller structural features have been mapped in detail in the Picher field and constitute important guides to the mining geologists who have to trace individual ore "runs". Minor folds, which have been depicted by structure contours drawn on the tops of L, N, and O beds, tend to parallel the trends of the "runs" (Fowler and Lyden, 1932; Fowler, 1938; McKnight and Fischer, 1970 and Lyden, 1950). Fractures, which are vertical or steeply inclined away from cores of dolomite and are discontinuous vertically and horizontally, commonly are present in the ore "runs". These fractures tend to be arcuate in plan view and follow the curving margins of the dolomitic masses. Their exposures in the working faces and in the back are mapped and utilized in predicting the extension of ore "runs" (Fowler and Lyden, 1932 and Lyden, 1950). This type of fracturing has been called *shearing* by Fowler and Lyden (1932), because slight movements along the fractures have produced slickensides on some of their surfaces.

GENERAL FEATURES OF THE ORE DEPOSITS

All of the ore deposits in the Tri-State district are characterized by three general features: (1) a close relationship to the surface of the earth; (2) a remarkable areal extent; and (3) no association with igneous rocks. The most striking feature of the ore deposits is their close relationship to the surface of the earth. The deepest deposits, those of the Picher field, are less than 145 m from the surface. The deposits are present over a remarkably large area which includes more than 5000 km^2 on the flank of the Ozark uplift. Within this district, there are many camps or ore fields of mineralization consisting of relatively low-grade, strata-bound, ore deposits. The igneous rocks in the district are Precambrian in age. Crystalline Precambrian rocks are transected in drill holes in and near the district. Crystalline rocks crop out in a knob at Spavinaw (Okla.), 65 km south of the district.

The ore bodies of the Tri-State district can be divided into three groups of different shapes commonly referred to in the district as: (1) runs; (2) circles; and (3) sheet deposits.

Runs

A "run" is an elongate, tabular body of mineralized chert breccias, which normally follows a stratum but does break through from one to another in places. "Runs" are common in the Boone Formation, except for the Grand Falls Member. The "runs" may be 1000 m or more long, 150 m wide, and more than 30 m high (Brockie et al., 1968), but most are of smaller dimensions.

In those areas where dolomite is present, most of the ore "runs" tend to follow closely

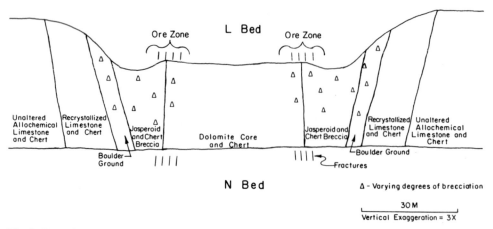

Fig. 2. Generalized section through the small-scale zoning in the M bed, eastern portion of the Picher field. (Modified after McKnight and Fischer, 1970 and Lyden, 1950.)

the edges of the irregularly elongate to roughly circular dolomitic masses. The runs, therefore, are of an arcuate shape in plan view, and their inner, concave walls are of dolomite and their outer, convex walls are of jasperoid as shown in Fig. 2, which is modified after Lyden (1950), Hagni and Saadallah (1965) and McKnight and Fischer (1970). Farther out from the convex side, there is an open zone, which is characterized by chert boulders, iron sulfide, and calcite crystals. In pull drifts, which have been driven between ore "runs", Hagni and Saadallah (1965) discovered a zone of limestone about 5–40 m wide outside the jasperoid and boulder zones. They determined that this limestone consists largely of sparry calcite. According to their interpretation, this orthochemical limestone zone had been altered by ore-forming solutions from the allochemical crinoidal limestone which is present at a greater distance from the ore bodies. In this regard, Pinckney and Rye (1972) have observed a similar type of alteration in Mississippian limestone wallrock that surrounds an ore body in the southern Illinois Fluorspar district. Here, they were able to observe progressive stages in the recrystallization of calcite oolites to sparry calcite. They interpreted the low $\delta^{18}O$ and $\delta^{13}C$ values, which they found in the altered, recrystallized limestone, to be a result of a partial exchange with the ore solutions.

Where the dolomitic masses are ovate to circular, the "runs" form circular patterns around them. In areas of intense mineralization around narrow, elongate, dolomitic cores, two ore bodies, one on each side of the core, are mined as a single "run" and the dolomite is left to serve as pillars in the mine (Lyden, 1950, fig. 6).

Circles

The term "circle deposits" has been applied to "runs" which form partial to nearly complete circular map patterns. Maps of mined areas in the Picher field (McKnight and Fischer, 1970, pl. I and V—X) contain numerous circular patterns which are particularly well shown near the margins of the field where the mineralization is less intense. The smallest circles have diameters of 100 m or so across, and the largest have diameters as large as 2.5 km. These deposits have the same spatial relationships of dolomite, jasperoid, limestone, and fracturing as the "runs".

Circle structures are common in fields to the east of the Picher field. The oldest and best known of these is the Oronogo Circle. The ore was concentrated in breccias at the base and margins of the sink structure, which contained a thickness of 94 ft. (about 29 m) of shale (Schmidt and Leonhard, 1874, pp. 494—497 and Winslow and Robertson, 1894, pp. 573—576) and was said to have been present vertically throughout most of the Boone Formation.

Blankets or sheets

Partly broken, mineralized, chert bodies, which are stratified, are referred to in the district as "sheet ground" deposits. These deposits are confined to the O, P, and Q beds of the Grand Falls Chert Member in the Boone Formation. They have a large horizontal extent as compared to their thickness. They commonly are 3.5—4.5 m thick and have an areal extent up to about 1 km² (Brockie et al., 1968). The chert is relatively unbroken, but where it is broken it forms crackle breccias in which the fractured fragments have suffered little movement. The deposits are confined to the northeast and southwest edges of the Picher field and are common to the east in the Missouri portion of the Tri-State district.

MINERALOGY AND PARAGENESIS

The important minerals of the Tri-State district are few in number and simple in composition. They include the sulfides (sphalerite, galena, chalcopyrite, pyrite, and marcasite) and the gangue minerals (quartz, dolomite, and calcite). In their simplicity, they are like other ore deposits of the Mississippi Valley type.

The Tri-State minerals are well crystallized, and the individual crystals are commonly large in size. Crystals of calcite two-thirds of a meter long have been recorded (Winslow and Robertson, 1894; Farrington, 1900; Bain et al., 1901 and Smith and Siebenthal, 1907). Crystals of galena, which measure nearly 1/3 m on an edge, are known (Weidman, 1932), and crystals of sphalerite over 1/3 m across have been collected from the district (Weidman, 1932). For this reason, the Tri-State district has been one of the most outstanding localities in the world for the collection of beautifully crystallized mineral specimens. It is appropriate that a large collection of such specimens is preserved in the Tri-State Mineral Museum, which is housed in a new building that was dedicated in August of 1973 at Schifferdecker Park in Joplin (Mo.).

TABLE II

Minerals in the Tri-State ores

Major minerals	
Sulfides	galena, sphalerite, chalcopyrite, pyrite, marcasite
Carbonates	calcite, dolomite
Silicates	quartz
Minor minerals	
Native elements	sulphur
Sulfides	bornite, wurtzite, greenockite, millerite, covellite
Sulfosalts	enargite, luzonite
Oxides	cuprite, hematite, pyrolusite, limonite
Carbonates	smithsonite, aragonite, cerussite, hydrozincite, aurichalcite, malachite, azurite, leadhillite
Sulfates	barite, anglesite, gypsum, starkeyite, chalcanthite, melanterite, epsomite, goslarite, linarite, jarosite, plumbojarosite, aluminite, copiapite, caledonite, szomolnokite, carphosiderite
Arsenates	picropharmacolite, mimetite
Phosphates	vivianite, apatite, pyromorphite, wavellite, diadochite
Silicates	hemimorphite, allophane, chrysocolla, kaolinite, glauconite

Unlike many of the lead—zinc deposits of the western United States, the minerals of the Tri-State district are not intimately intergrown, are not extensively replaced by one another, and commonly exhibit druses or encrusting fabrics. The minerals in the Tri-State district are relatively free from the contamination of associated minerals. Much of the ore consists of crystals which were deposited one upon another on the surfaces of open spaces. Each mineral has generally crystallized alone rather than with other minerals.

The important minerals of the ore are ubiquitous. They occur in all of the mines, and the general sequence of mineral deposition is the same throughout the district, although local repetitions and omissions of parts of the sequence occur.

Because the detailed observations on mineralogy and paragenesis have been recorded elsewhere (Hagni and Grawe, 1964; Brockie et al., 1968 and McKnight and Fischer, 1970), they will not be repeated here. A complete list of minerals found in the Tri-State ore deposits is given in Table II. A summary of the paragenetic sequence is given in Fig. 3.

HOST ROCK CHARACTER

Ore deposits in limestone host rocks exhibit features which have been produced by a complex interplay of sedimentary, diagenetic, tectonic, and solutional processes. The rock types, which host the Tri-State ore deposits, are limestone, chert, dolomite, jasperoid, and shale. The cherty limestones of the district were subjected to gentle folding and fracturing, to subsequent solutional activity associated with a karst topography, and

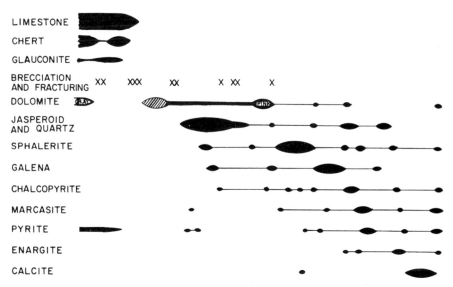

Fig. 3. Paragenetic diagram for the Tri-State ore deposits. (Modified after Hagni and Grawe, 1964.)

to replacement by dolomite, jasperoid, and the ores. Shales formed as facies of some limestones, as fillings in solutional cavities and depressions, and as insoluble residues from the solution of argillaceous limestone. If one is to understand the genesis of the Tri-State ore deposits and to conduct effective exploration for these deposits, he must have a better concept of the distribution and manner of development of their host rocks. The principal objective of this paper is to provide a summary of current knowledge pertaining to Tri-State host rocks and to give some detailed descriptions of their character.

Limestones

The limestones of the Boone Formation in the Tri-State district vary from light-gray, coarse-grained, crinoidal limestone to light-brown, fine-grained limestone. Beds K and M are particularly characteristic of the former and beds G—H of the latter. The limestone of bed R is dense and bluish-gray.

The depositional environment for the limestones is presumed to be characteristic of a warm, shallow, widespread sea that favored the growth of organisms, especially crinoids and bryozoans. The presence of small amounts of microcrystalline calcite in much of the Tri-State limestone suggests a lack of turbulence. This condition permitted lime mud to accumulate. The thin oolitic limestone at the base of the Baxter Springs Member, indicates the presence of shallow, agitated water at that time.

Bioherms

The discovery of reef structures in the Lead Belt of Southeast Missouri (Ohle and

Brown, 1954) and the realization of their importance in influencing the character and pattern of distribution of the sedimentary host rocks (Gerdemann and Meyers, 1972 and Larsen, 1973) and in controlling the localization of some of the mineralization in the Missouri Lead Belt (Snyder and Emery, 1956 and Snyder and Gerdemann, 1968), at Pine Point in the Northwest Territories, Canada (Campbell, 1967), and in some other Mississippi-Valley type deposits, has inspired some geologists to wonder whether reefs or bioherms might have played some part in the localization of the zinc–lead deposits of the Tri-State district. Although a definitive answer to this question remains for future investigation, a number of bioherms are known to occur in the lower part of the Boone Formation and correlative strata south of the district, and it is possible that chertified bioherms may be present in the ore-bearing horizons within the mineralized district.

Huffman (1958) has noted that large crinoidal reefs or bioherms are present in limestones which are approximately the equivalent of the Joplin and Grand Falls Members and located east of Prior, Oklahoma, about 24 km south of the Tri-State district.

Troell (1962) has described a large number of bioherms at the base of the Boone Formation in an area only 32 km south of the Granby field in the Tri-State district. He recognized 39 bioherms in the lower 1.2–3.6 m of the St. Joe Limestone. These bioherms are roughly circular in plan and measure up to 9 m thick and 90 m across. The calcitic mud, which constitutes 80% of the biohermal material, was apparently trapped by bryozoans which now comprise 8% of the structures.

Only a few kilometers south of the intensely mineralized Picher field, according to McKnight and Fischer (1970), even larger bioherms occur at the base of the overlying Reeds Spring Limestone. These structures are lens-shaped, have flat bases, and form circular patterns. Although their average thickness is about 9–18 m, some are more than 30 m thick. They are as much as 335 m across. These bioherms are composed largely of crinoidal debris and numerous bryozoan. This clastic material was presumably associated with and partially cemented by reef-building organisms, such as algae.

Certain textures of some of the cherts have led McKnight and Fischer (1970, pp. 41–42) to believe that bioherms are present within the confines of the Picher field. Much of the chert in the L bed exhibits a distinctive mottling that is locally referred to as "coach-dog" chert. McKnight and Fischer (1970) interpret the spots that form this mottling as chertified remains of bryozoans. They believe that the L bed "coach dog" chert masses are chertified bioherms and that the shaly limestone in the overlying K bed formed the sedimentary facies above and flanking these bioherms. Significantly, the K bed is one of the more important hosts for ore deposits in the Picher field. Additionally, if the chert masses are truly bioherms, they constitute further evidence for a shallow-water environment of deposition for the host limestones in the ore district.

To summarize, bioherms are known to occur in the lower Boone Formation near the district and probably within the ore district itself. The extensive effects of dissolution, dolomitization, and silicification, as evidenced by the chert and jasperoid, may have largely obliterated the evidence for the presence of these structures within the mineralized ore fields. It appears likely that additional bioherms have gone unrecognized in the district, and it is possible that such sedimentary structures may have had a role in determining the loci for subsequent dolomitization, fracturing, solution, and ore deposition.

Chert

Chert is an extremely abundant constituent in the rocks of the Tri-State district. The nature of its occurrence varies with its stratigraphic position: (1) massive beds, aggregating up to 9 m thick, constitute the L and N beds; (2) beds 5—15 cm thick are characteristic of the beds G—H and the sheet ground beds; and (3) nodules and lenses oriented with their longest dimensions parallel to the bedding are very common in the K and M beds.

Most Tri-State chert is white to light-gray and tan, but chert in the Reeds Spring Member (R bed) is dark-gray to bluish-black. Density, nature of fracture, and luster are other properties of these cherts which are fundamentally associated with their stratigraphic positions (Grohskopf and McCracken, 1949).

The chert consists primarily of microcrystalline quartz grains which are anhedral, equant, and intimately interlocked. The individual quartz grains are 3—15 μ across. Small amounts of chalcedonic quartz occur as radial microfibrous aggregates which form veinlets and replace sponge spicules and other organic remains.

In the chert, tiny inclusions of a substance which is brown in transmitted light and white in reflected light is believed by Folk and Weaver (1952) to be water-filled cavities and by McKnight and Fischer (1970) to represent organic matter. Other constituents, which are recognizable with the aid of the microscope, are pyrite, glauconite, and calcite. Pyrite is a constant, minor constituent especially in the darker cherts. It occurs as very small, usually anhedral grains, 3—300 μ in diameter. Glauconite occurs rarely in small rounded grains in some cherts. It is less abundant in the chert than in the jasperoid.

Anhedral calcite is abundant in some cherts, where it occurs as irregular patches. Such limy chert is referred to as "cotton rock" in the district. Surficial leaching of the limestone portion of the "cotton rock" has formed tripoli deposits which have been mined at Seneca (Mo.). Fossils, particularly crinoid-stem segments, are found in "cotton rock" and in the margins of chert nodules and lenses. The original calcite of such fossil material exhibits varying degrees of replacement by chert, from little or no replacement to complete conversion of the material to microcrystalline quartz.

The relative amounts of chert in association with the Tri-State ore deposits have been mapped (chertification maps) and used as an exploration guide in the past. The great abundance of chert in the mining district proper, the evidence for replacement of limestone, and an apparent systematic diminishing of chert abundance southward (Giles, 1935) have been interpreted by some geologists to indicate that the chert was introduced as an early phase of ore-forming solutions (Weidman, 1932; Bastin, 1933; Giles, 1935; Fowler et al., 1935 and Ridge, 1936). More recent interpretation favors the formation of chert during the deposition and diagenesis of the enclosing limestones (Hagni and Grawe, 1964; Brockie et al., 1968 and McKnight and Fischer, 1970). This interpretation is supported by the occurrence of chert in Mississippian limestones that lie far beyond the confines of the mining district and by a similar stratigraphic distribution of cherts of varied character outside the district. That the chert was formed during Meramecian time is indicated by the occurrence of chert pebbles in conglomerates of Chesterian age (Weidman, 1932). Residual chert breccias are very common at the contact between the Mississippian and Pennsylvanian formations in the Tri-State district. Part of the reason for the relatively large quantity of chert in the district is due to its accumulation as an insoluble residue from the partly dissolved host limestone (Hagni and Desai, 1966). The tendency to mistake some of the lighter-colored jasperoids for chert in drill cuttings has contributed to an excessive estimate of chert abundance (Brockie et al., 1968, p. 416—417). Nevertheless, chert is very abundant in the district and a source of abundant silica was required during Mississippian time.

KARST FEATURES

Underground solution of the Boone Formation in conjunction with erosion of the surface at various intervals since the end of the Meramecian period has had a pronounced effect upon the formation's soluble limestone. The erosional surface which forms the post-Mississippian, pre-Pennsylvanian unconformity is particularly important as a development of typical karst topography with its associated system of solution channels, caves, and sinks. The Pennsylvanian shales, which were deposited upon this very irregular surface, filled the solutional depressions. Further solutional activity and slumping of the shale during subsequent periods of erosion and solution accentuated the irregularity of this surface. In the Missouri portion of the Tri-State district, abundant outliers of Pennsylvanian shale attest to the processes. The close association of ore with the open ground at the margins of these shale areas in Missouri is shown on the Missouri Geological Survey "Shale Maps" (Missouri Geological Survey and Water Resources, 1942). In the Picher field, pockets of shale, mostly black, are common in the Boone Formation, particularly in the M bed. Most of the black shale in these pockets was deposited in Pennsylvanian time, and it partly slumped during subsequent limestone dissolution. Black shale from a cave in the Kenoyer mine has been dated as Pennsylvanian in age by its plant remains (McKnight and Fischer, 1970, p. 70). Some of the lighter-colored shales formed prior to the black shales were produced as insoluble residues from dissolution of the Boone limestone (McKnight and Fischer, 1970). Other shales present in the Boone Formation may represent shaly facies in that formation (Brockie et al., 1968, p. 428). Pipe slumps which bottom in the Reeds Spring Formation (Brockie et al., 1968) indicate active solution to at least that depth in the Mississippian.

Solutional activity associated with karst development has produced striking lateral variations in the thicknesses of the more soluble limestone beds, particularly that of the M bed, and has contributed to the development of the residual chert breccias, which are so characteristic of the Tri-State mines. In contrast to a thickness of 21 m or more at the margins of the Picher field, the M bed is less than 9 m thick in nearly all of the mined area of that field and locally may be as thin as 2.5 m (Hagni and Desai, 1968). The close association of the thin M bed with the ore deposits may be due partly to original deposi- tional differences and partly to dissolution by the ore-forming solutions, but much solu- tional activity was associated with the karst development when the chert breccias were formed and the host limestone conditioned with open spaces and channels. This condi- tioning, of course, facilitated the introduction of the subsequent mineralizing solutions. Some of the high-angle fracturing of the host rocks may have developed in association with the solutional activity in a manner similar to that observed at the Young mine in the east Tennessee zinc district (McCormick et al., 1969).

DOLOMITE

The distribution of coarsely crystalline, gray dolomite in the Tri-State district is close- ly associated with the positions of the ore bodies. This dolomite constitutes the single most important guide to ore, and its mode of formation has remained perhaps the most perplexing problem in the district.

The close association of the Tri-State ore deposits with areas of gray dolomite has been recognized for a long time (Smith and Siebenthal, 1907, p. 14). In fact, the early miners in the district said that "spar (dolomite) is the mother of jack (sphalerite)". Underground mapping by mining company and U.S. Geological Survey geologists has served to de- lineate the distribution of this dolomite for portions of the Picher field (McKnight et al., 1944), and its general distribution in the district has been discussed by McKnight (1950), Lyden (1950), Brockie et al., (1968), and McKnight and Fischer (1970).

Dolomite is common in the M and K beds. Although it may occur in other beds, it is uncommon in sheet ground (O, P, and Q beds). In plan view, the dolomitic masses or zones form irregular, elongate to circular areas. The dolomitic zones are comprised of massive gray dolomite, varying amounts of chert nodules, chert breccia, and jasperoid as shown in Fig. 4. Small quantities of pink dolomite in the form of saddle-shaped, rhom- bohedral crystals, occupy vugs and fractures in or near the masses of gray dolomite.

Determination of the lateral extent of the individual masses of dolomite is of utmost importance to mining geology in the Tri-State district. Although ore can occur in areas devoid of dolomite and gray dolomite is present in a few areas without accompanying mineralization, most ore bodies have a marked tendency to be located along one or both margins of any given mass (Fig. 2). Careful mapping along the mine walls of the contact between the dolomite and jasperoid zones usually provides an indication of the direction of an ore "run". Underground mapping of the dolomitic areas has led to the discovery of

Fig. 4. Dolomite (medium gray, speckled) wall typical of the ore "runs". Note that the chert lens (white) in the center of the photograph is broken, but yet it is nearly in place and cemented by jasperoid (black). M bed, Whitebird mine.

many bypassed portions of the ore deposits and has extended the life of the Picher field many years. Mineralization extends beyond the peripheries of the dolomite and in some localities reaches sufficient grades to warrant mining entirely through the cores of dolomite or even through areas devoid of dolomite.

Although it was recognized at an early date that coarse gray dolomite exhibits a spatial relationship to the ore deposits and was formed by replacement of limestone, the precise factors which determined the location of the dolomite still remain uncertain. One of the reasons for this uncertainty is that cross-sections through the central part of the wider, more developed dolomitic masses are rarely exposed by mining because they are barren of ore. In those places where pull drifts have been cut through the dolomitic cores or where sufficient mineralization has extended into the core of a dolomitic mass to warrant mining part of it, a variety of relationships are evident. Dolomite may extend across the entire core and the enclosed chert lenses appear to exhibit a somewhat lesser degree of brecciation than they do in the adjacent jasperoid areas. In other places, unreplaced remnants of limestone may be present in the dolomitic cores. Slump structures have been exposed in some of the cores in the Picher mines. McKnight (1950, p. 58) notes that barren cores examined by him are characterized by "a maximum concentration of residual chert, highly leached and closely compressed, with only a little interstitial dolomite, introduced Cherokee shale, or perhaps clay residual from the Boone formation."

The commonly accepted interpretation of the above relationships is that the gray dolomite was formed when the limestones were affected by ore solutions introduced along zones of fracturing and brecciation that have been associated with slight flexing of the beds (Fowler and Lyden, 1932) or with surface slump structures (Buckley and

Buehler, 1905). Support for this interpretation comes from the fact that it is difficult to find a specimen of dolomite that does not exhibit a little intermixed jasperoid, a matrix which was deposited with the ores. The interpretation does, however, admit a contradiction. If the plumbing system for the dolomitizing solutions was provided by the fractures, which have been exposed in the ore runs, very special circulation patterns would have been required so that the dolomitizing solutions could migrate in one direction while the silicifying solutions traveled in the other.

Brockie et al. (1968) have suggested that the gray masses of dolomite may have been formed by the selective replacement of some original sedimentary feature, that the replacement may have been prior to and unrelated to the introduction of the ore-forming solutions, and that the fracturing exposed in the mine walls could have subsequently developed at the margins of the dolomitized masses. The earliest time at which dolomitization could have occurred was after the solidification and fracturing of the chert, because chert breccia is common in much of the gray dolomite. The occasional occurrence of blocks of gray dolomite that have been cemented by sulfide-bearing jasperoid, as shown in Fig. 5, indicates that locally, at least, dolomitization has preceded the emplacement of the jasperoid.

If the masses of dolomite are indeed replaced sedimentary features, might the bioherms described earlier, or a related facies, be the replaced sedimentary structure? If the masses are dolomitized bioherms, they have thus far defied identification. The supposed destruction of original sedimentary features and fossils in the coarse-grained, gray dolomite has been a hindrance to possible identification.

Regardless of what factors were responsible for the localization of the masses of dolomite, it appears that, as in the case of the ore solutions, the solutions responsible for dolomitization were brines (Friedman and Sanders, 1967).

Fig. 5. Gray dolomite blocks (medium gray) with bottoms and sides fringed by pink dolomite (light gray) in a matrix of banded jasperoid (dark gray). Sphalerite (speckled gray) is concentrated in the jasperoid above the blocks. M bed, Lucky Jew mine.

JASPEROID

The jasperoid in the Tri-State district is a dark-gray or brown to nearly black micro-crystalline quartz. It is a very abundant constituent, and a knowledge of its occurrence is necessary to understand the genesis of the ore deposits.

Jasperoid forms a zone or halo around many of the cores of dolomite (Fig. 2). A typical mine exposure in the jasperoid zone in the M bed is shown in Fig. 6. Toward a core, jasperoid becomes interbanded with gray dolomite (Fig. 7); in the opposite direction, it may be interbanded with limestone (Fig. 8). Jasperoid is closely related to the district's sulfide mineralization in space, time of deposition, and genesis. Sulfide crystals are disseminated throughout nearly all the jasperoid and were crystallized when the jasperoid was deposited.

The question has been discussed as to whether jasperoid could simply have been a black, reducing, siliceous mud that formed at or near the sea floor. If this is true, it would have a significant bearing upon the genesis of the disseminated sulfides enclosed in the jasperoid. Although very rare features suggest that a part of the jasperoid went through a plastic or mushy state and other features show that locally the jasperoid was a chemical sediment which filled solution channels, most of it is a replacement of limestone and to a lesser extent of dolomite and shale.

Although jasperoid may appear surficially to be somewhat similar to chert, the two materials are different in several important respects (Table III). Jasperoid is more vitreous and more crystalline than chert. In many jasperoid specimens, small quartz crystals can be observed with the aid of a hand-lens. Jasperoid has an uneven fracture whereas that of

Fig. 6. Jasperoid wall typical of ore "runs" showing jasperoid (black) and chert (white). M bed, Whitebird mine.

Fig. 7. Horizontally banded gray dolomite (light gray) and jasperoid (dark gray). G–H beds, Piokee mine.

chert is smooth and sharp. Jasperoid commonly contains disseminated sphalerite, galena, marcasite, and pyrite, whereas chert characteristically lacks disseminated lead and zinc sulfides. Jasperoid is closely associated with the sulfide mineralization, whereas chert often occurs interbedded with limestone in barren areas between ore deposits. Micro-scopically, jasperoid is also quite distinct from chert. Most jasperoid quartz grains are 14–70 μ across, whereas the grains in chert are usually less than 14 μ across. Unlike

Fig. 8. Interbanding of jasperoid (dark gray) and limestone (light gray) at a gradational contact between the ore zone breccia and massive limestone. M bed, Kenoyer mine.

TABLE III

Comparison of Tri-State chert and jasperoid

Descriptive properties	Chert	Jasperoid
Color	white to tan; medium gray in some strata	dark-gray to black
Form	nodules, lenses, beds	cements matrix between brecciated fragments
Banding	crude, smooth, concentric; banding may follow margins of nodules and lenses, but most chert lacks distinct banding	thin, horizontal banding is characteristic; may be contorted locally
Fracture	smooth	uneven
Grain size	microcrystalline; all quartz grains too small to be seen with unaided eye; relatively uniform grain size, most grains $< 14\mu$	microcrystalline; some quartz crystals can be seen with aid of hand lens; variable grain size, most grains $14-70\mu$
Grain shape	mostly equant	commonly elongated
Rock types partially replaced	limestone	limestone and dolomite
Association with muds	no particular relationship; forms nodules, lenses, and brecciated fragments in muds	locally gradational with dark, semiconsolidated muds
Textural relationship to gray dolomite	present as nodules, lenses, and brecciated fragments in dolomite	commonly interbanded with gray dolomite; locally gray dolomitic brecciated fragments are enclosed by jasperoid
Textural relationship to sulfide mineralization	Zn–Pb–Cu sulfides confined to fractures; chert never contains disseminated Zn–Pb–Cu sulfides except where developed in tiny vugs; contains sporadic, disseminated, fine-grained pyrite	disseminated Zn–Pb–Cu sulfides are present in most jasperoid; contains abundant, disseminated, fine-grained pyrite and marcasite
Spatial relationship to sulfide mineralization	extends far beyond the confines of the district and its sulfide mineralization	lateral and vertical extent essentially coincident with sulfide mineralization

most chert, the jasperoid quartz grains have a strong tendency to be elongate in shape as shown in Fig. 9.

Chert may be crudely banded around the margins of nodules and lenses. Jasperoid, in contrast, commonly exhibits distinct, thin, mostly horizontal banding. This banding is due to small differences in quartz grain sizes and to variations in the amount of brown, opaque, organic matter, and disseminated sphalerite.

Fig. 9. Elongated quartz crystals in a thin section of jasperoid under crossed nicols. M bed, Netta White mine. (100×)

The overall character and distribution of jasperoid varies greatly depending upon its areal and stratigraphic position. In the G–H beds, there are striking examples of the replacement by jasperoid of thin limestone beds which contain interbedded chert. In unmineralized areas, the G–H beds consist of interbedded chert and limestone (Fig. 10), whereas the limestone is replaced by jasperoid in the ore zones (Fig. 11). Similar replacement of limestone occurs in thinner bedded host rocks, which comprise the sheet ground (beds O, P, and Q) and which typically form broad, flat deposits of ore like that shown in

Fig. 10. Horizontally banded and nodular chert (light gray) in limestone (medium gray) typical of G–H beds in unmineralized areas. Piokee mine.

Fig. 11. Horizontally banded chert (light gray) and jasperoid (black) typical of mineralized G–H beds. Sphalerite crystals (white spots) are present in jasperoid but absent from the chert. St. Joe mine.

Fig. 12. Sphalerite is disseminated in the dark jasperoid, and it increases in abundance upward to form crystals in the narrow, horizontal, open space provided beneath the overlying chert band. The chert is fractured and partially brecciated, and the fractures are partly healed by jasperoid. Sphalerite is not disseminated in the chert but occurs only along the fractures in the chert.

Fig. 12. Thin chert lenses (white) separated by jasperoid (black) containing sphalerite (gray specks) typical of mineralized "sheet ground". Jasperoid also fills fractures in the chert. The vertical edge of the number plate is about one centimeter long. Top of specimen is indicated by arrow. O bed, Blue Goose No. 1 mine.

Fig. 13. Sharp contact between massive limestone (lighter gray, on the left) and chert breccia (white) cemented by jasperoid (dark gray) typical of M bed. Blocks of limestone containing chert lenses, broken from the bedded limestone, are displaced to the right and surrounded by jasperoid. Lucky Jew mine.

The unmineralized portion of the M bed is typified by chert nodules and lenses which are disseminated throughout the unit's coarse-grained, readily soluble limestone. Although the contact between the unmineralized portion of the M bed and the silicified, mineralized portion is typically obliterated by the development of a solution-formed boulder zone, distinct contacts can be observed locally. In such instances, blocks of

Fig. 14. Chert breccia (white) cemented by jasperoid (dark gray to black) typical of M bed. Note the bending of the jasperoid bands beneath the large chert fragments at the left center portion of the photograph. Top of specimen is indicated by the arrow. M bed, Lucky Jew mine.

Fig. 15. Nodular and brecciated chert (white) typical of K bed, cemented by jasperoid (dark gray). Part of the chert is highly brecciated, and the horizontal, nodular, chert lenses in the upper part of the photograph have settled upon each other as a consequence of limestone dissolution. Crawfish mine.

broken limestone and chert are cemented by and partly replaced by jasperoid (Fig. 13). Irregular bands of the unbroken limestone wall can be seen to be darkened through incipient silicification. Farther from the limestone contact, where there is additional solution and replacement of the M bed limestone, all of the limestone is dissolved and replaced, and only the relatively insoluble chert breccia remains in a matrix of jasperoid (Fig. 14). Chert breccia of this type is very characteristic of the ore "runs" of beds K and

Fig. 16. Chert breccia containing fragments which exhibit a variety of textures: massive white, banded light gray, and mottled. The mottled fragments consist of chert (light gray) and limestone (medium gray). The fragments are cemented by jasperoid (dark gray). M bed, Lucky Jew mine.

Fig. 17. Mosaic chert breccia (white) in M bed, cemented by jasperoid (dark gray). Much of the breccia in the left portion of the photograph could be placed back together with the solid chert lens to the right. Note bands in jasperoid which bend over some chert fragments in the upper left portion of the photograph. Whitebird mine.

M in the Tri-State district (Fig. 15). Although a typical breccia consists of mixtures of chert fragments of varied character (Fig. 16), local mosaic breccias are comprised of fragments which have moved apart only short distances. Fig. 21 shows a mosaic breccia which was formed from the chert lens in the right center portion of the photograph. The fragments of the previously, highly fractured, chert lens had moved apart during the gradual subsidence, silicification, and solution of the originally interbedded limestone in the central portion of the photograph where they lacked the support of the underlying chert nodule. These fragments, as well as part of the chert breccia in Figs. 15 and 17 appear to "float" in the jasperoid host.

Sawkins (1969) postulated that some of the floating mosaic breccias that occur in several of the ore districts may have developed as a result of a chemical reaction between the host rocks and the ore-forming brines; a reaction that is similar in nature to the alkali-aggregate reactivity that is well known to civil engineers. If such a process was responsible for part of the brecciation of the Tri-State chert, it would have to have taken place at an early period before the chert had developed its present degree of crystallinity, because Missouri cherts in their present state are considered to be a nonreactive aggregate material.

Smaller features also demonstrate that much of the jasperoid originally replaced limestone. Fig. 18 shows a thin galena vein which transgresses at a high angle the horizontal bedding of bed B, which here consists of chert that contains abundant, small patches of limestone. For a distance of 15—20 cm around the galena vein, there is a halo in which the limestone patches in the chert have been replaced by jasperoid. Coarsely crystalline

Fig. 18. Alteration halo (dark gray) around a very thin galena (light gray; under hammer) vein transgressing the bedding. The galena vein and alteration halo terminated upward. The rim of the halo is marked by discontinuous patches of calcite (white). The alteration halo consists of chert mottled with jasperoid, whereas the surrounding country rock is chert (light gray) mottled with limestone. E bed, Netta mine.

calcite was subsequently deposited in small openings at the margin of the halo, perhaps as a result of reprecipitation of the earlier dissolved limestone.

Less commonly, jasperoid may replace host rocks other than limestone. Shales and shaly limestones may be partially to completely silicified. Fig. 19 shows contorted and

Fig. 19. Contorted and broken banded jasperoid. E bed, Tar Creek mine. One-half natural size.

Fig. 20. Contorted jasperoid bands. A few small scattered galena crystals (white specks) are present. E bed, Tar Creek mine. One-half natural size.

broken bands, which may have originally been shale but are now completely silicified to jasperoid. Very rarely, jasperoid bands, which are highly contorted, as shown in Fig. 20, indicate either that the jasperoid locally had gone through a mushy stage or that it had replaced a plastically deformed mud-like material. Jasperoid locally exhibits transitions that pass through partly silicified shale into areas of unsilicified and unmineralized shale (Smith and Siebenthal, 1907 and Brockie et al., 1968). Banded, jasperoidal dolomites, such as shown in Fig. 7, were formed, according to the interpretation of McKnight and Fischer (1970), by having the limestone bands of a banded dolomitic limestone selectively replaced by jasperoid. If this is true, small, microscopic, jasperoid, quartz crystals, which are found within the margins of some dolomite crystals, also may have been formed by replacement.

"Geopetal" fabrics are rarely observed in the Tri-State jasperoid. Jasperoid bands may be bent beneath some chert fragments, as shown by the large chert fragment marked "top" in Fig. 14. "Geopetal" fabrics involving sulfides are even less common. In Fig. 21, the bands of the jasperoid can be seen to bend over the large sphalerite crystal in the upper portion of the specimen and to bend beneath two, large, sphalerite crystals at the bottom of the specimen. The bands become thinner, pinch out against sphalerite crystals, and thicken between the two lower sphalerite crystals. Additionally, the base of the

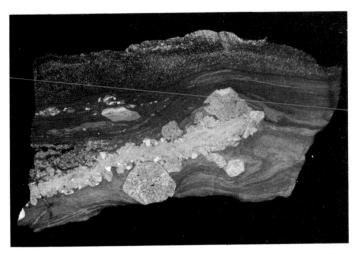

Fig. 21. "Geopetal" fabric showing jasperoid bands (dark gray and light gray) warped over and under sphalerite crystals (white). The top, broad, jasperoid band contains finely disseminated sphalerite (white specks). A few chert fragments (light gray and white, banded) are present near the center of the specimen. M bed, Netta mine. One-half natural size.

broad band of jasperoid forming the upper one-quarter of the specimen, which contains abundant fine sphalerite, bends over both the upper large sphalerite crystal and over the chert breccia. Although these features are rare and have not been described in the Tri-State district before, their interpretation is of interest in regard to the genesis of both the sphalerite and the jasperoid. Amstutz and Bubenicek (1967) believe that sulfide "geopetals", which occur in some southeast Missouri lead belt ores, indicate that they developed during the diagenesis of the enclosing sediments. In the Tri-State district, the development of such fabrics did not necessarily take place at an early time, because muds and softened rock developed during a number of periods subsequent to sedimentation and diagenesis; the evidence for this is present in the mines today. The rare "geopetals", which are present in Tri-State ores, probably developed as a result of limestone solution and vertical compression of the original section during the time of ore deposition. The vuggy character of the left side of the specimen in Fig. 21 is characteristic elsewhere of limestone that has been partly replaced by jasperoid, and the remaining limestone subsequently removed by solution. This character indicates that limestone replacement and solution had effected the development of this "geopetal" fabric.

Another unusual occurrence of jasperoid is shown in Fig. 22. Very distinct, nearly horizontal to slightly bowed bands or lenses of jasperoid are continuous laterally for about 6 m and pinch out at each end. These bands, unlike the jasperoid banding discussed above, exhibit sharp contacts with each other. The bands show distinct differences in grain size and sulfide content. Those bands with greater sulfide content exhibit larger grain sizes of both sphalerite and jasperoid quartz than do bands with lesser sphalerite

Fig. 22. Well-banded jasperoid (dark gray and black) deposited in a solution channel. The color variation is due to grain size and sphalerite content. The lower part of the banded jasperoid is covered by ore broken during mining. M bed, Webber mine.

content. Jasperoid occurrences, such as these, were probably formed by the local filling of open solution channels in contrast to more common occurrences where jasperoid replaces limestone and other rock types. The grains of sphalerite in these channel fills are crystals. They are not clastic grains like those from the east Tennessee zinc district that were described by Kendall (1960).

Jasperoid had not formed prior to Chesterian time, because jasperoid pebbles are not present with the chert pebbles in the conglomerates of Chesterian age.

GENESIS

The Tri-State district is located in a region that is unique for the confluence of structural features. The district is located on the flank of a major, positive, structural element (the Ozark uplift), a condition which is common to Mississippi Valley type deposits. The pinchout lines of two shales, which are older than the ore deposits, occur at or near the margins of the district. Siebenthal (1915, p. 198–199) emphasized the fact that the Chattanooga Shale of Late Devonian age pinches out at the southern margin of the district and is absent beneath most of the Tri-State ore fields (Fig. 1). Brockie et al. (1968) have shown that the Northview Shale of Early Mississippian age is commonly thinner than 3 m in the district and that its 3 m isopach approximately coincides with the northern limit of the field. Thus, most of the mineralization occurs above carbonate rock of pre-Late Devonian age.

Pennsylvanian shale unconformably overlies the Mississippian rocks containing the ore deposits. The porosity and permeability of the residual chert and solutional features developed in association with the unconformity were important in fostering the deposition of the ores in the Tri-State district. Callahan (1964) emphasized the importance of unconformities in relation to the location of a number of Mississippi Valley type ore districts. In addition, the present erosional edge of the Pennsylvanian shale is nearly coincident with the northwest margin of the district (Fig. 1). Most of the ore fields in the Tri-State district are aligned along a northeast trending axis, the position of which is very close to the present edge of the shale, particularly if one includes outliers of shale not shown in Fig. 1.

The district occurs in an area of intersecting major structural features. These include faults, throughs, and large folds many of which appear to reflect basement structures. Northwest trending structures, characteristic of the region to the north of the district, are intersected by northeast trending ones, which are common in the region to the south of the district. The individual ore fields are coincident with major folds and faults, e.g., the Miami trough and Bendelari monocline of the Picher field. Asymmetrical mineral relationships and fluid inclusion temperatures appear to support the concept that the Miami trough had an influence upon the circulation of the ore solutions in the Picher field. Mineral coatings on a single crystal or within a vug tend to be more concentrated on the side toward the Miami trough (Stoiber, 1946). Filling temperatures for fluid inclusions tend to be slightly higher in sphalerite specimens that have been collected near the trough as compared with those occurring farther from the structure (Schmidt, 1962). These major structural features have also played an important part in permitting ground water access to soluble limestone beds and in channeling at least part of the subsequent ore forming solutions in the Tri-State ore fields.

The Grand Falls Member is an especially permeable horizon in the Tri-State district. The crackle breccia character of its chert together with its bedding plane solution cavities provide it with a high degree of permeability. This horizon has been of great importance in promoting lateral movement of solutions in the district. It is the major shallow aquifer to which most of the water wells in the district have been drilled. The occurrence of broad, blanket-shaped, ore deposits in the Grand Falls Member indicates at least its local importance to the ore plumbing system. That the ore solutions probably travelled extensively in this member is indicated by the presence of minor sphalerite mineralization which can be detected over very wide areas in the district (Brockie, oral personal communication, 1973).

The smaller features in which individual ore bodies are localized, are of sedimentary, diagenetic, structural, and karst origin. The presence of crinoidal bioherms, some of which have associated shaly facies, suggests that original differences in the mineralogy and texture of these shallow-sea sediments may have made them susceptible to the effects of solution and slight tectonic activity and thereby prepared them for ultimate deposition of the ore. Those parts of the Boone Limestone, which were more soluble and more fractured or broken, served as the loci of solution, dolomitization, jasperoidization, and mineralization. Part of the dissolution of the Boone limestone and the accompanying accumulation of chert breccia, originated through ground-water action associated with the development of a karst topography on the post-Mississippian, pre-Pennsylvanian, erosion surface. This subsurface ground-water activity aided in preparing the host rocks with open spaces along the fractures and in developing additional solution-related fractures which facilitated the introduction of the subsequent mineralizing solutions. Much solution of the host limestone took place at the time of ore deposition as shown by the extensive replacement of limestone by jasperoid and sulfides. That solution of the limestone continued beyond the time of the deposition of jasperoid and most sulfides is shown by the presence of boulder zones at the outer edges of the ore runs. These zones, which consist of residual, partly broken, chert slabs, have many openings, the walls of which are lined only with the late minerals, calcite and marcasite. The boulder zones are not mineralized, despite their favorable "open ground" character, because dissolution of the limestones in these areas had not developed until after practically all of the ore had been deposited. The large calcite crystals deposited in the boulder zones and to a lesser extent in cavities associated with the ore deposits may simply represent redeposition of the calcium carbonate which earlier had been taken into solution from the limestone.

Provided the assumptions regarding fluid inclusions are correct, they appear to afford the best information on the character of the Tri-State ore-depositing solutions. Primary fluid inclusions in Tri-State sphalerite indicate that the ore-bearing solutions were strong brines with salinities as high as 30% salts that were dominated by sodium, calcium, and chlorine (Roedder, 1967; Chapter 4, Vol. 2). The brines were warm (about 80–120°C) during the deposition of the sphalerite and somewhat cooler (approximately 50–70°C) during the deposition of calcite, as determined by analyses of the filling temperatures for the fluid inclusions (Schmidt, 1962). Such brines have the capability of transporting significant quantities of metals in the form of chloride complexes (Helgeson, 1969).

There is no solid evidence that igneous magmas have taken any significant part in the development of the Tri-State ore deposits. White (1958) demonstrated that the chemical composition of the fluids trapped in the minerals of Mississippi Valley type ore deposits is strikingly similar to that of oil-field brines. A relationship to petroleum also is suggested by the common occurrence of viscous bitumen in some of the mines. Thus, the ore-bearing solution may have been generated in the adjacent sedimentary basin and subsequently moved upward along the flank of the Ozark uplift (see Wolf, 1976, for summary on compaction fluids as possible ore solutions). Clayton et al. (1966) found that the oxygen- and deuterium-isotope content of the deep, salty, basin waters, including those of the Illinois basin north of Illinois—Kentucky fluorspar district, is similar to that of the present day surface waters in the areas examined. They concluded that the brines were predominantly meteoric waters which had undergone deep circulation and modification. Circulation cells of meteoric ground water may be important in the development of many types of ore deposits (Taylor, 1973). The restriction of all of the ore deposits of the Tri-State district to positions near the earth's surface suggests that meteoric waters have had some roll in the development of these ores partly through the preparation of the host rocks and perhaps through mixing with brines.

Part or even all of the constituents for the minerals deposited in the Tri-State ores may have been derived from the sedimentary rocks which the brines traversed. Tiny, euhedral, apatite crystals disseminated in some jasperoids are believed by McKnight and Fischer (1970, p. 122) to have recrystallized from sedimentary collophane, because apatite occurs only in those limestone beds containing collophane. The fact that pink dolomite was deposited after gray dolomite, together with the restriction of the former to or near areas of the latter, has suggested to Brockie et al. (1968, p. 421) that pink dolomite is simply a remobilization of gray dolomite. The late, often large, calcite crystals deposited in boulder ground and throughout parts of the ore deposits may simply consist of calcium carbonate dissolved from the sedimentary limestone during the deposition of the ore deposits and redeposited at a subsequent time when the waters filling the openings were nearly devoid of metals and had acquired a cooler, more dilute character. One might further speculate that the silica, which formed the jasperoid that is so common to the Tri-State ore deposits, might have had its origin in the partial solution of the chert which is so abundant in the Mississippian limestones of the Tri-State region. If this is true, it could account for the puzzling contrast of intense silicification in the Tri-State ores as opposed to the almost complete absence of silicification from many other Mississippi Valley type deposits, such as the southeast Missouri Lead Belt ores, which occur in the noncherty Bonneterre Formation.

The source for the sulfur, which forms the sulfides of the Tri-State deposits, was the sedimentary rocks of the region. The sulfur isotope ratios of the sulfides are unlike those in sulfide deposits that are derived from magmatic sources or are of volcanic–exhalative origin. They are more characteristic of the type of sulfur that is generated by reduction of the sulfate that is present in metal-bearing brines (Jensen and Dessau, 1967). Various substances or organisms have been suggested for the sulfur reduction, such as methane and associated organic matter (Barton Jr., 1967) and sulfate-reducing bacteria (Bastin, 1926 and Jensen and Dessau, 1967). Black, viscous, bitumen percolated downward over the walls of some of the mines in the Picher field. A chemical analysis of this petroliferous material had yielded 1.1% sulfur. This amount has suggested to Skinner (1967) that the sulfur present as organic complexes in this material may have become unstable as a result of the temperature rise associated with the introduction of a metal-rich brine. An even closer associate of the ores is the solid organic material that is finely disseminated in the jasperoid and gives the jasperoid its dark color. Iron sulfide originally present in the host rocks may have provided some of the sulfur for the formation of the sulfide ores.

The metals themselves require no magmatic contribution and probably have simply been leached from the rocks of the region by the Na–Ca–Cl brines. Pyrite, marcasite, sphalerite, and galena, the four most abundant metallic sulfides in the Tri-State ores, occur commonly over a wide region in the Pennsylvanian shales as sedimentary or diagenetic constituents in the form of disseminated crystals, sulfide concretions, and ironstone septeria fillings. This illustrates the ability of these sulfides to be deposited during sedimentation and diagenesis of the enclosing Pennsylvanian shales. Although the sphalerite and galena from such occurrences appear to have some chemical characters which differ from that of the sulfide in the ore deposits, they conceivably could constitute a source for part of the metals. The sphalerites from such occurrences have different trace element populations of cadmium, gallium, and germanium, as contrasted to that of sphalerite crystals in the ore deposits. A single analysis of galena from one occurrence of this type northwest of the district indicates that it consisted of ordinary lead (Heyl et al., 1966) in contrast to the markedly radiogenic, J-type, lead isotopic ratios of the Tri-State ores (Cannon Jr. et al., 1961). Mixing of a smaller radiogenic lead component derived from a uranium-bearing source, such as that reported for brines northwest of the district (Gott and Hill, 1953), together with ordinary leads derived from the Carboniferous formations could produce lead isotopic contents like those of the Tri-State ores. Alternatively, the metals could have been leached from the Mississippian or possibly older formations. For example, minor amounts of galena and sphalerite have been reported by Lee (1933) from sedimentary rocks that were intersected in deep drill holes in central and western Kansas. Study of the conodonts from these rocks indicates that their stratigraphic position is equivalent to the M bed, which is the main ore horizon in the Tri-State district (Goebel et al., 1968).

The deposition of the metallic sulfides took place over a long period of time. This condition is attested to by the many repetitions in the paragenetic sequence (Hagni and Grawe, 1964) and is suggested by the wide range of lead isotopes recorded from the center to the edge of a single galena crystal (Cannon Jr. et al., 1963). The sulfides were even subjected to a period of corrosion between periods of deposition (McKnight and Fischer, 1970). The movement of the ore-depositing fluid was an exceedingly sluggish one in which gravity was the dominant force in determining the vertical component of mineral asymmetry (Stoiber, 1946).

Whether the metal-bearing brines that circulated through channels in the Mississippian rocks and deposited the Tri-State ore deposits had their ultimate origin in the adjacent sedimentary basin or whether they were surface waters that acquired a salty character as a result of evaporation during a period of hot, humid climate and/or by contact with evaporites and reached the sites of ore deposition through some type of underground circulation cell is a subject for interesting future research.

The characteristic features of the Tri-State district are summarized in Table IV. The principal characteristics for all Mississippi Valley type deposits, listed partly after Snyder (1968), are given for comparative purposes. The table emphasizes the similarities of the Mississippi Valley type deposits. The ore deposits of the Tri-State district, one of the principal Mississippi Valley districts, shares many of the characteristics common to the other districts. The main feature, which tends to distinguish the Tri-State district, is its greater development of solution collapse breccia and intensive silicification.

TABLE IV

Comparison of the characteristics of the Tri-State with other Mississippi Valley type deposits

Mississippi Valley type deposits	Tri-State deposits
Host rock character and alteration	
A. Shallow-water carbonates	Shallow-water Mississippian carbonates including biohermal, clastic and oolitic materials
B. Certain stratigraphic horizons are the principal producers, but mineralization extends over a large stratigraphic interval	Ore locally mined from overlying Pennsylvanian shales and stratigraphically lower Northview Formation
C. Unconformities beneath and within the mineralized stratigraphic interval influence sedimentary facies patterns	Effect of unconformities upon subsequent patterns of sedimentation needs further study
D. Dolomitization; most ore deposits are confined to dolomitized areas	Ore deposits occur in dolomitized rock, especially at the margins of dolomitic masses; very minor mineralization occurs locally in partly silicified limestone
E. Recrystallization; limestone contiguous to ore deposits may show recrystallization of fossils and oolites to sparry calcite	Limestones may be recrystallized for distances of about 40–80 feet beyond the edge of silicification
F. Silicification; developed to variable degrees	Intense silicification of limestone, dolomite, and muds to form jasperoid
Regional structure and ground preparation	
A. Districts occur on the flanks or above major positive features, especially domal structures	District is located on the flank of the Ozark dome at the margin of sedimentary basins to the northwest and southwest
B. Major faults and fault breccias are important to the localization of subdistricts or fields, but most individual deposits are controlled by local structures of sedimentary, tectonic, and dissolution origin	Although the intersection of the Miami trough and Bendelari monocline is located near the center of the very productive Picher field, only a small number of individual ore bodies have locations and orientations which could suggest that they might be directly related to the major structures
C. Permeable horizons allow extensive lateral movement of the ore solutions	The sheet ground is the principal shallow aquifer in the district; it has promoted extensive lateral migration of the ore solutions
D. Ground prepared by minor faults and fractures	Minor fractures provided access to soluble horizons for subsequent solutions
E. Ground further prepared by solution and collapse through the action of ground waters related to unconformities both above and within the mineralized stratigraphic intervals	Additional ground preparation through dissolution along fractures by ground waters related to erosion surfaces

TABLE IV (continued)

Mississippi Valley type deposits	Tri-State deposits
Local structural controls	
A. Sedimentary structures	Sedimentary structures
(1) Ridges or bars	(1) Bars have not been observed
(2) Reefs and reef-like structures	(2) Bioherms are present; their influence on ore deposition needs further study, but the intense silicification in the ore district renders difficult the study of these and other sedimentary features
(3) Submarine slide breccias	(3) Submarine slide breccias have not been observed
(4) Pinchouts	(4) Pinchouts of the Chatannooga and Northview shales beneath the ore deposits broadly outline the district
(5) Lateral facies changes	(5) Ore deposits mostly located near silicified dolomite–limestone contacts
B. Tectonic Structures	Tectonic structures
(1) Folds	(1) Minor flexing has been important in developing fractures in the cherty carbonate host rock; the fractures determine the ultimate loci for individual ore deposits
(2) Minor faulting and fracturing	(2) Minor fractures constitute a major part of the plumbing system and play an important part in the localization of individual deposits
(3) Fault breccias	(3) Fault breccias are of very minor importance in the localization of individual deposits
C. Solutional structures	Solutional structures
(1) Solutional collapse breccias are very important; many were initiated along fractures by ground water related to erosion surfaces, and the process was continued by ore fluids; solutional thinning of the underlying beds is a dominant factor in the development of many collapse structures; collapse structures may occur near and parallel to major faults; a great variety of shapes is exhibited by the resultant breccia bodies	(1) Solutional collapse breccias, developed by solution enlargement of fractures along dolomite–limestone contacts, are the dominant ore-bearing structures
Ore character	
A. Open space filling of vugs, fractures, and breccias is the principal manner of ore occurrence	Much of the ore is deposited as well developed crystals that line irregularly shaped vugs in breccias and host rocks
B. Disseminated replacements are associated with cavity-filling ores	Disseminated ore, developed by replacement and silicification of carbonate host rocks, is closely associated with cavity-filling ores
C. Veins are locally associated with predominantly stratiform deposits	True veins are of very rare occurrence in the district

TABLE IV (continued)

Mississippi Valley type deposits	Tri-State deposits
Ore character	
D. Size of ore deposits is controlled by the size of the local controlling structure	Size of most ore deposits is controlled primarily by the size of prepared bodies of breccia; size of sheet ground deposits determined mainly by the size of openings in a permeable horizon
E. Most deposits are relatively low grade	Grade has averaged mostly between 2 and 4% combined Zn and Pb
Mineralogy	
A. Principal minerals are restricted to a small number	Principal minerals restricted to eight: sphalerite, galena, chalcopyrite, marcasite, pyrite, quartz, dolomite, and calcite
B. Principal minerals have simple chemical compositions	Principal minerals are simple sulfides, carbonates, and silica
C. Few additional minor minerals	Enargite, luzonite, wurzite, and barite are locally present in very minor quantities
Mineral paragenesis	
A. Substantial portions of the mineralization were deposited as open space fillings in which crystals of one mineral coat those of another; these well crystallized minerals are particularly well suited for a study of their sequence of deposition	Minerals commonly deposited as well crystallized, open space fillings which are especially well suited for a study of mineral paragenesis
B. Minerals were deposited in a general sequence in which most of one mineral tends to be deposited before most of another mineral	Principal minerals deposited in the following general sequence: dolomite, quartz, sphalerite, galena, chalcopyrite, marcasite, pyrite, and calcite
C. Two minerals, whose periods of deposition are closely spaced in time, may exhibit partially overlapping relationships	Local overlapping deposition
D. Repetitive deposition of the principal minerals in many intervals is characteristic. Although many of the intervals beyond those of their main depositional periods are of minor importance, other repetitions are quantitatively significant and may represent separate pulses of the ore solutions	Repetitions common; chalcopyrite and pyrite were deposited during eight intervals, sphalerite and marcasite during six, and galena and quartz during five
Chemical Constituents in the Ore	
A. Major elements include: Zn, Pb, Cu, Fe, Si, Ca, Mg, Ba, F, S, C, and O	Major elements are: Zn, Pb, Cu, Fe, Si, Ca, Mg, S, C, and O
B. Minor elements present in the principal minerals: Fe, Ag, Cd, Ge, Ga, In, Co, and Hg, in sphalerite; Ag, Sb, Bi, and As in galena; Co, Ni, Ag, and Cu in pyrite and marcasite; Sr, Y, Ba, and Mn in calcite. Additional minor elements reported for these minerals, and perhaps part of the above elements are present partly to wholly as microscopic inclusions of separate phases	Traces of Fe, Ag, Ge, Ga, In, Co and Hg are present in sphalerite; Ag and Sb in galena; Ni, Ag, and Cu in marcasite; Ni in pyrite. The As-bearing minerals enargite, and luzonite, occur in trace quantities

TABLE IV (continued)

Mississippi Valley type deposits	Tri-State deposits
Chemical Constituents in the Ore	
C. Isotopic composition	Isotopic composition
(1) Lead: Wide range of values, J-type radiogenic character (for deposits of Mississippi Valley proper)	(1) Lead: Wide range of values even in a single crystal. District is type locality for J-type (Joplin) lead
(2) Sulfur: Wide range of values, characteristic of biogenic sulfur	(2) Sulfur: The relatively small number of analyses suggests a broad range of sulfur isotope values
Fluid inclusion character	
A. Mainly a very concentrated saline brine, dominated by Na, Ca, and Cl; becomes more dilute in some younger minerals	Brine with salinities as high as 30% salts
B. Filling temperatures range from about 40−150°C; showing a general decrease with time	Most sphalerite was deposited in the temperature range of about 80−120°C, and calcite deposited subsequently about 50−70°C
C. Density greater than normal water	Density perhaps as much as 1.1

REFERENCES

Amstutz, G.C. and Bubenicek, 1967. Diagenesis in sedimentary mineral deposits. In: G. Larsen and G.V. Chilingar (Editors), *Diagenesis in Sediments. Developments in Sedimentology, 8.* Elsevier, Amsterdam, pp. 417−475.

Bain, H.G., Van Hise, C.R. and Adams, G.I., 1901. Preliminary report on the lead and zinc deposits of the Ozark region. *U.S. Geol. Surv. Ann. Rep.,* 22 (part 2): 23−227.

Barton Jr., P.B., 1967. Possible role of organic matter in the precipitation of the Mississippi valley ores. In: J.S. Brown (Editor), *Genesis of Stratiform Lead−Zinc−Barite−Fluorite Deposits (Mississippi Valley Type Deposits) − A Symposium, New York. Econ. Geol. Monogr.,* 3: 371−377.

Bastin, E.S., 1926. A hypothesis of bacterial influence in the genesis of certain sulphide deposits. *J. Geol.,* 34: 773−792.

Bastin, E.S., 1933. Relations of cherts to stylolites at Carthage, Missouri. *J. Geol.,* 41: 371−381.

Bieber, C.L., 1955. The structural geology of southwestern Missouri. *Mo. Geol. Surv. Water Resour.,* unpublished.

Brockie, D.C., Hare Jr., E.H., and Dingess, P.R., 1968. The geology and ore deposits of the Tri-State district of Missouri, Kansas, and Oklahoma. In: J.D. Ridge (Editor), *Ore Deposits of the United States, 1933−1967.* Am. Inst. Min. Metall. Pet. Eng., New York, N.Y., pp. 400−430.

Buckley, E.R. and Buehler, H.A., 1905. The geology of the Granby area. *Mo. Geol. Surv. Rep.,* 4 (2nd ser.): 120 pp.

Callahan, W.A., 1964. Paleogeographic premises for prospecting for strata-bound base metal deposits in carbonate rocks. *CENTO Symposium on Mining Geology, Ankara, Turkey,* pp. 191−248.

Campbell, N., 1967. Tectonics, reefs and stratiform lead−zinc deposits of the Pine Point area, Canada. In: J.D. Ridge (Editor), *Ore Deposits of the United States, 1933−1967.* Am. Inst. Min. Metall. Pet. Eng., New York, N.Y., pp. 59−70.

Cannon Jr., R.S., Pierce, A.P. Antweiler, J.C. and Buck, K.L., 1961. The data of lead isotope geology related to problems of ore genesis. *Econ. Geol.*, 56: 1–38.

Cannon Jr., R.S., Pierce, A.P. and Delevaux, M., 1963. Lead isotope variations with growth zoning in a galena crystal. *Science,* 142: 574–576.

Clayton, R.N., Friedman, I., Graf, D.L., Mayeda, T.K., Meents, W.F. and Shimp, N.F., 1966. The origin of saline formation waters. *J. Geophys. Res.,* 71: 3869–3882.

Denison, R.E., Hetherington, E.A. and Otto, J.B., 1969. Age of basement rocks in northeastern Oklahoma. *Okla. Geol. Notes*, 29: 120–128.

Farrington, O.C., 1900. Crystal forms of calcite from Joplin, Missouri. *Publications of the Field Columbian Museum, Geol. Ser.,* I: 232–241.

Folk, R.L. and Weaver, C.E., 1952. A study of the texture and composition of chert. *Am. J. Sci.,* 250: 498–510.

Fowler, G.M., 1938. Structural control of ore deposits in the Tri-State zinc and lead district. *Eng. Min. J.,* 139: 46–51.

Fowler, G.M. and Lyden, J.P., 1932. The ore deposits of the Tri-State district. *Am. Inst. Min. Metall. Eng. Trans.,* 102: 206–251.

Fowler, G.M., Lyden, J.P., Gregory, F.E. and Agar, W.M., 1935. Chertification in the Tri-State mining district. *Am. Inst. Min. Metall. Eng. Trans.,* 115: 106–163.

Friedman, G.M. and Sanders, J.E., 1967. Origin and occurrence of dolostones. In: G.V. Chilingar, H.J. Bissell and R.W. Fairbridge (Editors), *Carbonate Rocks. Developments in Sedimentology, 9A.* Elsevier, Amsterdam, pp. 267–348.

Gerdemann, P.E. and Meyers, H.E., 1972. Relationship of carbonate facies patterns to ore distribution and to ore genesis in the southeast Missouri lead district. *Econ. Geol.,* 67: 426–433.

Giles, A.W., 1935. Boone chert. *Geol. Soc. Am. Bull.,* 46: 1815–1878.

Goebel, E.D., Thompson, T.L., Waugh, T.C. and Mueller, L.C., 1968. Mississippian conodonts from the Tri-State district, Kansas, Missouri, and Oklahoma. In: D.E. Zeller (Editor), *Short Papers on Research in 1967. State Geol. Surv. Kans. Bull.,* 191 (part 1): 21–25.

Gott, G.B. and Hill, J.W., 1953. Radioactivity in some oil fields of southeastern Kansas. *U.S. Geol. Surv. Bull.,* 988-E: 122 pp.

Grohskopf, J.G. and McCracken, E., 1949. Insoluble residues of some Paleozoic formations of Missouri, their preparation, characteristics, and application. *Mo. Geol. Surv. Water Resour. Rep. Invest.,* 10: 39 pp.

Hagni, R.D. and Desai, A.A., 1966. Solution thinning of the M bed host rock limestone in the Tri-State District, Missouri, Kansas, Oklahoma. *Econ. Geol.,* 61: 1436–1442.

Hagni, R.D. and Grawe, O.R., 1963. Tabular review of the genesis of Tri-State ores. Guidebook to the geology in the vicinity of Joplin, Missouri. *Assoc. Mo. Geol. Ann. Field Trip,* 10: 36–44.

Hagni, R.D. and Grawe, O.R., 1964. Mineral paragenesis in the Tri-State district, Missouri, Kansas, Oklahoma. *Econ. Geol.,* 59: 449–457.

Hagni, R.D. and Saadallah, A.A., 1965. Alteration of host rock limestone adjacent to zinc–lead ore deposits in the Tri-State district, Missouri, Kansas, Oklahoma. *Econ. Geol.,* 60: 1607–1619.

Helgeson, H.C., 1969. Thermodynamics of hydrothermal systems at elevated temperatures and pressures. *Am. J. Sci.,* 267: 729–804.

Heyl, A.V., 1972. The 38th parallel lineament and its relationship to ore deposits. *Econ. Geol.,* 67: 879–894.

Heyl, A.V., Delevaux, M.E., Zartman, R.M. and Brock, M.R., 1966. Isotopic study of galenas from the Upper Mississippi Valley, the Illinois–Kentucky, and some Appalachian Valley mineral districts. *Econ. Geol.,* 61: 933–961.

Huffman, G.G., 1958. Geology of the flanks of the Ozark uplift. *Okla. Geol. Surv. Bull.,* 77: 281 pp.

Ireland, H.A., 1955. Precambrian surface in northeastern Oklahoma and parts of adjacent states. *Bull. Am. Assoc. Pet. Geol.,* 39: 468–483.

Jensen, M.L. and Dessau, G., 1967. The bearing of sulfur isotopes on the origin of Mississippi Valley type deposits. In: J.S. Brown (Editor), *Genesis of Stratiform Lead–Zinc–Barite–Fluorite Deposits (Mississippi Valley Type Deposits) – A Symposium, New York. Econ. Geol. Monogr.,* 3: 400–408.

Kendall, D.L., 1960. Ore deposits and sedimentary features, Jefferson City mine, Tennessee. *Econ. Geol.*, 55: 985–1003.

Kisvarsanyi, E.B., 1974. Operation basement: buried Precambrian rocks of Missouri – their petrography and structure. *Bull. Am. Assoc. Pet. Geol.*, 58: 674–684.

Larsen, K.G., 1973. Depositional environments of the Bonneterre Formation in southeast Missouri – an example of epeiric sea sedimentation (abstract). *Geol. Soc. Am., North-Central Sect. Ann. Meet.*, 7: 330–331.

Lee, W., 1940. Subsurface Mississippian rocks of Kansas. *State Geol. Surv. Kans. Bull.*, 33: 114 pp.

Lyden, J.P., 1950. Aspects of structure and mineralization used as guides in the development of the Picher field. *Am. Inst. Min. Metall. Eng. Trans.*, 187: 1251–1259.

McCormick, J.E., Evans, L.L., Palmer, R.A., Rasnick, F.D., Quarles, K.G., Mellon, W.V. and Riner, B.G., 1969. Mine geology of the American Zinc Company's Young mine. In: *Papers on the Stratigraphy and Mine Geology of the Kingsport and Mascot Formations (Lower Ordovician) of East Tennessee. Tenn. Div. Geol. Rep. Invest.*, 23: 45–52.

McCracken, M.H., 1971. Structural features of Missouri: *Mo. Geol. Surv. Water Resour. Rep. Invest.*, 49: 99 pp.

McKnight, E.T., 1950. The Tri-State region. In: C.H. Behre, Jr., A.V. Heyl, and E.T. McKnight (Editors), *Zinc and lead deposits of the Mississippi Valley. Int. Geol. Cong. Rep.*, 18 (part 7): 52–58.

McKnight, E.T. and Fischer, R.P., 1970. Geology and ore deposits of the Picher field, Oklahoma and Kansas. *U.S. Geol. Surv. Prof. Pap.*, 588: 165 pp.

McKnight, E.T. et al., 1944. *Maps Showing Structural Geology and Dolomitized Areas in Part of the Picher Zinc–Lead Field, Oklahoma and Kansas*; based on Field Maps prepared by McKnight, E.T., Fischer, R.P., Addison, C.C., Bowie, K.R., Thiel, J.M., Owens, M.F., Jr., and Wells, F.G. In: Tri-State zinc–lead investigations, Preliminary maps 1–6, U.S. Geol. Surv., Wash., D.C.

Missouri Geological Survey and Water Resources, 1942. Joplin district (Jasper and Newton Counties, Mo.), Geological maps showing mining and mineralized areas, 6 sheets.

Ohle Jr., E.L. and Brown, J.S., 1954. Geologic problems in the southeast Missouri lead district. *Geol. Soc. Am. Bull.*, 65: 201–221, 935–936.

Pierce, W.G., 1935. Contour map of the base of the Cherokee Shale in the zinc and lead district of southeastern Kansas. *Dep. of the Interior, Memorandum for the Press*, 4 pp.

Pinckney, D.M. and Rye, R.O., 1972. Variation of O^{18}/O^{16}, C^{13}/C^{12}, texture, and mineralogy in altered limestone in the Hill mine, Cave-in-Rock District, Illinois. *Econ. Geol.*, 67: 1–18.

Ridge, J.D., 1936. The genesis of the Tri-State zinc and lead ores. *Econ. Geol.*, 31: 298–313.

Roedder, E., 1967. Environment of deposition of stratiform (Mississippi Valley-type) ore deposits, from studies of fluid inclusions. In: J.S. Brown (Editor), *Genesis of Stratiform Lead–Zinc–Barite–Fluorite Deposits (Mississippi Valley Type Deposits) – A Symposium, New York. Econ. Geol. Monogr.*, 3: 349–361.

Sawkins, F.J., 1969. Chemical brecciation, an unrecognized mechanism for breccia formation. *Econ. Geol.*, 64: 613–617.

Schmidt, R.A., 1962. Temperatures of mineral formation in the Miami-Picher district as indicated by liquid inclusions. *Econ. Geol.*, 57: 1–20.

Schmidt, A. and Leonhard, A., 1874. The lead and zinc regions of southwest Missouri. *Mo. Geol. Surv. Rep. for 1874*. pp. 380–734.

Siebenthal, C.E., 1908. Mineral resources of northeastern Oklahoma. *U.S. Geol. Surv. Bull.*, 340: 187–230.

Siebenthal, C.E., 1915. Origin of the zinc and lead deposits of the Joplin region – Missouri, Kansas, and Oklahoma. *U.S. Geol. Surv. Bull.*, 606: 283 pp.

Siebenthal, C.E., 1925. Contour map of the surface of the beds underlying the Cherokee shale in a portion of the Picher district, Okla., showing relations of ore bodies to the surface contoured. *Dep. of the Interior, Memorandum for the Press*.

Skinner, B.J., 1967. Precipitation of Mississippi Valley-type ores: A possible mechanism. In: J.S. Brown (Editor), *Genesis of Stratiform Lead–Zinc–Barite–Fluorite Deposits (Mississippi Valley Type Deposits) – A Symposium, New York. Econ. Geol. Monogr.*, 3: 363–369.

Smith, W.S.T. and Siebenthal, C.E., 1907. Description of the Joplin district in Geologic Atlas of the United States. *U.S. Geol. Surv. Folio,* 148: 20 pp.

Snyder, F.G., 1968. Geology and mineral deposits, midcontinent, United States. In: J.D. Ridge (Editor), *Ore Deposits of the United States, 1933–1967.* Am. Inst. Min. Met. Pet. Eng., New York, N.Y., pp. 257–286.

Snyder, F.G., 1970. Structural lineaments and mineral deposits, eastern United States. In: D.O. Rausch and B.C. Mariacher (Editors), *World Symposium on Mining and Metallurgy of Lead and Zinc.* Am. Inst. Min. Metall. Pet. Eng., New York, N.Y., 76–94.

Snyder, F.G. and Emery, J.A., 1956. Geology in development and mining, southeast Missouri lead belt. *Am. Inst. Min. Metall. Eng. Trans.,* 208: 1216–1224.

Snyder, F.G. and Gerdemann, P.E., 1968. Geology of the southeast Missouri lead district. In: J.D. Ridge (Editor), *Ore Deposits of the United States, 1933–1967.* Am. Inst. Min. Metall. Pet. Eng., New York, N.Y., pp. 326–358.

Stoiber, R.E., 1946. Movement of mineralizing solutions in the Picher field, Oklahoma–Kansas. *Econ. Geol.,* 41: 800–812.

Taylor Jr., H.P., 1973. The application of oxygen and hydrogen isotope studies to problems of hydrothermal alteration and ore deposition (abstract). *Geol. Soc. Am., Abstr. 1973 Ann. Meet.,* pp. 834–835.

Troell, A.R., 1962. Lower Mississippian bioherms of southwestern Missouri and northwestern Arkansas. *J. Sediment. Petrol.,* 32: 629–664.

Weidman, S., 1932. The Miami-Picher zinc–lead district, Oklahoma. *Okla. Geol. Surv. Bull.,* 56: 177 pp.

White, D.E., 1958. Liquid of inclusions in sulfides from Tri-State (Missouri–Kansas–Oklahoma) is probably connate in origin (abstract). *Geol. Soc. Am. Bull.,* 69: 1660–1661.

Winslow, A. and Robertson, J.D., 1894. Lead and zinc deposits. *Mo. Geol. Surv. Rep.,* 7: 763 pp.

Wolf, K.H., 1976. Ore genesis influenced by compaction. In: G.V. Chilingarian and K.H. Wolf (Editors), *Compaction of Coarse-Grained Sediments, II. Developments in Sedimentology,* 18 B. Elsevier, Amsterdam, in press.

Chapter 11

APPALACHIAN ZINC—LEAD DEPOSITS

ALAN D. HOAGLAND

INTRODUCTION

Appalachian zinc—lead deposits and the classic Mississippi Valley deposits of the mid-continent are similar in many respects. The Appalachian deposits, however, are found in rocks which have undergone deformation by one or more of the orogenic events of the region. For many years the ore bodies in this mountain province were believed to have been formed from hydrothermal solutions which travelled from a deep source into zones in favorable carbonate horizons. Both the paths through which the solutions travelled and the sites of deposition were presumed to have been prepared by tectonic structures related to the mountain-building process. During the last two decades, in a period of extensive exploration and mine development, particularly in Tennessee, but also in Virginia, Pennsylvania, and Canada, old evidence has been re-evaluated and new evidence examined. A pre-orogenic age of the mineralization is now generally accepted and dependence on tectonic structure for ore control is no longer mandatory. Ore deposition was accomplished fairly close to the surface while the strata were still in their essentially horizontal position just as were those of the Mississippi Valley. The similarity of the ore deposits in the Appalachian mountain province with those of the undeformed mid-continent region becomes more obvious when the folds and great thrust faults of the Appalachians are assumed not to have been formed and the original attitude of ore districts is reconstructed. A fresh look at the history of sedimentation, environment, and physiography of the region is generating concepts of the genesis of Appalachian deposits parallel to, and to a considerable extent dependent on, the new ideas related to the character and genesis of the classic Mississippi Valley-type deposits in the mid-continent region.

From the perspective of Appalachian geology, there appear to be three main classes of deposits (Figs. 1 and 2).

(*1*) The most important and distinctive of these is the essentially lead-free zinc deposits represented by those of east Tennessee; central Tennessee; Timberville (Va.); Friedensville (Pa.) and probably Newfoundland zinc in the Great Northern Peninsula. The distinctive characteristics of these deposits, in addition to the essential absence of lead, are comparable sedimentary and paleophysiographic environments in Lower Ordovician dolomite. Members of this class are grouped in Table I under the caption "Lower Ordo-

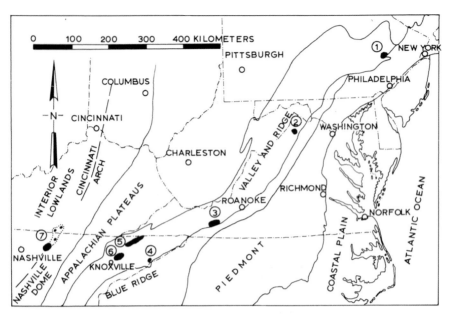

Fig. 1. Major Appalachian-type deposits in eastern United States.
1 = Friedensville, Pa.: zinc in Lower Ordovician Beekmantown Formation; *2* = Timberville, Va.: zinc in Lower Ordovician Beekmantown Formation; *3* = Austinville–Ivanhoe, Va.: zinc–lead in Lower Cambrian Shady Formation; *4* = Embreeville, Tenn.: zinc–lead in Lower Cambrian Shady Formation; *5, 6* = east Tennessee districts, Copper Ridge and Mascot–Jefferson City, respectively; *7* = central Tennessee. In these east and central Tennessee districts the zinc ore is in Lower Ordovician Mascot and Kingsport Formations.

vician Host". The east Tennessee deposits belonging to this class, as described later in the text, are identified in Table I as Mascot–Jefferson City and Copper Ridge.

(*2*) A second class of major economic importance is represented by the Austinville–Ivanhoe deposit in Virginia which is characterized by zinc with subordinate but economically important lead mineralization. Austinville, according to Callahan (1964), is localized at a facies change in breccia and reef material.

(*3*) The third class represented by Gays River (N.S.) is now being intensively explored and studied. The zinc–lead mineralization in this district occurs in a reef complex which developed on a major unconformity in or at the edge of a large evaporite basin. Gays River mineralization in a carbonate host of Mississippian age may be the youngest of economically important Appalachian deposits as well as the only one clearly associated with an evaporite sequence. The lead-isotope characteristics of Gays River are normal, similar to the Pine Point (N.W.T.) lead. These are distinctively different from Appalachian leads which form a rather coherent slightly anomalous group between the "J" leads of the Mississippi Valley deposits and normal leads which both Gays River and Pine Point resemble (Table II). Indeed, the Gays River environment and mineralization have much more in common with the Pine Point district than with any of the other Appalachian deposits.

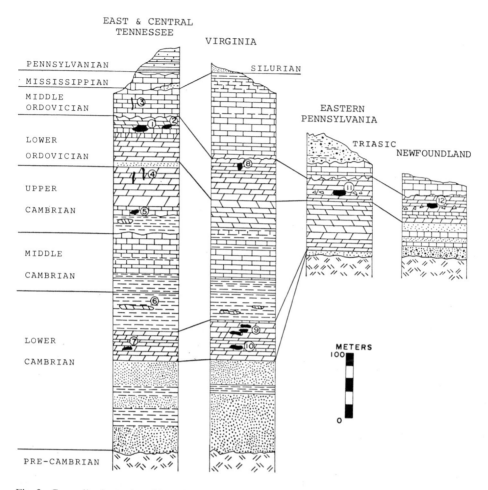

Fig. 2. Generalized stratigraphic relationships in principal Appalachian districts from Tennessee to Newfoundland. In east Tennessee, the Mascot–Jefferson City and Copper Ridge ore deposits (1) are confined to the Kingsport Formation and the lower part of the Mascot Dolomite. In central Tennessee (2) ore is predominately in the Middle and the Lower Mascot Dolomite. Uneconomic veins containing fluorite, barite, sphalerite, and galena, (3) occurring in Middle Ordovician limestone in central Tennessee are probably genetically related to the deeper ore in that district. Veins in the Copper Ridge Dolomite, (4) in the Powell River district are probably genetically related to the deeper stratiform deposits in the Maynardville Formation (5) in the same district. Uneconomic zinc–lead mineralization in the Evanston district, (6) occurs in dolomite lenses in the predominately shaly Rome Formation. At Embreeville oxidized zinc–lead ore, (7) occurs in deeply weathered Shady Dolomite. The Timberville, Virginia zinc deposit, (8) was developed in the Beekmantown Dolomite the stratigraphic equivalent of the Mascot and Kingsport Formation of east Tennessee. The Austinville Mine (9) and Ivanhoe Mine (10) are in the Shady Dolomite (Austinville and Ribbon Members respectively) in a strongly developed reef facies. The Friedensville Zinc Mine (11) in the Beekmantown of eastern Pennsylvania is correlated with the deposits in the Mascot–Jefferson City and Copper Ridge districts in east Tennessee as well as the Newfoundland Zinc deposits (12) occurring in the St. George Formation in the Great Northern Peninsula.

TABLE I

Characteristic \ District	Austinville, Va. (Lower Cambrian Host)	Embreeville, Tenn.	Gays River, Nova Scotia (Mississippian Host)	Powell River District (Upper Cambrian Host)	Friedensville District (Lower Ordovician Host)	Timberville, Pa.	*Copper Ridge, Va.	*Mascot–Jeff City	Newfoundland Zinc	Central Tennessee
District Rank	2	4	NR	4	3	4	3	1	NR	NR
Shallow marine carbonate	■	■	■	■	■	■	■	■	■	■
Dolomite-limestone transition zone	■	■	■	■		■	■	■	■	■
Open space filling predominate	■	■	■	■	■	■	■	■	■	■
Predominately dolomite gangue	■	■	■	■	■	■	■	■		■
Fe in sphalerite < 1%	■	■	■	■	?	■	■	■	■	■
Pyrite + marcasite < 5%	■	■	■	■	?	■	■	■	■	■
Barite + Fluorite < 0.5%	■	■	■	■	■	■	■	■	■	■
Galena rare or absent	■	■	■	■	■	■	■	■	■	■
Karst phenomena important	■	■	■	■	■	■	■	■	■	■
Deposited from brine solutions	?	?	?	?	■	■	■	■	■	■
Deposition between 80°C - 180°C	?	?	?	?	?	■	■	■	■	■
Slightly anomalous lead	■	■		?		■	?	■	?	■
Hydrocarbons associated	■			■		■	■	■	■	■
Reef facies important	■	■			■	■	■	?	?	■
Sphalerite with subordinate galena	■	■	■	■	■	■	■	■	■	■
Evaporite facies associated	■	■		■		■	■	■	■	■
Fissure veins associated	■	■	■		■	■	■	■	■	■
Predominately calcite gangue	■	■	■	■	■	■	■	■	■	■

The filled squares indicate strong correlation between characteristics shown horizontally with districts represented by vertical columns. Question marks indicate possible or questionable correlation while open spaces indicate absence of correlation. *Copper Ridge and Mascot–Jefferson City are the principal zinc districts in east Tennessee. **First rank: estimated more than $3.5 \cdot 10^6$ metric tons of metal; second rank: estimated between $1.5 \cdot 10^6$ and $3.5 \cdot 10^6$ metric tons of metal; third rank: estimated between $0.5 \cdot 10^6$ and $1.5 \cdot 10^6$ metric tons of metal; fourth rank: estimated less than $0.5 \cdot 10^6$ metric tons of metal; NR: unranked new district; present evidence indicates 3rd rank or higher.

Shallow-water marine carbonate host rocks

All of the important zinc and zinc–lead deposits of the Appalachians occur in shallow-water marine carbonate sediments. Further, it appears that many, if not all, are located within or close to the transition zone between the dominantly dolomite facies, believed to be of diagenetic origin, and the dominantly limestone facies on or near the edge of the continental shelf. Ore is commonly found in dolomitized limestone at or adjacent to the interface of limestone and dolomite. The dolomite in the latter case is generally attributed to dolomitization by the ore-related solutions although it is not clear whether the

TABLE II

Comparison of lead-isotope values in Mississippi Valley and Appalachian deposits[1]

	204	206	207	208	206/204	207/204	208/204	Σ/204[2]	206/207	208/206	208/207	208 / (206+207)
Wisconsin	1.230	27.25	19.86	51.66	22.15	16.15	42.00	80.30	1.37	1.89	2.60	1.10
Joplin–Picher	1.240	27.36	19.97	51.43	22.06	16.11	41.48	79.65	1.37	1.88	2.58	1.09
New Pb belt, Mo.	1.263	27.08	20.32	52.35	21.44	16.09	40.66	78.19	1.33	1.90	2.53	1.08
Old Pb belt, Mo.	1.283	26.73	20.51	51.48	20.83	15.99	40.13	76.94	1.30	1.93	2.51	1.09
Illinois–Kentucky	1.301	26.13	20.47	52.10	20.08	15.73	40.05	75.86	1.28	1.99	2.55	1.12
Central Kentucky	1.302	26.11	20.71	51.88	20.05	15.91	39.85	75.80	1.26	1.99	2.51	1.11
Mascot, Tenn.[3]	1.316	25.74	20.74	52.19	19.56	15.77	39.66	74.95	1.24	2.03	2.52	1.12
Friedensville, Pa.[3]	1.323	25.46	20.74	52.47	19.24	15.68	39.66	74.58	1.24	2.06	2.53	1.14
Flat Gap, Tenn.[4]	1.326	25.38	21.02	52.28	19.14	15.85	39.43	74.42	1.21	2.03	2.52	1.13
Embreeville, Tenn.[5]	1.324	25.76	20.93	51.98	19.46	15.81	39.26	74.52	1.23	2.02	2.48	1.11
Austinville, Va.[6]	1.329	25.51	21.08	52.07	19.19	15.85	39.17	74.22	1.21	2.04	2.47	1.12
Bamford, Pa.[5] (L–C)	1.348	25.26	21.24	52.15	18.74	15.76	38.69	73.18	1.19	2.06	2.46	1.12
Birmingham Mine, Pa.[5]	1.349	25.11	21.20	52.34	18.61	15.72	38.80	73.13	1.18	2.08	2.47	1.13
Central Tenn. (Veins)[7]	1.349	26.20	20.97	51.67	19.42	15.41	38.30	73.13	1.26	1.97	2.49	1.10
Central Tenn. (Knox)[7]	1.359	26.12	20.82	51.68	19.22	15.32	38.03	72.57	1.25	1.98	2.48	1.10
Califon, N.J.[5] (L–C)	1.356	25.33	21.21	52.06	18.68	15.64	38.39	72.71	1.19	2.06	2.45	1.12
Pine Point, N.W.T.	1.354	25.01	21.45	52.19	18.47	15.84	38.54	72.86	1.17	2.09	2.43	1.12
Gays River, N.S.	1.360	24.88	21.43	52.33	18.29	15.76	38.48	72.53	1.16	2.10	2.44	1.13

[1] Except as otherwise noted, data from J.S. Brown (1962); [2] Σ/204 = 206 + 207 + 208/204; [3] There is doubt respecting these data as galena is exceedingly rare at Mascot and has been unreported from Friedensville in the last 40 years; [4] Crawford and Hoagland (1968); [5] Heyl et al. (1966); [6] Brown and Weinberg (1968); Kohls, D.W. (1970) New Jersey Zinc Co., private report.

criteria used to distinguish diagenetic dolomite from hydrothermal dolomite are in all cases diagnostically valid.

The Lower Ordovician sediments particularly the Mascot, Kingsport and Beekmantown Formations which are important host strata for the Tennessee and Timberville (Va.) ores are perhaps unique in the great lateral persistence of distinctive strata and combinations of strata which constitute a remarkable series of key beds. Subtle features such as a few centimeters of floating sand grains in dolomite, chert-nodule zones, texture- or color-variants of beds, especially when such features occur in characteristic sequences, have been traced for several tens of kilometers and similar overlapping features and zones have been traced for even greater distances (Oder and Miller, 1945; Stagg and Fischer, 1970; Harris, 1973). These sediments were formed in an epi-continental sea under sub-tidal conditions. In contrast to the remarkable persistence and extent of thin key beds which are characteristic of the Lower Ordovician of Tennessee and Virginia, the Austinville facies of the Lower Cambrian Shady Formation exhibits a high degree of lateral variability in lithologic character. The prominent reef structures and other sharply changing lithologies typical of reef-complex environments make detailed correlation of strata difficult or impossible. The localization of ore within this facies at Austinville has genetic significance and is moreover an important exploration guide. The shallow-water reef environment in the Gays River district (N.S.) shows a comparable variety of sedimentary lithologies. At Gays River, as noted above, the mineralized reefs were developed along the edge of a large evaporite basin, whereas at Austinville the off-reef facies is probably limestone of the open sea.

Paleokarst and paleoaquifer environments

Much attention has been given to the paleoaquifer system which developed in the Lower Ordovician carbonate sediments in the Appalachians during the long period of uplift and erosion prior to deposition of Middle Ordovician sediments. A symposium on this subject and its relationship to Appalachian zinc deposits was held at the University of Tennessee at Knoxville and was recorded as "A paleoaquifer and its relation to economic mineral deposits: The Lower Ordovician Kingsport Formation and Mascot Dolomite" (*Econ. Geol.*, 66(5): 695–810). In east and central Tennessee and at Friedensville (Pa.) zinc ore was deposited in debris-filled dissolution zones as much as 240 m below the unconformity. The unconformity and extensive karst phenomena related to it have been observed from the Great Northern Peninsula of Newfoundland southwestward along the Appalachian trend to the Mississippi embayment in Alabama. West of the embayment, in Oklahoma and Texas, comparable karst phenomena are developed in similar Lower Ordovician sediments. Fresh meteoric water, responsible for the dissolution of these rocks, must have completely replaced the connate water in the rocks where karst features were extensively developed. Dolomite beds in the eroding rocks appear to have been relatively well-lithified as evidenced by the angularity and competence of the dolomite-breccia fragments. Limestone is, however, rarely preserved as fragments in the karst breccias indicating, perhaps, that limestone strata were relatively less lithified at the time of the pre-Middle Ordovician erosion period.

Although large debris-filled zones of dissolution were formed, it appears doubtful that there were at any time great open caverns such as are common in modern caves, and most of the dissolution with simultaneous accumulation of fine and coarse debris must have taken place below the water table. These karst-related breccia zones are prominent features of the mineralized areas in east Tennessee and central Tennessee. They are much more extensive in distribution than ore, but they appear to have been a necessary feature of the system which brought the mineralizing solutions into the depositional sites. Roedder (1971; see also Vol. 2, Chapter 4) has shown that the brines carrying the zinc were at temperatures ranging from 70°C to 170°C and they were later than the meteoric waters which developed the karst. The later hydrothermal solutions utilized this permeable system and to a degree, at least, enlarged and modified it. Callahan (1968) states that the ore at Friedensville occurs in karst-breccia zones related to the post-Lower Ordovician unconformity, and it is tempting to speculate that other deposits which share this stratigraphic and paleophysiographic environment such as Timberville (Va.) and Newfoundland zinc bear a similar relationship.

The importance of several unconformities in the Mississippian and in the pre-Pennsylvanian interval in the formation of dissolution breccias and the preparation of the ground for ore in the Tri-State district is, to some extent, comparable to the karst-breccia system in Tennessee and elsewhere in the Appalachians.

Mineralization

A distinctive quality of Appalachian zinc and zinc—lead deposits is the relative purity of the sphalerite (Table III). The low iron is notable, averaging about 0.5%. The sphalerite from the lead-free deposits of the Lower Ordovician is especially low in iron, particularly those from central Tennessee, the Mascot—Jefferson City distrct, and Friedensville. Copper Ridge district sphalerite shows the greatest variation with a range from about 0.3% to 4.5% Fe averaging, however, close to the Appalachian average. Cadmium content shows considerable range from Friedensville's 0.05% to 0.3% in east Tennessee and 1% in the non-commercial veins of central Tennessee. Silver, copper, and lead are low as excepted. Galena, too, is relatively free of minor elements.

Appalachian deposits are also characterized by a simple suite of gangue minerals of which dolomite is the most abundant and prominent. (The central Tennessee deposits with a dominantly calcite gangue are exceptions unless, as could be argued, these deposits lying just west of the Appalachian structural province do not belong in this classification at all.) Calcite (except in central Tennessee) is subordinate in quantity and is later than the sulfide mineralization. Silica is inconspicuous but is commonly present as thin silicified selvages along limestone—ore contacts, as local intergrowths of cryptocrystalline

TABLE III

Semi quantitative spectrographic analyses of sphalerites from selected ore deposits in eastern United States[1]

Elements	Central Tennessee Veins	Central Tennessee, Knox	East Tennessee	Friedensville, Pa.[2]	Austinville, Va.
Si	3	0.08	0.03	0.4	0.7
Fe	0.2	0.3	0.5	0.4	0.7
Mg	0.03	0.01	0.02	0.005	0.2
Ca	0.3	0.05	0.07	0.005	0.2
Ti	0.001	0.0005	0.0003	ND	0.001
Mn	0.0003	0.0003	0.004	0.0001	0.02
Ag	0	0.0001	0.0002	ND	0.001
Ba	0.4	0.004	0.0004	ND	0.01
Cd	1	0.6	0.3	0.05	0.1
Cu	0.2	0.1	0.03	0.0001	0.003
Ge	0.02	0.03	0.008	0.001	0
Ga	0.02	0.06	0.0005	0.005	0.0001
Pb	0.04	0.03	0.05	0.0001	0.02
Sr	0.04	0	0	ND	0

[1] Kohls, D.W., 1968, New Jersey Zinc Co. Private report; analyses by U.S. Geol. Surv.
[2] Callahan (1968).

black jasper, and as bands or patches within the ore breccias. Introduced silica, however, probably seldom constitutes more than 5% or 10% of the ore. Primary chert nodules are local components of the stratigraphic section and are common constituents of the breccias increasing the total silica content of the ore.

Pyrite and more rarely marcasite range from essentially absent or rare in the northeastern part of the Mascot–Jefferson City district to as much as 10% or more in local atypical pyritic zones at Austinville. Average pyrite content at Austinville, Friedensville, and Flat Gap is less than 5%.

Galena at Austinville is the third most abundant sulfide after pyrite. Although it has been reported from the old Mascot Mine in the Mascot–Jefferson City district, to the writer's knowledge, this mineral is extremely rare or absent from the mines in that district. At Friedensville a lead-isotope analysis is reported by Brown (1962, p. 682), but no galena has been observed since systematic geologic studies of this deposit were begun in the nineteen thirties. It may be stated, therefore, that Friedensville and Mascot–Jefferson City are lead-free districts. Minor amounts of galena are present at the Flat Gap Mine in the Copper Ridge district. Other ore bodies (as yet undeveloped) in this district are, however, similar to Mascot–Jefferson City district mineralization, and the Copper Ridge district considered as a whole also belongs in the lead-free category. The lead-isotope characteristics of Appalachian zinc and zinc–lead deposits fall into a slightly anomalous group between the "J" leads of the upper Mississippi Valley (Wisc.), Tri-State (Joplin–Picher), Lead Belt (southeast Mo.) deposits and the more normal leads from, for example, Pine Point and Gays River (Table II).

Barite and fluorite are rare accessory minerals in Appalachian deposits. Gypsum and anhydrite occur locally, but they too are rare. Finally, although chalcopyrite has been observed in most of the Appalachian districts, it is a rare mineral in this environment.

The important minerals, then, are sphalerite, locally galena as at Austinville, dolomite (and calcite in central Tennessee), and to a lesser degree, silica. Pyrite is commonly present in minor amounts in most of the Appalachian deposits. Sphalerite and dolomite are prominent as open-space filling. These minerals coat the breccia fragments, fill the open spaces between the fragments, the cracks of the inflation breccias, and the fine interstices of the breccia matrix. Although significant dolomite replacement has taken place in limestone areas and in breccia matrix, sphalerite replacement of matrix and limestone is subordinate, and replacement of dolomite fragments by sphalerite is minute and insignificant. Silica replacement of limestone and dolomite is a common ore-associated phenomenon but it is not quantitatively very significant.

AUSTINVILLE (VIRGINIA)

The Austinville Mine in the Valley and Ridge physiographic province of southwestern Virginia has been an important source of zinc and lead since colonial time and is estimated to have produced in excess of 1.2 million tons of these metals. The primary ore minerals are sphalerite and galena having an average zinc/lead ratio of about 5/1. The ore is localized in a shallow-water facies of the Lower Cambrian Shady Dolomite which at Austinville is about 550 m thick and is characterized by prominent archeocyathid reefs. Commercially important mineralization is confined to about 400 m of the formation including the upper 130 m of the Ribbon Member and nearly all of the overlying Austinville Member. The lowest mineralization stratigraphically is within 200 m of the base of the Paleozoic carbonate sedimentary section which has a total thickness of about 4,000 m.

The Austinville structural environment is typical of the southern Appalachian Valley and Ridge province with N 60°E fold axes and parallel imbricate thrust faults overriding to the northwest. Klippen of Lower Cambrian clastics overlying later Paleozoic formations are impressive evidence of the magnitude of these structures. Fold axes are tilted to the southeast with the southeast limbs of the folds dipping at relatively moderate attitudes while the northwest limbs are steep or overturned (Brown and Weinberg, 1968).

The Austinville–Ivanhoe ore zones extend over a distance of about 10 km nearly parallel to the N 60°E regional strike. By far the largest part of the ore occurs in the upper 200 m of the ore zone on the southeast limb of the Austinville anticline. In the Ivanhoe Mine, which is to the southwest and contiguous with the Austinville Mine, the ore is principally confined to the lowest 100 m of the mineralized zone and lies on the north-dipping limb. Complex faulting and rapidly changing facies make the precise structural interpretations uncertain. The ore bodies are pencil shaped with lens-like cross sections. The base of the ore tends to be conformable with the underlying strata while the upper surface is generally convex upward. The long dimension of the ore zones is roughly parallel to the regional Appalachian trend with lengths of as much as 2,000 m or more. Dip length may be as great as 120 m, and thickness normal to bedding may exceed 30 m locally. The beds underlying the ore appear to be normal unbrecciated dolomite. The base of the ore is commonly the richest and is characterized by sphalerite rosettes. The rosette zone is overlain by rubble breccia which in turn gives way upward to mosaic breccia in which mineral content is weakest.

The most abundant gangue mineral is white dolomite which occurs as augen or discontinuous bands along stratification and as filling between breccia fragments with vague ghost-like contacts merging imperceptibly into coarsely crystalline stratified dolomite. There is a direct quantitative relationship between ore and the coarsely crystalline white dolomite (Brown and Weinberg, 1968). Other gangue minerals, such as barite, fluorite, gypsum, and anhydrite are rare. Pyrite is the second most abundant sulfide in the ore. Galena is an important minor constituent, and chalcopyrite is very rare.

Both the original sedimentary variants and later-imposed structures are called on by Brown and Weinberg (1968, p. 185) to form the channels for the hydrothermal solutions which deposited the ore and gangue minerals. They cite the vertical stacking of ore lenses near fault zones; a general halo of coarsely crystalline dolomite surrounding the ore; tectonic breccias as ore hosts; and the pattern of metal ratios as supporting the postdeformational age of the mineralization. They believe the mineralization was accomplished during or subsequent to the Appalachian orogeny in Late Pennsylvanian or Permian time. Callahan (1964), who studied Austinville early in his career, has emphasized the significance of the unique sedimentary environment of this deposit and recently (1974) suggested the possible syngenetic origin of deposits of the Austinville type.

Features such as the dolomite halos and the "stacking" of ore lenses appear to have been formed or modified from sedimentary structures related to reef development. The breccias, attributed to tectonism by Brown and Weinberg (1968), are, the writer suggests, similar to fore-reef breccias which have been modified by solution collapse; and the overlying mosaic breccias are inflation structures typical of the roof zones of dissolution breccia bodies.

The stratigraphic discontinuities of the reef complex localized minor faulting, fracturing, and related structures as the trough of carbonate sedimentation subsided. These subsidence structures probably contributed to the permeability of the reef complex and influenced the flow of mineralizing solutions at a somewhat later time. The relatively great permeability of the reef was not at once destroyed by the deposition of ore sulfides and gangue minerals but was sustained by contemporaneous dissolution of limestone and alteration of limestone to dolomite. A synthesis of available evidence, in the writer's opinion, suggests that the mineralization had been essentially completed in the Early Paleozoic, probably before the end of the Cambrian Period.

The Appalachian orogeny superimposed a regional structural overprint which modified the early sedimentary structures. Numerous faults of diverse type and magnitude are present in the Austinville—Ivanhoe Mine and the adjacent region. Some are in close spatial association with the ore bodies, but many others are not. On the whole, there appears to be a random or chance association of faulting with ore rather than a genetic relationship. As noted earlier, ore at Austinville lies on the south limb of an anticline, and the ore has a gentle northeasterly pitch. At Ivanhoe, the ore is on the north limb and has a moderate pitch to the west. Apparently the Austinville ore, although modified and deformed by late orogenic activity, received its essential form and structure during deposition of the unique Austinville archeocyathid reef facies of the Shady Dolomite in Early Cambrian time. There is no recognized evidence bearing on the nature or source of the ore-bearing solutions.

FRIEDENSVILLE (PENNSYLVANIA)

The Friedensville Mine lies in a small Lower Paleozoic re-entrant in the Precambrian crystalline rocks which border the Appalachian Valley to the southeast. New York City is only 120 km to the east, Philadelphia 65 km to the south and The New Jersey Zinc Company Palmerton smelter is but 50 km to the north. The ore averaging slightly more than 6% Zn is lead-free with a sphalerite/pyrite ratio of about 3/1. It occurs in a breccia zone about 30 m thick near the base of the Lower Ordovician Beekmantown Dolomite. An important unconformity separates the Beekmantown from the overlying Jacksonburg Formation of Middle Ordovician age. This stratigraphic environment is comparable to that of the east Tennessee zinc deposits with which Friedensville is in many ways similar. It is unique, however, in being the most strongly deformed of the major Appalachian zinc deposits.

The most comprehensive and authoritative published description of the Friedensville deposit was written by Callahan (1968), who was for many years manager of exploration for The New Jersey Zinc Company. Friedensville has been studied by a number of geologists, but the modern coordinated view and detailed understanding of this deposit is to a remarkable degree the result of Callahan's work; and the following description is based largely on his published papers, private reports, and personal communication. In the preamble to his description of the geology Callahan (1968, pp. 97—98) cautioned the reader as follows: ". . . regarding the features associated with ore and its environment, it must be remembered that much of the data are biased, having been obtained in the immediate vicinity of the mineralization. This situation is not uncommon in mineralized areas where effort is directed to avoid barren ground. To what extent these features are common in the same geologic environment remote from mineralization is indeterminate because of lack of outcrops, drill holes, and mine openings. This being so, there is a tendency to try to interpret all features in the mineralized environment in terms of their relation to the ore-forming process whereas it is probable that very few of these features operated to cause or control that process." He suggested that many features, neutral in relation to the ore-forming process, are distractions and that: "Much of the controversy regarding ore genesis stems from this difficulty of distinguishing in the mineralized environment, the unique from the ubiquitous and the temporal aspect of features now spatially associated."

Stratigraphy

The zinc ore lies in a breccia zone about 850 m above the Precambrian crystalline basement and some 300 m below the erosional unconformity which separates the Lower Ordovician Beekmantown Dolomite from the overlying Jacksonburg of Middle Ordovician age. The latter, consisting of a basal limestone member overlain by a thin-bedded argillaceous limestone, provides the source raw material for the important cement industry of Lehigh County.

Underlying the ore zone is a dark thin-bedded shaly dolomite or shaly limestone, the Evans marker, 30–40 m thick. This basal member of the Beekmantown Dolomite overlies nearly 800 m of Cambrian dolomite of the Allentown and Tomstown dolomites.

The upper 300–340 m of the Beekmantown has a variable lithology with ribbony, cherty, sandy, conglomeratic, and breccia beds interstratified with more massive beds. Limestone is more common in the upper portion of the formation. The Lower Beekmantown is darker, more dolomitic and is characterized by thicker more massive bedding. A few key beds with floating sand grains have been found to be useful, but the formation has a paucity of good marker horizons which is a significant handicap in determining subtle but possibly significant structural relationships. Callahan (1968, p. 100) described the ore zone as follows: "From the standpoint of mineralization, the most important unit... is designated *sedimentary breccia* ... The term *sedimentary breccia* is used to emphasize that the breccia did not result from deformation but rather is composed of fragments derived from solution-collapse cemented by finer detritus derived from the solution process. The breccia was formed as a consequence of a karst topography and sub-surface drainage system during uplift and erosion during Late Beekmantown time. The mineralization generally is restricted to the sedimentary matrix of the breccia and fractures in the fragments are post mineralization." It should not be assumed that the ore is everywhere co-extensive with the breccia as this structure is not ubiquitously mineralized, nor is the *sedimentary breccia* uniformly developed as a planar stratigraphic feature of the formation.

There are four main ore zones exposed on the surface at Friedensville, each of which contributed to a significant production of oxidized zinc ore during the latter part of the last century. Each is at essentially the same stratigraphic horizon, but each occupies a different position in the structural framework of the area being distributed two on the gentle south limb, one near the axis but plunging away from the axis with depth, and another on the nearly vertical north limb. All of these may have been parts of a single interconnecting ore zone before the beds were folded and erosion truncated the structure.

The main ore zone lies on the south limb with a long axis at an acute angle to the axial plane so that with increasing depth the ore moves away from the anticlinal axis. Callahan commented on the structure (1968, p. 103) as follows: "It is the writer's opinion that structure per se did not control the localization of the ore inasmuch as ore deposits are present on all elements of the fold and are not related to faulting. At most, tectonics merely deformed horizontally-disposed ore bodies and determined their subsequent pattern and attitude of occurrence. Because the Taconic revolution is the earliest post-Beekmantown tectonic event in the area and the sulfide minerals themselves show evidence of deformation as well as indifference to structural setting, it is suggested that the mineralization is pre-Taconic; that is, Ordovician in age."

Butler (1935) described the primary ore at Friedensville as occurring: (1) in small replacement veinlets; (2) in large masses which are probably metasomatic; and (3) as open-space fillings. Regarding the latter, he said (p. 898): "Filling of open spaces generally resulted in crystalline, somewhat banded ore, the bands due to repeated deposition of pyrite and sphalerite. The structure is mostly conformable to projections of country rock* on which the minerals have been deposited. Sphalerite continued to form after pyrite ceased. Etching of the sphalerite brings out twinning and marginal granulation in some specimens, although others show little evidence of deformation." Butler found no evidence of inheritance of deformed textures from the country rock, and he concludes that there is abundant microscopic evidence of the deformation of the Friedensville deposit.

A suite of typical specimens from the Friedensville main ore zone were studied by William C. Kelly and Figs. 3–6 inclusive are his photomicrographs and descriptions of the more strongly deformed material. Kelly (written communication, 1973) observes that the degree of deformation is not the same in all specimens. Massive bands of relatively coarsely crystalline sphalerite appear to be only slightly deformed as does the very fine-grained "flinty" sphalerite. Fine-grained sphalerite which appears consistently in all specimens along grain boundaries and in microfractures or faults present is recrystallized. Regarding the deformed ore, Kelly comments, ". . .breccia ore from the main ore zone shows the most extreme and spectacular plastic deformation of the sphalerite and the features are similar to those developed in experimental deformation where the sphalerite is deformed to high total strains.**" The dolostone constituents in the breccia are fragmented and deformed and the small rhombs making up this rock show no internal plastic deformation. This is expected as dolostone (particularly when fine-grained) would behave

---

* Butler's "projections of country rock" are probably fragments of country rock which are the megascopic components of the ore breccia, ADH.
** As reported by Clark and Kelly (1973).

Fig. 3. Polished section, photomicrograph and description by W.C. Kelly. Deformed breccia ore from main ore zone, Friedensville Mine, Pa. Typical of deformed ore showing fractured pyrite (white) distorted sphalerite grains (bent primary twins and abundant deformation twins) and the usual late, undeformed, fine-grained sphalerite along grain boundaries. In no case does deformed sphalerite appear squeezed into fractures in broken pyrite.

Fig. 4. Polished section, photomicrograph and description by W.C. Kelly. Blow up of small area in photo Fig. 3 showing bending of primary twins and close up of intergranular, undeformed, fine-grained sphalerite. Such folding of primary twins is probably accomplished by translation gliding on 111.

Fig. 5. Polished section, photomicrograph and description by W.C. Kelly, from same specimen as Figs. 3 and 4. Deformation twins where sphalerite apparently dragged along pyrite fragments (white area, right). Some twins are faulted. The usual fine-grained sphalerite occurs along faults and also on contact of pyrite with deformed sphalerite. Some chert is intergrown with this late sphalerite.

Fig. 6. Polished section, photomicrograph and description by W.C. Kelly, of deformed breccia ore, main ore zone, Friedensville Mine. Extreme folding contortion is typical. Note fine-grained, late undeformed sphalerite at base of photo.

Fig. 7. Photograph of polished specimen of deformed ore. Main ore zone, Friedensville Mine, Pa. Matrix foliation and shearing of sphalerite, fracturing of dolostone fragments (black); light and medium gray areas are mixture of light colored sphalerite and darker rock matrix; white areas are crystalline dolomite.

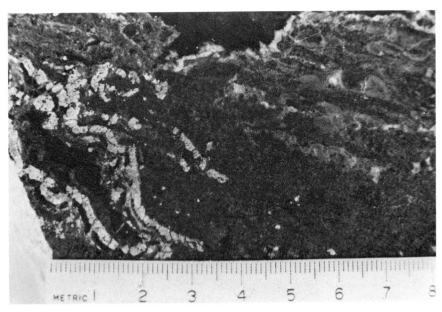

Fig. 8. Photograph of polished specimen of deformed ore. Main ore zone, Friedensville Mine, Pa. In lower left fractured pyrite around elongated breccia fragment oriented in axial plane. Sheared sphalerite and matrix in central area and irregular shapes of very fine-grained sphalerite at upper right.

as a relatively strong and brittle material under conditions that would produce a ductile behavior in sphalerite (Clark and Kelly, 1973). An obvious megascopic feature of the Friedensville ore is the fractured and deformed pyrite (Figs. 7–8). Kelly's study shows that fine-grained chalcedonic quartz which is commonly intergrown with carbonate gangue, occurs as a laminated filling along with carbonate in the fractures of shattered pyrite. Curiously, some of the chert layers filling these fractures appear to be folded and the quartz shows undulose extinction suggesting that the chert itself has been deformed. This suggests repetitive stages of deformation which is compatible with Callahan's conclusion that the ore was subjected to the Taconic as well as the Appalachian orogeny. It is puzzling, however, that sphalerite migration into the fractures of the shattered pyrite was not observed.

EAST TENNESSEE

The principal zinc and zinc–lead deposits of east Tennessee are found in the Appalachian Valley and Ridge province (Fig. 9) which is bounded on the southeast by the Blue Ridge and on the northwest by the Appalachian Plateau provinces. The Valley is an area of moderate relief underlain by carbonate rocks modified by long northeasterly-trending ridges underlain by the more resistant clastic formations. It ranges from 20 to 30 km in width and is characterized by an imbricate series of thrust faults which dip at low to moderate angles to the southeast. The boundary with the Blue Ridge province is sinuous in response to a complex of branching low-angle faults, some of which override the Valley and Ridge to the northwest for as much as 10 km or more. A thick sequence of Paleozoic sediments ranging in age from Cambrian to Pennsylvanian underlie the Appalachian Valley. Upper Cambrian and Lower Ordovician rocks are dominantly dolomite, whereas limestone constitutes Middle Ordovician and part of the Upper Mississippian. Clastic rocks predominate in the Lower Cambrian, Silurian, and Pennsylvanian systems.

The dominant structural feature of the Tennessee Valley is a series of seven or more southeasterly-dipping shingles each of which has moved northwestward on thrust faults overriding younger rocks. Typically these breaks occur along asymmetrical anticlinal fold axes which are strongly tilted towards the broad, gently-dipping southeast limbs. The steeply-dipping or overturned northwest limbs have markedly subordinate expression.

Appalachian-type zinc and zinc–lead deposits are well represented in the carbonate sediments of the Ordovician and Cambrian systems in Tennessee (Fig. 9). Among these are the economically important deposits of the Mascot–Jefferson City, the Copper Ridge, and the central Tennessee districts which will be described in some detail following a brief comment on several lesser deposits of academic if not economic interest.

Embreeville

The Embreeville zinc–lead deposit is located at Bumpass Cove at the foot of the Blue

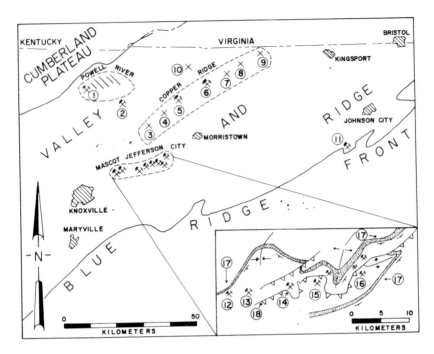

Fig. 9. Appalachian deposits of east Tennessee. Inset of Mascot–Jefferson City district showing anomalous minor anticlinal and synclinal structures. *1, 2* = the New Prospect and Straight Creek Mines, respectively, in Powell River district with NW trending vein structures; *3, 4* = the Luttrel and Puncheon Camp prospects; *5, 6* = the Idol and Flat Gap Mines; *7, 8, 9* = the Shiloh, Eidson, and Independence prospects, all in the Copper Ridge district; *10* = the Evanston district, *11* = Embreeville. Inset: *12* = Mascot Mine; *13* = Immel Mine; *14* = Young Mine; *15* = New Market Mine; *16* = Jefferson City Mine; *17* = outcrop belts of the Lower Ordovician Kingsport Formation; *18* = Rocky Valley thrust fault. All of the important ore in the district lies in the northern Kingsport belt in the footwall of the Rocky Valley fault.

Ridge Mountains on the extreme southeast edge of the Valley and Ridge province. Oxidized zinc and lead ores have been mined during periods of high metal prices prior to 1945 from deeply weathered Shady Dolomite of Early Cambrian age. Only small patches of the primary ore, too small for exploitation, have been found. In close spatial association with the secondary ores of lead and zinc are residual limonite and residual manganese deposits. The limonite appears to be derived mainly from a persistent pyritic zone from near the top of the Lower Cambrian clastic section and there is no genetic relationship with the zinc and lead. The Embreeville structural and stratigraphic environment is similar to that of Austinville although the typical Austinville facies is not developed. It is probable that even before weathering, the ore which was formed at Embreeville was on a significantly smaller scale than at Austinville and Ivanhoe; yet the two districts have a similar genesis.

Powell River district

The Straight Creek Mine and the New Prospect Mine are small high-grade deposits which were moderately productive seventy to eighty years ago. A very small tonnage was gleaned during the two world wars, but significant new reserves were not found and the mines have since been inactive. The New Prospect Mine lies on the Powell Valley anticline, a broad nearly flat structural anomaly along the western edge of the Valley and Ridge province. The Straight Creek Mine is on the east side of this dome and is involved in the strong folding which is characteristic of portions of the Valley and Ridge. Both of these ore bodies are composed of sphalerite with subordinate galena in brecciated and dolomitized Maynardville Formation of Late Cambrian age. Dolomite is the principal gangue mineral, and pyrite, a minor constituent, averages less than 5%. Both of these deposits are considered to be localized and controlled by fault zones intersecting a favorable limestone horizon (Brokaw et al., 1966). At the Straight Creek Mine, where tectonism is intense, the ore is deformed and crushed in the fault zones. In view of the newer concepts relating to the pre-orogenic age of the ores in the Mascot–Jefferson City and the Copper Ridge districts, consideration of a similar relationship at Straight Creek and New Prospect appears to be appropriate.

A series of sub-parallel vertical faults in the Copper Ridge Dolomite (Upper Cambrian) also occurs on the Powell Valley anticline. The faults are characterized by horizontal movement and extend for lengths of from several 100 m to 5 or 6 km. Sphalerite, galena, and dolomite mineralization occurs locally along these structures as small veins of a few thousand to a few tens of thousand tons. Only the Bunch Hollow vein has had any recorded production and this amounted to only a few hundred tons of concentrate produced during World War II (Brokaw, 1948). This vein terminates abruptly at a depth of about 100 m on a bedding plane, and it is probable that this is characteristic of a lack of vertical continuity of the structures in this district. The attention that has been given to the Powell River district appears, in retrospect, to be out of proportion to its importance, but the notion persists that deposits of significant size and grade may be present in the Maynardville Formation in bodies similar to that of the New Prospect Mine some 400–500 m below the surface.

Evanston district

The Evanston district is 10 km north of the Flat Gap Mine from which it is separated by a major thrust fault. Stratiform bodies of zinc and lead mineralization have been prospected in a dolomite bed in the Rome Formation of Early Cambrian age. Past efforts to commercialize these deposits have been unsuccessful (Secrist, 1924), and the occurrences appear to be too small and low-grade to be of economic value in the foreseeable future. They are, however, of interest because of their close association to shale in the sedimentary environment and because of the absence of notable permeable conduits which appear capable of bringing mineralizing solution into this environment. The Rome Formation in east Tennessee is composed of sandstone, silt, and shale with local dolomite lenses. At Evanston the upper 3 m of a 40 m dolomite bed is locally mineralized directly below a thin-bedded shale. The width of the mineralized zones is only a few meters, and the combined zinc and lead content is less than 5%, generally much less. The mineralization is predominantly sphalerite with minor galena, pyrite, and traces of chalcopyrite in breccia with dolomite gangue. Disseminated grains of sphalerite and galena have been observed locally in the shale above the mineralized dolomite. The mineralization appears to be uninfluenced by the local structural setting and may be predominantly controlled by the sedimentary environment.

Mascot–Jefferson City and Copper Ridge districts

By far the most important deposits of zinc in Tennessee are those of the Mascot–Jefferson City district and of the Copper Ridge district. These two districts, though separated by some 15 km (and the Saltville thrust fault of 5 km or more throw), are similar in character and genesis, and they were probably parts of a single large district before the Appalachian orogeny and subsequent erosion disturbed their original geographic relationship. It will be convenient, however, to mention at the outset the several differences that, apart from the geographic discontinuity, make them distinguishable.

The first of these is the structural setting. The Mascot–Jefferson City district (Fig. 9) is located on a gentle cross anticlinal fold giving the outcrop path of the Kingsport Formation a pronounced warp. A minor low-angle thrust fault has a horizontal offset of about 2,700 m at the Jefferson City Mine and dies out about 8 km to the northeast and the southwest. All of the ore northeast of the Young Mine is in the footwall within about 400 m of this fault. The prominence of this and other faults, both normal and reverse, together with numerous minor folds and warps subsidiary to and modifying the larger structures, have been carefully mapped in order to guide mining development and exploration. It was generally believed that these structures played a major role in preparing the ground and providing the access for the mineralizing solutions which deposited the ore. By 1960, however, district geologists (Kendall, 1960; Oder and Ricketts, 1961) were moving away from this position towards the concept that ore deposition pre-dated major faulting and probably was completely pre-orogenic.

The Copper Ridge district, in contrast, is a long unbroken monocline without major flexures or other structures. The few faults and minor flexures are unimpressive compared to those of the Mascot–Jefferson City district. The dip of the beds, however, changes gradually and imperceptibly from about 50° near the Virginia border to about 35° at Idol some 60 km to the southwest. Footwall and hangingwall thrust faults are far distant from the ore horizon of the Copper Ridge district.

As noted above, genetic concepts by 1960 were beginning to recognize the possibility of a pre-orogenic age to mineralization in east Tennessee and doubts on this question

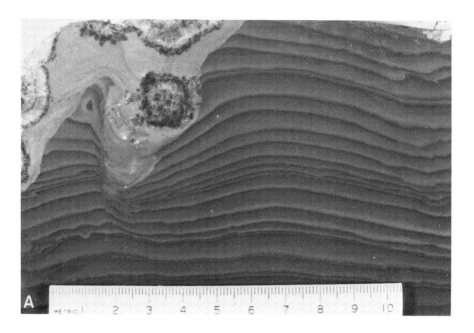

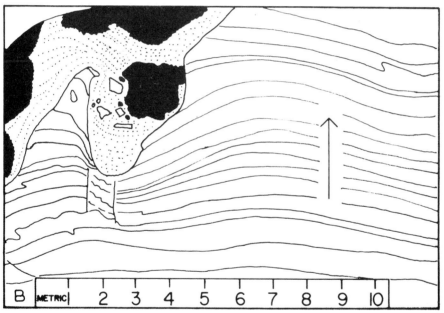

Fig. 10. A. Post-ore silt and sand composed of clasts of dolomite and of sphalerite, Jefferson City Mine (Tenn.). Two generations, the earlier varved silt is disturbed by soft-rock deformation, light colored varves are rich in sphalerite. Irregular zone in upper left is late clastic filling which is more than 50% sphalerite. Similar features have been described from the Cracow–Silesian region by Bogacz et al. (1973).

B. Key to Fig. 10A: the arrow indicates top; black areas in upper left = sphalerite; stippled area = late detritus with fragments of sphalerite and of dolostone in poorly sorted matrix of sphalerite and dolomite; banded area = early varved detritus of silt size composed of sphalerite and dolomite clasts.

were dispelled when Kendall (1960) at the Jefferson City Mine discovered and correctly interpreted the significance of pockets of clastic dolomite and sphalerite sand with incorporated fragments of ore (Figs. 10 and 11). Very similar features from the Muschelkalk of the Cracow–Silesian region are described and their origin discussed by Bogacz et al. (1973). These clastic deposits in the east Tennessee districts occupy vugs which remained open in the ore zones after the deposition of sphalerite had been completed. Well-developed stratification is these post-ore clastic deposits is parallel to the host rock Kingsport dolomite. The pre-orogenic age of the mineralization being so conclusively demonstrated, it was no longer possible to attribute the breccia bodies with which the ore is associated to tectonic activity (Kendall, 1961). It is now generally accepted that the significant structures with which the ore is associated are related to sedimentary features of the marine environment and to the extensive system of karst features which were developed in a prolonged and widespread period of erosion between Early and Middle Ordovician time.

Fig. 11. Jefferson City Mine (Tenn.), from the same vug as Fig. 10. The top of the figure is up. Two generations of post-ore silt and sand are composed of clasts of dolomite and sphalerite. Note original wall of vug with thin band of white crystalline dolomite in upper right hand corner. Thick band of sphalerite with rosette structures lines the vug coating the earlier white dolomite. The early finely laminated silt composed of dolomite and sphalerite detritus with prominent sphalerite-rich layers has been greatly deformed and eroded in the soft-rock stage by deposition of later, very poorly sorted material composed of fragments of ore, gangue dolomite, dolostone and a rich matrix of fine sphalerite. This vug, as with most vugs in the Jefferson City Mine, was completely filled by this post-ore detritus, quite effectively reducing to a very low level the permeability of the once extremely permeable ore zone.

Although mining of zinc ores began on a small scale at Jefferson City before the Civil War, the period of large-scale operations was initiated in 1913 when the American Zinc, Lead and Smelting Company took over the Mascot Mine and placed in operation a mill for treatment of sulfide ore with a 900 metric ton per day capacity. Other companies entered the field, and by 1963 four mills with a combined daily capacity of nearly 11,000 metric tons of ore were operating. The grade of the ore ranges from less than 3% Zn to 5% Zn or more, the average value in recent years approximating 3½% Zn or slightly less. Sphalerite in this district is notably low in iron with a minimum of other contaminating elements. Concentrates produced contain from 1 to 3 kg per ton of cadmium and 62% Zn or more. The ore is essentially lead-free.

The ore environment

The mineralization is found over a stratigraphic range of about 200 m from a few meters below the U bed up well into the Mascot dolomite (Fig. 12). The predominance of commercial ore is found in the 60 m zone made up of the U bed, S bed, R bed, and the lower 20–30 m of the Mascot dolomite. The U, S, and R beds are fine-grained dense limestone horizons and the ore occurs in dolomitized and brecciated equivalents of these beds as well as in brecciated dolomite horizons immediately overlying them.

The ore bodies are irregular in shape both in, and perpendicular to the stratification. Distinctive types of ore structures are described as *coarse rock matrix breccias, fine rock matrix breccias,* and *inflation breccias.* Coarse rock matrix breccias are developed in the limestone horizons of the Kingsport Formation, whereas the other breccias are formed from the fine-graine dolomite beds of the Kingsport and the overlying Mascot Formations.

Coarse rock matrix breccias

The coarse rock matrix breccias are typified by ore bodies in the U bed which reach their fullest development in the eastern part of the district in the vicinity of Jefferson City. These structures are associated with interesting features which are described by Fulweiler and McDougal (1971, p. 765) as "bedded ore structures". Kendall (1960, pp. 995–996) first described these features tentatively attributing their origin to algal activity and referring to them as "reefs" (Fig. 13). They occur as discrete elongated bodies of light-colored crystalline dolomite interbanded with dark siliceous dolomite which form a capping on long, narrow ridge-like protuberances of fine-grained primary dolomite a meter or two high. These string-like structures are flanked by limestone which commonly appears to be warped downward around the dolomite core. A band of sphalerite in some places as thick as 25 or 30 cm characteristically caps the banded dolomite. These long, narrow structures lead into the major U bed ore zones and indeed appear to be the base upon which the ore zones have been developed. The "reef-like" banded dolomite, de-

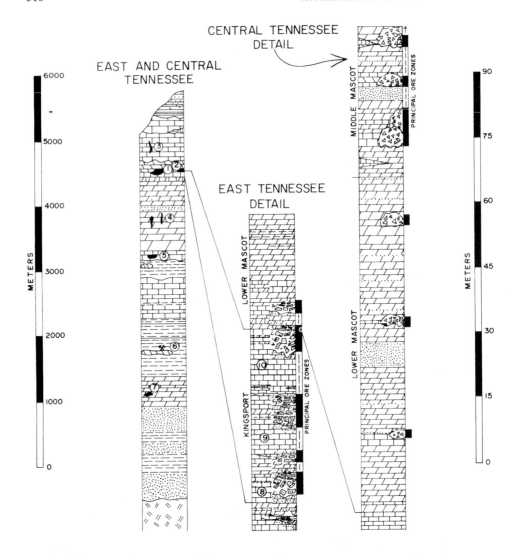

Fig. 12. Stratigraphic relationships in east and central Tennessee compared. *1* = Mascot–Jefferson City and Copper Ridge districts; *2* = Central Tennessee: note the east Tennessee ores are concentrated in the Kingsport Formation and the Lower Mascot Formation to a lesser degree. In central Tennessee the ore zones appear to be concentrated in the Middle Mascot with lesser amounts in the Lower Mascot. Ore in the Kingsport in the central Tennessee district appears to be of relatively minor importance. Central Tennessee vein deposits (*3*) are not represented in east Tennessee. Powell River district (*4*) and (*5*), Evanston (*6*) and Embreeville (*7*) are in strata that are too deep to explore in central Tennessee under present economic conditions. The U Bed (*8*), the S Bed (*9*) and the R Bed (*10*) in east Tennessee are limestone horizons except that where they are mineralized they are invariably crystalline dolomite. These contain the richest ore and are the most productive horizons in east Tennessee.

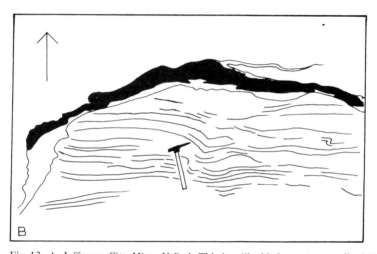

Fig. 13. A. Jefferson City Mine, U Bed. This is a "bedded ore structure" of Fulweiler and McDougal (1971) which Kendall (1960) first suggested might be algal structures. The writer, at first skeptical of Kendall's interpretation, now accepts these and similar structures in the Kingsport and to a lesser degree in the Mascot Formations as algal stromatolites. The alternate bands are dolomitic the darker ones being high in silica. The light gray above with the white dolomite intergrowths is the typical sphalerite cap. The overlying dark gray rock is aphanitic U Bed limestone. Below the stromatolitic zone is a pedestal of fine-grained dolomite which is concealed by muck in this picture. These features are only a meter or two wide. When followed, they lead to typical U Bed ore zones and clearly play a significant role in localization of the ore. (Reproduced from *Econ. Geol.*, 1971, Vol. 66, p. 766.) B. Key to Fig. 13A: the arrow indicates top; white area at top of picture = U Bed limestone with local veinlets of white dolomite and calcite; black and white band = sphalerite (black) with minor gangue dolomite; the irregularly banded area below = the algal stromatolite zone; the white base of the picture is muck covering a pedestal of fine-grained dolomite.

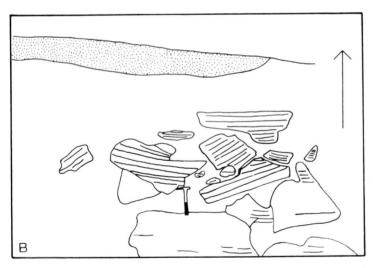

Fig. 14. A. Jefferson City Mine: coarse rock matrix breccia of the U Bed. Note the sub-angular slabs and partly rounded boulders of crystalline dolostone with white sparry dolomite bands and augen. It appears that these structures were formed before the brecciation and the banded and augen dolomite rock was an anomalous sedimentary facies of the U Bed limestone. This facies is closely related in space and genesis with the "bedded ore structures" (algal stromatolites) of Fig. 13. (See also, Fig. 15.) B. Key to Fig. 14A: the arrow indicates top; upper white area = basal T Bed dolomite; stippled area = upper U Bed clastic dolomite. Below in bottom three fourths of picture are sub-angular to rounded blocks, variously disoriented from the horizontal, in a coarsely crystalline dolomite matrix which is locally rudely stratified.

scribed above, appear to be of sedimentary origin. They are intimately associated with U Bed ore zones, which zones are characterized by a high degree of stratigraphic control (Fig. 15).

The ore zones which have been developed in the limestone horizons of the U, S, and R beds are characterized by a coarse rock matrix breccia of sub-angular to partially rounded fragments of coarse dolomite with abundant almond-shaped eyes of white crystalline dolomite which are elongated parallel to the stratification. Locally these augen structures coalesce to form bands. The matrix of this breccia is composed of clastic grains of rudely-stratified dolomite (Fig. 14). The upper part and the flanks of these breccias have suffered considerable dissolution and are characteristically well mineralized with bands and disseminated interstitial sphalerite. The limestone walls for a distance of a few tens of centimeters are characteristically silicified and contain disseminated rosettes of sphalerite. The silicification and the sphalerite rosettes represent the clearest example of replacement found in the district, but by far the predominant part of the sphalerite in the ore was deposited along with gangue dolomite as a filling of open space.

A fine-grained dolomite bed (T bed) overlies the U bed and ore commonly extends upwards into the lower meter or two of the T bed dolomite in well-developed inflation or founder breccias. Slabby breccia fragments of T bed dolomite are often found in the U bed as much as a meter below their normal position. T bed fragments are coated with coarsely crystalline sphalerite, and sphalerite fills the cracks of the overlying inflation breccia.

It is postulated that the augen and related banded structures in the coarse-crystalline dolomite are inherited from structures which were formed during sedimentation. They are profoundly different in character from the typical aphanitic limestone of the Kingsport Formation. U bed breccias which consist of the sub-angular to rounded fragments of these augen rocks appear to have been formed by the sedimentary process, pre-dating the deposition of the overlying T bed. Comparable breccias in the S and R beds would have a comparable genesis (Fig. 15). Although dolomitization was in part, at least, later and contemporaneous with ore deposition, an early dolomitization of these augen breccias of the Kingsport Formation appears to have taken place. Brecciation of the overlying T bed took place much later and resulted in foundering of T fragments in the upper part of the U bed and inflation or crackle breccia of the lower T bed. Deposition of sphalerite on a significant scale was contemporaneous with foundering of the T bed.

The U bed ore was developed in zones within the strata of relatively great permeability. This anomalous permeability was a legacy preserved from the reef-like structures of the sedimentary period. Later after the development of the post-Lower Ordovician karst system, the sphalerite mineralization was introduced.

Fine rock matrix breccias

It is significant that the Kingsport Formation is dominantly composed of fine-grained

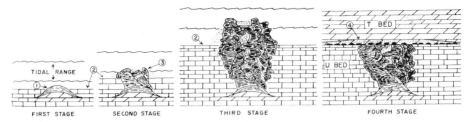

Fig. 15. Schematic cross-sections depicting the development of coarse matrix breccia zones. *1* = "Bedded ore structure" or algal stromatolite of Fig. 13; *2* = lime mud; *3* = reef-like structures building above the surrounding lime mud with moderate break-up producing large fragments and coarse carbonate sand; *4* = truncation of reef to base level of T Bed at horizon of nodular chert zone (chert nodules, black). A lense of carbonate detritus derived from breakdown of reef marks the top of the U Bed. Similar structures, common in the U Bed, are present in other horizons of the Kingsport Formation. The richest ore in the district is associated with coarse matrix breccia zones.

aphanitic limestone or its coarsely crystalline dolomitized stratigraphic equivalent. The age relationships of the alteration, principally dolomitization, of the Kingsport limestones have not been satisfactorily resolved. Conflicting evidence suggests an early diagenetic age related to sedimentary conditions such as reefs and, on the other hand, a later hydrothermal origin related to the ore-bearing solutions. It is probable that both early dolomitization of reefs, such as those of the U bed described above and late dolomitization by hydrothermal solutions, were operative, the latter working through and enlarging the areas of diagenetic coarse-grained dolomite.

In contrast to the coarse-crystalline dolomites which are extensively developed in the Kingsport Formation, are the fine-grained dolomite beds (such as the T bed in the Kingsport) in the dominantly dolomite strata of the overlying Mascot Dolomite. These dolomites are of regional extent and are considered to be of sedimentary or early diagenetic origin. They are referred to as "primary dolomites" by Tennessee geologists.

The second distinctive ore-body type is developed in breccia bodies in which the fragments are composed entirely or at least dominantly of the fine-grained primary dolomite. Large bodies of rubble breccia have been developed in the Mascot Dolomite extending from the unconformity at its top downward into the Kingsport Formation to depths of 250–300 m. These breccias are composed of angular fragments of dolomite ranging up to several meters in major dimension imbedded in a dolomitic rock matrix which is itself a micro-breccia (Fig. 16). They have been described as "dry breccias" (because they are barren of ore) and by the more descriptive term, "fine rock matrix breccias". The breccia fragments are randomly disoriented and have been displaced downward, some as much as 20 m or more from their normal stratigraphic position. Several places have been found where fine rock matrix breccia bodies are associated with paleo-sink holes at the unconformity on top of the Mascot Formation, such as in outcrop at Luttrell and at Shiloh in the Copper Ridge district and in drill holes in the Mascot–Jefferson City district. The mineralized breccias at the Lost Creek Barite Mine described by

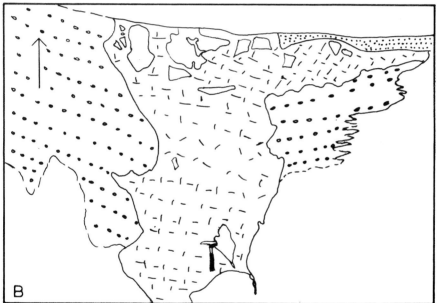

Fig. 16. Jefferson City Mine: fine rock matrix breccia in the U Bed. Note the base of T Bed dipping gently to the right at top of picture and irregular boundary of fine matrix breccia coming nearly to point at bottom. Hammer head points to breccia fragment. Limestone and breccia boundary sharply defined in lower part. Upper part of breccia mass is in contact with banded dolostone similar to the fragments of Fig. 14, but here an integral part of U Bed. The fine matrix breccia is believed to be near the bottom of an intricate system of debris-filled (and locally mineralized) solution openings related to the karst development in the early Middle Ordovician. A thin zone of ore occurs in the zone just below the T Bed.

B. Key to Fig. 16A: the arrow indicates top; white area at top = fine-grained dolomite of the basal T Bed. Stippled area = medium coarse-grained detrital dolomite, upper member of the U Bed; circle symbol = banded coarse-grained dolostone of the U Bed; irregular hachures in central area = fine-grained dolostone matrix with dolostone fragments (uncolored); white areas in lower left and right = aphanitic, unaltered U Bed limestone.

Carpenter et al. (1971) may occupy a sinkhole structure which has been modified by later solution collapse associated with barite mineralization.

Fine rock matrix breccias are developed on a wide scale in east Tennessee and give evidence of the extensive karst system which was developed in the post Early Ordovician–pre-Middle Ordovician erosion interval. Dissolution breccias extend from the unconformity at the top of the Mascot Dolomite downward as much as 250 m into the Kingsport Formation. No single drill hole reveals the full extent of this phenomena but many holes, combined with information in the mines, have indicated the ubiquity of the complex system of karst-related structures. These features developed with both conformable zones parallel to the strata and cross-cutting zones at high angles to the strata. The conformable structures, however, are much greater in lateral extent and in the mineralized districts are best developed in the strata near the top of the Kingsport Formation and the base of the Mascot Dolomite.

In these areas, the fine rock matrix breccias are commonly highly modified by the mineralizing solutions which have caused an extensive second stage of limestone dissolution. The result of this dissolution is to cause a profound re-organization of the fine rock matrix breccia zones letting them move downward and laterally into space created by the dissolution and alteration of limestone horizons. The fine matrix material may be partially or entirely removed by settling downward especially from the upper parts of the breccia zones creating large volumes of open space and zones of high permeability. Sphalerite and dolomite deposition has resulted in productive ore zones along the tops and sides of the breccia bodies. The zones which extend down into the limestone horizons of the Kingsport Formation are the most productive.

Inflation breccias

Overlying the fine rock matrix breccias and the ore breccias, a zone of inflation breccias is characteristically developed. These, locally referred to as crackle or mosaic breccias, are developed in relatively thick-bedded, fine-grained dolomite strata from which the underlying support has been removed by dissolution (Fig. 17). They are characterized by blocky fragments only slightly disoriented resembling crude brick work with a tendency to droop into the central zone of the underlying breccia. The inflation breccias merge with the rubble breccias below and upward into undisturbed strata. The inter-fragment areas are filled with white dolomite gangue and disseminated sphalerite. Commonly the lower part of the zone is sufficiently mineralized to be ore, but the grade becomes leaner upward as the size of the fragments increases and the space between them decreases.

Ore breccias

The three breccia types (i.e., coarse rock matrix breccia, fine rock matrix breccia, and

Fig. 17. Flat Gap Mine, Copper Ridge district, Tenn. Inflation breccia above mineralized zone much thinned by dissolution. Note small reverse displacement on TRIO marker bed, flexure due to uneven downward movement, and extensive bedding plane inflation with related transverse fragmentation. White is gangue dolomite with disseminated sphalerite.

inflation breccia) may be mineralized, but they are not invariably so. Indeed, on a regional basis they are much more likely to be barren or only sparsely mineralized. Ore areas are rare, whereas the apparently favorable breccia structures and alteration features are relatively common. In mineralized areas, the three breccia types are closely related and usually occur as pairs. Both the fine and coarse rock matrix types are characteristically capped by inflation breccias, and in most cases, the coarse rock matrix zones have been more or less intermixed with fine rock matrix and rubble. Ore breccia containing fragments of ore with later generations of superimposed sphalerite deposition testify to the continuity of brecciation and permeability within the ore zone during a prolonged period of mineralization. Readjustment and settling downward of the ore breccia contemporaneous with sphalerite deposition is explained by essentially continuous dissolution of limestone substrate by the hydrothermal solution.

CENTRAL TENNESSEE

Zinc was discovered in Smith County near the axis of the Nashville Dome in Lower Ordovician dolomite by The New Jersey Zinc Company in 1968. The Elmwood Mine now being developed was expected to go into production during the fall of 1974 (Winslow and Hill, 1973). The full significance of this discovery is not yet clear as exploration of a large mineralized area in northeast central Tennessee and an adjacent area in Kentucky is at an early stage. The central Tennessee district is in the Interior Lowland physiographic province and, strictly construed, should not be included with deposits of the Appalachian region. It lies, however, very close to the western border of the Cumberland Plateau in the Central Basin of Tennessee along the edge of the Highland Rim (Fig. 1). Besides its proximity, it has many features of an Appalachian-type district. It is of particular interest because the absence of tectonic structures in central Tennessee deposits emphasizes the similarity and linkage between the mid-continent deposits of the Mississippi Valley and those of Appalachia. There has been very little published about this newly discovered district, and a brief description here may, therefore, be appropriate.

The ore-bearing strata are essentially horizontal and are correlated by Stagg and Fischer (1970) with the Mascot Dolomite of east Tennessee. This Lower Ordovician ore host is unconformably overlain by two hundred to more than four hundred meters of Middle Ordovician limestone beds. Fissure veins in Middle Ordovician limestones containing barite, fluorite, sphalerite, and galena in varying amounts have been known in the district for more than 100 years (Jewell, 1947). Several attempts to mine the strongest of these have been unsuccessful and the veins are commercially unimportant. The sphalerite of the veins and of the typical Appalachian-type ore bodies in the underlying Lower Ordovician rocks have strong chemical similarity (Table III) and are believed to be genetically related. A close spatial relationship between the vein deposits and the deeper ore, if it exists at all, is obscure.

Dependable criteria for the selection of areas favorable for the occurrence of deep ore based on surface exposures have not been established, and geophysical techniques do not appear to be useful to distinguish ore-associated phenomena. Exploration, therefore, depends to a very high degree on diamond drilling and the evaluation of sedimentological, stratigraphic, paleophysiographic, paleostructural, and ore-related alteration and mineralogic features. The central Tennessee ore (Fig. 18) appears to be largely confined to a stratigraphic zone in the Middle and Lower Mascot dolomite approximately 100—200 m below the same erosion surface which was the significant factor in ground preparation at Friedensville (Pa.), the Mascot—Jefferson City and Copper Ridge districts in east Tennessee, and probably Timberville (Va.). As in east Tennessee, the greatest concentration of the ore appears to lie within a 60—100 m stratigraphic zone. The vertical range of permeable ground which was formed by the early karst development and enhanced by later brine movement is considerably greater than the vertical range of ore deposition.

The dominant structural features of the Lower Ordovician strata are extensive solution breccias formed by the great karst system which was developed before the deposition of the Middle Ordovician sediments (Gilbert and Hoagland, 1970). These solution breccias were enlarged by further dissolution during subsequent periods of the Paleozoic by enriched connate brines carrying zinc as the principal base metal. The movement of the zinc-bearing brines through the permeable paleoaquifer in the Lower Ordovician carbonate beds was accompanied by dolomitization on a wide scale and minor silicification, but only in local areas where the conditions were favorable for the deposition of zinc sulphide (Fig. 19).

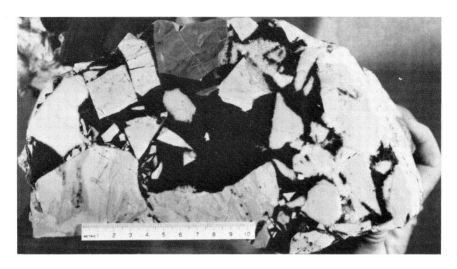

Fig. 18. Elmwood Mine, central Tenn. Polished specimen of ore breccia. Angular fragments of light and medium gray dolostone are cemented by dark reddish brown sphalerite. Note absence of gangue minerals and sharp contacts between fragments and sphalerite. Calcite gangue, however, is both common and abundant in the Elmwood Mine whereas dolomite, the common gangue mineral of east Tennessee, is rare. Fluorite, barite, and galena are noticeably developed locally; but on average, these minerals are sparse constituents of the ore. Pyrite and marcasite are uncommon and rare.

The mineralogy of the ores differs from that of east Tennessee as calcite rather than dolomite is the principal gangue mineral; and galena, fluorite, and barite are more common accessories. Small blebs of hydrocarbon attached to sphalerite and associated minerals lining vugs in the ore zones are not unusual.

GENESIS

Noble (1963) proposed that Mississippi Valley lead—zinc deposits may have been formed by the expulsion of formation fluid from compaction of thick piles of argillaceous sediments in adjacent basins of deposition (a mechanism similar to that which he has proposed for vanadium—uranium deposits of the Colorado Plateau). Discussing the character of ore deposits formed by water of compaction, he said (p. 1150): "Water from compaction may have been, at different places and at different times, an ore forming fluid, a cementing agent, a solvent of rock constituents, or an agent of alteration." For a summary of this subject, see Wolf (1976).

Roberts (1967, p. 196) in speaking of the compaction of basinal shales, said: "Assuming that this pore space is water filled and the rock is lithifying to a shale, the water must be trapped under conditions of increasing pressure and temperature, in close contact with all the mineral grains of the rock. The situation would be ideal for solution to take

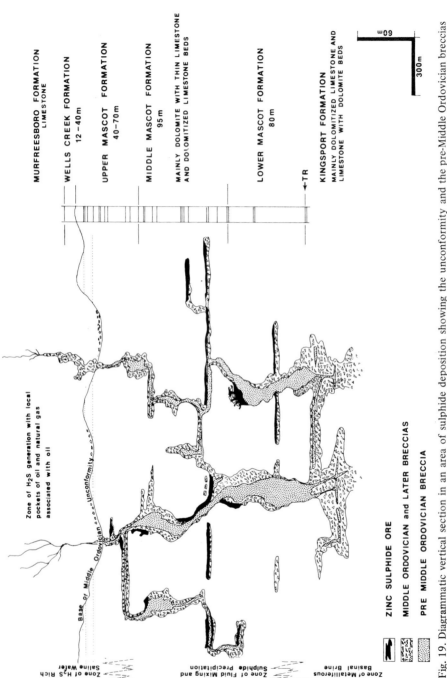

Fig. 19. Diagrammatic vertical section in an area of sulphide deposition showing the unconformity and the pre-Middle Ordovician breccias related to it. Middle Ordovician and later breccias were produced by dissolution and dolomitization of limestone by the pregnant brines which followed the permeable zones developed by the pre-Middle Ordovician karst. Ore deposition tended to favor the upper part of these breccia zones. Above the denser metal-bearing brine was a layer of saline water carrying H_2S related to indigenous petroleum in Knox and Middle Ordovician carbonate rocks. Between these two fluids was a zone of mixing and sulphide deposition (modified from Gilbert and Hoagland, 1970. Courtesy of Geological Society of America).

place, and if the sediment is rich in adsorbed metals and sulfur, a potent ore solution could develop. . ."

The Appalachian basin adjacent to the east Tennessee and central Tennessee zinc districts has more than 150,000 km^3 of shale of Middle Ordovician age. The compaction and lithification of this rock must have generated immense volumes of connate brines rich in metal. It is here suggested that a significant part of this pregnant brine must have found its way into the huge paleoaquifer of the Upper Knox (Fig. 20) to be driven through this system to areas of lower hydraulic potential along the Appalachian shelf (Fig. 19).

Jackson and Beales (1967) and, more recently Beales and Onasick (1970), have developed a model for the localization of Mississippi Valley-type deposits from the migration of metal-rich basinal fluids into the zone of precipitation in carbonate rocks on the flanks of the basins. They suggest precipitation was effected through the mixing of metal-carrying fluids with sulphur-carrying fluids, the latter having been generated in the carbonate host rocks at, or at least relatively close to, the site of deposition.

The similarity of the deposits of Appalachian and Mississippi Valley-types of ores to each other and their widespread distribution throughout the world, suggests that many significant genetic factors are shared. This has been emphasized by Ohle (1959) who argued that consideration of the group as a whole offered the most promising approach toward reaching a better understanding of genesis. Kendall's discovery (1960) of pre-orogenic age for east Tennessee ore deposition opened the way for the general acceptance of this fact and has dramatically illustrated the structural similarity which existed at the time of ore emplacement between the deposits of Appalachian and Mississippi Valley-type. The essential requirements for the formation of deposits such as those of east and central Tennessee (and perhaps more generally for deposits of this type in other parts of the world) appear to have been:

(1) A carbonate host-rock environment with sufficient open space to permit the necessary volume of ore mineral to be deposited.

(2) A through-going permeable channel extending from source areas to sub-areal or sub-marine discharge. Discharge areas are believed to be relatively close to the sites of deposition.

(3) A source area for the enrichment of connate brines in zinc and other metal ions to feed the channel systems referred to above. The source areas are believed to be the large basins of argillaceous rocks which have physical access to the strata containing the channel system (Fig. 20). They may have been tens to hundreds of kilometers distant from the depositional sites.

(4) Conditions which can produce and sustain a hydraulic gradient sufficient to drive the metal-rich brines from the source area through the channel system to surface or sub-marine discharge. These could be produced largely by sediment loading combined with basin settling, diagenesis, and crystallization. Early orogenic movements in the basin may also have played a significant role in the development of the hydraulic system.

(5) Physicochemical relationships between the channel system and the enriched brines

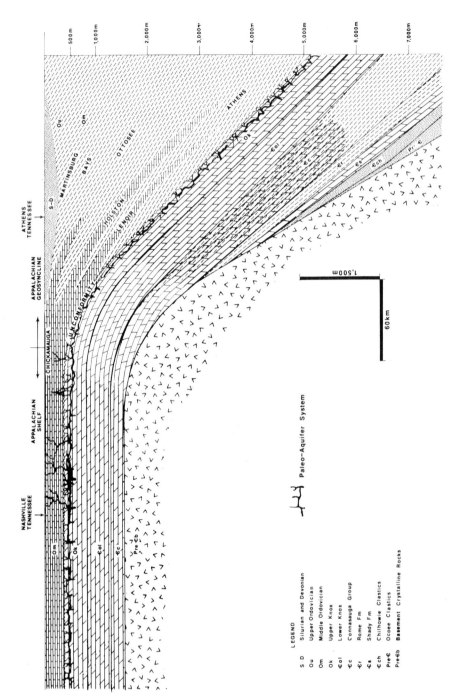

Fig. 20. Diagrammatic reconstruction of pre-Mississippian Appalachian shelf and geosyncline and conceptual depiction of ore genesis, showing the regional and pervasive extent of karstification in Upper Knox beds below the unconformity. Note also the transition from Middle Ordovician limestone strata of moderate thickness on the shelf to the great thickness of argillaceous sediments in the geosyncline. Metal-bearing brines were forced by the compaction and lithification of the shales into the karst structures and upward to areas of lower hydrostatic potential on the shelf.

which would maintain the permeability of the channels. This may have been accomplished by an equilibrium between the dissolution activity of the brines in the channels and the rate of deposition from the brines reacting with the precipitating agent. An excess of the dissolution activity enhances the permeability, whereas an excess of deposition is restrictive. Long-lasting channels are more likely to produce large ore deposits than are short-lived channels. Limestone is more readily attacked by the pregnant solutions than dolomite, and channels in strata with interbedded limestone and dolomite have a tendency to enlarge and perpetuate themselves until the accessible limestone has been removed or altered to dolomite or silica.

(6) A deposition agent such as H_2S, is required in a second fluid system which will mix with the pregnant brines to precipitate ZnS, as suggested by Jackson and Beales (1967). The H_2S may have been generated through the decay of organic material indigenous to the carbonate rocks of the shelf area and the biogenic reduction of sulphur associated with the evolution of petroleum (Skinner, 1967).

(7) A favorable coincidence of time and space was required. It was necessary for the two fluid systems to mix within the channel system for a considerable period of time to produce significantly large concentrations of base metal sulphides.

REFERENCES

Barton Jr., P.B., 1967. Possible role of organic matter in the precipitation of the Mississippi Valley ores. In: J.S. Brown (Editor), *Genesis of Stratiform Lead–Zinc–Barite–Fluorite Deposits. Econ. Geol. Monogr.* 3: 371–378.

Beales, F.W. and Onasick, E.P., 1970. Stratigraphic habitat of Mississippi Valley-type ore bodies. *Inst. Min. Met. Trans.,* 79: B145–B154.

Billings, G.K., Kesler, S.E. and Jackson, S.A., 1969. Relation of zinc-rich formation waters, northern Alberta, to the Pine Point ore deposit. *Econ. Geol.,* 64: 385–391.

Bogacz, K., Dzulynski, S. and Haranczyk, C., 1973. Caves filled with clastic dolomite and galena mineralization in disaggregated dolomites. *Ann. Soc. Geol. Pol.,* 43:59–76.

Bridge, J., 1956. Stratigraphy of the Mascot–Jefferson City zinc district, Tennessee. *U.S. Geol. Surv. Prof. Pap.,* 277: 76 pp.

Brokaw, A.L., 1948. Geology and mineralogy of the east Tennessee zinc district. In: K.C. Dunhan (Editor), *Symposium on the Geology, Paragenesis and Reserves of the Ores of Lead and Zinc. 18th Int. Geol. Congr.,* London, pp. 60–67.

Brokaw, A.L., Rodgers, J., Kent. D.F., Laurence, R.A. and Behre Jr., C.H., 1966. Geology and mineral deposits of the Powell River area, Clairborne and Union Counties, Tennessee. *U.S. Geol. Surv. Bull.,* p. 1222-c.

Brown, J.S., 1962. Ore leads and isotopes. *Econ. Geol.,* 57: 673–720.

Brown, W.H., 1935. Quantitative study of ore zoning, Austinville Mine, Virginia. *Econ. Geol.,* 30: 425–433.

Brown, W.H. and Weinberg, E.L., 1968. Geology of the Austinville–Ivanhoe district, Virginia. In: J.D. Ridge (Editor), *Ore Deposits of the United States, 1933–1967.* AIME, New York, N.Y., pp. 169–186.

Butler, R.D., 1935. Mylonitic sphalerite from Friedensville, Pennsylvania. *Econ. Geol.,* 30: 890–904.

Callahan, W.H., 1964. Paleophysiographic Premises for Prospecting for Strata-bound Base Metal Mineral Deposits in Carbonate Rocks. *Cento Symposium on Mining Geology and Base Metals,* Ankara, pp. 191–248.

Callahan, W.H., 1967. Some spatial and temporal aspects of the localization of Mississippi Valley–Appalachian type ore deposits. In: J.S. Brown (Editor), *Genesis of Stratiform Lead–Zinc–Barite–Fluorite Deposits. Econ. Geol., Monogr.,* 3: 14–19.

Callahan, W.H., 1968. Geology of the Friedensville zinc mine Lehigh County, Pennsylvania. In: J.D. Ridge (Editor), *Ore Deposits of the United States, 1933–1967.* AIME, New York, N.Y., pp. 95–107.

Callahan, W.H., 1974. Syngenesis versus epigenesis of Mississippi Valley–Appalachian type base metal mineral deposits. *Econ. Geol.* 69: 123–124.

Cannon, R.S. and Pierce, A.P., 1967. Isotopic varieties of lead in stratiform deposits. In: J.S. Brown (Editor), *Genesis of Stratiform Lead–Zinc–Barite–Fluorite Deposits. Econ. Geol., Monogr.,* 3: 427–434.

Carpenter, R.H., Fagan, J.M. and Wedow Jr., H., 1971. Evidence on the age of barite, zinc, and iron mineralization in the lower Paleozoic rocks of east Tennessee. *Econ. Geol.,* 66: 792–798.

Clark, B.R. and Kelly, W.C., 1973. Sulfide deformation studies, I. Experimental deformation of pyrrhotite and sphalerite to 2,000 bars and 500°C. *Econ. Geol.* 68: 332–352.

Crawford, J. and Hoagland, A.D., 1968. The Mascot–Jefferson City zinc district, Tennessee. In: J.D. Ridge (Editor), *Ore Deposits of the United States, 1933–1967,* AIME, New York N.Y., pp. 242–256.

Currier, L.W., 1935. Zinc and lead region of southwestern Virginia. *Va. Geol. Surv. Bull.,* 43 pp.

Fulweiler, R.E. and McDougal, S.E., 1971. Bedded-ore structures, Jefferson City Mine, Jefferson City, Tennessee. *Econ. Geol.,* 66: 763–769.

Gilbert, R.C. and Hoagland, A.D., 1970. Paleokarst phenomena and the post-Knox unconformity of middle Tennessee. *Geol. Soc. Am. Abstr.,* 2 (7): p. 558.

Hanshaw, B.B., Back, W. and Deike R.G., 1971. A geochemical hypothesis for dolomitization by ground water. *Econ. Geol.,* 66: 710–724.

Harris, L.D., 1971. A lower Paleozoic paleoaquifer; the Kingsport Formation and Mascot dolomite of Tennessee and southwest Virginia. *Econ. Geol.,* 66: 735–743.

Harris, L.D., 1973. Dolomitization model for Upper Cambrian and Lower Ordovician carbonate rocks in the eastern United States. *J. Res. U.S. Geol. Surv.,* 1: 63–78.

Heyl, A.V. 1967. Some aspects of genesis of stratiform zinc–lead–barite–fluorite deposits in the United States. In: J.S. Brown (Editor), *Genesis of Stratiform Zinc–Lead–Barite–Fluorite Deposits. Econ. Geol. Monogr.,* 3: 20–32.

Heyl, A.V., Delevaux, M.H., Zartman, R.E. and Brock, M.R., 1966. Isotopic study of galenas from the upper Mississippi Valley, the Illinois–Kentucky, and some Appalachian Valley mineral district. *Econ. Geol.,* 61: 933–961.

Hill, W.T., 1969. Mine geology of The New Jersey Zinc Company's Flat Gap mine and Treadway in the Copper Ridge district In: *Papers on the Stratigraphy and Mine Geology of the Kingsport and Mascot Formations (Lower Ordovician) of East Tennessee. Tenn. Div. Geol. Rep.,* 23: 76–90.

Hill, W.T. and Wedow Jr., H., 1971. An early Middle Ordovician age for collapse breccias in the east Tennessee zinc districts as indicated by compaction and porosity features. *Econ. Geol.,* 66: 725–734.

Hill, W.T., Morris, R.G. and Hagegeorge C.G., 1971a. Ore controls and related sedimentary features at the Flat Gap mine, Treadway, Tennessee. *Econ. Geol.,* 66: 748–756.

Hill, W.T., McCormick, J.E. and Wedow Jr., H., 1971b. Problems on the origin of ore deposits in the Lower Ordovician formations of east Tennessee. *Econ. Geol.,* 66: 799–804.

Hoagland, A.D., 1967. Interpretations relating to the genesis of east Tennessee zinc deposits. In: J.S. Brown (Editor), *Genesis of Stratiform Lead–Zinc–Barite–Fluorite Deposits. Econ. Geol. Monogr.,* 3: 52–58.

Hoagland, A.D., 1971. Appalachian strata bound deposits: their essential features, genesis, and the exploration problem. *Econ. Geol.,* 66: 805–810.

Hoagland, A.D., Hill, W.T. and Fulweiler, R.E., 1965. Genesis of the Ordovician zinc deposits in east Tennessee. *Econ. Geol.*, 60: 693–714.

Jackson, S.A. and Beales, F.W., 1967. An aspect of sedimentary basin evolution: the concentration of Mississippi Valley-type ores during late stages of diagenesis. *Bull. Can. Pet. Geol.*, 15: 383–433.

Jewell, W.B., 1947. Barite, fluorite, galena, sphalerite veins of middle Tennessee. *Tenn. Div. Geol. Bull.*, 51: 114 pp.

Kendall, D.L., 1960. Ore deposits and sedimentary features, Jefferson City mine, Tennessee. *Econ. Geol.*, 55: 985–1003.

Kendall, D.L., 1961. Ore deposits and sedimentary features in Tennessee. *Econ. Geol.*, 56: 1137–1138.

Laurence, R.A., 1939. Origin of the Sweetwater, Tennessee barite deposite. *Econ. Geol.*, 34: 190–200.

Laurence, R.A., 1944. An early Ordovician sink hole deposit of volcanic ash and fossiliferous sediments in east Tennessee. *J. Geol.*, 52: 235–249.

Laurence, R.A., 1968. Ore deposits of the southern Appalachians. In: J.D. Ridge (Editor), *Ore Deposits of the United States, 1933–1967*, AIME, New York, N.Y., pp. 242–256.

Laurence, R.A., 1971. Evolution of thought on ore controls in east Tennessee. *Econ. Geol.*, 66: 696–700.

Maher, S.N., 1971. Regional distribution of mineral deposits beneath the pre-Middle Ordovician unconformity in the southern Appalachians. *Econ. Geol.*, 66: 744–747.

McCormick, J.E., Evans, L.L., Palmer, R.A. and Rasnick, F.D., 1971. Environment of the zinc deposits of the Mascot–Jefferson City district, Tennessee. *Econ. Geol.*, 66: 757–762.

Miller, B.L., 1924. Lead and zinc ores of Pennsylvania. *Pa. Geol. Surv., 4th Ser. Bull.*, M-5: 60 pp.

Noble, E.A., 1963. Formation of ore deposits by water of compaction. *Econ. Geol.*, 58: 1145–1156.

Oder, C.R.L. and Miller, H.W., 1945. Stratigraphy of the Mascot–Jefferson City district. *Am. Inst. Min. Met. Eng. Tech. Publ.*, 1818: 9 p., also in: *AIME Trans.*, 1948, 178: 223–231.

Oder, C.R.L. and Ricketts, J.E., 1961. Geology of the Mascot–Jefferson City district, Tennessee. *Tenn. Div. Geol. Rep.*, 12: 29 pp.

Ohle, E.L., 1959. Some considerations in determining the origin of ore deposits of the Mississippi Valley type. *Econ. Geol.*, 54: 769–789.

Ohle, E.L., 1967. Origin of ore deposits of Mississippi Valley type. In: J.S. Brown (Editor), *Genesis of Stratiform Lead–Zinc–Barite–Fluorite Deposits. Econ. Geol. Monogr.*, 3: 33–39.

Ohle, E.L. and Brown, J.S., 1954. Geologic problems in the southeast Missouri lead district. *Bull. Geol. Soc. Am.*, 65: 201–221, 935–936.

Roberts, W.M.B., 1967. Sulphide synthesis and ore genesis. *Miner. Deposita*, 2(3): 188–199.

Rodgers, J., 1948. Geology and mineral deposits of Bumpas Cove, Unicoi and Washington Counties, Tennessee. *Tenn. Div. Geol. Bull.*, 54: 82 pp.

Roedder, E., 1967. Environment of deposition of stratiform (Mississippi Valley type) ore deposits, from studies of fluid inclusions. In: J.S. Brown (Editor), *Genesis of Stratiform Lead–Zinc–Barite–Fluorite Deposits. Econ. Geol. Monogr.*, 3: 349–362.

Roedder, E., 1971. Fluid inclusion evidence on the environment of formation of mineral deposits of the southern Appalachian Valley. *Econ. Geol.*, 66: 777–791.

Secrist, N.H., 1924. Zinc deposits of east Tennessee. *Tenn. Div. Geol. Bull.*, 31: 165 pp.

Skinner, B.J., 1967. Precipitation of Mississippi Valley-type ores: a possible mechanism. In: J.S. Brown (Editor), *Genesis of Stratiform Lead–Zinc–Barite–Fluorite Deposits. Econ. Geol. Monogr.*, 3: 363–370.

Snyder, F.G., 1968. Geology and mineral deposits. Mid-continent United States. In: J.D. Ridge (Editor), *Ore Deposits of the United States, 1933–1967*. AIME, New York, N.Y., pp. 257–286.

Snyder, F.G. and Odell, J.H., 1958. Sedimentary breccias in the southeast Missouri lead district. *Bull. Geol. Soc. Am.*, 69: 899–926.

Stagg, A.K. and Fischer, F.T., 1970. Upper Knox stratigraphy of middle Tennessee. *Geol. Soc. Am. Abstr.*, 2(7): p. 693.

Wedow Jr., H., 1971. Models of solution collapse structures from drilling statistics: an aid to explora-
    tion. *Econ. Geol.,* 66: 770–776.
Winslow, K.R. and Hill, W.T., 1973. The Elmwood project. *Min. Congr. J.,* March: 1
Wolf, K.H., 1976. Ore genesis influenced by compaction. In: G.V. Chilingarian and
    Compaction of Coarse-Grained Sediments, 2. Elsevier, Amsterdam, in press.

Chapter 12

THE McARTHUR* ZINC–LEAD–SILVER DEPOSIT: FEATURES, METALLOGENESIS AND COMPARISONS WITH SOME OTHER STRATIFORM ORES

I.B. LAMBERT

INTRODUCTION

There are a number of extremely large stratiform Pb–Zn–Ag deposits of Middle Proterozoic (Carpentarian) age in Australia. The most important of these are the unmetamorphosed zinc-rich McArthur deposit in the Northern Territory, the low-grade metamorphosed Mount Isa and Hilton deposits in northwest Queensland and the highly metamorphosed Broken Hill deposit in western New South Wales.

The keys to an increased understanding of any such group of broadly similar ores are most likely to be found in the deposits which have been least modified by post-lithification processes. Comprehensive studies of these should increase the chances of being able to establish relatively narrow limits for the geological and physicochemical conditions of ore formation and, in addition, should provide reference points for assessment of metamorphosed and deformed deposits. With these aims in mind, this chapter reviews the features and genesis of the McArthur deposit and compares it with Mount Isa, Hilton, Broken Hill and, very briefly, Sullivan and Rammelsberg. It also presents a working classification scheme which relates these "McArthur-type" deposits to other general types of stratiform ores of volcano–sedimentary associations.

The McArthur deposit consists of a series of tuffaceous shale orebodies separated by relatively metal-poor, inter-ore beds. It was delineated following the discovery of a hemimorphite-rich gossan in 1955, but its features are not widely known internationally because the fine grain size of the ore presents milling problems which have held up the commencement of mining. However, the deposit can be sampled extensively from crosscuts and diamond drill cores, and it has been the subject of wide-ranging investigations in recent years.

---

* Also known as the H.Y.C. deposit.

SCOPE

The first section outlines the regional geology of the northern Australian province in which the McArthur deposit is located. This is followed by descriptions of the local stratigraphy, depositional environments and structures.

The next sections describe the general features, mineralogy and geochemistry of the McArthur deposit, the minor deposits in its vicinity, the mineralogical and geochemical dispersion haloes in the country rocks, and the available isotope information. These data are then used as a basis for considerations of ore genesis.

The final section discusses the other Pb–Zn-rich deposits, and briefly considers classification of stratiform base metal sulphide deposits of volcano–sedimentary associations.

REGIONAL GEOLOGY, NORTHERN AUSTRALIA

The McArthur deposit occurs within a major metalliferous province of Lower–Middle Proterozoic age in northern Australia. This province also contains the Mount Isa, Hilton

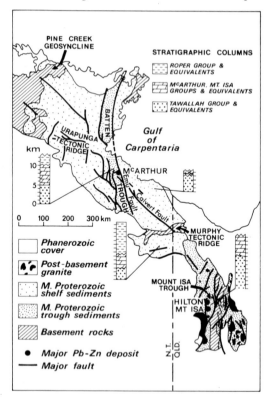

Fig. 1. Regional geology, central northern Australia, based on Plumb and Derrick (1975). *N.T.* = Northern Territory, *QLD* = Queensland. See text for discussion.

and many smaller Pb—Zn—Ag deposits, the Mount Isa and many smaller Cu deposits, and a number of U deposits. Cover rocks of the Great Artesian Basin obscure the relationships between this northern Australian region and the Willyama province, some 1400 km to the south, which contains the Broken Hill and some smaller Pb—Zn—Ag deposits, and a few minor Cu deposits.

The regional geology of northern Australia has been discussed by Plumb and Derrick (1975). The main features are summarized in Fig. 1.

The McArthur deposit occurs towards the eastern margin of the Batten Trough (Fig. 1), which contains up to 5.5 km of essentially unmetamorphosed Carpentarian dolomites, siltstones, shales and tuffites. Much thinner sequences of sediments are present on either side of this fault-bounded trough.

The trough sequences to the south of the Murphy Tectonic Ridge (Fig. 1) are generally similar to those of the Batten Trough, but are variably metamorphosed, and considerably more deformed (Bennett, 1965; Smith, 1969; Mathias et al., 1973). The metamorphic grade at Mount Isa is lower greenschist facies.

Radiometric age data cited by Plumb and Derrick (1975) indicate deposition of the ore-bearing McArthur Group sediments between 1600—1400 m.y. B.P., and penecontemporaneous deposition of the Mount Isa Group between 1540—1400 m.y. B.P.

Middle Proterozoic igneous activity was more intense in the Mount Isa region than in the Batten Trough (Bennett, 1965; Plumb and Derrick, 1975). Thin tuffite beds (Croxford, 1964) appear to be more abundant in the Mount Isa Group than they are in the McArthur Group. Thick piles of basalts, the Eastern Creek Volcanics, occur unconformably under the Mount Isa Group. The complex Sybella Granite was intruded, and some rhyolites were extruded, in the interval between the eruption of the Eastern Creek basalts and the sedimentation of the Mount Isa Group. A few dolerite dykes intrude the Mount Isa Group sediments.

The Carpentarian sediments of this northern Australian province were deposited on continental crust. No pre-Carpentarian basement rocks are exposed within about 150 km of the McArthur deposit; the nearest are the metasediments and granites of the Murphy Tectonic Ridge, which Plumb and Derrick (1975) tentatively correlate with the Lower Proterozoic rocks of the Pine Creek geosyncline in the northwest of the Northern Territory (see Fig. 1). The Urapunga Tectonic Ridge (Fig. 1) is overlain by a very thin sequence of sediments, and basement rocks could have been exposed there during deposition of part of the McArthur Group. This ridge could well have extended west to divide the Batten Trough into two separate depositional basins.

STRATIGRAPHY AND DEPOSITIONAL ENVIRONMENTS, McARTHUR AREA

The stratigraphic succession in the McArthur area is summarized in Table I, and in the diagrammatic cross-section through the eastern portion of the Batten Trough, Fig. 2. The terminology used is that of Plumb and Brown (1973).

TABLE I

Generalized stratigraphic succession in the McArthur area, based on Plumb and Brown (1973).

System	Group	Sub-Group	Formation	Member		Thickness (m)		Main rock types	
Adelaidean (Upper Proterozoic)	Roper							Sandstones and siltstones (some feldspathic, glauconitic or ferruginous); minor conglomerates	
		Batten				~1000		Sandstones and siltstones; minor dolimitic sediments and chert, tuffites	
			Reward Dolomite			30-1000?		Cherty dololutite, intraclast and pelletal dolarenite, dolomitic shale, coarse graded arenites and breccia beds, tuffites	
			Barney Creek Formation	H.Y.C. Pyritic Shale	Cooley Dolomite	0-500+	0-550+	Carbonaceous, dolomitic, tuffaceous shales and siltstones; pyrite, sphalerite-galena; carbonate-rich arenite and breccia interbeds, tuffites	Brecciated dolomite, minor carbonaceous shale, non-stratiform Pb-Zn(-Cu) mineralization
				W-fold Shale		15-150		Tuffaceous, dolomitic shales and siltstones	
Carpentarian (Middle Proterozoic)	McArthur	Umbolooga	Teena Dolomite	Coxco Dolomite		15-80		Thin bedded dololutites with frequent thin interbeds of shales, tuffites, gypsum pseudomorphs	
						6-30		Cherty dololutites, intraclast dolarenites and flake breccias, rare halite casts	
			Emmerugga Dolomite	Mitchell Yard Dolomite		15-120		Dololutite, intraclast dolomite flake breccias, tuffites	
				Mara Dolomite		100-500		Cherty dololutite, dololutite, intraclast dolomite flake breccias and dolarenites, halite casts, tuffites	
			Tooganinie Formation	Myrtle Shale		30-240		Red dolomitic siltstone, dololutite, halite casts	
				Leila Sandstone		~140		Dolomitic sandstones, siltstones	
			Tatoola Sandstone			~140		Quartz and dolomitic arenites and sandstones	
			Amelia Dolomite			90-240		Dolomitic arenites and siltstones, cherts, gypsum pseudomorphs[1]	
			Mallapunyah Formation			30-750		Dolomitic arenites and siltstones, cherts	
	Tawallah							Sandstones, siltstones, conglomerates, minor basic and felsic volcanics	

[1] M.D. Muir (personal communication).

Carbonate in the McArthur Group sediments is predominantly dolomite. Calcite fragments occur in sedimentary breccias at the base of the H.Y.C. Pyritic Shale Member, and minor amounts of calcite have been recorded in some of the dolomite units. Dolomites in the vicinity of the McArthur deposit are iron-rich. Halite casts and dolomite pseudomorphs after gypsum indicate high salinities during deposition and/or diagenesis of several of the dolomitic members. The evidence for bedded gypsum in rocks of this age is particularly noteworthy.

There are a number of thin tuffite beds and variable amounts of pyroclastic debris in the shales and siltstones of the ore-bearing McArthur Group sediments. The underlying

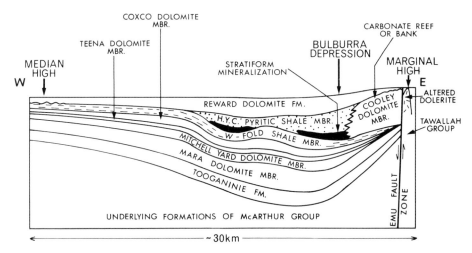

Fig. 2. Schematic E–W section through the McArthur area during deposition of the Reward Dolomite, based on Brown et al. (1975) and Murray (1975). All units thin as they approach the Emu Fault zone and to a lesser extent the median uplift area. The major period of subsidence of the Bulburra Depression commenced during deposition of the Coxco Dolomite Member. The boundary between the H.Y.C. Pyritic Shale and the Cooley Dolomite is quite sharp to the east of the McArthur deposit.

Tawallah Group (Table I) contains minor proportions of basic and felsic volcanics. Small, highly altered, basic intrusions occur in the Emu Fault zone near the McArthur deposit, and minor basalts and felsic porphyries occur in the relatively thin shelf sequence to the east of the Emu Fault.

The features of the various units from the Myrtle Shale Member up to the Reward Dolomite Member are summarized below, with emphasis on the ore-bearing H.Y.C. Pyritic Shale Member. Most of the information presented has been drawn from Brown et al. (1975) and Murray (1975). The initial part of this sequence of sediments appears to have formed during a major transgression. This produced a change from mainly subaerial siltstones, through intertidal and shallow marine carbonates to deeper-water, shaley carbonates, tuffaceous mudstones, and local graded arenite and breccia beds. Subsequent regression resulted in deposition of shallow-water and intertidal carbonates. All units thin towards the Emu Fault zone, indicating that this was a penecontemporaneous structure.

Myrtle Shale

This member consists mainly of red dolomitic siltstone, which appears to represent a subaerial deposit composed largely of wind-transported dust. Thin dololutite and silty dolomite interbeds, often with halite casts, increase in abundance near the top. These are indicative of early stages of marine transgression, with intermittent evaporative conditions. The latter led to precipitation of halite at the surface and/or interstitially. Later incursions dissolved the salt crystals and the voids were filled by sediment.

Mara Dolomite

The boundary between the Myrtle Shale and the overlying Mara Member is transitional over about 30–40 m. It is marked by the appearance of abundant cherty dololutite with algal stromatolites, interbedded with red dolomitic siltstone and silty dololutite with halite casts. The dolomite in this transition zone is commonly brecciated; fragments from a few millimetres to a few metres across sit in a matrix of red dolomitic siltstone. The breccia fragments appear to have been generated by collapse of dolomite beds following solution of halite-rich layers, as has been postulated for Devonian carbonates in Western Canada (Beales and Oldershaw, 1969). The stromatolites are mainly small laterally-linked hemispheroidal and flat laminated types. Present-day stromatolites of these types form in highly saline intertidal zones sheltered from wave action, with restricted access to the open ocean (Logan et al., 1964).

In the upper few hundred metres of the Mara Member, chert-free dololutite increases in importance, intraformational dolomite flake breccias and intraclast dolarenites are fairly common and there are some beds of oolitic dolarenite. The stromatolites change towards the top of the unit with the incoming of conical (*Conophyton*), ellipsoidal (oncolites) and basinal types. These features indicate intervals of shallow-marine and intertidal conditions of sedimentation.

Sequences of near-uniform dololutites in the Mara Members are interrupted by several structureless to well-laminated, conformable, chert-like beds, usually between 2 and 100 cm thick. The main constituent of these beds is finely crystalline, anhedral, potash feldspar. They also contain highly angular quartz grains, many of which have embayed margins, and minor amounts of clays and ferroan dolomite. The quartz grains are most likely derived from crystals of felsic volcanic origin, and this, together with their mode of occurrence suggests that these beds are tuffites. Similar beds occur throughout the overlying units of the McArthur Group and will be discussed further in the section dealing with the H.Y.C. Pyritic Shale.

Mitchell Yard Dolomite

The disappearance of stromatolites from the sequence is arbitrarily taken by Brown et al. (1975) as the boundary between the Mitchell Yard Dolomite Member and the underlying Mara Dolomite Member. The Mitchell Yard Dolomite is a relatively thin unit, and is absent near the Emu Fault zone. It appears to have formed mainly in a shallow-marine environment. It varies from uniform dololutite to intraformational breccia consisting of broken laminae (flakes) or blocks of dololutite in a fine grained, unlaminated dolomite matrix. There is some silicification of the fragments.

Several tuffite beds, similar to those in the Mara Member, have been recorded in the Mitchell Yard Dolomite.

Teena Dolomite

This formation can be divided into two members. The lower member is thin, but laterally persistent, and marks a single brief cycle of transgression and regression. It consists of cherty dololutite containing some stromatolites of similar types to those in the upper Mara Member. There are some shaley beds with rare halite casts, oolitic and intraclastic doloarenites, dolomite flake breccias and sandy doloarenites.

The upper member, the Coxco Dolomite, has similarities with the Mitchell Yard Member, but contains frequent thin beds of potash-rich mudstone, which probably have a considerable tuffaceous component. There are also a few of the chert-like tuffite beds. Slump folding is present towards the top of the unit. Deposition of the Coxco Dolomite appears to have been accompanied by differential subsidence in the eastern part of the Batten Trough, the subsidence being greatest in the Bulburra Depression. Williams (1974) reported dolomite pseudomorphs after gypsum in this member.

W-Fold Shale

This member was deposited over a wide area, but in general is thicker and has a lower dolomite component in the Bulburra Depression, where water depths were presumably greatest. The contact with the underlying Coxco Member is usually gradational over a metre or so, but in places carbonate-rich and carbonate-poor sediments alternate for up to about 10 m.

Near the McArthur deposit, the W-Fold Shale consists of green and red mudstones and dolomitic mudstones, which have significant tuffaceous components. Elsewhere it is generally greyish.

H.Y.C. Pyritic Shale

This member is not widely distributed but is economically the most important unit in that it contains the McArthur deposit and some smaller Pb—Zn deposits (see section dealing with minor mineralization). It consists largely of black and dark-grey pyritic shales and siltstones, which are thin-bedded and widely disturbed by preconsolidational slumping, sliding, scouring, load casting, and de-watering structures. It also contains intervals of less-pyritic carbonaceous shales and siltstones, dololutites, graded dolarenites, slump and turbidite breccia beds, and tuffites. Tuff contribution reached a maximum during deposition of this unit. The H.Y.C. Pyritic Shale was deposited mainly in the Bulburra Depression (Fig. 2) which was the area of greatest subsidence, but thinner sequences of apparently equivalent carbonaceous shales occur in other basins of lesser subsidence.

Hamilton and Muir (1974) reported abundant biomorphic cellular structures in insoluble organic matter from the McArthur deposit.[1]

---

[1] J. Oehler (personal communication, 1975) has recorded a diverse and locally abundant assemblage of organically preserved unicellular and filamentous blue-green algae and filamentous bacteria from chert nodules in ore. He considers the microfossil assemblage represents a predominantly deep-water bacterial biota (i.e., deeper than the photic zone) with most of the algae being detrital.

The McArthur deposit occurs near the base of a 500 m section of the H.Y.C. Pyritic Shale; it is separated from the underlying W-Fold Shale by some 20 m of carbonaceous dolomitic shale and breccia beds.

The fine-grained beds in the H.Y.C. Pyritic Shale Member differ from most other tuffites in the McArthur Group sediments in that they are greyish-green. In thin section it can be seen that they generally have higher abundances of vitroclastic debris (Fig. 3), pyrite, sphalerite, galena, arsenopyrite, and organic matter than the stratigraphically lower tuffites. Croxford and Jephcott (1972) recognized six major tuffite beds, between 8 and 30 cm thick, within the 55 m cross-cut they studied through the McArthur deposit. The original glass shards have now been altered largely to potash feldspar (adularia), but in some places they are partly replaced by sulphide minerals.

The unusually high potash contents (up to 13% K_2O), which characterize the tuffite beds through the McArthur Group, must have been generated during diagenesis. Unstable high-temperature phases in the volcanic ashes (particularly glass) may have reacted with saline pore waters, in the manner proposed by Iijima and Hay (1968) for the Green River Formation, Wyoming. Potassium could have been enriched in the pore waters partly as a result of dolomitization of illitic limestones (Swett, 1968) and/or solution of evaporites.

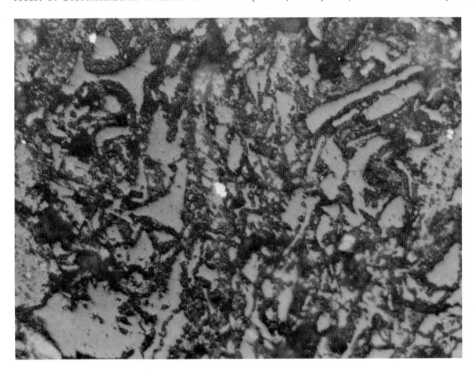

Fig. 3. Photomicrograph of tuffite band in H.Y.C. Pyritic Shale Member, showing vitroclastic texture. Original glass has been altered largely to K-feldspar. Width of field is 500 μ; reflected light. (Ref. No. Te115/60.)

The fine-grained nature of the tuffites in the McArthur Group suggests that the volcanic ashes were erupted on land, or possibly in shallow water, and transported for some distance by wind currents before being deposited in the sedimentary basin.

There are textural and mineralogical gradations between the tuffites and the carbonaceous shales (Fig. 4), indicating that the latter have variable tuff components. In general, the shales have higher contents of dolomite, organic matter, quartz and clays.

The dolomite-rich arenites and sedimentary breccia beds occur in the relatively thick sequences of H.Y.C. Pyritic Shale Member in the Bulburra Depression. These were deposited rapidly, presumably as a result of slumping and turbidity currents triggered off by volcanic activity and subsidence.

The dolarenites are usually from 2.5 to 5 cm thick, and are graded and poorly sorted. They are quite numerous; Croxford and Jephcott (1972) recorded about 350 of them through the McArthur deposit. The rock fragments are largely dolomitic, but there are lesser proportions of sandstone, siltstone, shale and pyritic bituminous shale. Some of the dolomite clasts are partially mineralized, and others consist of variably silicified oolites, oncolites and stromatolite material.

Fig. 4. Photomicrograph of relatively low-grade McArthur ore showing interbanded metal sulphides and tuffaceous shale. Note that pyrite (white) occurs as fine grained crystals and as coarser subspherical grains. Sphalerite (mid-grey) contains small inclusions of galena and chalcopyrite. Width of photomicrograph is 500 μ; reflected light. (Ref. No. Te115/20.)

The sedimentary breccia beds are thicker, less common and less widely distributed than the dolarenites. They generally vary in thickness from about 25 cm up to 15 m and in most cases wedge out towards the west, indicating they were derived largely from the eastern edge of the Bulburra Depression. There are at least fifteen such beds through the H.Y.C. Member. They contain clasts of the same lithologies as those in the arenites, and occasional fragments of igneous rock similar to the intrusions in the Emu Fault zone. The clasts are poorly sorted, angular to sub-angular; they are often between 1 and 5 cm across, but may exceed 30 cm. In some cases there are lateral transitions from breccias to graded turbidites and thence to dolarenites.

The matrix in the breccia beds is usually dolomitic silt, but in some cases the clasts have settled into underlying soft carbonaceous muds. Breccia beds above the McArthur deposits contain minor amounts of coarse-grained sulphides similar to those found in the inter-ore beds.

The two lowest breccia beds are interesting in that their clasts are mainly Fe-poor calcite. These appear to be undolomitized, or dedolomitized, Cooley Dolomite (see below).

Cooley Dolomite

At the eastern edge of the Bulburra Depression, the H.Y.C. Pyritic Shale Member gives way fairly rapidly to the Cooley Dolomite Member (Figs. 2, 6). This member consists largely of brecciated Fe-rich dolomite with fine-grained, sometimes sulphidic, dolomite-rich matrix. The individual fragments are thin-bedded to laminated dololutite. Stromatolitic structures occur but are not very abundant except in some of the eastern sections. There are some minor carbonaceous shale intervals in the western part, and some dolarenites in the eastern part.

The Cooley Dolomite has been interpreted as an algal reef complex (Murray, 1975), and as a near-shore carbonate bank (Brown et al., 1975). At the eastern side of the McArthur deposit the transition from H.Y.C. Pyritic Shale Member to Cooley Dolomite occurs over a distance of only 250 m.

The Cooley Dolomite extends eastwards to the Emu Fault zone, where upfaulted Tawallah Group rocks (Table I) were probably close to sea level during the build-up of this carbonate formation.

Reward Dolomite

This formation consists mainly of dololutite and pelletal or intraclastic doloarenite and usually contains abundant chert nodules. It contains some interbedded dolomitic shale, coarse dolarenites and breccias, particularly in the vicinity of the Bulburra Depression which ceased subsiding during deposition of this unit. Unconsolidated slump folds are common, and columnar stromatolites, *Conophyton* and oncolites are abundant in some thinner sequences.

A variety of depositional environments are indicated for the Reward Dolomite. Lateral variations in thickness and rock types and local unconformities with the underlying shales suggest deposition on a warped surface. Relatively thin sequences of intertidal sediments containing land-derived sand grains accumulated in shallow-water areas, and thicker sequences of muddy sediments with interbedded breccia beds formed in deeper-water basins.

STRUCTURE

The sediments in the Bulburra Depression have been affected by two series of broad basinal folds, one trending $070°$ and the other $020°$ (Cotton, 1965). From north to south through the McArthur deposit, the surface dips of the mineralized beds vary from about 85 to $20°$E, and the change from steep to shallow dips occurs across a monoclinal flexure.

The major faults in the McArthur area (Fig. 1) trend approximately N–S (e.g., Emu Fault) and NW–SE (e.g., Calvert Fault).

Earth Research Technology Satellite (ERTS) imagery of the McArthur area contains prominent lineaments in these two fault directions. The lineaments parallel to the Calvert Fault cross the Emu Fault zone; they partly represent faults, and partly strong jointing in the Cambrian Buckalara Sandstone which unconformably overlies the Carpentarian shelf sediments. Another lineament system apparent on the ERST imagery trends $0.70°$. This possibly reflects relatively minor faults and the fold trend mentioned above. The general $040°$ trend of the McArthur River may reflect faulting.

The McArthur deposit, at its closest point is about 2.5 km from the Emu Fault. It is located at a zone on intersection of the Calvert, $020°$, $040°$ and $070°$ lineament systems.

GENERAL FEATURES OF THE McARTHUR DEPOSIT

The general features of the McArthur deposit have been described by Cotton (1965), Croxford and Jephcott (1972) and Murray (1975), and the information presented in this section is largely from these publications.

Size

The McArthur deposit contains an estimated 200 million tons of ore with approximately 10% Zn, 4% Pb, 45 ppm Ag (1.5 oz/ton), and 0.2% Cu. It extends over an area of about 1.5 km^2 and is generally in the vicinity of 55 m thick, but increases to about 130 m thick towards its eastern edge (Fig. 6).

Fig. 5. Slab of high-grade McArthur ore, showing black and dark grey tuffaceous shale laminae and abundant bands of pyrite, sphalerite and galena. Note preconsolidational slumping, scouring and dewatering structures, and concretionary pyrite (light grey patches). Specimen is 10 cm across.

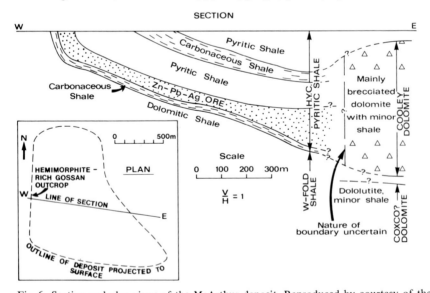

Fig. 6. Section and plan views of the McArthur deposit. Reproduced by courtesy of the Carpentaria Exploration Company.

Orebodies and inter-ore beds

The McArthur deposit contains seven mineralized shale orebodies separated by relatively metal-poor, dolomite-rich inter-ore beds (Fig. 7). At the north-end of the deposit, a relatively small area of stratiform mineralization is known underneath the lowest of these orebodies, but this is not considered here.

Orebodies. The orebodies consist essentially of thin laminae of fine-grained sulphides and tuffaceous shales (Figs. 4, 5). In high-grade parts, very few non-sulphide laminae are discernable (Fig. 8). Coarse-grained lenses and irregular to rounded patches of pyrite are scattered through the ore (Figs. 5, 9), and there are occasional small discordant sulphide-carbonate veinlets. These veinlets often contain some barite.

There are several relatively carbonate-rich intervals within the two lowest orebodies. These are characterized by rounded or angular nodules of dolomite, which may be from less than a millimetre to greater than a centimetre across, surrounded by sulphide minerals and carbonaceous tuffaceous shale (Fig. 10). Some of these appear to be micro-concre-

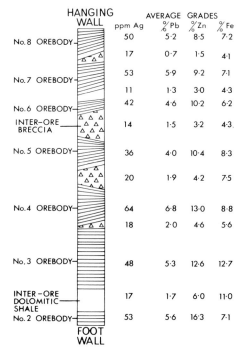

Fig. 7. Section through the McArthur deposit showing the seven orebodies, the inter-ore beds and the metal grades. The orebodies are numbered 2–8; a small area of stratiform mineralization has been located under the number 2 orebody in the northern part of the deposit. (Reproduced from Murray (1975), by courtesy of Carpentaria Exploration Company and Australasian Institute of Mining and Metallurgy.)

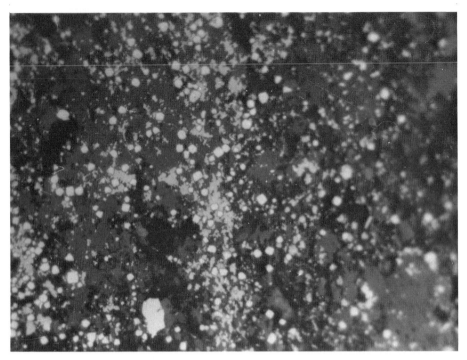

Fig. 8. Photomicrograph of high-grade McArthur ore showing galena-rich band (light grey) and fine grained pyrite crystals (whitish) in a matrix of sphalerite. The sphalerite matrix is fine-grained and contains small irregular blebs of galena and chalcopyrite. Black patches are carbonaceous shale. Reflected light; 200 μ across field. (Ref. No. Te115/19.)

tions which forced aside the enclosing sulphidic muds, but others have the appearance of residuals after partial dissolution of carbonate laminae, possibly by acidic metalliferous solutions.

Thin, poorly-mineralized beds occur within each of the orebodies. These include graded dolarenites, and less common tuffites (Fig. 3) and nodular chert beds.

There are many structures within the orebodies that indicate the fine-grained conformable sulphides bands must have formed prior to consolidation of the sediments (refer to Fig. 5). Slump, slide and scour structures are widespread, and are particularly abundant in the eastern part of the deposit, adjacent to the Cooley Dolomite. The sulphide and shale bands are also frequently broken up by unmineralized micro-faults which are interpreted as early-diagenetic dewatering structures. The textural relationships in the nodular dolomite intervals are complicated, because carbonates can accumulate relatively rapidly and can undergo a variety of diagenetic processes near, and at, the sediment surface which can lead to their lithification before associated detrital sediments. The writer considers that the textures are consistent with syngenetic–early diagenetic mineralization, which partly involved replacement of slightly earlier formed carbonate coupled with some subsequent recrystallization.[1] The uncommon sulfide–carbonate veinlets in the ore are fair-

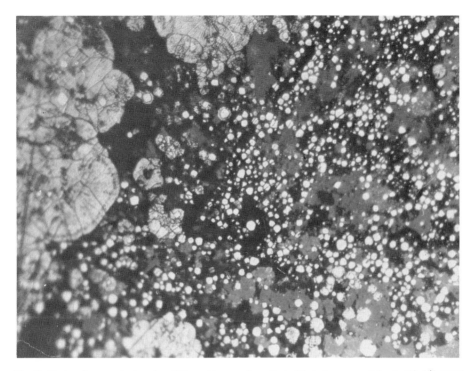

Fig. 9. Photomicrograph showing different types of pyrite in McArthur ore; etched with 1/1 HNO$_3$ for 45 sec. Fine-grained crystals of pyrite are unaffected by the acid, but many of the coarser grains contain rims of more reactive pyrite. The large concretionary grains contain nuclei of fine-grained pyrite. Mid-grey mineral is sphalerite, black is carbonaceous shale. Reflected light; 200 μ across field. (Ref. No. Te115/26.)

ly straight and must have formed when compaction of the sediments was essentially completed.

Inter-ore beds. The relatively metal-poor intervals within the McArthur deposit are generally 1–3 m thick, except in the eastern part where they are several-fold thicker.

The inter-ore beds between the two lowest ore bodies consist essentially of bituminous dolomitic shales with some nodular dolomite. They contain abundant pyrite and significant, but sub-economic, amounts of fine-grained sphalerite and galena.

In contrast, the beds between the higher orebodies are mainly dolomitic arenites and breccias. They contain some minor amounts of coarse-grained sphalerite, galena and pyrite, with lesser chalcopyrite, marcasite and arsenopyrite. These sulphides occur within dolomite clasts and as discrete grains. The inter-clast matrix material in these beds is most commonly dolomitic silt which may contain minor fine-grained pyrite, but there are some areas of mineralized carbonaceous mud containing fine-grained pyrite, sphalerite and galena.

---

[1] It is noteworthy that the high quality of preservation of their microfossils points to formation of the chert nodules in the ore during the very early stages of diagenesis (J. Oehler, personal communication, 1975).

Fig. 10. Photomicrograph of nodular dolomite within the McArthur deposit. Fine-grained pyrite (whiteish) and patches of sphalerite (light grey) and galena (slightly greyer than pyrite) occur between dolomite "nodules" (mid-greys). Reflected light; 1000 μ across field. (Ref. No. Te115/26n.)

MINERAL RELATIONSHIPS IN THE OREBODIES

The tuffaceous shale laminae within the orebodies consist mainly of quartz, K-feldspar, ferroan dolomite, illite, chlorite, kaolin, disseminated sulphides, and carbonaceous matter. The presence of illite and kaolin may be indicative of slightly acid conditions during deposition/diagenesis of the shales.

The main sulphide minerals in the McArthur deposit are pyrite, sphalerite and galena. Sulphides present in minor amounts include chalcopyrite, arsenopyrite, marcasite, chalcocite and covellite.

Significant proportions of the metal sulphide bands in the orebodies are dominated by one mineral, i.e., they are essentially monomineralic. No systematic ordering has been detected amongst these bands.

The sulphide mineral relations summarized below are based on Reinhold (1961, quoted by Cotton, 1965), Croxford (1968) and Croxford and Jephcott (1972), as well as some unpublished data obtained by the present author.

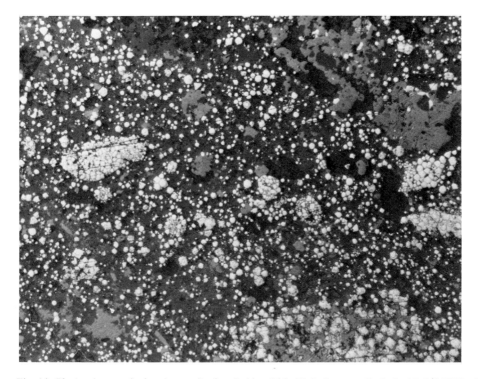

Fig. 11. Photomicrograph showing pyrite framboids within McArthur ore. Etched with 1/1 HNO₃ for 45 sec. They consist of clusters of fine-grained pyrite crystals and are developed locally within the ore and country rocks. Reflected light; 500 μ across field. (Ref. No. Te115/26f.)

Pyrite

This is the most abundant sulphide mineral in relatively low-grade ore, but it can be slightly less abundant than sphalerite in high-grade ore.

There are several varieties of pyrite within the McArthur orebodies (Figs. 9, 11), and these indicate several generations of formation of this mineral.

The first-formed pyrite is banded to disseminated and consists of very fine-grained, euhedral to rounded crystals, generally between 2 and 5 μ across (see Figs. 4, 8—11). It is the most abundant variety, constituting roughly 50—60% of the total pyrite. It is largely unaffected by 1/1 HNO₃.

There are coarser-grained varieties of pyrite in the ore which contain nuclei of the fine-grained crystals, surrounded by variable proportions of later formed pyrite. They can be subdivided in a somewhat arbitrary fashion into three main types.

Spherical and subspherical pyrite grains, often in the size range 5—20 μ, are disseminated through the ore and sometimes form distinctive bands (see Figs. 4, 9). Etching of this pyrite with 1/1 HNO₃ generally shows up rims of relatively reactive pyrite around the fine-grained unreactive cores. Overall, this variety constitutes roughly 30—40% of the total pyrite.

Minor proportions of large, rounded, irregular or anhedral concretions of pyrite are scattered through the ore (Figs. 5, 9). Many of these are visible in hand specimen and some examples are up to several millimetres across (Fig. 5). They typically form discontinuous layers within the ore and have forced aside surrounding materials in a manner which suggests these were unconsolidated during growth of the concretions. Etching with HNO_3 reveals relatively few fine-grained acid-resistant crystals surrounded by rather complex rims and mosaics of variably more reactive pyrite.

Framboidal pyrite is relatively uncommon overall, but there are some local concentrations. These consist of spherical to lens-shaped clusters of the fine-grained crystals or, rarely, the larger spherical grains, and individual framboids are up to 200 μ across. The inter-crystal areas are usually filled with pyrite which is more reactive to HNO_3, but sphalerite and galena occur within a minority of these structures.

A distinctive but minor occurrence of pyrite in the ore is replacing fragments and matrix within some of the tuffites.

It is noteworthy that there is little evidence for embayed margins and other features indicative of extensive solution or replacement of pyrite in the ore. It is also interesting that very little pyrite occurs in the rare sulphide—carbonate veinlets.

Sphalerite

In general sphalerite is the most abundant mineral in high-grade ore, but is less abundant than pyrite in low-grade ore. It occurs mainly as fine-grained almost monomineralic bands (Figs. 4, 8) up to a millimetre or so thick, and as irregular patches from a few microns to a few hundred microns across.

Relatively minor proportions of coarse-grained sphalerite occur in the nodular dolomite ore, with some of the concretionary pyrite, and in the small sulphide-carbonate veinlets. Very thin apophyses of sphalerite occasionally cut through the concretionary pyrite, filling cracks in the latter mineral. Therefore, this sphalerite formed after crystallization of the concretionary pyrite (which, as mentioned above, is considered to have formed before consolidation of its enclosing sediments).

The conformable sphalerite bands contain variable proportions of fine-grained, euhedral to rounded pyrite, coarser-grained subspherical pyrite, and smaller irregular blebs of galena and chalcopyrite (Figs. 4, 8). The coarser-grained sphalerite is characterized by minute spindles of chalcopyrite aligned along crystallographic directions.

Sphalerite in thin sections of high-grade ore is often opaque. This is possibly because of inclusions of chalcopyrite, many of which may be sub-microscopic. The iron content of the sphalerite is low, but variable. Transparent samples vary in colour from light straw colour to reddish-brown.

Galena

This mineral is significantly less abundant than pyrite and sphalerite. It is closely associated with sphalerite, occurring mainly as small irregular grains within the sphalerite bands and as concordant, essentially monomineralic bands and schlieren (Fig. 8). Minor amounts of later-formed galena occur in cross-cutting veinlets, in nodular dolomite ore, in association with some concretionary pyrite, and as subhedral to auhedral crystals up to about 0.06 mm across scattered through the mineralized shales.

Minor sulphide minerals

Chalcopyrite is present mainly as irregular and spindle-shaped blebs within sphalerite (Fig. 8), but also occurs as occasional large grains. It is significantly more abundant in high-grade ore in the bottom four orebodies. Arsenopyrite is widespread in minor amounts. Small acicular crystals of this mineral are scattered through some tuffite bands. Marcasite occurs in minor amounts in some of the concretionary pyrite layers and in sulfide—carbonate veinlets. Chalcocite and covellite are rare and probably are secondary after chalcopyrite.

GEOCHEMISTRY OF THE OREBODIES[1]

The data summarized in this section are from Coxford and Jephcott (1972), Lambert and Scott (1973), Corbett et al. (1975) and Murray (1975).

The order of presentation is as follows. The metals which are most highly enriched in the ore are considered first, in the order Zn and Pb, Fe, Ag, Cu, As, Cd, Hg, Sb and Tl. Other elements of interest are then considered in alphabetical order: Al, Ba, carbonate (CO_2), Cl, Co and Ni, Mn, organic C, K and Na, Si, Se, Sr and V. A considerable number of additional elements have been determined, *viz.,* Au, B, Bi, Cr, Ga, La, Li, Mo, P, Rb, Sc, Th, Ti, U, Y and Zr. These are not discussed here because they are not significantly enriched in the ore relative to unmineralized sediments.

Zinc and lead

High-grade ore contains up to 24% Zn and 12% Pb and the overall average ore grade is 10% Zn and 4% Pb. The seven orebodies have variable Zn/Pb ratios within the fairly narrow range 2.9—1.6 (see Fig. 7). There is no systematic trend in this ratio with stratigraphic level, or with orebody thickness. However, near the lateral extremities of the orebodies, Pb decreases more rapidly than Zn.

---

[1] For a general treatment of trace element composition of orebodies, see Chapter 1 by Mercer, Vol. 2.

In the lower orebodies, these highest-grade portions are roughly central with respect to their plan views and there is a general contraction in the areal extent of the orebodies upwards accompanied by an eastwards shift of the highest-grade ore.

Iron

The iron contents of ore samples are quite variable, with most samples falling in the range 4–25%. The average content is around 10% Fe. Pyrite accounts for most of the iron, but there is typically 0.5–2% Fe which is soluble in hydrochloric acid, and this is largely in ferroan dolomite.

Silver

The silver contents are proportional to the Pb grades. The highest value recorded is 120 ppm (approximately 4 oz/ton).

Copper

This metal is most abundant in the lower and middle orebodies reaching a peak of around 0.5% in No. 4 orebody. The overall average Cu content is about 0.2%.

Arsenic

The orebodies have high arsenic contents, mainly in the range 1000–2000 ppm and their mean As/Fe ratio is in the vicinity of 0.01.

Cadmium

This metal correlates well with Zn and is present in amounts between 20 and 550 ppm.

Mercury

A relatively small number of analyses are available for this element. The indicated range of Hg contents in the orebodies is from 770 to 1740 ppb.

Antimony

This element has been recorded in amounts up to 130 ppm, and averages about 50 ppm through the orebodies.

Thallium

This element varies considerably in abundance through the orebodies, but the major proportion of the values falls in the range 50–120 ppm.

Other elements

Alumina contents of the mineralized shales are generally in the range 4–10% (average about 7%) and SiO_2/Al_2O_3 ratios are within the range 3.5–5. Carbonate-rich interbeds have lower alumina contents and higher SiO_2/Al_2O_3 ratios.

Barium is low through the ore, with only rare values exceeding 220 ppm.

Carbonate (CO_2) contents in the orebodies are typically less than 10% and do not vary in a systematic fashion with ore grade. The relatively low grade, nodular dolomite ore intervals have up to about 20% CO_2 w/w.

Chlorine contents of the ore shales are mainly in the range 40–220 ppm. Values in the carbonate rich inter-ore beds range up to around 500 ppm, presumably reflecting the presence of abundant brine inclusions in the dolomites.

Cobalt and nickel are not high in the orebodies. Maximum recorded values are 150 ppm Co and 90 ppm Ni. The Co/Ni ratios are mainly in the range 0.5–3, and average around 1.5.

Manganese is moderately high in the orebodies where it is commonly between 0.2 and 0.6%. It is largely bound up in dolomite and can be almost entirely dissolved out with hydrochloric acid.

Organic matter is widespread through the mineralized shales, but the highest recorded organic carbon content in the ore is only 0.8% w/w (Lambert and Scott, 1975; Corbett et al., 1975). Much higher values quoted by Lambert and Scott (1973) were subsequently found to be in error because of SO_2 contamination of the analyses caused by the very high pyrite contents of the samples. An organic matter separate from the ore shales has been analysed by Saxby (1970) who found it to have between 72 and 84% C, 5% H, 1% N, between 6.5 and 16.5% O and between 2.5 and 5.4% S.[1]

Potassium is quite high and sodium very low in the mineralized shales. Most ore samples contain between 1 and 4% K_2O and the average content is close to 2%; the carbonate-rich inter-ore beds generally have between 0.7 and 2.7% K_2O. Potassium contents correlates with alumina contents. In contrast, Na_2O values rarely exceed 0.3%.

Silica contents of the mineralized shales are generally in the range 20–35%. The carbonate-rich inter-ore beds often have SiO_2 contents in the same range, but a significant proportion have between 10 and 20% SiO_2.

Selenium was not detected at 2 ppm in all except one of the available analyses; the anomalous value was 15 ppm.

Strontium is extremely low, rarely exceeding 30 ppm in the mineralized shales or carbonate-rich inter-ore beds.

Vanadium is not enriched in the ores, where the highest recorded value is 100 ppm. The average value for the ore shales is around 70 ppm, and that for the carbonate-rich inter-ore beds is around 30 ppm.

---

[1] See Chapter 5 by Saxby, Vol. 3, and especially his table III and fig. 4.

MINOR MINERALIZATION IN THE McARTHUR AREA

H.Y.C. Pyritic Shale

Several stratiform Zn deposits of significantly lower grade and smaller tonnage than the McArthur deposit occur within sub-basins of H.Y.C. Pyritic Shale Member *viz*., the W-Fold, Teena and Mitchell Yard deposits. These are respectively 8 km to the west-south-west, 11 km to the west-southwest and 6 km to the south-southwest of the McArthur deposit. The Teena and Mitchell Yard deposits are essentially pyritic shales containing only 1–2% Zn and traces of Pb, but the W-fold deposit is much more important.

The W-Fold prospect is 30–40 m thick and appears to be more extensive laterally than the McArthur deposit. Folding and faulting are relatively intense in this deposit. It generally has 2–3% Zn, but there is a 2–3 m interval of about 10% Zn near the base. It is some 22–27 m above the base of the H.Y.C. Pyritic Shale Member and is possibly stratigraphically equivalent to the basal part of the McArthur deposit. It contains rare dolarenite interbeds, but sedimentary breccias like those in the McArthur deposit are absent. Bedding is often mildly disturbed and slumped.

It is interesting that pyrite is not abundant in the W-fold deposit, only becoming prominent above the high-grade zinc interval. Galena is a minor constituent throughout. Sphalerite occurs in thin, fine-grained bands and is generally similar to that in the McArthur deposit. In some places, however, it has been recrystallized along with en-closing dolomite-rich beds and forms coarser-grained bands and concretions which often have centres of grey dolomitic silt material.

Minor Pb mineralization with subordinate Zn and Cu occurs in a slump breccia at the Reward prospect, some 13 km west-southwest of the McArthur deposit. This may be stratigraphically equivalent to the H.Y.C. Pyritic Shale.

The W-fold, Teena and Reward deposits occur in a general west-southwesterly direction from the McArthur deposit, within the 0.70° trending lineament system visible on ERTS imagery.

Cooley Dolomite

Minor mineralization is encountered in most drill holes in the Cooley Dolomite, and four significant deposits have been discovered in this unit.

The Ridge 1 and 2 deposits occur near the western margin of the Cooley Dolomite and are known to extend to within about 300 m of the McArthur deposit. Ridge 1 consists of discordant veins of coarse-grained sphalerite, with minor galena and chalcopyrite. Ridge 2, which is predominantly a Pb deposit, is characterized by two contrasting styles of mineralization. The eastern part consists of discordant veins in brecciated dolomite and the western part consists of broadly concordant mineralization in contorted pyritic shale which is intercalated with dolomite breccia.

The Cooley 1 and 2 deposits occur in the eastern part of the Cooley Dolomite. Both are discordant vein deposits and Cooley 1 is Pb-rich, whilst Cooley 2 is Cu-rich. They contain coarse-grained galena, sphalerite, chalcopyrite, marcasite, pyrite and minor tetra-hedrite. The sulphides are associated with ferroan dolomite, quartz and minor barite. The mineralization appears to be in open-space structures varying from minute veinlets to veins that occasionally exceed 10 cm thick.

Other dolomite units, Batten Trough

A number of small, fissure-filling/replacement Pb deposits are scattered through dolomitic units in the McArthur Group. Several occur along the Emu Fault zone to the south-southeast of the McArthur deposit, and there are a few others at variable distances in a general westerly direction from the McArthur deposit. The stratigraphic positions of a number of these deposits are uncertain, but some are definitely within the Emmerugga Dolomite and a few are tentatively regarded as being in the Reward Dolomite. All contain minor sphalerite, and barite is present in some of them.

There are some small Cu deposits in the Emu Fault zone and dolomites and siltstones of the Amelia Dolomite and Tooganinie Formation.

Shelf sequences to east and west of Batten Trough

A number of small Pb deposits and some minor Cu deposits occur in the relatively thin sequences of McArthur Group equivalents to the east of the Emu Fault, and along the Urapunga Tectonic Ridge.

MINERAL DISPERSIONS AROUND McARTHUR DEPOSIT

Sulphides

Pyrite. This mineral is most abundant within the H.Y.C. Pyritic Shale. Very highly pyritic strata are particularly characteristic of the hanging-wall shales for several hundred metres above the McArthur deposit, but also occur locally at distances up to at least 23 km from this deposit.

Euhedral to rounded pyrite crystals typically a few microns across, constitute the dominant form of pyrite in the H.Y.C. Pyritic Shale country rocks (Fig. 12). This is very similar to the common fine-grained pyrite in the McArthur deposit, and likewise is not differentially etched by 1/1 HNO_3. Framboidal clusters of the fine-grained crystals are scattered through the shales and are locally quite abundant. They are occasionally cemented by pyrite that is more reactive to HNO_3.

Larger, spherical pyrite grains, often of the order of 20 μ across, are locally abundant

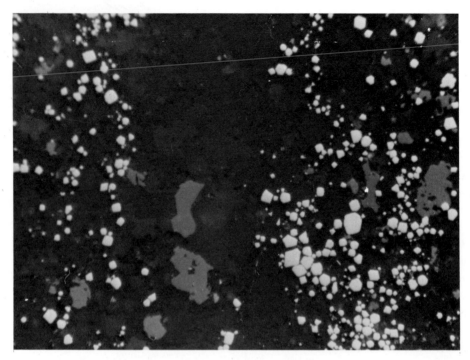

Fig. 12. Photomicrograph of carbonaceous, dolomitic pyritic shale from the H.Y.C. Member, about 7 km from the McArthur deposit. This specimen contains significant amounts of sphalerite (mid-grey) and traces of galena. Pyrite is very similar to the fine-grained variety in the McArthur deposit; it is largely unaffected by etching with 1/1 HNO_3. Reflected light, 200 μ across field (Ref. No. W.H.1.)

in the shales above the McArthur deposit (Fig. 12). In contrast with the rather similar variety of pyrite in the ore, etching of these hanging-wall pyrites with HNO_3 only rarely reveals more reactive rims.

Concretionary pyrite layers are very uncommon in the country rock shales. Etching these with HNO_3 reveals complex mosaic and concentric structures, and considerable variations in reactivity. In many cases, pyrite grains within the concretions have grown around carbonate or silicate nuclei, rather than the fine-grained unreactive pyrite nuclei which characterize the generally similar variety of pyrite in the ore.

The W-Fold Shale and the Reward Dolomite, which respectively underlie and overlie the H.Y.C. Pyritic Shale, generally contain only minor amounts of pyrite. This is mainly in the form of small disseminated crystals.

Other sulphides. Minor amounts of sphalerite, galena and arsenopyrite, generally in that order of abundance, are scattered through the pyrite-rich beds of the H.Y.C. Pyritic Shale Member. The sphalerite and galena occur in several forms: as small essentially mono-mineralic schlieren; as distinct subhedral to irregular grains usually between 15 and 100 μ across; and as infillings in clusters of pyrite crystals. The arsenopyrite is typically in the form of fine-grained euhedral crystals, less than 15 μ long.

Non-sulphides

There is a dispersion halo of ferroan dolomite above, below and beside the McArthur deposit, which is readily apparent from the rusty brownish colours of weathered samples.

In contrast with many other stratiform ores there are no recorded siliceous or clay-rich alteration zones around the mineralization.

GEOCHEMICAL DISPERSION HALOES AROUND McARTHUR DEPOSIT

Lambert and Scott (1973) carried out detailed geochemical investigations of about 130 drill core samples of foot and hanging-wall rocks, and stratigraphically equivalent units at various distances up to about 20 km from the McArthur deposit. They determined all major elements, CO_2, organic C, S, Ag, As, Au, B, Cd, Co, Cr, Ga, Ge, Hg, La, Li, Mo, Ni, Pb, Sb, Sc, Se, Sr, Tl, Te, U, V, Y, Zn and Zr. Their analyses are tabulated in Corbett et al. (1975).

Brown et al. (1975) presented partial analyses (*viz.* acid insolubles, Ca, Mg, Fe, K, Mn, P, Zn, Pb, Cu, Co and Ni) for over 300 outcrop samples of McArthur Group sediments collected at distances from a few hundred metres to about 60 km from the McArthur deposit.

Elements with dispersion haloes

The only elements for which significant dispersion haloes have been delineated are Zn, Pb, Fe, Mn, As, Hg and Tl.

Zinc and lead. The W-Fold Shale underneath the McArthur deposit has Zn contents in the range 100—2600 ppm and a maximum recorded Pb content of 50 ppm. Similar metal concentrations persist in the lowermost strata of the H.Y.C. Pyritic Shale, but there are significantly higher concentrations for up to 10 m or so beneath the lowest orebody.

The top of the McArthur deposit is fairly well-defined, with Zn grades decreasing from greater than 5% to less than 2% over an interval of 5 m or so, and Pb values decreasing accordingly. However, anomalous Zn contents up to 1.7% and Pb contents up to 0.75% occur intermittently in the hanging wall H.Y.C. Pyritic Shale for distances of at least 275 m above the ore.

Anomalously high Zn and Pb contents occur within the H.Y.C. Pyritic Shale Member in each of the drill cores investigated by Lambert and Scott (op. cit.). At distances of about 100 m from the McArthur deposit, Zn and Pb levels in the pyritic shales are commonly of the other of 2% and 0.25%, respectively. The contents of these metals decrease gradually with increasing distance from the ore deposit, but some pyritic shales from 20 km away still have around 0.5% Zn and 0.1% Pb.

The outcrop samples of Brown et al. (1975) included only one specimen of H.Y.C. Pyritic Shale, because this member weathers very readily. This specimen, collected about 200 m from the McArthur deposit, was heavily stained with Fe oxides. However, it retained anomalously high Zn and Pb contents (0.5% Zn and 0.1% Pb).

The data of Brown et al. (op. cit.) for fresh outcrop samples indicated no significant Zn–Pb anomalies in the other lithologies from the Myrtle Shale up to the Reward Dolomite.

Acid soluble iron. This is the predominant form of Fe in all units except the H.Y.C. Pyritic Shale.

The Fe contents of the dolomites show a general decrease with increasing distance from the McArthur deposit (Lambert and Scott, 1973; Brown et al., 1975). Within and closely adjacent to the McArthur ore deposit the dolomites frequently contain 2–4% Fe. This Fe is generally in the reduced state, occurring within the dolomite lattice; the reddish colouration in the dolomitic W-Fold Shale under the McArthur deposits, however, is caused by apparently primary iron oxide. Anomalous iron contents have been recorded through the stratigraphic interval from the Mara Dolomite to the Reward Dolomite. They appear to be distributed over a roughly semi-circular area, with a radius of approximately 15–20 km, extending westwards from the Emu Fault zone; outside this area, the Fe contents in the dolomites are less than 1%.

Pyritic iron. Pyrite is the major Fe-bearing phase only in the H.Y.C. Pyritic Shale. It is not a very abundant constituent in the foot-wall shales of this member, where Fe is generally 4–5%, or less. However, the hanging-wall shales frequently have higher pyrite contents than the orebodies; their Fe contents are commonly on the range 20–30%. The so-called carbonaceous shales in the hanging wall (Fig. 6), appear to be relatively poor in pyrite in hard specimen, but can contain significant amounts of this mineral; their Fe contents are generally between 2 and 12%.

Pyrite is a major constituent of the shales for a few hundred metres laterally away from the McArthur deposit; 25–30% Fe is not uncommon in these rocks. However, Fe contents have decreased to 5–10% in the pyritic shales at distances of 2 km from the deposit, and such levels persist in the more distant shales.

Manganese. This element is most abundant (up to 1.9% Mn) immediately beneath the McArthur deposit where it is concentrated in both dolomitic and shaley beds of the W-Fold Shale and basal H.Y.C. Pyritic Shale. The Mn contents of these units decrease gradually with increasing distance from the ore; a few kilometres away they average around 0.5%.

Arsenic. This element rarely exceeds 50 ppm except in the H.Y.C. Pyritic Shale, where it shows a reasonably good correlation with Fe. The As/Fe ratios in the pyritic shales of this member are considerably lower than within the McArthur ore. Within a few hundred metres above and laterally away from the ore, the pyritic shales frequently contain at least 1000 ppm As; at distances of 15–20 km from the ore they generally contain 150–300 ppm As.

Mercury. Background values of this element in units other than the H.Y.C. Pyritic Shale are around 50–150 ppb. The pyritic shales typically contain 200–700 ppb Hg, with most, but not all, of the higher values being within about 7 km of the ore.

Thallium. This element was studied by Smith (1973). It is concentrated in the pyritic shales immediately above the McArthur deposit, where he reported a maximum value of 830 ppm. Tl contents drop off rapidly above the ore; pyritic shales some 200 m up into the hanging wall contain only about 10 ppm. Anomalous Tl contents (>10 ppm) occur in the pyritic shales near the base of the H.Y.C. Member for at least 7 km from the ore.

Other elements

Copper does not have a significant dispersion halo. One sample of pyritic shale 100 m from the ore had 150 ppm Cu, but all other samples analysed from the H.Y.C. Pyritic Shale contained less than 100 ppm. There are some minor copper anomalies (150–300 ppm) in the Teena and Cooley Dolomites, and these are not associated with Zn and Pb anomalies.

Silver contents of most of the pyritic shales outside the McArthur deposit are less than 5 ppm, but samples from the hanging wall within a few metres of the top of the ore contain around 10 ppm. Ag contents of close to 5 ppm are associated with the above mentioned Cu anomalies in the Teena and Cooley Dolomites.

Cobalt rarely exceeds 30 ppm in the country rock shales and is significantly less abundant in the carbonates.

Nickel contents of the shales average around 30 ppm and the carbonates contain significantly lower amounts of this element.

Antimony is below 5 ppm in the vast majority of the country rock shales and carbonates.

Selenium is below 2 ppm in the analysed pyritic shales, with the exception of two samples (15 ppm and 2 ppm).

Alumina contents of the country rock shales and carbonates are quite variable. The average Al_2O_3 in the pyrite-poor shales is around 10% with very few recorded values above 14%. The majority of the highly pyritic shales contain between 3 and 7% Al_2O_3. The carbonates rarely contain more than 8% Al_2O_3 and generally have much less.

Silica averages about 50% in the sulphide-poor country rock shales. The pyritic shales have lower contents which tend to be inversely proportional to their sulphide contents. The carbonates have variable silica contents which rarely exceed 35%.

Potassium does not decrease progressively with distance for the McArthur deposit. It is high (often >5% K_2O) in shales from the H.Y.C., W-Fold and Teena Members. Sodium on the other hand is very low in most of the analysed rocks from the McArthur area, rarely exceeding 0.3% Na_2O.

Barium rarely exceeds 400 ppm and 250 ppm, respectively, in the shales and carbonates.

Strontium rarely exceeds 50 ppm in the shales and carbonates.

Carbonate contents of the country rock shales are highly variable and average out at around 12% CO_2 w/w.

Organic carbon contents up to 5% have been recorded in carbonaceous shales of the H.Y.C. Pyritic Shale Member, but the average value for these rocks is around 1% C w/w. The highly pyritic shales commonly contain less than 0.6% C. The insoluble carbonaceous matter in these country rocks is compositionally indistinguishable from that in the ore (J. Saxby, personal communication, 1974). Shales within the W-Fold and Teena Members have very low organic carbon contents.

Vanadium is most abundant in the carbonaceous shales, which commonly contain between 50 and 250 ppm of this element. The majority of the carbonates contain less than 50 ppm V.

ISOTOPE DATA[1]

Sulphur isotopes

McArthur deposit. Sulphur isotope ratios in the McArthur deposit were investigated by Smith and Croxford (1973). These authors analysed a suite of samples from the cross-cut through the ore described by Croxford and Jephcott (1972). They separated the coexisting sulphides by chemical dissolution techniques.

Their results are summarized in Fig. 13. The $\delta^{34}S$ values for sphalerite (+3.3 – +8.9‰) and galena (−1.2 – +5.7‰) show no systematic trends with stratigraphic position. In contrast, the $\delta^{34}S$ values for pyrite vary over a much wider range (−3.9 – +14.3‰) and become progressively heavier from the base to the top of the deposit. It is clear that there is a rough approach to isotopic equilibrium between sphalerite and galena, and that the pyrite sulphur is distinctly different.

Smith and Croxford (1973) concluded that there were dual sulphur and metal sources involved in the formation of the McArthur deposit. They considered that Fe was introduced separately from Zn–Pb, that the sulphur in the pyrite was generated by bacterial reduction of sulphate in a fairly closed system, and that the sulphur in the sphalerite and galena was brought in with the Zn–Pb ore solutions. In addition, they pointed out that the differences between the $\delta^{34}S$ values for coexisting sphalerite and galena are indicative of a wide range of temperatures – from 100 to 270°C on the temperature scale of Kajiwara and Krouse (1971). This may well be an apparent range, reflecting incomplete isotopic equilibration during low-temperature ore formation and diagenesis, but the possibility of non-systematic temperature fluctuations of this magnitude during ore formation cannot be completely dismissed.

The dual sulphur source model proposed by Smith and Croxford (1973) satisfactorily explains the existing sulphur isotope data, but four alternative interpretations of these data merit further consideration:

(1) Fe was introduced separately from Zn–Pb, and was precipitated as iron monosul-

---

[1] For general treatments of isotope data, see Chapters 7, 8, and 9 in Vol. 2 by Fritz, Sangster, and Köppel and Saager, respectively.

phides by non-biogenic sulphur; pyrite formed during early diagenesis by reaction with biogenic sulphur.

(2) Non-biogenic iron monosulphides precipitated from Fe–Zn–Pb–S solutions, and were subsequently converted to pyrite by reaction with biogenic sulphur.

(3) A sulphur-poor, Zn–Pb brine dissolved some of the earlier formed pyrite from throughout the ore interval, producing an isotopically homogeneous solution with a $\delta^{34}\Sigma S$ close to +6‰, from which the sphalerite and galena precipitated (Williams and Rye, 1974).

(4) All the sulphide minerals incorporated non-biogenic sulphur, but the sphalerite and galena precipitated at fairly constant physicochemical conditions, whilst the pyrite formed separately under progressively changing conditions. Ohmoto (1972) showed that the isotopic compositions of sulphide minerals are strongly controlled oxygen fugacity, pH and temperature conditions in the ore forming environment.

Model (1) is a variation on the interpretation of Smith and Croxford which cannot be ruled out by the existing data. Model (2) was considered by Smith and Croxford (1973) who did not favour it. It appears incompatible with the Pb isotopic data of Gulson (1975 – see discussion in next section). Model (3) does not seem likely on the basis of the arguments presented below in the discussion of the genesis of the McArthur deposit. Model (4) cannot be assessed further until detailed information is forthcoming on the

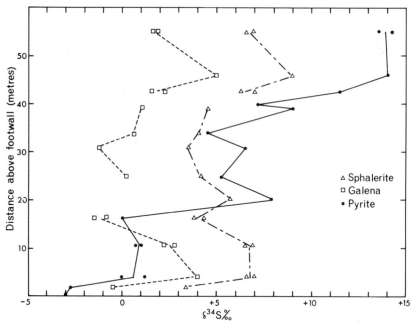

Fig. 13. Sulphur isotope data through the McArthur deposit, after Smith and Croxford (1973). Note that the pyrite has a distinctly different trend from sphalerite and galena.

Eh–pH conditions under which the pyrite was generated, and until it is established whether equilibrium calculations like those of Ohmoto (op. cit.) are applicable at the relatively low temperatures which are thought to have characterized the precipitation of the McArthur sulphides. Delineation of the physicochemical conditions of pyrite formation are complicated by the fact that this mineral could have formed directly from Fe-rich solutions, or by diagenetic sulphidization of iron monosulphides or iron oxide–hydroxides.

Country rocks. Smith and Croxford (1975) determined the sulphur isotope ratios of pyrite in the shales below and above the McArthur deposit. The δ^{34}S values of the foot-wall pyrite vary within the range $+13 - +18‰$, whilst the values for the hanging-wall pyrite fall in the range $-5 - +5‰$. Therefore, there are marked sulphur isotope discontinuities at the bottom and top of the ore.

Smith and Croxford (1975) believed that biogenic sulphur was involved in pyrite formation throughout the H.Y.C. Pyritic Shale Member and that the different ranges of δ^{34}S values reflect changing water depth, sulphate availability and degree of reduction of the sulphate reservoir.

Lead isotopes

J.R. Richards (1975) has determined isotope ratios of HCl-soluble leads from the McArthur deposit and several of the minor deposits around it. Gulson (1975) compared isotopic ratios of leads in galena, sphalerite and pyrite from the McArthur deposit, and pyrite from the country rocks.

The galena leads in the McArthur area appear to fall into two isotopic categories. The first is highly homogeneous, and includes samples from the McArthur deposit and the stratiform Mitchell Yard deposit (the other stratiform deposits have not been investigated). The isotopic ratios of the samples analysed fall within the following ranges: 206/204 = 16.07–16.16; 207/204 = 15.37–15.48; 208/204 = 35.57–35.90. They lie close to, but significantly above the "average ore-lead growth curve" (Richards, 1968). Their homogeneity suggests either a very homogeneous source, or thorough mixing of the Pb before ore deposition. Their model galena age is approximately 1600 m.y.

The second category of galena leads comprises the more coarsely crystalline galenas from the non-stratiform deposits in the Emmerugga and (?) Reward Dolomites; the mineralization in the Cooley Member has not been investigated. These display a typically "anomalous" pattern, which suggests partial admixture of lead from at least two sources. The spread of isotopic data allows that one component of such a mixture could be lead similar to what is found in the McArthur deposit. It does not seem possible, however, that the stratiform McArthur and Mitchell Yard deposits contain merely a better-mixed average of the leads found in the non-stratiform deposits; at least some of the lead in the fissure-filling and replacement deposits has to be derived from a different, more radiogenic source.

Gulson (1975) found that the trace amounts of lead in sphalerites from the McArthur deposit are isotopically identical with the lead in coexisting galenas. However, leads in pyrite from the H.Y.C. Pyritic Shale Member within and around the McArthur deposit are isotopically more variable and relatively radiogenic (206/204 = 16.26—16.49; 207/204 = 15.42—15.58; 208/204 = 35.80—36.31). He concluded that distinct Fe and Zn—Pb solutions must have been derived from different sources and must have migrated separately to the Bulburra Depression. This is in good accord with the model advocated by Smith and Croxford (1973) on the basis of their sulphur isotope data.

Gulson's (1975) data allow that some of the lead in the pyrite orebody could have been derived from the Zn—Pb—Ag ore-forming solutions. On this basis some of the pyrite rim structures in the ore could have precipitated from the Zn—Pb—Ag ore-forming solutions, thus incorporating less radiogenic Pb than the predominant fine-grained pyrite crystals.

Carbon and oxygen isotope data

Smith and Croxford (1975) found little variation in the ^{13}C content of organic matter throughout the H.Y.C. Pyritic Shale Member, with $\delta^{13}C$ values ranging from −28 to −31‰ P.D.B.

They also found that carbonates from the Bulburra Depression have $\delta^{13}C$ values between 0 and −3‰ and $\delta^{18}O$ values in the range −6 − −9‰ P.D.B., and interpreted these almost invariant isotopic compositions as indicating the predominantly marine nature of the basin.

GENESIS OF McARTHUR DEPOSIT

Most workers who have studied the geology of the McArthur deposit have concluded that it formed on the sea floor from metalliferous exhalations (e.g., Cotton, 1965; Croxford, 1968; Brown et al., 1975; Murray, 1975). They based this conclusion on the conformable nature of the ore, its sedimentary and early diagenetic structures and its association with tuffaceous sediments.

This section attempts to establish a rather more detailed metallogenetic picture by considering the very recent geochemical data (Croxford and Jephcott, 1972; Lambert and Scott, 1973, Corbett et al., 1975) and isotopic data (Smith and Croxford, 1973; Gulson, 1975) along with salient geological features. It discusses the possibility of a replacement origin for the deposit, the source of metals and sulphur, the migration and entrapment of the ore solutions, and the temperatures of ore formation and diagenesis.

Arguments against a replacement origin

The McArthur deposit contains relatively minor amounts of sulphides which must have formed after deposition of the sediments (e.g., sulphides replacing dolomite and tuffaceous debris, concretionary sulphides, cross-cutting sulphide-bearing veinlets). However, it should be emphasized that these do not necessarily indicate that the whole of the ore was introduced after the sediments were laid down — several generations of sulphides could have been formed by normal post-depositional processes acting on an essentially syngenetic ore (e.g., Lambert, 1973). It must also be borne in mind that so-called paragenetic sequences in ore deposits could, in many cases, have been generated merely by differing behaviours of coexisting minerals during recrystallization (e.g., Stanton, 1964; Roberts, 1965).

The geochemical data (Croxford and Jephcott, 1972; Lambert and Scott, 1973, Corbett et al., 1975) do not support selective replacement of alumino-silicate or carbonate components of the H.Y.C. Pyritic Shale of sufficiently large scales to be quantitatively important in the formation of the McArthur Zn–Pb–Ag mineralization. This is demonstrated, for instance, by the fact that the average Al_2O_3, SiO_2 and CO_2 contents of the ore shales, after subtraction of the ZnS and PbS components and recalculation to 100%, are similar to those in the country-rock shales. Therefore the bulk of the sphalerite and galena must have been *added to* the sediments. It remains possible, though, that reactions of the ore solutions with alumino-silicate or carbonates could have played a role in generating suitable conditions for metal precipitation.

The possibility that the McArthur deposit was formed by extensive dissolution and replacement of early-formed pyrite by later Zn–Pb–Ag solutions was suggested by Williams and Rye (1974), and this model warrants detailed consideration. A mechanism of this type has previously been invoked for some other stratiform sulphide deposits, most notably the White Pine Mine (Brown, 1971).

In terms of such a pyrite dissolution model, the comformable sphalerite and galena bands and the early diagenetic structures which affect them could only be satisfactorily explained if the early-formed pyrite bands, or the thin-bedded shales, exercised an intricate control over precipitation of these minerals. The essentially monomineralic sphalerite and galena bands could conceivably have formed as a result of different migration rates of the metals through the sediments (e.g., Lambert and Bubela, 1970).

However, there are a number of arguments against formation of the McArthur ore by dissolution or replacement of pyrite. The most important are:

(1) The ore textures do not suggest large-scale dissolution or replacement of pyrite — the predominant fine-grained pyrite within the McArthur deposit is virtually identical to that in the shales above, below and beside it. Furthermore, any such process would produce a general inverse relationship between ore-grade and abundance of fine-grained pyrite, but no consistent relationship of this type has been recorded.

(2) Any such process should generate a discordant fringe zone around the ore, similar to that at the White Pine Mine, but no evidence for this has been recorded at McArthur.

(3) It is very doubtful whether sphalerite and galena could be precipitated in such perfectly conformable bands by a two-stage process of this kind; certainly the White Pine ores are not so conformable.

(4) If galena and sphalerite precipitation followed circulation of a metalliferous solution throughout the whole ore interval, as the sulphur isotope data demand of the pyrite dissolution model, the base metal ratios in each of the orebodies should be similar, or perhaps vary in a systematic fashion. Furthermore, there should be widespread mineralization in the carbonate-rich inter-ore beds.

(5) It is unlikely that there could have been extensive circulation of ore solutions through the compacted fine-grained sediments.

So, whilst a pyrite dissolution model could well have operated in the case of the White Pine and some other stratiform deposits, there is no good evidence for its importance at McArthur. The data available for the McArthur deposit appear more in accord with separate but penecontemporaneous, syn-sedimentary to early diagenetic precipitation of Fe and Zn–Pb. This could have occurred if the bulk of the Fe or/and Zn–Pb precipitated before, or as, the solutions discharged into the sedimentary basin. The precipitated metal phases would probably have formed fine-grained suspensions which settled gradually to the bottom of the depositional basin.

On this basis, the low metal contents in the turbidite and breccia inter-ore beds are consistent with rapid deposition of these units, and the essentially monomineralic metal sulphide bands in the ore can be explained by (1) differential precipitation of sphalerite and galena from solutions containing both Zn and Pb; (2) differential settling of suspended ZnS and PbS precipitates, or, (3) short-term fluctuations in the relative abundances of Zn and Pb in the ore solutions.

Consideration of volcano-exhalative metal sources

The different lead isotopic compositions of pyrite and sphalerite–galena indicate that the Fe and Zn–Pb–Ag were derived from different sources. It appears that the best means of assessing what these different sources could have been would be to investigate the Pb-isotope ratios in a representative suite of sedimentary, pyroclastic and igneous rocks from the McArthur region. Until such isotopic "tracer" data are available any discussions of the metal sources must be based on circumstantial evidence and, therefore, can only be speculative.[1]

The high-metal grades provide one line of circumstantial evidence in favour of the involvement of igneous activity in some way in the formation of the McArthur deposit. Its average Fe + Zn + Pb content is almost 25 wt%, which is much greater than the metal grades of stratiform deposits in sedimentary sequences with no volcanic components (e.g., the ores in the Kupferschiefer of Germany–Poland and the Copperbelt of Africa).[2]

---

[1] Since this chapter was written, Gulson (1975) has analysed Pb from two dolomite samples in the McArthur deposit; both contained Pb that is distinctly more radiogenic than that in the ore galena.

[2] Cf. Chapter 7 by Jung and Knitzschke, Vol. 6, on the Kupferschiefer and Chapter 6 in this volume, by Fleischer et al., on the African Copperbelt.

Iron. Lambert and Scott (1973) pointed out that the period of Fe enrichment in the sediments around the McArthur deposit (i.e., the stratigraphic interval containing pyrite and/or ferroan dolomite) was a period of tuffaceous activity. This suggests the possibility of influx of Fe-bearing exhalations similar to those in a number of present-day volcano–sedimentary environments. In many cases these are essentially sea or meteoric waters which became heated as they descended along favourable structures, leached Fe and relatively minor amounts of other metals, and returned to the surface (e.g., Ferguson and Lambert, 1972).[1] At McArthur, such waters could have become highly saline as they circulated through the evaporative sediments.

The relatively high radiogenic components in the trace amounts of Pb that were transported along with the Fe can possibly be ascribed to the fact that the northern Australian region is a U province. Many of the granitic basement rocks have high U contents and quite high Th contents (Heier and Rhodes, 1966; Heier and Lambert, unpublished data), and there are some U deposits within the Proterozoic sediments. A small component of Pb from such radioactive rocks should be sufficient to provide the observed enrichments of radiogenic Pb isotopes (Gulson, 1975).

The Fe in the sediments is dominantly in the reduced state, except for apparently primary iron oxides in the W-fold Shale under the McArthur deposit. Therefore, the Fe must have been introduced into the sedimentary basin largely in the reduced state, or as iron oxides–hydroxides which were reduced after deposition.

The iron oxides which give the reddish colours to the stratigraphically lower Myrtle Shale Member and Mallapunyah Formation are much more widespread and are unlikely to be of exhalative origin.

Zinc–lead–silver. The stratiform Zn–Pb–Ag deposits in the McArthur area are in the stratigraphic interval with the highest proportion of tuffaceous debris and a significant proportion of this debris has vitroclastic textures. Therefore, it appears likely that felsic magmas from which the pyroclastic materials were derived, could have been directly or indirectly involved with generation of the ore-forming fluids. The igneous activity could have generated favourable physicochemical conditions for leaching Zn–Pb–Ag from nearby rocks, or it could have led to release of juvenile metal-rich fluids. If magmatic emanations were involved, it is likely that these would have mixed quite extensively with sediment brines as they migrated towards the surface.

The relatively unradiogenic nature of the galena Pb is consistent with (1) leaching of unexposed igneous intrusives related to preceeding tuffaceous activity, (2) leaching of any strata in the McArthur and underlying groups which have suitably unradiogenic Pb, or (3) largely juvenile Pb.

It is probable that the Zn–Pb–Ag solutions also carried a certain amount of Fe which

---

[1] See Chapter 4 by Degens and Ross, Vol. 4, on the Red Sea area.

precipitated in the ore as the pyrite rims around the predominant fine-grained pyrite crystals. Furthermore, these solutions must have introduced the bulk of the Cu, As, Tl and Hg.

Consideration of metal sources unrelated to volcanism

Whilst weathering and erosion processes acting on metalliferous terrains have been implicated in the formation of some stratiform ores in sedimentary sequences, this does not appear to be feasible for the McArthur deposit. Factors which mitigate against the importance of such processes here are the apparent lack of a suitable source terrain, the high metal grades in the ore, and the evidence for dual metal sources.

However, a number of authors have proposed that significant proportions of the metals incorporated into sediments during sedimentation can be released to pore solutions during diagenesis, mainly as a result of desorption, solution and/or mineralogical changes (e.g., Davidson, 1966; Jackson and Beales, 1967; Billings et al., 1969; Lambert, 1973; Roberts, 1973; Carpenter et al., 1974; Wolf, 1976). Organic matter, clays and carbonates all have the ability to concentrate metals during sedimentation and these metals could be subsequently released to form soluble metal complexes in the pore solutions. Tuffaceous debris, particularly glass, could also release metals as it altered to phases stable at low temperatures.

Jackson and Beales (op. cit.) proposed that desorption of metals from clays and organic matter in shaley sediments could lead to the generation of metal-enriched formation waters, and Billings et al. (op. cit.) and Carpenter et al. (op. cit.) subsequently documented metalliferous oil-field brines which they considered were formed in this manner. On the other hand, Davidson (op. cit.) believed that brines could leach major quantities of metals from carbonates, and Roberts (op. cit.) argued for release of metals during dolomitization of precursor carbonate minerals (cf. also several chapters on the Mississippi Valley-type ores in this multi-volume publication).

Metal-enriched formation waters have been implicated by the above authors in the genesis of the Phanerozoic Mississippi Valley-type Pb—Zn deposits. Such deposits are never stratiform to the same degree as the McArthur deposit, and mainly occur as open-space fillings in their host sediments. In addition, they are not usually closely associated with igneous or pyroclastic rocks; they are characterized by isotopically inhomogeneous Pb which can be highly radiogenic, and they have low Ag contents.

It would appear, therefore, that the McArthur deposit could not have precipitated from formation waters similar to those postulated for the Mississippi Valley-type mineralization. However, at the present state of our knowledge, we cannot entirely rule out the possibility that somewhat different formation brines — having relatively homogeneous, unradiogenic Pb, and high Ag contents — evolved during diagenesis of the McArthur Group sediments.

It is obvious that distinction between "formation waters" and "volcanic exhalations" can become a problem in semantics. For instance, igneous activity could heat pore waters, thereby facilitating their ascent to the surface and possibly also enhancing their ability to extract metals from the sediments through which they pass. For a summary on the problems on ore genesis by compaction fluids, see Wolf (1976).

Sources of non-biogenic sulphur

It appears most likely that the sulphur in the sphalerite and galena came in with the Zn–Pb–Ag. Smith and Croxford (1973) proposed that the Zn and Pb were introduced into the basin as sulphides, or sulphide complexes in brines. It also is feasible that distinct sulphur species were transported in significant amounts in the same solutions as Zn and Pb chloride complexes (e.g., Helgeson, 1964; Nriagu and Anderson, 1970), and that metal sulphide precipitation occurred in response to physicochemical changes as the solutions emanated at the sea floor.

Whatever its form in the ore solution, the sulphur could have been generated by non-bacterial reduction of sulphate, it could have been dissolved from the rocks permeated by the ore-forming brines, or it could have been juvenile (i.e., given off by igneous magmas). If the latter, it would have to have undergone isotopic exchange with sulphate. A minor proportion of sulphur could also have been released during degradation of organic matter.

Significance of structures and organic matter in transport and entrapment of metals

The McArthur deposit is localized near the Emu Fault zone and other lineament systems (see section on structure). These structures could have been important in the migration of the metalliferous solutions into the Bulburra Depressions. The Fe and base metal anomalies around the McArthur deposit appear to emanate from the region of the Emu Fault, suggesting that the Fe and Zn–Pb–Ag solutions ascended to the submarine surface via this zone and/or faults branching from it. Stratiform ore deposits apparently formed where the metalliferous solutions, and/or precipitates from them, were trapped in deeps in the sedimentary trough. The metal anomalies in the hanging-wall shales could indicate either (1) intermittent minor influxes of Zn–Pb solutions for a considerable period after the formation of the stratiform ores, or (2) continued availability of sufficient amounts of metals for ore formation, but absence of suitable basins for trapping them.

Organisms and organic matter can concentrate metals in sedimentary environments and metals can be transported as soluble organo-metallic complexes (see Chapters 5 and 6, Vol. 2, by Saxby and Trudinger, respectively). However, the low organic C contents and high metal grades of the Zn–Pb–Ag ore provide no support for the importance of these processes in the formation of the McArthur deposit. It is likely that the carbonaceous matter in the deposit merely reflects the existence of quiet, relatively deep, reducing basins suitable for ore accumulation.

Temperature limits

No fluid inclusion temperature[1] data are available because of the extremely fine-grained nature of the McArthur deposit, but low-temperature ore accumulation is suggested by other lines of evidence:

(1) If bacterial reduction of sulphate was involved in pyrite formation, the temperatures at the sites of biogenic activity must have been less than about 100°C.

(2) The low rank of the carbonaceous matter within the ore indicates that it has never been subjected to temperatures exceeding approximately 150°C (Taylor, 1971).

(3) The sulphur isotope data for coexisting sphalerite and galena can be interpreted in terms of a poor approach to isotopic equilibrium at low temperatures of precipitation.

Consideration of low-copper content

The McArthur deposit has a much lower Cu/Pb + Zn ratio than is characteristic of any common rock type. Hence, it appears that there must have been separation of Cu from Zn and Pb during derivation or migration of the ore solution. One feasible explanation for this involves the relative instabilities of Cu complexes in low-temperature, sulphide-bearing brines (e.g., Helgeson, 1964; Nriagu and Anderson, 1970). Thus, if the ore solution was generated by a fairly low-temperature leaching process (up to roughly 200°C), it may never have contained a significant proportion of Cu. Alternatively, if it was derived at some moderate to high temperature, it may have incorporated a considerable amount of Cu, the bulk of which precipitated as the brine was cooled and/or diluted during migration to the Bulburra Depression.

There is no good evidence for the initial nature of the metalliferous brines. However, the existence of minor Cu-bearing deposits in the Emu Fault zone, the Cooley Dolomite and formations stratigraphically beneath the McArthur deposit could be interpreted as supporting the possibility that the metalliferous solutions originally contained significant amounts of Cu in addition to Zn and Pb.

GENESIS OF MINOR DEPOSITS

The widespread mineralization in the Cooley Dolomite suggests that Zn–Pb–Ag–S ore solutions which formed the McArthur deposit could, in part at least, have migrated through permeable zones in the Cooley Dolomite, before discharging into the immediately adjacent Bulburra Depression. The Cooley and Ridge 1 deposits formed entirely within this carbonate reef or bank, but the Ridge 2 deposit could have formed partly within it,

---

[1] For a general chapter on fluid inclusions, see Chapter 4 by Roedder, Vol. 2.

partly within muds draped on it, and partly on the sea floor. Further assessment of this discharge zone hypothesis for the Ridge 2 deposit must await the availability of relevant Pb and S isotope data.

It appears, on the other hand, that at least some of the non-stratiform mineralization in units other than Cooley Dolomite could not have formed from solutions similar to those which generated the stratiform mineralization. The Pb-isotope data of Richards (in preparation) indicate that the coarse-grained non-stratiform deposits he analysed contain admixtures of Pb from at least two sources. One component of their Pb could be the same as that in the galena of the stratiform ore and another could have been derived from similar source rocks to the Fe-rich solutions which gave rise to the more radiogenic pyrite in the carbonaceous shales. No sulphur-isotope data are available for these deposits. Their sulphur could have been carried in the same solutions as the metals, or it could have been generated in organic-rich zones in the carbonates by biogenic or non-biogenic reduction of sulphate, or by degradation of sulphur-bearing organic matter.

COMPARISONS WITH OTHER STRATIFORM ORES OF VOLCANO-SEDIMENTARY ASSOCIATIONS

"McArthur-type" Pb—Zn—Ag deposits of Australia

Mount Isa and Hilton. The Mount Isa and Hilton Pb—Zn—Ag deposits are situated about 20 km apart (Fig. 1) in the Urquhart Shale Formation. Both are basically similar to the McArthur deposit, but have higher Pb contents and have been significantly recrystallized and mobilized as a result of the low-grade regional metamorphism. The Hilton deposit is not yet being mined, but developmental work is at an advanced stage.

Detailed descriptions of the geology of the Mount Isa Mine have been given recently by Bennett (1965, 1970) and Mathias and Clark (1975), and of the Hilton deposit by Mathias et al. (1973) and Mathias and Clark (1975). Cross-sections through these deposits are given in Figs. 14 and 15.

The Pb—Zn—Ag orebodies are thinly banded within pyritic tuffaceous and dolomitic shales and siltstones, which are considered to be time equivalents of the H.Y.C. Pyritic Shale (Plumb and Derrick, 1975). Thin, highly potassic tuff beds occur through the Urquhart Shale and many contain abundant vitroclastic debris (Croxford, 1964). A few tuffite beds have been recorded in the underlying and overlying dolomitic siltstone formations.

The main sulphide minerals are galena, sphalerite, pyrite and pyrrhotite, and there are minor amounts of chalcopyrite, arsenopyrite, marcasite and tetrahedrite. The pyrrhotite could have formed from pyrite during metamorphism (e.g., Lambert, 1973). Pyritic shales, with minor sphalerite and galena, are widespread outside the orebodies.

The main non-sulphide minerals in the Urquhart Shale are ferroan dolomite, K-feldspar, quartz, muscovite, chlorite and carbonaceous matter. The latter is more graphitic than at McArthur, reflecting the regional metamorphism (Saxby, 1970). (Cf. also Chapter 5 by Mookherjee, Vol. 4, on metamorphism of ores.)

The fourteen Pb–Zn–Ag orebodies at Mount Isa occur within an approximately 1100 m thick section of Urquhart Shale and they have a total published size (ore mined plus reserves) of around 100 million tons and an average grade of 7.8% Pb, 6.0% Pb and 130 ppm Ag. The seven orebodies at Hilton occupy a stratigraphic interval of some 300 m and contain at least 35 million tons of 7.7% Pb, 9.6% Zn and 125 ppm Ag. In both cases there is evidence for some repetition of the stratigraphic section by faulting, large-scale pre-consolidational sliding and folding; isoclinal folds with amplitudes up to 200 m have been recorded in the Mount Isa Mine. However, tectonical thickening seems insignificant, and it follows that the mineralization at Mount Isa and Hilton formed over longer periods than at McArthur, where the maximum thickness of the deposit is only 130 m. The inter-ore beds at Mount Isa and Hilton are mainly dolomitic and tuffaceous shales and siltstones containing sub-economic sphalerite and galena.

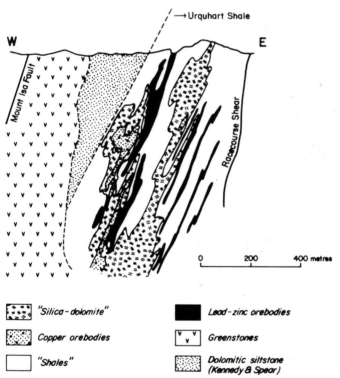

Fig. 14. Composite cross-section of the northern part of the Mount Isa mine, after Bennett (1965). At the southern end of the mine there is more Cu and less Pb–Zn–Ag, and sheared contacts have been exposed between the Urquhart Shale and underlying greenstones. (Reproduced by courtesy of Mount Isa Mines and Australasian Institute of Mining and Metallurgy.)

Sulphur isotope ratios in the Mount Isa orebodies were measured by Solomon (1965) who recorded wide δ^{34}S ranges for pyrite (+7 − +31‰) and pyrrhotite (+8 − +29‰) and slightly more restricted ranges for sphalerite (+10 − +23‰) and galena (+3 − +15‰). His data do not show trends of the type found by Smith and Croxford (1973) at McArthur, but he did not analyse an analogous series of samples from bottom to top of the deposit; furthermore, the presence of pyrrhotite could have modified pyrite trends. Solomon favoured a biogenic origin for all the sulphide minerals in the ore, but it must remain a possibility that non-biogenic sulphur was involved in the formation of the sphalerite and galena, as is indicated by the more systematic sampling of the McArthur deposit. The δ^{34}S values at Mount Isa are generally heavier than at McArthur, which suggests that these orebodies could have formed under significantly different physicochemical conditions.

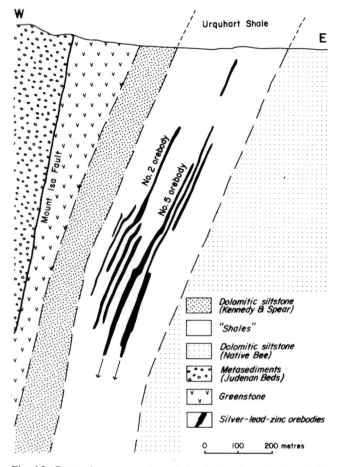

Fig. 15. Composite cross-section of the Hilton deposit after Mathias et al. (1973). (Reproduced by courtesy of Mount Isa Mines and Bureau of Mineral Resources.)

Galena lead isotope ratios at Mount Isa were measured by Ostic et al. (1967). They are isotopically homogeneous and similar to those at McArthur: 206/204 = 16.212–16.232; 207/204 = 15.590–15.615; 208/204 = 36.218–36.253. The model galena age is approximately 1600 m.y.

The obvious similarities between the Pb–Zn–Ag deposits at Mount Isa, Hilton and McArthur strongly suggest that closely analogous processes were involved in their formation, There are a number of penecontemporaneous faults in the Mount Isa–Hilton area which could have acted as channel-ways for the ore solutions.

Immediately adjacent to the Pb–Zn–Ag orebodies at Mount Isa, there is a huge non-stratiform Cu deposit (Fig. 14). This occurs in a complex "silica dolomite" facies of the Urquhart Shale. The main sulphide minerals in the Cu ores are pyrite, pyrrhotite and chalcopyrite, and these occur in irregular veinlets and patches. Bennett (1965) noted that the "silica dolomite" contains four basic rock types in intimate association: (1) medium to coarse grained dolomite: (2) irregularly brecciated dolomitic shale in a crystalline carbonate–quartz matrix; (3) partly recrystallized shale; and (4) fractured and brecciated shales with quartz veining. No "silica dolomite", or Cu orebodies, have been reported at Hilton.

The "silica dolomite" is conformable with the undeformed shales in many recorded contacts, but as a whole it broadly transgresses the bedding in the shales. There is commonly a gradual increase in the degree of recrystallization of the dolomite at the contacts and in some cases "ghosts" of shale and tuff beds can be traced for some distance into the "silica dolomite". The Pb–Zn–Ag mineralization never continues into the "silica dolomite", but in a few instances can be traced virtually to the contact, where there are local increases in the pyrrhotite content. The silica dolomite has a sheared basal contact with basic greenstones.

There are two schools of thought concerning the genesis of the Cu deposit at Mount Isa. Bennett (1965, 1970) and Stanton (1972) consider that the Cu ore formed syngenetically in a near-shore, algal reef–reef breccia environment which was subsequently heavily sheared and recrystallized. The other school of thought considers that the Cu ore formed epigenetically. The latter is supported by geochemical evidence presented by Smith and Walker (1971), who concluded that the Cu was derived from the underlying greenstones during diagenesis, metamorphism and/or hydrothermal activity. If the Cu was introduced epigenetically there has to be a reason for the restriction of the Cu mineralization to the Urquhart Shale, when dolomitic siltstones and shales from adjacent formations are also underlain by the greenstones. An explanation could be that the mineralized zone was a relatively permeable, brecciated reef zone in the Urquhart Shale but, equally importantly, the presence of pyrite would probably have been essential for "fixing" the Cu. A trace-Pb isotope study could well resolve the syngenetic/epigenetic controversy. The isotope ratios in the minor amounts of Pb in the chalcopyrite should be compared with those of the Pb in the greenstones, the country rock pyrite and the Pb–Zn–Ag orebodies. If the chalcopyrite of the Cu deposit has isotopically similar Pb to that in the galena of the Pb–Zn–Ag deposit, this would be strong evidence that the Cu precipitated differentially from the same syn-sedimentary ore solution as the Pb–Zn–Ag. On the other hand, different Pb isotope ratios would be indicative of different modes of formation of these deposits, and consideration should then be given to whether the data support the epigenetic greenstone source/pyrite trap model for the Cu solutions.

The absence of significant Cu mineralization at Hilton and McArthur can be explained in terms of the lack of greenstone source rocks in the immediate vicinities of the pyrite shales.

Broken Hill. Detailed descriptions of the geology of this deposit have been given recently by Carruthers (1965), Lewis et al. (1965), Pratten (1965), Johnson and Klingner (1975), and in Chapter 6 by Both and Rutland, Vol. 4. See also the discussion by King (Chapter 5, Vol. 2) on the "evolution" of genetic hypotheses.

It is the richest deposit for the size in the world, totalling roughly 200 million tons of ore with as average grade of around 12% Pb, 12% Zn and 115 ppm Ag. There are six separate lodes at the southern end of the deposit (Fig. 16), and the upper three have distinctly higher Zn/Pb ratios. At the northern end, the upper three lodes are not economic.

The Broken Hill deposit differs in a number of respects from the McArthur, Mount Isa and Hilton deposits. The ore is coarse-grained and massive. It is conformable within a rather complex lode horizon which contains sillimanite gneisses, blue quartzites, garnet quartzite, garnet sandstone and concordant pegmatites. The main rock types outside the lode horizon are quartzo-feldspathic gneisses with or without garnet, sillimanite-garnet-biotite gneisses, quartzites, mica schists, amphibolites and banded iron formations. None of the rocks contain significant amounts of carbonate minerals or carbonaceous matter. Some of the gneisses are geochemically indistinguishable from felsic volcanics, but there is no conclusive proof of their pre-metamorphic nature. Thus, although it is widely assumed that there was volcanism during accumulation of the ore-bearing strata at Broken Hill, the major textural and mineralogical reconstitutions that have taken place make it impossible to estimate just how important this could have been.

The granulite facies metamorphism at Broken Hill has been dated by Rb–Sr isochron methods at approximately 1700 m.y. (Pidgeon, 1967; Shaw, 1968). The ore was affected by this metamorphism and therefore must be older than the major Carpentarian Pb–Zn–Ag deposits of northern Australia.

The mineralogy of the Broken Hill deposit has been described by Stillwell (1959). The major ore minerals are coarse-grained galena and sphalerite (marmatite). Pyrrhotite is locally abundant but overall is a fairly minor constituent. Chalcopyrite is another minor mineral, and there are traces of arsenopyrite, lollingite and tetrahedrite. Pyrite is present in small amounts but is probably secondary after pyrrhotite.

The main gangue minerals are quartz, calcite, manganiferous garnet, rhodonite, bustamite and fluorite, and each orebody has its own characteristic gangue assemblage. The lode horizon contains minor pyrrhotite for considerable distances away from the orebodies, but there is no iron sulphide-rich envelope analogous to that around the previously described deposits.

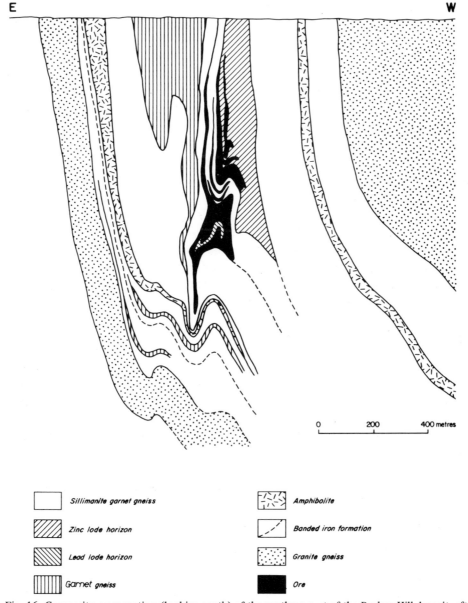

Fig. 16. Composite cross-section (looking south) of the southern part of the Broken Hill deposit, after The Zinc Corporation. The upper three lodes are not economic at the northern end of the deposit. Note that the orebodies are in a syncline–anticline drag fold structure. (Reproduced by courtesy of the Zinc Corporation.)

The $\delta^{34}S$ values measured for bulk sulphide samples from the Broken Hill ore fall within the range $-2.2 - +4.7‰$ (Stanton and Rafter, 1966). The mean of approximately $+1‰$ and the restricted spread of the values suggests that a major proportion of the sulphur incorporated in the ore minerals could have been juvenile. It is equally feasible, though, that there was originally a significantly wider spread of $\delta^{34}S$ values, analogous to those at McArthur and Mount Isa, which was narrowed down during the high-grade metamorphism.

The Pb in the Broken Hill galena is isotopically homogeneous and lies on the average growth curve (Ostic et al., 1967). The available data fall in the following ranges: 206/204 = 16.109–16.131; 207/204 = 15.526–15.548; 208/204 = 36.007–36.155. The model age of the galena is slightly greater than 1600 m.y.

The paucity of carbonates, the absence of an iron sulphide-rich envelope and the older age of the mineralization distinguish the Broken Hill deposit from the McArthur, Mount Isa and Hilton deposits. The other different features of the Broken Hill mine could merely be functions of its metamorphic grade.

Because of the complexity of the Broken Hill deposit there is divergence of opinion concerning its origin, with various syngenetic and epigenetic origins being advocated. The writer favours syngenetic–early diagenetic ore formation involving moderately low-temperature solutions that were fairly similar to those which generated the McArthur, Mount Isa and Hilton deposits. However, the environment of ore formation was evidently characterized by relatively little carbonate precipitation and organic activity.

Possible "McArthur-type" deposits of other countries

Sullivan. The large Sullivan deposit of British Columbia (e.g., Freeze, 1966) is here tentatively classified as a "McArthur-type" Pb–Zn–Ag ore. It consists of banded and massive ore in Upper Proterozoic argillites, siltstones and quartzites which have been regionally metamorphosed to greenschist facies. No definite tuffs or lavas have been recognized amongst the country rocks in the vicinity of the ore, but there are numerous sills and some dykes of intermediate to acid composition in the underlying strata. Igneous activity which gave rise to these intrusives could have led to the generation of the Sullivan ore solution, or yielded juvenile, metal-rich fluids.

The Sullivan deposit has a number of features which are not characteristic of the Australian deposits described above. In particular, it has a relatively high Sn content, and it is associated with zones of tourmalinization, chloritization and albitization. When the full significance of these differences is understood, it may well be that the Sullivan deposit has to be classified into a distinct Pb–Zn–Ag sub-group.

Rammelsberg. The Rammelsberg deposit of Germany can also be provisionally classified with the "McArthur-type" deposits. It consists of several ore lenses which are conformable within a Devonian slate formation containing intercalated sandstones, limestones and a few tuff horizons.

The Pb–Zn ore at Rammelsberg consists of finely interbanded sulphides and slate. It grades down into Cu-rich, massive ore and up into barite-rich ore. Similar vertical metal zoning occurs in the "Kuroko-type" Zn–Cu–Pb–Ag–Au deposits. However, Rammelsberg cannot be classified as "Kuroko-type" because it differs from such ores in a number of important respects. For instance:

(1) Rammelsberg lacks the felsic tuff, tuff breccia and lava footwall rocks which are features of "Kuroko-type" mineralization.

(2) Rammelsberg has a lower overall Cu/Pb + Zn ratio than is characteristic of the "Kuroko-type" ores.

(3) Sulphur in Rammelsberg pyrite is not in isotopic equilibrium with that in coexisting sphalerite, galena and chalcopyrite (Anger et al., 1966), in contrast with the situation for the "Kuroko-type" ores.

(4) There are abundant thin beds of detrital sediment within the Rammelsberg ore lenses, but not in the "Kuroko-type" ores.

Certain "Alpine-" or "Mississippi Valley-type" ores. Some of the Phanerozoic Pb–Zn ores in the European Alps and elsewhere are stratiform in sedimentary sequences which contain tuffaceous horizons and/or porphyry intrusions (e.g., Kostelka and Petrascheck, 1967; Maucher and Schneider, 1967). Whilst these are usually referred to as "Alpine-" or "Mississippi Valley-type" deposits, they do have affinities with the "McArthur-type deposits.

Other ore types

The spectrum of stratiform base metal sulphide deposits[1] of volcano–sedimentary associations can be divided into several general types on the basis of metal ratios, country-rock associations and, to a certain extent, ages. The writer favours the five-fold working classification scheme presented in Table II which, at the present state of our knowledge, seems to provide an adequate basis for both scientific investigation and exploration.

The Pb–Zn–Ag or "McArthur-type" deposits clearly fall near the sedimentary end of this spectrum. The other four ore types recognized in Table II are associated with relatively high proportions of volcanic rocks. They are smaller and more massive than the main "McArthur-type" ores, and presumably formed relatively rapidly. All contain abundant pyrite and/or pyrrhotite.

The Zn–Cu–Pb–Ag–Au deposits are here named "Kuroko-type" after the Japanese ores which are the youngest known, least modified examples of this group. These Kuroko deposits have been reviewed recently by Ishihara (1974) and Lambert and Sato (1974). Ores of this group formed in association with felsic volcanism in mature island arcs and

---

[1] For a spectrum of ore types composed of transitional varieties, see Chapter 4 by Gilmour, Vol. 1.

TABLE II

Working classification scheme for stratiform base metal sulphide deposits of volcano-sedimentary associations, based in part on Stanton (1972) and Hutchinson (1973)

Ore types	Minor metals	Main rock associations	Examples	Ages
Pb–Zn–Ag "McArthur-type"	Cu	tuffaceous mudstones and siltstones	McArthur, Mount Isa, Hilton, Broken Hill (Australia), Sullivan (Canada), Rammelsberg (Germany)	Proterozoic Palaeozoic
Zn–Cu–Pb–Ag–Au "Kuroko-type"		felsic tuffs and lavas of calc-alkaline suite, "volcanic" sediments, mudstones	Kuroko (Japan); New Brunswick, etc. (Canada/U.S.A.); Captains Flat, Woodlawn, Rosebery (Australia); Skellefte deposits (Sweden)	Proterozoic Phanerozoic
Zn–Cu–Ag–Au "Superior-type"	Pb	andesitic to rhyolitic tuffs and lavas of calc-alkaline (?) suite, "volcanic" sediments	Noranda, etc. (Canada), West Shasta (U.S.A.)	Archaean Phanerozoic
Cu–Au "Cyprus-type"	Zn–Ag	tholeiitic basalts (pillowed), chalks and marls, mudstones	Skouriotissa, etc. (Cyprus); Ergani Maden, etc. (Turkey); Besshi, etc. (Japan); Atlantis II muds (?) (Red Sea)	Phanerozoic
Ni "Yilgarn-type"	Cu	ultramafic lavas and/or shallow intrusives, "volcanic" sediments	Kambalda, etc. (Australia)	Archaean

active continental margins. They are typically zoned from Cu-rich, through Pb–Zn-rich, up to barite-rich. Their pyrite is in isotopic equilibrium with coexisting sulphide minerals, and the isotopic and fluid inclusion data indicate precipitation of the ores at around 200–250°C from ore solutions with a major seawater component. Some deposits of this group, for example Rosebery, have somewhat lower Cu/Pb + Zn ratios than is characteristic for the Kuroko deposits, a feature which the writer interprets as indicating that they formed at slightly lower temperatures than the Japanese ores.

The Zn–Cu–Ag–Au ores are here named "Superior-type" after the numerous Archaean massive sulphide deposits which occur in the Superior province of Canada (e.g., Sangster, 1972). They occur in association with mafic–felsic lavas and tuffs and are characteristically zones from Cu-rich up to Zn-rich. Their low Pb contents are compatible with derivation of their ore solutions predominantly from andesitic to basaltic rocks, and/ or with derivation of the ore solutions from felsic rocks followed by fractionation of Pb from Zn–Cu before or during ore formation.

The group of Cu–Au deposits is here termed "Cyprus-type" after the famous examples of such ores on that island. These have been reviewed recently by Constantinou and Govett (1973). The "Cyprus-type" deposits occur in association with tholeiitic basalts. They usually contain significant amounts of Zn in their upper parts. It is thought that they formed on the ocean floor in the vicinity of accreting plate margins, and were subsequently incorporated into continental crust by overthrusting of oceanic lithosphere. Some examples, however, may have formed in association with basaltic volcanism in young island-arc environments. This could be the case for Besshi and similar deposits of Japan (Kanehira and Tatsumi, 1970). These are relatively deformed and metamorphosed, but are basically similar to the Cyprus ores. One noteworthy point of difference is that the Japanese deposits have relatively low Au contents. The Recent Red Sea Zn–Cu-rich muds (e.g., Degens and Ross, 1969) can probably also be included in this broad group, although sedimentary rocks appear to have played a more significant role in the derivation of their metals than was the case for the other basalt-associated deposits.

The so-called "Kuroko-", "Superior-" and "Cyprus-type" deposits are all widely considered to have formed sub-aqueously from thermal exhalations which discharged during the waning stages of major volcanic cycles (e.g., Hutchinson, 1973). Zones of ascent of the hydrothermal solutions are indicated by the network mineralization and altered rocks which occur beneath the stratiform ores of each of these groups.

The Archaean Ni deposits are here termed "Yilgarn-type" after the craton of this name in Western Australia where they are particularly well developed (e.g., Ewers and Hudson, 1972). In general, they consist of massive ore which grades up to disseminated ore, and they occur at the base of ultramafic flows and/or shallow intrusions. Available evidence (Ewers and Hudson, op. cit.) supports their formation at high temperatures by gravitational separation of immiscible sulphide liquids during emplacement of ultramafic magmas. Minor mineralization in adjacent sediments could be of submarine exhalative origin.

CONCLUSIONS

The unmetamorphosed, stratiform McArthur Zn–Pb–Ag deposit occurs near the edge of a fault-bounded trough containing over 5 km of Middle Proterozoic dolomites, shales, siltstones and minor tuffites.

The deposit consists of seven mineralized shale orebodies separated by relatively metal-poor, dolomite-rich shales, arenites and breccia beds. The sulphide minerals are mostly very fine-grained and occur in thin conformable bands. Pyrite, galena and sphalerite are the major sulphides and chalcopyrite, arsenopyrite and marcasite are minor constituents. The shales are composed essentially of potash feldspar, ferroan dolomite, quartz, illite, chlorite, kaolin and carbonaceous matter. Dolomitic strata below and above the mineralized shale are iron-rich and contain occasional thin potash-rich tuffite beds.

The available information points to syngenetic–early diagenetic formation of the sulphide bands. It is probable that the metalliferous solutions were derived in some way as a result of the igneous activity which gave rise to the tuffaceous debris in the enclosing sediments.

Sulphur and lead isotope data suggest dual metal and sulphur sources. The bulk of the Fe and the bulk of the Zn–Pb–Ag must have been derived from different sources and it appears that there could not have been significant mixing of the two solutions prior to precipitation of the metals from either or both of them. The sulphur in the sphalerite and galena is non-biogenic, but there appears to be a biogenic sulphur component in the pyrite.

The Mount Isa and Hilton Pb–Zn–Ag deposits are basically similar to the McArthur deposit. The Broken Hill deposit is older and has several significantly different features, but its size and Cu-poor nature characterizes it as a "McArthur-type" ore. Important overseas Pb–Zn-rich deposits with "McArthur-type" affinities include Sullivan, Rammelsberg, and certain stratiform "Alpine-" or "Mississippi Valley-type" ores.

Four other broad groups can be recognized amongst stratiform base metal sulphide ores of volcano–sedimentary associations, each of which is associated with relatively high proportions of volcanic rocks. These groups are the Zn–Cu–Pb–Ag–Au ("Kuroko-type") deposits, the Zn–Cu–Ag–Au ("Superior-type") deposits, the Cu–Au ("Cyprus-type") deposits and the Ni ("Yilgarn-type") deposits.

ACKNOWLEDGEMENTS

The following people are thanked for providing pre-prints of their forthcoming papers and for valuable discussion: M.C. Brown, B.L. Gulson, W.J. Murray, K.A. Plumb, J.R. Richards and J.W. Smith. Constructive comments by E.M. Bennett, K.M. Scott and G.H. Taylor led to worthwhile improvements in the chapter.

The Baas Becking Laboratory is supported by the Australian Mining Industry Research

Association, the Bureau of Mineral Resources and the Commonwealth Scientific and Industrial Research Organization.

REFERENCES

Anger, G., Nielsen, H., Puchelt, H. and Ricke, W., 1966. Sulphur isotopes in the Rammelsberg ore deposit (Germany). *Econ. Geol.*, 61: 511–536.
Beales, F.W. and Oldershaw, A.E., 1969. Evaporite-solution brecciation and Devonian carbonate reservoir porosity in western Canada. *Am. Assoc. Pet. Geol. Bull.*, 53: 503–512.
Bennett, E.M., 1965. Lead–zinc–silver and copper deposits of Mount Isa. *Commonw. Min. Metall. Congr., 8th, Melb.*, 1: 1233–1246.
Bennett, E.M., 1970. History, geology, and planned expansion of Mount Isa properties. Paper presented to World Symposium on the Mining and Metallurgy of Lead and Zinc. *Am. Inst. Min. Eng., St. Louis.*
Billings, G.K., Kesler, S.E. and Jackson, S.A., 1969. Relation of zinc-rich formation waters, Northern Alberta, to the Pine Point ore deposit. *Econ. Geol.*, 64: 385–391.
Brown, A.C., 1971. Zoning in the White Pine copper deposit, Ontonagon County, Michigan. *Econ. Geol.*, 66: 543–573.
Brown, M.C., Claxton, C. and Plumb, K.A., 1975. The Proterozoic Barney Creek Formation and some associated carbonate units, McArthur Group, N.T. *Bur. Miner. Resour. Aust. Rec.*, in preparation.
Carpenter, A.B., Trout, M.L. and Pickett, E.E., 1974. Preliminary report on the origin and chemical evolution of lead and zinc-rich oil field brines in central Mississippi. *Econ. Geol.*, 69: 1191–1206.
Carruthers, D.S., 1965. An environmental view of Broken Hill ore occurrence. In: J.McAndrew (Editor), *Geology of Australian Ore Deposits. Commonw. Min. Metall. Congr., 8th, Melb.*, 1: 339–351.
Constantinou, G. and Govett, G.J.S., 1973. Geology, geochemistry and genesis of Cyprus sulphide deposits. *Econ. Geol.*, 68: 843–858.
Corbett, J.A., Lambert, I.B. and Scott, K.M., 1975. Results of analyses of rocks from the McArthur area, Northern Territory. *C.S.I.R.O. Miner. Res. Lab. Tech. Commun.*, 57.
Cotton, R.E., 1965. H.Y.C. lead–zinc–silver deposit, McArthur River. In: J. McAdrew (Editor), *Geology of Australian Ore Deposits. Commonw. Min. Metall. Congr., 8th. Melb.*, 1: 197–200.
Croxford, N.J.W., 1964. Origin and significance of volcanic potash-rich rocks from Mount Isa. *Trans. Inst. Min. Metall.*, 74: 33–43.
Croxford, N.J.W., 1968. A mineralogical study of the McArthur lead–zinc–silver deposit. *Proc. Aust. Inst. Min. Metall.*, 226: 97–108.
Croxford, N.J.W. and Jephcott, S., 1972. The McArthur lead–zinc–silver deposit, N.T. *Proc. Aust. Inst. Min. Metall.*, 243: 1–26.
Davidson, C.F., 1966. Some genetic relationships between ore deposits and evaporites. *Inst. Min. Met. Trans.*, 75: B216–B225.
Degens, E.T. and Ross, D.A. (Editors), 1969. *Hot Brines and Recent Heavy Metal Deposits in the Red Sea.* Springer, New York, N.Y., 600 pp.
Ewers, W.E. and Hudson, D.R., 1972. An interpretive study of a nickel–iron sulphide ore intersection, Lunnon Shoot, Kambalda, Western Australia. *Econ. Geol.*, 67: 1075–1092.
Ferguson, J. and Lambert, I.B., 1972. Volcanic exhalations and metal enrichments at Matupi Harbour, New Britain, T.P.N.G. *Econ. Geol.*, 67: 25–37.
Freeze, A.C., 1966. On the origin of the Sullivan orebody Kimberley, B.C. In: *Tectonic History and Mineral Deposits of the Western Cordillera.* Can. Inst. Min. Metall., 8 (Spec. Vol.).
Gulson, B.L., 1975. Differences in lead isotopic compositions in the stratiform McArthur zinc–lead–silver deposit. *Miner. Deposita*, 10: 277–286.
Hamilton, L.H. and Muir, M.D., 1974. Precambrian microfossils for the McArthur River lead–zinc–silver deposit, N.T., Australia. *Miner. Deposita*, 9: 83–86.

Heier, K.S. and Rhodes, J.M., 1966. Thorium, uranium and potassium concentrations in granites and gneisses of the Rum Jungle Complex, N.T., Australia. *Econ. Geol.*, 61: 563—571.

Helgeson, H.C., 1964. *Complexing and Hydrothermal Ore Deposition*. Macmillan, New York, N.Y., 128 pp.

Hutchinson, R.W., 1973. Volcanogenic sulphide deposits and their metallogenic significance. *Econ. Geol.*, 68: 1223—1246.

Iijima, A. and Hay, R.L., 1968. Analcime composition in tuffs of the Green River Formation of Wyoming. *Amer. Mineral.*, 53: 184—200.

Ishihara, S. (Editor), 1974. *Geology of Kuroko Deposits. Min. Geol., Spec. Issue*, 6, Tokyo, 435 pp.

Jackson, S.A. and Beales, F.W., 1967. An aspect of sedimentary basin evolution: the concentration of Mississippi Valley-type ores during late stages of diagenesis. *Bull. Can. Pet. Geol.*, 15 (4): 383—433.

Johnson, I.R. and Klingner, G.D., 1975. The Broken Hill ore deposit and its environment. In: C.L. Knight (Editor), *Economic Geology of Australia and Papua New Guinea — Metals*. Aust. Inst. Min. Metall., Melb.

Kajiwara, Y. and Krouse, H.R., 1971. Sulphur isotope partitioning in metallic sulphide systems. *Can. J. Earth Sci.*, 8: 1397—1408.

Kanehira, K. and Tatsumi, T., 1970. Bedded cupriferous iron sulphide deposits in Japan: a review. In: T. Tatsumi (Editor), *Volcanism and Ore Genesis*. Univ. Tokyo Press, Tokyo, 448 pp.

Kostelka, L. and Petrascheck, W.A., 1967. Genesis and classification of Triassic Alpine lead—zinc deposits in the Austrian region. In: J.S. Brown (Editor), *Genesis of Stratiform Lead—Zinc—Barite Deposits. Econ. Geol., Monogr.*, 3: 443 pp.

Lambert, I.B., 1973. Post-depositional availability of sulphur and metals, and formation of secondary textures and structures in stratiform sedimentary sulphide deposits. *J. Geol. Soc. Aust.*, 20: 205—215.

Lambert, I.B. and Bubela, B., 1970. Banded sulphide ores: the experimental production of monomineralic sulphide bands in sediments. *Miner. Deposita*, 15 (2): 97—102.

Lambert, I.B. and Sato, T., 1974. The Kuroko and associated ore deposits of Japan: a review of their features and metallogenesis. *Econ. Geol.*, 69: 1215—1236.

Lambert, I.B. and Scott, K.M., 1973. Implications of geochemical investigations of sedimentary rocks within and around the McArthur zinc—lead—silver deposit, Northern Territory. *J. Geochem. Explor.*, 2: 307—330.

Lambert, I.B. and Scott, K.M., 1975. Carbon contents of sedimentary rocks within and around McArthur zinc—lead—silver deposits, Northern Territory. *J. Geochem. Explor.*, in press.

Lewis, B.R., Forward, P.S. and Roberts, J.B., 1965. Geology of the Broken Hill lode, reinterpreted. In: J. McAndrew (Editor), *Geology of Australian Ore Deposits. Commonw. Min. Metall. Congr., 8th, Melb.*, 1: 319—332.

Logan, B.W., Rezak, R. and Ginsburg, R.N., 1964. Classification and environmental significance of algal stromatolites. *J. Geol.*, 72: 68—83.

Mathias, B.V. and Clark, G.J., 1975. Mount Isa copper and silver—lead—zinc orebodies — Isa and Hilton Mines. In: C.L. Knight (Editor), *Economic Geology of Australia and Papua New Guinea — Metals*. Aust. Inst. Min. Metall., Melbourne.

Mathias, B.V., Clark, G.J., Morris, D. and Russell, R.E., 1973. The Hilton deposit — stratiform silver—lead—zinc mineralization of the Mount Isa type. *Bur. Miner. Resour. Aust. Bull.*, 141: 33—58.

Maucher, A. and Schneider, H.J., 1967. The Alpine lead—zinc ores. In: J.S. Brown (Editor), *Genesis of Stratiform Lead—Zinc—Barite Deposits. Econ. Geol., Monogr.*, 3: 443 pp.

Murray, W.J., 1975. McArthur River H.Y.C. lead—zinc and related deposits. In: C.L. Knight (Editor), *Economic Geology of Australia and Papua New Guinea — Metals*. Aust. Inst. Min. Metall., Melbourne.

Nriagu, J.O. and Anderson, G.M., 1970. Calculated solubilities of some base metal sulphides in brine solutions. *Inst. Min. Met. Trans., Sect. B.*, 79: B208—B212.

Ohmoto, H., 1972. Systematics of sulfur and carbon isotopes in hydrothermal ore deposits. *Econ. Geol.*, 67: 551—578.

Ostic, R.G., Russell, R.D. and Stanton, R.L., 1967. Additional measurements of the isotopic composition of lead from stratiform deposits. *Can. J. Earth Sci.*, 4: 245–69.

Pidgeon, R.T., 1967. A rubidium–strontium geochronological study of the Willyama Complex, Broken Hill, Australia. *Petrol.*, 8: 283–324.

Plumb, K.A. and Brown, M.C., 1973. Revised correlation and stratigraphic nomenclature in the Proterozoic carbonate complex of the McArthur Group, Northern Territory. *Bur. Miner. Resour. Aust. Bull.*, 139: 103–115.

Plumb, K.A. and Derrick, G.M., 1975. Geology of the Proterozoic rocks of Northern Australia. In: C.L. Knight (Editor), *Economic Geology of Australia and Papua New Guinea – Metals*. Aust. Inst. Min. Metall., Melbourne.

Pratten, R.D., 1965. Lead–zinc–silver ore deposits of the Zinc Corporation and New Broken Hill Consolidated Mines, Broken Hill. In: J. McAndrew (Editor), *Geology of Australian Ore Deposits. Commonw. Min. Metall. Congr., 8th, Melb.*, 1: 33–335.

Reinhold, J.J., 1961. Notes on the microscopic examination of the H.Y.C. lead–zinc ore. Mount Isa Mines Limited. *Microsc. Tech. Rep.*, 436, unpublished.

Richards, J.R., 1968. "Primary" leads. *Nature*, 219 (5151): 258–259.

Richards, J.R., 1975. Lead isotope data on three Australian galena localities. *Miner. Deposita*, 10: 287–301.

Roberts, W.M.B., 1965. Recrystallization and mobilization of sulphide at 2000 atmospheres and in the temperature range 50–145°. *Econ. Geol.*, 60: 168–180.

Roberts, W.M.B., 1973. Dolomitization and the genesis of the Woodcutters lead–zinc prospect, Northern Territory, Australia. *Miner. Deposita*, 8: 35–56.

Sangster, D.F., 1972. Precambrian volcanogenic massive sulphide deposits in Canada: A review. *Geol. Surv. Can., Pap. 72–22*, 44 pp.

Saxby, J.D., 1970. Technique for the isolation of kerogen in sulphide ores. *Geochim. Cosmochim. Acta*, 34: 1317–1326.

Shaw, S.E., 1968. Rb–Sr isotopic studies of the mine sequence rocks at Broken Hill. In: M. Radmanovich and J.T. Woodcock (Editors), *Broken Hill Mines. Aust. Inst., Min. Metall., Monogr.*, 3: 185–198.

Smith, W.D., 1969. Penecontemporaneous faulting and its likely significance in relation to Mount Isa ore deposition. *Geol. Soc. Aust., Spec. Publ.*, 2: 225–235.

Smith, R.N., 1973. *Trace Element Distributions in Some Major Stratiform Ore Bodies.* B.Sc (Hons) Thesis, Univ. Melbourne, unpublished.

Smith, J.W. and Croxford, N.J.W., 1973. Sulphur-isotope ratios in the McArthur lead–zinc–silver deposit. *Nat. Phys. Sci.*, 245: 10–12.

Smith, J.W. and Croxford, N.J.W., 1975. An isotopic investigation of the environment of deposition of the McArthur mineralization. *Miner. Deposita*, 10: 269–276.

Smith, S.E. and Walker, K.R., 1971. Primary element dispersions associated with mineralization at Mount Isa, Queensland. *Bur. Miner. Resour. Aust. Bull.*, 131: 80 pp.

Solomon, R.J., 1965. Investigation into sulphide mineralization at Mount Isa, Queensland. *Econ. Geol.*, 60: 737–765.

Stanton, R.L., 1964. Textures of stratiform ores. *Nature*, 202: 173–174.

Stanton, R.L., 1972. *Ore Petrology.* McGraw-Hill, New York, N.Y., 713 pp.

Stanton, R.L. and Rafter, T.A., 1966. The isotopic constitution of sulphur in some stratiform lead–zinc ores. *Miner. Deposita*, 1: 16–29.

Stillwell, F.L., 1959. Petrology of the Broken Hill Lode and its bearing on ore genesis. *Proc. Aust. Inst. Min. Met.*, 190: 1–84.

Swett, K., 1968. Authigenic feldspars and cherts resulting from dedolomitization of illitic limestones: a hypothesis. *J. Sediment. Petrol.*, 38: 128–135.

Taylor, G.H., 1971. Carbonaceous matter as a guide to the genesis and history of ores. *Soc. Min. Geol. Japan, Spec. Issue*, 3: 283–288.

Williams, N., 1974. Epigenetic processes in the stratiform lead–zinc deposit at McArthur River, Northern Territory, Australia. *Abstr. Geol. Soc. Am. Meet., Miami*, 1006.

Williams, N. and Rye, D.M., 1974. Alternative interpretation of sulphur isotope ratios in the McArthur lead–zinc–silver deposits. *Nature*, 247: 535–537.

Wolf, K.H., 1976. Ore genesis influenced by compaction. In: G.V. Chilingarian and K.H. Wolf (Editors), *Compaction of Coarse-Grained Sediments, 2.* Elsevier, Amsterdam, in press.